Table of Atomic Weights

Element	Symbol	Atomic Number	Atomic Weight	Element	Symbol	Atomic Number	Atomic Weight
Actinium	Ac	89	227.028	Neodymium	Nd	60	144.24
Aluminum	Al	13	26.98154	Neon	Ne	10	20.179
Americium	Am	95	(243)[a]	Neptunium	Np	93	237.0482
Antimony	Sb	51	121.75	Nickel	Ni	28	58.71
Argon	Ar	18	39.948	Niobium	Nb	41	92.9064
Arsenic	As	33	74.9216	Nitrogen	N	7	14.0067
Astatine	At	85	(210)	Nobelium	No	102	(255)
Barium	Ba	56	137.34	Osmium	Os	76	190.2
Berkelium	Bk	97	(247)	Oxygen	O	8	15.9994
Beryllium	Be	4	9.01218	Palladium	Pd	46	106.4
Bismuth	Bi	83	208.9804	Phosphorus	P	15	30.97376
Boron	B	5	10.81	Platinum	Pt	78	195.09
Bromine	Br	35	79.904	Plutonium	Pu	94	(244)
Cadmium	Cd	48	112.40	Polonium	Po	84	(210)
Calcium	Ca	20	40.08	Potassium	K	19	39.098
Californium	Cf	98	(251)	Praseodymium	Pr	59	140.9077
Carbon	C	6	12.01115	Promethium	Pm	61	(147)
Cerium	Ce	58	140.12	Protactinium	Pa	91	231.0359
Cesium	Cs	55	132.9054	Radium	Ra	88	226.0254
Chlorine	Cl	17	35.453	Radon	Rn	86	(222)
Chromium	Cr	24	51.996	Rhenium	Re	75	186.2
Cobalt	Co	27	58.9332	Rhodium	Rh	45	102.9055
Copper	Cu	29	63.546	Rubidium	Rb	37	85.4678
Curium	Cm	96	(247)	Ruthenium	Ru	44	101.07
Dysprosium	Dy	66	162.50	Samarium	Sm	62	150.4
Einsteinium	Es	99	(254)	Scandium	Sc	21	44.9559
Erbium	Er	68	167.26	Selenium	Se	34	78.96
Europium	Eu	63	151.96	Silicon	Si	14	28.086
Fermium	Fm	100	(257)	Silver	Ag	47	107.868
Fluorine	F	9	18.99840	Sodium	Na	11	22.98977
Francium	Fr	87	(223)	Strontium	Sr	38	87.62
Gadolinium	Gd	64	157.25	Sulfur	S	16	32.06
Gallium	Ga	31	69.72	Tantalum	Ta	73	180.9479
Germanium	Ge	32	72.59	Technetium	Tc	43	98.9062
Gold	Au	79	196.9665	Tellurium	Te	52	127.60
Hafnium	Hf	72	178.49	Terbium	Tb	65	158.9254
Helium	He	2	4.00260	Thallium	Tl	81	204.37
Holmium	Ho	67	164.9304	Thorium	Th	90	232.0381
Hydrogen	H	1	1.00797	Thulium	Tm	69	168.9342
Indium	In	49	114.82	Tin	Sn	50	118.69
Iodine	I	53	126.9045	Titanium	Ti	22	47.90
Iridium	Ir	77	192.22	Tungsten	W	74	183.85
Iron	Fe	26	55.847	Unnilennium[b]	Une	109	(266)
Krypton	Kr	36	83.80	Unnilhexium	Unh	106	(263)
Lanthanum	La	57	138.9055	Unnilpentium	Unp	105	(262)
Lawrencium	Lr	103	(256)	Unnilquadium	Unq	104	(261)
Lead	Pb	82	207.19	Unnilseptium	Uns	107	(262)
Lithium	Li	3	6.941	Uranium	U	92	238.029
Lutetium	Lu	71	174.97	Vanadium	V	23	50.9414
Magnesium	Mg	12	24.305	Xenon	Xe	54	131.30
Manganese	Mn	25	54.9380	Ytterbium	Yb	70	173.04
Mendelevium	Md	101	(258)	Yttrium	Y	39	88.9059
Mercury	Hg	80	200.59	Zinc	Zn	30	65.38
Molybdenum	Mo	42	95.94	Zirconium	Zr	40	91.22

[a] Value in parentheses is the mass number of the most stable or best-known isotope.
[b] These unnil-names are temporary. Official names and symbols for these elements have not been agreed upon.

BASIC
CONCEPTS
OF
CHEMISTRY

BASIC CONCEPTS OF CHEMISTRY

Third Edition

Leo J. Malone

Saint Louis University
St. Louis, Missouri

WILEY

John Wiley & Sons

New York   Chichester   Brisbane   Toronto   Singapore

Library of Congress Cataloging in Publication Data:

Malone, Leo J., 1938–
    Basic concepts of chemistry / Leo J. Malone. — 3rd ed.
        p.   cm.
    Bibliography: p.
    Includes index.
    ISBN 0-471-84930-8
    1. Chemistry.   I. Title.
QD31.2.M344   1989
540 — dc19

Printed in the United States of America

10 9 8 7 6 5 4 3 2 1

PREFACE

Basic Concepts of Chemistry is not only an introduction to chemical concepts, but is also an introduction to the way chemistry is studied. The level, content, and sequence of topics have been chosen with a sensitivity towards the students who have had little or no background in chemistry or who have had a significant interruption since their first exposure to this topic. The text is designed primarily for students who wish to obtain the background and confidence needed to pursue a main sequence chemistry course. However, this text is also used quite successfully in the general chemistry part of a course for the health professions or in a one-semester, terminal course.

Previous editions of this text are known for their friendly, conversational style. It is hoped that this edition will improve on that reputation. Chemistry need not be intimidating. The sequence of topics in the text provides a step-by-step construction of the science, with one topic building logically on the previous one. However, since few instructors agree exactly on the most desirable sequence of topics, the text has been designed with considerable flexibility. For example, some may prefer to discuss moles and stoichiometry as early as possible. Hence, the instructor is advised to proceed directly from Chapter 3 (after Section 3-5) to Chapter 8 and then to Section 9-1 (equations) and 9-3 (stoichiometry). The author feels that it is preferable to present the topic of moles in a thorough manner, as is done in Chapter 8, instead of a superficial introduction early, with the supporting explanations and examples presented later.

There is obviously more material in this text than can be easily covered in a one-semester course. Generally, Chapters 1 through 10 can be covered, but there can be much variation, depending on the topics that the instructor wishes to emphasize. The nuclear discussion in Chapter 3 is completely optional, discussion and problems are separated from the rest of the chapter so that they can be included at any time. This topic is included here because it follows logically from the discussion of the nature of the nucleus. Many instructors appreciate its inclusion at this point because the mystique of the topic is effective in building interest early in the course. Other instructors prefer its inclusion at a later point. In any case, it can easily be omitted or included later. The discussions of orbitals and box diagrams in Chapter 4 are also optional and may be deleted without prejudice in later discussions.

By its very nature chemistry is a quantitative science. It is understood that many, if not most, students using this text require some preparation or review of the extensive mathematical concepts applied to chemical systems. This text has been written and revised with that very important fact in mind. The mathematical tools of measurements and conversions are presented early in Chapter 2 and are supplemented by detailed reviews in the appendixes. The next five chapters introduce many basic chemical concepts that are not primarily mathematical in nature. However, in Chapter 8 the quantitative nature of chemistry becomes apparent with the introduction of the counting unit of chemistry known as "the mole." In between Chapters 2 and 8, the students, with the direction of their instructor, are given an opportunity to improve their mathematical, algebraic, and problem-solving skills. There are separate appendixes on basic math (Appendix A), basic algebra (Appendix B), and scientific notation (Appendix C). These appendixes include discussions, worked-out examples, and drill problems (with answers provided). They are specially designed for application to the types of problems encountered in elementary chemistry. These three appendixes also include self-diagnostic tests so that students can easily determine the extent of review that they need.

The first three appendixes stress the basic tools needed for calculations in chemistry. The next (Appendix D) provides assistance in putting these tools to use in solving problems. The typical student beginning the study of chemistry needs more than a one- or two-page introduction to a technique of problem solving known as the factor-label or dimensional analysis method. Almost all introductory and general chemistry textbooks now use this method in problem solving. Appendix D builds on the introduction to problem solving in Chapter 2 by reviewing the origin and use of conversion factors and by many examples of one-step and multi-step problems. In addition to worked out examples, many exercises contain answers provided at the end of the appendix.

The next three appendixes provide aid to the student with specialized topics. Appendix E contains a discussion of logarithms. Although it is now easy to bypass a knowledge of the meaning of a logarithm by use of a hand calculator, many instructors prefer that students have an understanding of the basic concepts. Appendix G contains a discussion of graphs. Since much data in chemistry is conveniently represented in graphical form, a discussion of graphing techniques is appropriate in the study of basic chemistry. Appendix G discusses the use of hand calculators. Although instructions accompany the calculator, a review of the use of the functions most applicable to chemistry problems is of help to many students using this instrument for the first time in science.

Many other features of this text are designed to help students understand and organize the topics. The following is a list of these special features that are retained in this edition of the text.

1   Simple analogies are used that relate the concrete to the abstract. Analogies that are themselves easily understood can be useful in making new concepts seem real.

2   Introductions to the chapters provide a connection between what is to come and what has been previously discussed. This approach emphasizes the continuity of chemistry. Also included in the introduction are specific topics that are particularly relevant and that should be reviewed.

3   At the beginning of each chapter, a one sentence chapter purpose sets the general emphasis of the chapter. Chapter objectives follow and refer to more specific skills and concepts to be mastered.

4   At the end of each chapter, a review summarizes the important concepts of the chapter in a unique way using tables, diagrams, or flow charts when possible.

5   Numerous example problems are worked out in the text in a step-by-step manner. There are usually two or three examples of each type of problem with careful explanations of procedures.

6   End-of-chapter problems are assigned in the margins of the text after a particular topic has been discussed and sample problems worked out. This is designed to give students direction for the immediate reinforcement of a concept without affecting the continuity of the discussion.

7   End-of-chapter problems are numerous and of varying difficulty. They are categorized by topic. About 60 percent of the answers are provided in Appendix I. Many of the quantitative problems also include solutions.

8   New terms are introduced in boldface type. The definitions of the terms are in italics.

9   A comprehensive glossary of terms (Appendix H) provides easy reference to the definitions and terms used throughout the text.

10 Comprehensive review tests are included after Chapters 3, 7, 10, and 14. These are designed to integrate the material in the intervening chapters.

The main emphasis in this revision has been on improved flow and clarity within the chapters. This required the rearrangement of some topics and extensive rewriting of the discussions. Those who are familiar with previous editions will find the style more colorful and descriptive. Special consideration has been given to logical transitions between concepts. Most of the additional pages in this edition are a result of expanded discussion of concepts rather than addition of new topics. There are about fifteen percent more end of chapter problems, and many of the problems whose answers are not given in the text have been changed. A few new topics have been added such as a discussion of the origin of the names of the elements in Chapter 3, the experiments of Lord Rutherford in Chapter 4, and thermochemical equations in Chapter 9.

The Study Guide to accompany this text has also been revised to provide close support for the text. In the Study Guide, related sections within a chapter have been grouped for discussion and self-testing. In most Study Guides the chapter is reviewed in its entirety. Fifteen topics of current interest (e.g., The Antarctic Ozone Hole) have been included in the Study Guide to emphasize the role of chemistry in related sciences. The laboratory manual that accompanies the text has been revised by Professors Steve Murov and Brian Stedjee.

ACKNOWLEDGMENTS

Writing a chemistry text is certainly not an individual project. In particular, I wish to thank my colleagues in the chemistry department of Saint Louis University for their many helpful comments, especially during the preparation of the first edition. I owe special thanks to Dr. Judith Durham, who was particularly helpful. I am also grateful to my family for their assistance in proofreading, especially my wife Sharon, and my daugher, Katie. I thank my editor at Wiley, Dennis Sawicki, for his help, suggestions, and encouragement. Finally, the following people reviewed the manuscript and offered many useful comments and suggestions:

Prof. William Huggins
Pensacola Junior College

Richard Monson
California State University

Neil Coley
Chabot College

Philip Kint
University of Toledo

William Wasserman
Seattle Central Community College

James Petrich
San Antonio College

Frank Hoggard
Southwest Missouri State University

Steven Murov
Modesto Junior College

George Schenk
Wayne State University

The late Prof. Thomas Radin
San Joaquin Delta College

CONTENTS

Chapter 3   THE ATOM, THE STRUCTURE OF MATTER, AND NUCLEAR REACTIONS   63

Chapter 4 THE ARRANGEMENT OF THE ELECTRONS IN THE ATOM   103

Chapter 5 THE PERIODIC NATURE OF THE ELEMENTS   131

Chapter 6 THE NATURE OF THE CHEMICAL BOND   157

BASIC
CONCEPTS
OF
CHEMISTRY

FIRE
Of all the animals on earth only humans have dared to tame and
even understand this powerful and mysterious force.

CHEMISTRY: MATTER, CHANGES, AND ENERGY

1

PURPOSE OF CHAPTER

In Chapter 1 we introduce the discipline of chemistry and discuss the properties and changes that distinguish the various types of matter.

OBJECTIVES FOR CHAPTER

(Each objective is followed by the appropriate section in the text.) After completion of this chapter, you should be able to:

1. Describe what is meant by the scientific method. (1-2)
2. Define chemistry and matter. (1-2)
3. Distinguish between the types of pure matter—elements and compounds. (1-3)
4. Describe the three states of matter and give examples of each. (1-4)
5. Give examples of the physical properties of a substance and the physical changes it undergoes. (1-4)

6   Give examples of the chemical properties of a substance and the chemical changes it undergoes. (1-5)

7   Classify a sample of matter as either homogeneous or heterogeneous. (1-6)

8   Distinguish between the properties of mixtures and pure substances. (1-6)

9   Name and describe the various forms of energy. (1-7)

10   Classify a chemical or physical change as either exothermic or endothermic. (1-7)

11   Distinguish between kinetic and potential energy. (1-7)

12   Describe the significance of the laws of conservation of mass and energy. (1-8)

Fire—Fascinating! Powerful! Mystical! Onc can become mesmerized for hours just by gazing at the flames of a campfire as they dance and leap about. It is easy to understand why this awesome force was the source of so much mystery throughout the ages. For example, in Greek mythology there existed a god named Prometheus. Prometheus supposedly gave to animals the special tools they needed to survive in a hostile world. He incurred the wrath of the other gods, however, when he got to humans, because he gave them a tool reserved for the gods themselves. This tool was, of course, fire.

Archaeological evidence indicates that fire has been used by humans for at least 400,000 years. Fire must have been the great equalizer in prehistoric times. The raccoon has sharp vision at night, but fire allowed humans to bring light to the darkness. We have no warm, protective fur like the deer, but the heat of the fire made the long winter bearable. We do not have sharp teeth or powerful jaws like the lion, but fire made meat tender. We are not as strong or fast as many of the fierce predators, but fire held off these animals because they all dreaded this unpredictable force. It was obviously a monumental event when our ancestors overcame their natural fear of fire. In fact, it may have been the most significant event that marked our species as "human" and truly superior to other species. Other animals can communicate with each other and some can even use simple stone tools. Only humans dared to use fire, the tool of the gods.

Fire, earth, air, and water were considered by the ancient Greeks to be the basic elements of nature. For these people, fire was the transforming element that caused one substance to change into another. We now realize that fire itself is a result of the hot gases and energy associated with a transformation. As we will study later in this chapter, the transformation of one substance into another is called a **chemical reaction.** It is thus logical to state that our species became "human" thousands of years ago when it first put to use the science of chemistry in the form of fire.

Since the taming of fire, many other significant advances have been based on the observance and use of a chemical phenomenon. At first, people shaped tools from stone, but about 10,000 years ago a different naturally occurring substance gained value because it could be pounded into various shapes and sharpened to a much finer edge than ordinary stone. This substance was the metal copper. Still, this metal was relatively rare in its natural state, so it could not be utilized to a large extent. What was probably an accidental discovery opened the door to the age of metals. About 5000 B.C., someone in ancient Persia discovered that fire from the glowing coals of an earthen furnace could transform the green stone malachite into the red metal copper. Imagine the commotion that must have occurred as some prehistoric citizen insisted that a common rock could be changed into a valuable metal! The conversion of ores to metals is now part of the chemical science known as *metallurgy.* When first discovered, however, it must have been considered an example of the magic of fire.

The evidence that chemistry has been a major factor in our progress is impressive. About 3000 B.C., the Egyptians could dye cloth and embalm their dead. They were so good at these chemical processes that we can still tell what caused the death of some of the mummies from these times and even what diseases they may have had during their lifetimes. Despite their use of chemistry, however, the early Egyptians actually had no idea of why any of these procedures occurred.

Around 400 B.C. the Greeks went in another direction. They thought and argued a great deal about why things occurred in the world around them, but they didn't really put their ideas to practical use or even check out their ideas by experimentation. Still, some of their thoughts have proved to be consistent with modern developments.

The early centuries of the Middle Ages (500–1600 A.D.) are usually referred to as the Dark Ages in Europe because of the lack of art and literature and the decline of central governments in this period. In fact, there was a general loss of much of the civilization that Egypt, Greece, and Rome had previously built. In the case of chemistry, however, considerable advancement occurred. Some of the chemists of the time, especially those in Europe, were known as *alchemists* and were thought to possess magical powers. Among other things, alchemists searched for ways to transform cheaper metals such as lead and zinc into gold, which was thought to be the perfect metal.

Obviously, they never accomplished the impossible, and many good alchemists were executed for their failure. However, they may have been pleased to know that they did establish certain important chemical procedures, such as distillation and crystallization. They also discovered and prepared many previously unknown chemicals that we now refer to as elements and compounds.

Modern chemistry had its foundation in the late 1700s when the use of the analytical balance became widespread. At that time chemistry became a quantitative science, and from then on theories used to describe chemistry had to be checked and correlated with the results of experiments performed in the laboratory. From this came the modern atomic theory, first proposed by John Dalton around 1803. This theory, which in a slightly modified form is still accepted today, gave chemistry the solid base from which it could grow to serve humanity on an impressive scale. Actually, most of our understanding of chemistry has evolved in the last 100 years. In a way this makes chemistry a very young science. However, if we mark the beginning of "human" behavior as the use of fire, it is also the oldest science.

1-1 THE STUDY OF CHEMISTRY

Curiosity is defined as the desire to know and to learn about things. We all know how that can get you into trouble, but it is also why science has had such a significant impact on our daily lives. Curiosity has brought us airplanes, automobiles, plastics, synthetic fibers (Orlon, Dacron, and nylon), chemicals for treatment of cancer and other diseases, microchips for computers, and genetic engineering. The modern advancements of science make 100 years ago look almost like the stone age. At this rate, it is hard to imagine how things will be 100 years from now. In any case, like all other sciences, chemistry attracts the profoundly curious person, one who never outgrew the childhood habit of constantly asking the question "why?" As children, most scientists probably drove their parents crazy with their continual how's and why's.

Why study chemistry? The obvious answer is that it is required. But maybe you can appreciate *why* it is required. Chemistry is *the* fundamental natural science. Biology, physics, geology, as well as all branches of engineering and medicine, are based on an understanding of the chemical substances of which matter is composed. It is the beginning point in the course of studies that eventually produces all scientists, engineers, and physicians.

Besides scientists and engineers, all intelligent citizens need a foundation of scientific knowledge to make intelligent decisions concerning the future of our fragile environment. What is air and water pollution? (See Figure 1-1.) How did it get there and how do we get it out? What about our need for energy and its relation to the pollution problem? These questions and others can't be discussed intelligently without some knowledge about the nature of the matter involved. Neither should these questions be left solely to scientists. Everyone has a big stake in the answers. Thus, it seems reasonable to expect that a fundamental knowledge of chemistry should be a prerequisite not only for a specific course of study but for life in confusing and complicated times.

Chemistry can be studied and appreciated by the average person. It need

Figure 1-1 SMOG. To fully appreciate the origin and control of pollution, citizens need a basic understanding of chemistry.

not be feared. Chemistry is orderly, predictable, and entirely reasonable, which gives the study of this science an advantage over that of the social sciences. Psychology, for example, deals with the whims of human nature. The nature of humans is constantly changing. On the other hand, once we know the facts about chemicals, we can be assured that the nature of the matter that we have studied will never change.

If you enjoy mastering concepts in your mind or find a sense of satisfaction in solving a problem successfully, then you will find the study of chemistry challenging. Keep an open mind; it may be fun. The study of chemistry is much like the study of a language. As such, it requires special study habits and academic self-discipline. What follows is a checklist of questions dealing with the academic self-discipline needed in the study of this science.

1 Can You Budget Time on a Regular Basis to Study and Work Problems?

Like basketball or the piano, chemistry requires practice to be mastered. The practice should be done very soon after the topic is covered and not right before the test. One wouldn't wait until the night before the big game to first practice jump shots or the night before the recital to first practice the sonata. This text uses the margins to assign problems that emphasize a certain concept. It is wise to stop at that point and practice before you proceed. Each chapter and each topic within a chapter is like constructing a building. You can't put up the second floor until you have built a firm foundation with the first. The night before a test should be reserved for *review* and a good night's sleep—it is not the time to break new ground.

2   Are You Willing to Read Ahead?

If you were captain of a football team you would probably want to come out onto the field early just to get a look at the other team warming up. It may give you a feeling of what lies ahead. Likewise, reading ahead makes you aware of some of the concepts that you will be discussing. You may not completely understand the concepts, but that just makes you listen more intently when they are covered in lecture. Reading ahead can save time in note taking since you'll know that certain tables and definitions are available in the book.

3   Will You Attend Lectures Regularly and Take Good Notes?

You will notice (if you haven't already) that the successful students in college miss very few classes. It is almost impossible to stay up with the lecture unless you're there to hear it. A text such as this is meant to help you understand; it is not meant to replace your instructor.

Once you are there at the lecture you need to take orderly notes. Remember, you will need these notes to know what your instructor emphasizes and to help organize your review. When a problem is worked out on the board, copy the problem so that you can review it later and rework it yourself.

4   Can You Ask Questions in Class or Afterward?

You can't let a basic concept go by without owning it. If the class is such that asking questions during class is impossible (or embarrassing), ask later. Your instructor is available, and you owe yourself the answers. Step forward; this is your life, and your question is not "dumb."

5   Can You Keep at It Even if You Are Disappointed on the First Test?

An "A" is a great way to start, but what if that doesn't happen? In that case you'll have to reanalyze your study habits and try again. Don't be afraid to ask your instructor for advice in this matter. For most people, mastering chemistry takes perseverance. Give the course a chance with your best shot. You will know eventually whether you should "haul up the flag."

6   Can You Memorize?

Some definitions, names, and formulas have to be memorized. Some people say, "I don't want to memorize, I want to understand." To be sure, understanding is what you are after, but memorization is often necessary before understanding develops. In this respect studying chemistry is like studying a foreign language. Before you can learn to speak or write a new language, such things as vocabulary and verb declensions must be committed to memory.

7 Can You Do Problems Systematically and Neatly?

It is amazing how good performance in chemistry courses usually correlates with neat and orderly notes, papers, and tests. Order on paper shows an orderly mind. That helps in chemistry.

If you are able to show by your habits positive answers to these questions, you just may surprise yourself with how well this material comes to you and, as a result, just how fascinating chemistry can be.

1-2 THE NATURE OF CHEMISTRY

Humans have been strolling this good earth for hundreds of thousands of years. Yet only in the last 100 years has the vast explosion of scientific knowledge and technology (the practical application of science) occurred. After all of that time, why just now? Probably the best reason for the recent strides is the way scientists have learned to interpret the data from observations and experiments. In what is known as the **scientific method,** scientists examine data for trends or regularities in the results. From these trends a tentative explanation called a **hypothesis** is proposed to explain what is already known and to predict results of future experiments. In fact, experiments are specifically devised to test the hypothesis. If a hypothesis withstands the test of time and trial it gains the stature of a **theory.** A modern scientific theory is supported by a great deal of evidence. In addition to hypotheses and theories, which *explain* behavior, we have **laws,** which *describe* behavior with some concise mathematical relationship or statement. Under appropriate conditions, laws are not violated.

Modern scientific theories have come a long way since the time of the ancient Greeks when a theory did not have to correspond to reality. Current theories are not considered "scientific" unless they are supported by a great deal of information and are not directly contradicted by any available data. Occasionally, we hear of people who consider a modern scientific theory as little more than a guess. This is not the case. Still, new information and data are not always consistent with a particular theory. The theory must then either be adjusted to include the new data, be limited in its scope, or possibly discarded altogether.

Hypotheses, theories, and laws help the study of chemistry by allowing the classification of phenomena into categories. For example, the *periodic law* allows the chemistry of certain elements to be studied as a group rather than individually. You will find that this is a welcome simplification.

Let us now turn our attention specifically to the definition of chemistry. **Chemistry** *is that branch of science that deals with the nature of matter and the changes that matter undergoes.* It seems like a simple definition but two key words, "matter" and "changes," cover a lot of territory. We will spend the rest of the chapter on these two words. First, we look at the types of matter that constitute our world. **Matter** *is defined as anything that has mass and occupies space.* Matter is composed of substances. **A substance** *is just a particular kind of matter* such as water, a type of rock, or copper wire.

1-3 TYPES OF MATTER: ELEMENTS AND COMPOUNDS

All of the matter in and on the earth, in the sun and the moon, and in all the stars and planets of the night sky is made up of only about 90 unique substances called elements. **Elements** *cannot be broken down into simpler substances by chemical means and so are the most basic forms of matter that exist under ordinary conditions.* Many of these elements are very familiar to us in their free states. The shiny gold in a ring, the life-supporting oxygen in air, and the light but sturdy aluminum composing a can of soda are all examples of elements. The known elements are listed inside the front cover of this text. We will have more to say about the names and symbols of these elements in Chapter 3.

Our world is composed mostly of substances more complex than just the free elements. On earth, most elements are not found in their free state but are combined with other elements in compounds. **A compound** *is a substance composed of two or more elements that are chemically combined.* Thus, most ordinary substances with which we are familiar are actually compounds. For example, water is a compound composed of the elements oxygen and hydrogen; table salt is a compound composed of sodium and chlorine; and sugar is a compound composed of carbon, hydrogen, and oxygen. Although nature is composed of relatively few elements, various chemical combinations of these elements make up millions of known compounds. (See Figure 1-2.)

■
*SEE PROBLEMS
1-1 THROUGH
1-6.*

We are now ready to examine the properties of a specific element or compound. **Properties** *describe the particular characteristics or traits of a substance.* There are two kinds of properties: *physical* and *chemical.* Physical properties are discussed in the next section, which is followed by a discussion of chemical properties.

1-4 THE PHYSICAL PROPERTIES OF MATTER

The most obvious property that we would first observe about a substance is its physical state. The three **physical states** *of matter are solid, liquid, and gas.* In a later chapter we discuss how all substances are composed of fundamental particles. In the solid state, the particles remain close together in relatively fixed positions. Movement of the particles is confined mostly to vibrations about the fixed positions. Because of the fixed positions of the fundamental particles, **solids** *have a fixed shape and a fixed volume.* The particles in liquids are also close together but are free to move past one another. Thus, **liquids** *have a fixed volume but not a fixed shape.* Liquids flow and take the shape of the lower part of a container. In gases, however, the particles are in random motion and move freely in all three dimensions. Thus, **gases** *have neither fixed volume nor shape* and fill a container uniformly. (See Figure 1-3.)

Elements

gold

copper

helium

Compounds

water

butane

sugar

Figure 1-2 ELEMENTS AND COMPOUNDS. A few elements can be present on earth in their free state but most are chemically combined in compounds.

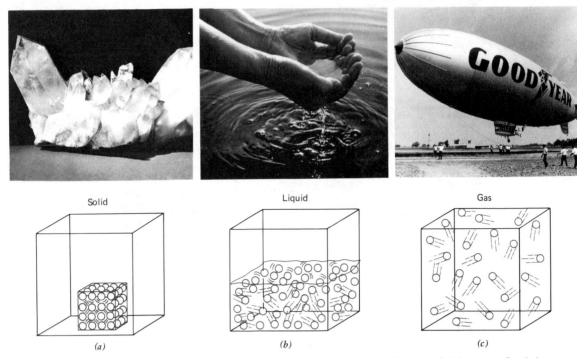

Solid   Liquid   Gas

(a)   *(b)*   *(c)*

Figure 1-3   THE THREE PHYSICAL STATES OF MATTER. *(a)* Solids have a fixed shape and volume. *(b)* Liquids have a fixed volume but an indefinite shape. *(c)* Gases have an indefinite shape and volume.

We are already familiar with many examples of all three physical states. Ice, rock, salt, and steel are substances that exist as solids; water, gasoline, and alcohol exist as liquids; ammonia, natural gas, and the components of air are present in nature as gases. In actual fact, we should specify the temperature of these substances because any one may be a gas at a high temperature (e.g., water steam), a solid at a low temperature (e.g., water ice), and a liquid in between (e.g., normal water).

The physical state of a substance at a particular temperature is an example of a physical property. **Physical properties** *can be observed without changing the substance into another substance.* Odor, color, and density or specific gravity (both relate to the ratio of mass to volume) are other examples of physical properties. Your physical properties, listed on your driver's license, may include height, weight, sex, and the color of your eyes and hair.

Another important physical property of elements and compounds is the temperature at which they change from one physical state to another. *The temperature at which a particular element or compound changes from the solid state to the liquid state is known as its* **melting point.** (The reverse is the **freezing point** and is the same temperature.) *The temperature at which an element or compound in the liquid state begins to bubble and change to the*

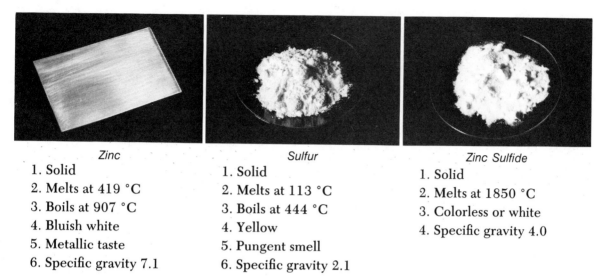

Zinc	Sulfur	Zinc Sulfide
1. Solid	1. Solid	1. Solid
2. Melts at 419 °C	2. Melts at 113 °C	2. Melts at 1850 °C
3. Boils at 907 °C	3. Boils at 444 °C	3. Colorless or white
4. Bluish white	4. Yellow	4. Specific gravity 4.0
5. Metallic taste	5. Pungent smell	
6. Specific gravity 7.1	6. Specific gravity 2.1	

Figure 1-4   PHYSICAL PROPERTIES. Zinc, sulfur, and zinc sulfide all have distinct and different physical properties.

gaseous state is known as its **boiling point.** (The reverse is the **condensation point,** also the same temperature.) Melting and boiling are examples of physical changes that a substance can undergo. **A physical change** *in a substance does not involve a change in the composition of the substance but is simply a change in physical state or dimensions.* Water, ice, and steam are all states of the same compound.

Elements and compounds are both forms of matter known as pure substances. **Pure substances** *have definite compositions with definite and unchanging properties.* Zinc and sulfur are elements and zinc sulfide is a compound; all are pure substances. In Figure 1-4 some of the physical properties of each are listed. Note particularly that although zinc sulfide is composed of two elements, zinc and sulfur, it has properties that are entirely unique and distinct from those of both of its parent elements.

Taken as a whole, the properties of each pure substance serve as a fingerprint allowing positive identification. A sample of zinc from the planet Mars would have the same physical properties as one from our own world.

1-5 THE CHEMICAL PROPERTIES OF MATTER

A substance also has chemical properties. **A chemical property** *relates to the ability or tendency to change into other substances by a chemical reaction.* Chemical properties are like a personality. One observes how the substance responds to heat or the presence of other substances to see what "its chemistry" is like. A chemical property describes a profound change in a substance known as a chemical change. *When a* **chemical change** *occurs in a substance, the chemical reaction produces other unique substances.* The decay of vegeta-

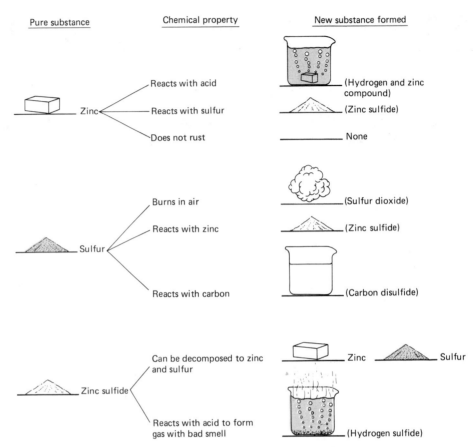

Figure 1-5   CHEMICAL PROPERTIES AND CHANGES OF SUBSTANCES.

tion, the burning of coal, and the rusting of iron all involve chemical changes. The rusting of iron, for example, involves the chemical reaction of the element iron with the element oxygen from the atmosphere. The rust (a compound, iron oxide) does not spontaneously change back into iron and oxygen. In Figure 1-5 some of the chemical properties of zinc, sulfur, and zinc sulfide are listed. Like their physical properties, the chemical properties of these pure substances can serve as a fingerprint to reveal their identity.

■
*SEE PROBLEMS
1-7 THROUGH
1-17.*

1-6 MIXTURES OF ELEMENTS AND COMPOUNDS

If we pick up a chunk of soil and look closely, it is obvious that we have a mixture of materials. We can see bits of sand, some black matter, and perhaps pieces of vegetation. On the other hand, if we mix sugar and water the sugar seems to disappear into the water. The fact that we have a mixture is certainly not obvious to the eye.

The sugar water is a **homogeneous mixture** known as a **solution. Homogeneous matter** *is the same throughout and contains only one phase. A* **phase** *is one*

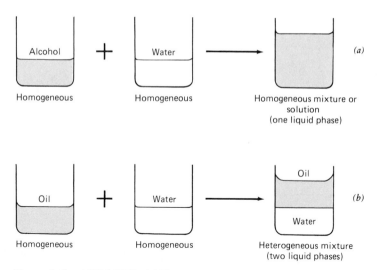

Figure 1-6 MIXTURES. *(a)* Water and alcohol form a homogeneous mixture (solution). *(b)* Water and oil form a heterogeneous mixture.

physical state (solid, liquid, or gas) with distinct boundaries and uniform properties. Note that pure matter, elements and compounds, discussed earlier, are also forms of homogeneous matter (if they are present in only one physical state). *Homogeneous matter may be either a pure substance or a solution.* Copper pennies, air, sugar, water, an alcohol – water solution [see Figure 1-6a] are all examples of homogeneous matter.

The dividing line between a pure substance and a homogeneous mixture is not always sharp. Complete purity is difficult and sometimes even impossible to achieve. For example, drinking water contains some dissolved solids and gases. Even rain water contains dissolved air. The gold in a ring contains other metals to give it strength. What is considered "pure" by the chemist depends on the application. For example, if a sample of matter is composed of 99% one element or compound it may be considered pure for most purposes. On the other hand, silicon must be "superpure" to be used in computer chips. In this case, it is composed of more than 99.9999% silicon.

Heterogeneous matter, like our handful of soil, *is a nonuniform mixture containing two or more phases with definite boundaries between the phases.* Examples of heterogeneous mixtures are salt and sand (two solid phases), oil and water (two liquid phases (see Figure 1-6b), and shaving cream foam (liquid and gas phase). Sometimes it is not easy to tell whether matter is homogeneous or heterogeneous. Creamy salad dressing and fog both appear uniform and thus homogeneous to the naked eye. If we were to slightly magnify each, however, the truth would become apparent. The salad dressing has little droplets of oil suspended in the vinegar (two liquid phases), and the fog has droplets of water suspended in the air (liquid and gas phases).

Heterogeneous mixtures can usually be separated into homogeneous components by simple laboratory procedures. For example, a solid suspended in a

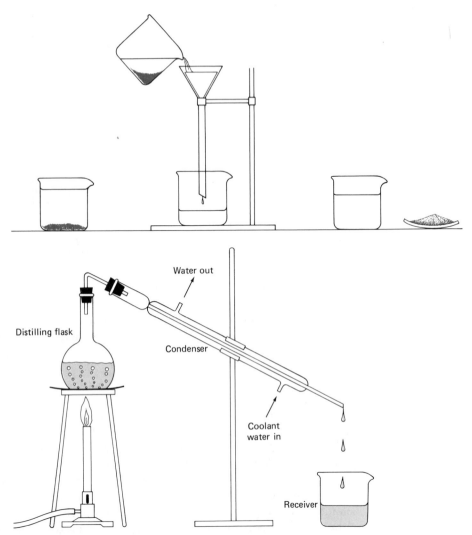

Figure 1-7   LABORATORY PROCEDURES USED TO SEPARATE MIXTURES. (a) A heterogeneous mixture of a solid in a liquid can be separated by filtration. (b) A homogeneous mixture of a solid dissolved in a liquid can be separated by distillation.

liquid can be separated by *filtration*. (See Figure 1-7a.) The solid remains on the filter and the liquid passes through. Separation of components of a homogeneous mixture (solution) can also be accomplished by laboratory procedures although the process can be somewhat more involved. For example, when table salt is mixed with water it disperses completely (all the way to the fundamental particles of the compound), making filtration impossible. In this case, a procedure known as *distillation* is used whereby the water is boiled away and condensed, eventually leaving the dry table salt in the flask. (See Figure 1-7b.)

Usually, we can distinguish a heterogeneous mixture from pure matter

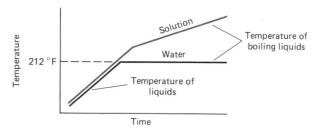

Figure 1-8 BOILING TEMPERATURES OF PURE WATER AND A SOLUTION. When pure water boils the temperature remains constant; when a solution boils the temperature slowly rises.

(elements and compounds) by visual observation. But it is not so easy with a homogeneous mixture. How then can we distinguish a homogeneous mixture from a pure substance? The answer is that we must examine the physical properties. Mixtures have properties that vary with the proportion of the components; elements and compounds have definite and unchanging properties. For example, solid table salt (sodium chloride) melts at exactly 1083 °C and water freezes at exactly 0 °C (32 °F). A solution of salt in water freezes anywhere from −18 °C to just under 0 °C depending on the amount of salt dissolved.

A mixture not only melts and boils at different temperatures than its components, but it does so over a range of temperatures. For example, if pure water is boiled until it is all gone, we would notice that the temperature remains at exactly 100 °C (212 °F) from start to finish. When a solution boils, however, the temperature of the boiling solution changes as the liquid boils away (except in some unusual cases). (See Figure 1-8.)

Another variable property of solutions that we can appreciate is the taste of sweetened coffee or tea. The more sugar that is dissolved, the sweeter it tastes, although there is a limit to how much sugar can be dissolved in a given amount of water.

A solution is much like a mixture of colored paints. If you mix red and blue you get violet. The shade of violet depends on the proportions of the two components. The violet color produced is simply a blend of the two and retains characteristics of the two original colors.

Sometimes, even advanced students of chemistry confuse mixtures of elements with compounds. For example, in Figure 1-9 we show a heterogeneous mixture of two elements, zinc and sulfur, on the left. Both solid phases are clearly visible and, if necessary, we could easily separate the two components. If this mixture is ignited with a hot flame, however, a vigorous chemical reaction occurs, forming the compound zinc sulfide. Through the magic of chemistry the mixture was transformed into a unique pure substance. *The zinc sulfide was formed from a mixture but is not itself a mixture.* It would take some very involved chemical procedures to decompose the compound back to its elements.

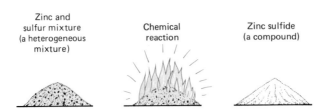

Figure 1-9   FORMATION OF ZINC SULFIDE. In this chemical reaction a mixture of elements is changed into a compound.

Formation of a compound from a mixture of elements is much like baking a cake. We mix eggs, flour, and other ingredients and allow the mixture to heat. What comes out is a unique substance that no longer looks like, tastes like, or actually is the mixture we put in the oven. The mix of ingredients has been transformed into a new homogeneous substance (we hope).

■
*SEE PROBLEMS
1-18 THROUGH
1-27.*

**1-7 ENERGY
CHANGES IN
CHEMICAL
REACTIONS**

When a log burns, it is obvious that more than just gas and ashes are formed. Large amounts of heat are also released by the combustion process. *Chemical reactions that release heat are called* **exothermic,** *and those that absorb heat are called* **endothermic.** An example of an endothermic process involves an "instant cold pack." When the compounds ammonium nitrate (a solid) and water are brought together in a plastic bag, a solution is formed. The endothermic solution process causes enough cooling to make an ice pack useful for treating sprains and minor aches for athletes. (See Figure 1-10.)

Heat is only one form of energy. But what is energy? When we have no energy left at the end of the day, the implication is clear: We have no capacity to do work. That is exactly how energy is defined. **Energy** *is the capacity or the ability to do work.* Energy exists in several forms, and chemistry is vitally concerned with any changes in energy involving chemical reactions. Almost all of our energy on earth originates from the sun in the form of *radiant* or *light energy.* (Light energy will be discussed in Chapter 4.) By a process called photosynthesis, a living tree can transform the solar, radiant energy into *chemical energy.* In this process, energy-poor compounds in the environment are transformed into energy-rich compounds in the tree. When logs from the tree

Figure 1-10   INSTANT COLD PACK. When capsules of ammonium nitrate are broken and mixed with water, a cooling effect results.

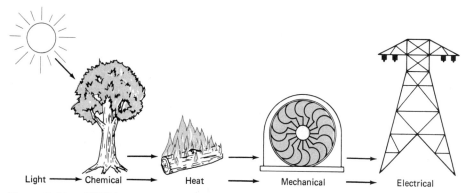

Figure 1-11   ENERGY. Energy is neither created nor destroyed but can be transformed.

burn, the chemical energy is transformed into *heat energy*, which we observed earlier. In this process, energy-poor compounds are also formed and returned to the environment. If the heat energy is used to produce steam that is used to turn a turbine, the energy is converted into *mechanical energy*, which is then converted into *electrical energy*. (See Figure 1-11.) The chemical reaction that occurs in a car battery, however, is used to convert chemical energy directly into electrical energy.

In addition to the forms of energy, there are two *types* of energy depending on whether the energy is available but not being used or actually in use. For example, a weight suspended above the ground has energy available because of its suspended position and the attraction of gravity for the weight. *Energy that is available because of position or composition is known as* **potential energy.** Other examples are water stored behind a dam and a coiled spring. In our example, when the weight is released, the energy can be put to use by the downward motion. *Energy resulting from motion is called* **kinetic energy.** To

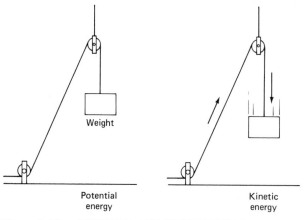

Figure 1-12   POTENTIAL AND KINETIC ENERGY. Potential energy is energy of position; kinetic energy is energy of motion.

move the weight back into position, energy must be supplied. (See Figure 1-12.) In future chapters, potential and kinetic energy will play a part in discussions of some chemical systems.

1-8 THE CONSERVA-TION OF MASS AND ENERGY

In the discussion of mixtures (Section 1-6) we illustrated the chemical reaction of a mixture of zinc and sulfur to produce zinc sulfide. (See Figure 1-9.) If the proportions of zinc and sulfur are just right, only zinc sulfide remains after the reaction. Furthermore, the mass of the zinc sulfide is exactly the same as that of the zinc and sulfur mixture from which it was formed. Not more than 200 years ago, however, scientists were still puzzled when wood burned and most of the mass seemed to disappear. At the time, this was explained by various theories. Our current understanding of chemical changes is based on the **law of conservation of mass,** *which states that matter is neither created nor destroyed in a chemical reaction.* Two hundred years ago the involvement of gases in chemical reactions was not fully understood. It is now known that the mass of the wood plus the mass of the oxygen in the air used in combustion equals the mass of the gases formed plus the mass of the ashes. (See Figure 1-13.) The gases involved are all invisible but still have mass.

In Chapter 8, chemical reactions will be introduced from a quantitative point of view. You will notice that many of the calculations are based on the law of conservation of mass.

In the discussion of energy changes in chemical reactions (Section 1-7) we illustrated the various changes in energy from one form to another in Figure 1-11. In all of these energy transformations the total energy did not change. Some transformations, such as mechanical to electrical, are inefficient because some of the energy is lost as heat, but again the total energy is constant. The **law of conservation of energy** *states that energy cannot be created nor destroyed but only transformed from one form to another.*

For our purposes in chemistry, the laws of conservation of mass and energy are completely valid and can be applied to an understanding of the nature of matter and its changes. In actual fact, the laws turn out to be less exact than was originally thought. In 1905 Albert Einstein proposed the now well-known relationship between mass and energy:

$$E = mc^2$$

$$\text{where } E = \text{energy}$$
$$m = \text{mass}$$
$$c = \text{speed of light}$$

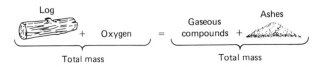

Figure 1-13 COMBUSTION OF WOOD. In a chemical reaction mass is conserved.

This amazing theory tells us that, in fact, energy results from a loss of mass. Because a small amount of mass produces a tremendous amount of energy, the mass lost or gained in a chemical reaction is far too small to be detectable. Therefore, such changes can be ignored. In nuclear reactions, however, there is a significant mass loss, which results in a vast release of energy. Nuclear energy is another form of energy and is the source of the energy of the sun. Nuclear energy is discussed in Chapter 3.

■
SEE PROBLEMS 1-28 THROUGH 1-35.

1-9 CHAPTER REVIEW

Chemistry is a branch of science that deals with the nature of matter and the changes it undergoes. Chemistry, like all sciences, is systematized to a large extent by hypotheses, theories, and laws.

All of the earth as well as the rest of the universe is composed of a relatively few basic substances called elements. On earth, most elements are not found in the free state but are chemically combined with other elements to form compounds. Both elements and compounds are known as pure substances because they have definite, unchanging physical and chemical properties. They also undergo definite and particular physical and chemical changes.

	Physical	Chemical
Property	Describes properties that do not result in a change into another substance	Describes types of chemical reactions the substance undergoes
Change	Changes in dimensions or physical state	Changes into other substances

When we look about us, the elements and compounds we observe are usually not found in the pure state but are often present as mixtures. Even drinking water contains some dissolved solids and air. Heterogeneous mixtures contain two or more distinct phases; homogeneous mixtures are the same throughout and exist in one phase. In contrast to pure substances, the properties of mixtures are variable and depend on the ratio of the components. In the diagram on the next page we start with a heterogeneous mixture, the most complex type of impure substance, and proceed step-by-step to the simplest pure substance, an element.

Energy is also involved in chemical reactions. The energy involved can be classified as to form and type.

Type		Form
Kinetic (energy of motion) ⎫ Potential (energy of position) ⎭ ——→ ENERGY ←——		⎧ Chemical ⎪ Heat ⎨ Light ⎪ Electrical ⎩ Mechanical

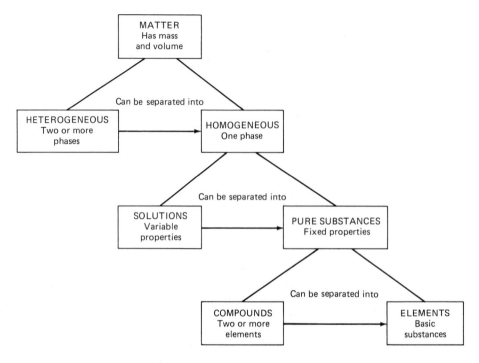

One form of energy can be converted into another. An important conversion in chemistry is between chemical and heat energy. When chemical energy is converted into heat energy, as when a log burns, the reaction is said to be exothermic. When heat energy is converted into chemical energy, the reaction is said to be endothermic.

In all transformations, whether between forms of energy or between substances in a chemical reaction, both mass and energy are conserved. Einstein's law tells us that there is a relationship between mass and energy, but the interchange between mass and energy in chemical reactions is negligible.

■ EXERCISES

Throughout the text, answers to all problems except those with the problem number in color are given in Appendix I.

ELEMENTS AND COMPOUNDS

1-1 List the elements that you think occur in the free state on earth. (Refer to the list of elements inside the front cover.)

1-2 What is the difference between an element and a compound?

1-3 Two centuries ago, water was thought to be an element. Why do you think scientists decided that it is a compound?

1-4 Identify the following as either an element or a compound. Refer to the list of elements inside the front cover if necessary.
(a) carbon monoxide
(b) hydrogen
(c) iron
(d) titanium dioxide
(e) potash
(f) sodium bicarbonate

1-5   Identify the following as either an element or a compound. Refer to the list of elements inside the front cover if necessary.
(a) ammonia   (d) mercury
(b) hydrogen   (e) stannous
    peroxide   fluoride
(c) aspirin   (f) uranium

1-6   A compound can be decomposed into two other substances. Are the two substances both elements?

PHYSICAL AND CHEMICAL PROPERTIES AND CHANGES

1-7   Which of the following describes the liquid phase?
(a) It has a fixed shape and a fixed volume.
(b) It has a fixed shape but not a fixed volume.
(c) It has a fixed volume but not a fixed shape.
(d) It has neither a fixed shape nor a fixed volume.

1-8   Which state of matter is compressible? Why?

1-9   A fluid is a physical state that flows so that it can be poured. What state or states can be classified as fluids?

1-10   Identify the following as either a physical or a chemical property.
(a) Diamond is the hardest known substance.
(b) Carbon monoxide is a poisonous gas.
(c) Soap is slippery.
(d) Silver tarnishes.
(e) Gold does not rust.
(f) Carbon dioxide freezes at -78 °C.
(g) Tin is a shiny, gray metal.
(h) Sulfur burns in air.
(i) Aluminum has a low density.

1-11   Identify the following as either a physical or a chemical property.
(a) Sodium burns in the presence of chlorine gas.
(b) Mercury is a liquid at room temperature.
(c) Water boils at 100 °C at sea level.
(d) Limestone gives off carbon dioxide when heated.
(e) Hydrogen sulfide has a pungent odor.

1-12   Identify the following as a physical or a chemical change.
(a) frying of an egg
(b) vaporization of dry ice
(c) boiling of water
(d) burning of gasoline
(e) breaking of glass

1-13   Identify the following as a physical or a chemical change.
(a) tanning of leather
(b) fermentation of apple cider
(c) compression of a spring
(d) grinding of a stone

1-14   Can an element be distinguished from a compound by examination of only its physical properties?

1-15   When table sugar is heated to its melting point it bubbles and turns black. When it cools it remains a black solid. Describe the changes.

1-16   A pure substance is a green solid. When heated, it gives off a colorless gas and leaves a brown, shiny solid that melts at 1083 °C. The shiny solid cannot be decomposed to simpler substances but the gas can. List all of the properties given and tell whether they are chemical or physical. Tell whether each substance is a compound or an element.

1-17   A pure substance is a greenish-yellow, pungent gas that condenses to a liquid at -35 °C. It undergoes a chemical reaction with a certain substance to form a white solid that melts at 801 °C and a brown, corrosive liquid with a specific gravity of 3.12. The white solid can be decomposed to simpler substances but the gas and the liquid cannot. List all of the properties given and tell whether they are chemical or physical. Tell whether each substance is a compound or an element.

MIXTURES AND PURE SUBSTANCES

1-18   Carbon dioxide is a compound that is not a mixture of carbon and oxygen. Explain.

1-19   When a teaspoon of solid sugar is dissolved in a glass of liquid water, what phase or phases are present after mixing?
(a) liquid only
(b) still solid and liquid
(c) solid only

1-20 Identify the following as homogeneous or heterogeneous matter.
  (a) gasoline
  (b) dirt
  (c) smog
  (d) alcohol
  (e) a new nail
  (f) vinegar
  (g) ice cubes in water
  (h) aerosol spray
  (i) air

1-21 Identify the following as homogeneous or heterogeneous matter.
  (a) a cloud
  (b) dry ice
  (c) whipped cream
  (d) bourbon
  (e) natural gas
  (f) a grapefruit

1-22 In what physical state or states does each of the substances listed in Problem 1-20 exist?

1-23 In what physical state or states does each of the substances listed in Problem 1-21 exist?

1-24 Tell whether each of the following properties describes a heterogeneous mixture, a solution (homogeneous mixture), a compound, or an element.
  (a) a homogeneous liquid that when boiled away leaves a solid residue
  (b) a cloudy liquid that after a certain time seems more cloudy toward the bottom
  (c) a uniform red solid with a definite and sharp melting point that cannot be decomposed into simpler substances
  (d) a colorless liquid that boils at one unchanging temperature and can be decomposed into simpler substances
  (e) a liquid that at first boils at one temperature but as the boiling continues, the temperature of the boiling liquid slowly rises (There is only one liquid phase.)

1-25 Tell whether each of the following properties describes a heterogeneous mixture, a solution (a homogeneous mixture), a compound, or an element.
  (a) a nonuniform powder that when heated first turns mushy and continues to melt as the temperature rises
  (b) a colored gas that can be decomposed into a solid and another gas (The entire sample of gas seems to have the same chemical properties.)
  (c) a sample of colorless gas, only part of which reacts with hot copper

1-26 List the following "waters" in order of increasing purity: ocean water, rain water, and drinking water. Explain.

1-27 Iron is attracted to a magnet but iron compounds are not. How could you use this information to tell whether a mixture of iron and sulfur forms a compound when heated?

MASS AND ENERGY

1-28 When a log burns, the ashes have less mass than the log. When zinc reacts with sulfur, the zinc sulfide has the same mass as the combined mass of zinc and sulfur. When iron burns in air, however, the compound formed has more mass than the original iron. Explain in terms of the law of conservation of mass.

1-29 From your own experiences tell whether the following processes are exothermic or endothermic.
  (a) decay of grass clippings
  (b) melting of ice
  (c) change in an egg when it is fried
  (d) condensation of steam
  (e) curing of freshly poured cement

1-30 A car battery can be recharged after the engine starts. Trace the different energy conversions from gasoline to the battery.

1-31 Windmills are used to generate electricity. What are all of the different forms of energy involved with generation of electricity by this method?

1-32 Identify the principal type of energy (kinetic or potential) exhibited by each of the following.
  (a) a car parked on a hill
  (b) a train traveling 60 miles/hr
  (c) chemical energy
  (d) an uncoiling spring
  (e) a falling brick

1-33 Identify the following as having either potential or kinetic energy or both.
  (a) an arrow in a fully extended bow
  (b) a baseball traveling high in the air

(c) two magnets held apart

(d) a chair on the fourth floor of a building

1-34 When you apply your brakes to a moving car, the car loses kinetic energy. What happens to the lost energy?

1-35 When a person plays on a swing, at what point in the movement is kinetic energy the greatest and at what point is potential energy the greatest? Assume that once started, the person would swing to the same height each time without an additional push. At what point is the total of the kinetic energy and the potential energy the greatest?

MEASUREMENTS
Chemistry is a quantitive science requiring many careful measurements.

MEASUREMENTS
IN CHEMISTRY

PURPOSE OF CHAPTER

In Chapter 2 we explore certain physical properties that are described by measurements. Included are the significance of the numerical quantity of the measurement and the units and techniques used to convert between units.

OBJECTIVES FOR CHAPTER

After completion of this chapter, you should be able to:

1 Distinguish between the precision and accuracy of a measurement. (2-1)

2 Determine the number of significant figures in a measurement. (2-1)

3 Express a result of a mathematical operation to the proper decimal place (addition and subtraction) or to the proper number of significant figures (multiplication and division). (2-2)

4 Convert a number to scientific notation. (2-3)

5   Carry out mathematical operations with numbers expressed in scientific notation. (2-3, Appendix C)

6   Name the basic metric units and their symbols. (2-4)

7   Write the relationships of desired units having the prefixes **kilo, centi,** and **milli.** (2-4)

8   Distinguish between mass and weight. (2-4)

9   Make conversions among metric units and between English and metric units using the factor-label method. (2-5)

10   Convert among the Celsius, Kelvin, and Fahrenheit temperature scales. (2-6)

11   Calculate the density of a substance and use it as a conversion factor between mass and volume. (2-7)

12   Distinguish among density, buoyancy, and specific gravity. (2-7)

13   Define the units for the amount of heat: the calorie and the joule. (2-8)

14   Calculate the specific heat of a substance given appropriate experimental data and use it as a conversion factor among heat, temperature change, and mass. (2-8)

This world in which we live seems to be getting more quantitative by the *minute*. Whether we are eating "quarter-*pounders*," paying two *dollars* for a "*foot*-long hot dog," or getting 23 *miles* to a *gallon* of gasoline it seems that we are always having to deal with units of measurement. **Measurements** *determine quantity, dimensions, or extent of something, usually in comparison with a standard.* Standards range from the length of a long dead king's foot (one foot) to the distance light travels in a vacuum in 1/299,792,458th of a second (one meter). Obviously, the latter standard is much more precise and dependable than a king's foot.

In the last chapter, we discussed how a substance can be identified by both its physical and chemical properties. In this chapter, we concentrate on how some of the physical properties are described. Many physical properties of a substance are qualitative, such as color, odor, and physical state. Others, however, are quantitative. These include mass, dimensions, volume, temperature, density, and specific heat. All of these properties require a numerical quantity and a unit of measurement (e.g., 3.65 lb). You are probably most familiar with English units but less so with the units of science which are basically all from the metric system. However, this unfamiliarity can be used to our advantage. We can use this as an opportunity to introduce an extremely useful method of problem solving for converting from one system to another. This method is called the factor-label method and it will serve us well here and in later chapters where conversion problems are common.

It is hard to imagine how a mechanic could repair an automobile engine without being able to use such tools as wrenches and screwdrivers. It is also hard to picture how one could grasp the meaning of measurements without the ability to use mathematics. "Why is there so much math in chemistry?" is a question often heard from the student new to this science. The answer is that chemistry, like physics and astronomy, is a **physical science**, that is, *a science concerned with the natural laws of matter other than those laws concerned with life (biology, botany)*. As we will see throughout this text, most of these natural laws are based on quantitative, reproducible measurements.

The math involved in this course and in most general chemistry courses at the college or university level is not at all awesome. In fact, no more than a year of high school algebra is needed. Still, many students need a review of mathematics. As you know, being in good physical shape allows physical exertion in sports to be relaxing, challenging, and just downright fun. If you're not in shape, exertion is not only painful but unsatisfying. Being in good mathematical shape has a similar effect on the study of chemistry. The normal sequence of chapters in this text will allow you time to review and strengthen your math background before the quantitative aspects of chemistry are introduced in force in Chapter 8. Students with specific needs can take advantage of this time to "get in shape."

At this point, you should notice that a fair portion of the back of this text is devoted to the appendixes. These provide self-tests to aid you in assessing your needs in addition to reviews, solved problems, and some exercises with answers. These appendixes are not in the back of the book because they are any less important than material present in the regular chapters. However, there is a wide variation in both the amount and type of material that each student must review. Therefore, it is not necessary to interrupt the flow of topics each time a mathematical procedure is encountered. When it is advisable to assess your mathematical needs you will be directed to do so by a notation in the margin. You are strongly urged to check your needs at this point and to undertake a comprehensive review if necessary. Almost everyone using this text requires at least some review, so don't feel like it is unusual to need some basic math, algebra, or problem-solving work.

Before we look into the quantitative properties discussed above, we will discuss the numerical quantities that are used. The topics center on how we express the number properly (Section 2-1), how we express a number that results from a mathematical operation (Section 2-2), and then how we conveniently express the very large and the very small numbers that we are required to use in chemistry (Section 2-3).

2-1 THE NUMERICAL VALUE OF A MEASUREMENT

How meaningful is the numerical part of a measurement? Perhaps, we can use an example to illustrate this question. In most large cities there is one official who is an expert at estimating the size of a crowd. Let's say that this official estimates the size of a certain gathering at 9000 people, and then notices that eight people leave. Should the new estimate be changed to 8992? No, of course not. The original number wasn't that good. The three zeros in the estimate of 9000 were not really measured numbers (as opposed to the zero in 108), but were there merely as fillers to indicate the magnitude of the number or, in other words, to locate the decimal point. In this example, the crowd would have to diminish by at least 500 before the official might feel justified in changing the estimate of the crowd to 8000. In the number 9000 only one of the numbers, the 9, is "significant." *The number of* **significant figures** *in a measurement is simply the number of measured digits and refers to the precision of the measurement.* **Precision** *refers to how close repeated measurements come to each other.* For example, repeated estimates of the crowd of 9000 may range from 8000 to 10,000, indicating an uncertainty of ± 1000. If the same crowd were seated in the bleachers of a stadium instead of milling about, a more *precise* estimate would be possible. In this case, repeated estimates might range from 9400 to 9600 with an average of 9500. Uncertainty has been reduced to ± 100. If the crowd went through a turnstile, an even more precise number could be given. In any case, in a measurement the significant figure farthest to the right is considered to be estimated.

In another example, the precision of a measurement is illustrated in Figure 2-1. The clock in the tower measures time to the nearest minute. Note that even so, the actual minute must be estimated. The nondigital wrist watch measures time more *precisely*, since the seconds as well as the minute can be read. Again, the last digit of the second hand is estimated. Finally, the stop-

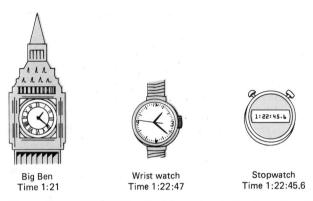

Big Ben
Time 1:21

Wrist watch
Time 1:22:47

Stopwatch
Time 1:22:45.6

Figure 2-1 PRECISION. Precision refers to the "exactness" of a measurement.

watch used to time track events is even more precise, since it can measure time to tenths of a second.

In Figure 2-1, the actual time can be compared with the standard, known as Greenwich time, which is established at an observatory in Greenwich, England. In the example the Greenwich time is 1:22:45.83. Note that the stopwatch time is not only the most precise, since its value has the largest number of significant figures, but it is also the most accurate. **Accuracy** *refers to how close the value of the measurement is to the true value.* In scientific measurements the most precise number is generally (but certainly not always) the most accurate.

As mentioned, the number of significant figures in a measurement is simply the total number of measured numbers. For example, 76.3 has three significant figures, and 4562 has four significant figures. That would be all there is to it except for the number zero. Zero serves two functions: It can be a legitimate number indicating "none," or simply a filler to locate a decimal point (like the zeros in the crowd of 9000 we had earlier). It is unfortunate that centuries ago when numbers were invented no one invented a symbol that could be used in place of a zero that is not significant. Since we don't have a separate symbol, we have some rules that tell us whether a zero is a significant figure or not.

1 When a zero is between other digits, it is significant (e.g., 7 0 9 has three significant figures).

2 Zeros to the right of a digit and to the right of the decimal point are significant (e.g., 8.0 has two significant figures as does 7.9 and 8.1; 5.7 0 0 has four significant figures).

3 Zeros to the left of the first digit are not significant. They are used to locate the decimal point (e.g., 0.0 0 7 8 has two significant figures, 0.0 4 0 6 0 has four significant figures).

4 Zeros to the left of a decimal point when no decimal point is shown may or may not be significant. It is usually assumed that they are not (e.g., 9 0 0 0 has one significant figure and <u>66</u>00 has two). What if a number such as 890 is actually known to three significant figures? That is, the zero is really a zero. This is a tough question. Some texts use a line over the zero to indicate that it is significant i.e., 89$\overline{0}$). As we will see shortly, scientific notation gives us a way to solve the problem. In most calculations in this text, measurements are expressed with three significant figures. Therefore, in calculations where other numbers have three significant figures we will assume that numbers such as 890 also have three significant figures.

EXAMPLE 2-1

How many significant figures are in the following measurements? (a) 1508 cm (b) 300.0 ft (c) 20.003 lb (d) 0.00705 gal

ANSWERS

(a) four (rule 1) (b) four (rule 2) (*Note:* Since the zero to the right of the decimal point is significant, the other two zeros are also significant because they lie between significant figures.) (c) five (rule 1) (d) three (rules 1 and 3)

SEE PROBLEMS 2-1 THROUGH 2-5.

2-2 SIGNIFICANT FIGURES AND MATHEMATICAL OPERATIONS

REFER TO APPENDIX G FOR DISCUSSION OF CALCULATORS.

What happens when we have to add, multiply, or divide measurements of various degrees of precision? Obviously, it is necessary to express the answer to the proper number of significant figures. This exercise has become even more important in recent years because of the use of the inexpensive hand calculator. These little tools are immense time savers but they are not as precise as they may initially appear. For example, 7.8 divided by 2.3 reads 3.33913043 on the display of a calculator. But if the original numbers had units indicating that they were measurements (e.g., 7.8 kg and 2.3 mL) the calculator's answer is much too precise. The calculator just assumes that all numbers punched into it have an unlimited number of significant figures. Thus, we must know how to express the answer properly regardless of what the calculator reads.

First, we consider how to express an answer when measurements are being added and subtracted. *When numbers are added or subtracted, the answer should be rounded off to the place farthest to the right in which all numbers have a significant figure.* Since most of the numbers expressed in this text are in decimal fraction form, we will look to the place farthest to the right of the decimal common to all numbers. This rule is illustrated by the following summation.

$$10.68$$
$$0.473$$
$$\underline{1.32}$$
$$12.473 = 12.47$$

Note that the "3" in the addition cannot be expressed in the answer, because two numbers are not known to that many decimal places. In fact, since the last significant figure in a measured number is estimated, it is assumed that there is an uncertainty of ± 1 in that number. Therefore, the number above is "rounded off" to 12.47. It is understood, however, that the true value may actually range from 12.46 to 12.48.

The rules for **rounding off** a number are as follows (the examples are all expressed in three significant figures):

1   If the digit to be dropped is less than 5, simply drop that digit (e.g., 12.44 rounds off to 12.4).

2   If the digit to be dropped is 5 or greater, increase the preceding digit by one (e.g., 0.3568 rounds off to 0.357 and 13.75 rounds off to 13.8).

In addition and subtraction we must also be careful with manipulations involving numbers containing nonsignificant zeros. The final answer must be rounded off to show the same number of nonsignificant zeros as in the original number. This can be illustrated with the crowd estimated at 9000 people discussed earlier. In this case, only the nine was a significant figure.

$$
\begin{array}{ccc}
9000 & 9000 & 9000 \\
-8 & -80 & -800 \\
\hline
8992 = \underline{9000} & 8920 = \underline{9000} & 8200 = \underline{8000}
\end{array}
$$

REFER TO APPENDIXES A AND B FOR A REVIEW OF BASIC MATH AND ALGEBRA.

EXAMPLE 2-2

Carry out the following calculations, rounding off the answer to the proper decimal place.

$$
\begin{array}{ccc}
7.56 & 14{,}000 & \\
+0.375 & +580 & 0.0327 \\
+14.2203 & +75 & -0.00068 \\
\hline
22.1553 = \underline{22.16} & 14{,}655 = \underline{15{,}000} & 0.03202 = \underline{0.0320}
\end{array}
$$

In multiplication and division the concern does not center on the location of the decimal point. Instead, *the answer must be expressed with the same number of significant figures as the multiplier, dividend, or divisor with the least number of significant figures.* In other words, the answer is only as precise as the least precise part of the problem (i.e., the chain is only as strong as its weakest link).

EXAMPLE 2-3

Carry out the following calculations, rounding off the answer to the proper number of significant figures.

(a) 2.34 in. $\times$ 3.225 in.

The answer on the calculator reads 7.5465. Since the first multiplier has three significant figures and the second has four, the answer should be expressed to three figures. The answer is rounded off to 7.55 in.^{2}

(b) 11.688 ft/4.0 sec

The answer, 2.922, should be rounded off to two significant figures. The answer is 2.9 ft/sec.

(c) 148.6 cm $\times$ 0.224 cm $= 33.2864$ cm$^2 = $ 33.3 cm^2

(d) $\dfrac{875.6 \text{ cm}^2}{62 \text{ cm}} = 14.122$ cm $= $ 14 cm

The planet Saturn is 10^{12} meters away at its closest approach to Earth.

2 meters tall

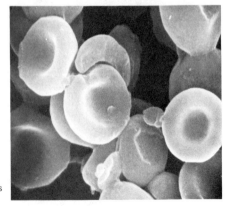

The diameter of one of these cells is about 10^{-6} meters

Figure 2-2 POWERS OF TEN. In astronomy extremely large numbers are encountered. In cell biology, very small numbers are used.

Before leaving the topic of significant figures, we should mention the effect of *exact* numbers on a calculation. An exact number is one that is either defined exactly or originates from a count. For example, 1 gal = 4 qt is an exactly defined relationship. As such it has unlimited significant figures (i.e., 1.00 . . . etc. gal = 4.00 . . . etc. qt). Counted numbers are also exact and have unlimited significant figures. For example, if there are 32 students in a certain classroom, the number 32 is considered exact (i.e., 32.00 . . . etc.). As we will see in this chapter, an exact relationship does not limit or affect the number of significant figures in a calculation. Relationships that originate from a measurement are not exact and limit the number of significant figures in a calculation as described. You will become more familiar with exact numbers as we proceed.

SEE PROBLEMS 2-6 THROUGH 2-13.

2-3 EXPRESSING LARGE AND SMALL NUMBERS: SCIENTIFIC NOTATION

Chemistry as well as many other sciences (see Figure 2-2) uses some huge numbers and some very small ones as well. For example, we will soon deal with an important number so large that it challenges our imagination to deal with it. This number is

$$602,000,000,000,000,000,000,000$$

With so many zeros this number is extremely awkward. A much more convenient way of both writing and reading the above number is

6.02×10^{23} ("six point oh two times ten to the twenty-third")

The latter number is expressed in scientific notation. **Scientific notation *expresses zeros used to locate the decimal point in powers of 10.* A number expressed in scientific notation has two parts: *the* coefficient *is the number that is multiplied by 10 raised to a power, and the* exponent *is the power to which 10 is raised.*

$$6.02 \times 10^{23}$$

Exponent

Coefficient

The following are some powers of 10 and their equivalent numbers.

$$10^0 = 1 \qquad\qquad 10^{-1} = \frac{1}{10^1} = 0.1$$

$$10^1 = 10 \qquad\qquad 10^{-2} = \frac{1}{10^2} = 0.01$$

$$10^2 = 10 \times 10 = 100 \qquad\qquad 10^{-3} = \frac{1}{10^3} = 0.001$$

$$10^3 = 10 \times 10 \times 10 = 1000 \qquad\qquad 10^{-4} = \frac{1}{10^4} = 0.0001$$

$$10^4 = 10 \times 10 \times 10 \times 10 = 10,000 \qquad \text{etc.}$$

etc.

The *standard* method of expressing numbers in scientific notation is with one digit to the left of the decimal point (e.g., 4.56×10^6 rather than 45.6×10^5).

EXAMPLE 2-4

Express each of the following numbers in scientific notation with one digit to the left of the decimal point.

(a) 47,500 (b) 5,030,000 (c) 0.0023 (d) 0.0000470

SOLUTION

(a) The number 47,500 can be factored as $4.75 \times 10,000$. Since $10,000 = 10^4$, the number can be expressed as

$$4.75 \times 10^4$$

A more practical way to transform this number to scientific notation is to count to the left from the *old* decimal point to where you wish to put the new decimal point. The number of places counted to the left will be the positive exponent of 10.

$$4 \quad 7 \quad 5 \quad 0 \quad 0 = 4.75 \times 10^4$$

(b)

$$5, \; 0 \quad 3 \quad 0, \; 0 \quad 0 \quad 0 = 5.03 \times 10^6$$

(c) 0.0023 can be factored into 2.3×0.001. Since $0.001 = 10^{-3}$, the number can be expressed as

$$2.3 \times 10^{-3}$$

A more practical way is to count to the right from the *old* decimal point to where you wish the new decimal point. The number of places counted to the *right* will be the negative exponent of 10.

$$0. \; 0 \quad 0 \quad 2 \quad 3 = 2.3 \times 10^{-3}$$

(d)

$$0. \; 0 \quad 0 \quad 0 \quad 0 \quad 4 \quad 7 \quad 0 = 4.70 \times 10^{-5}$$

EXAMPLE 2-5

Add the following numbers: 1.24×10^{-6}, 8.12×10^{-7}, 1.11×10^{-5}. Express the answer to the proper decimal place.

SOLUTION

To add coefficients, the exponents of 10 must all be the same. In this case, the most convenient exponent is -6, but -5 or -7 could also be used. Changing all numbers to the same exponent of 10 and adding we have

$$
\begin{array}{r}
1.24 \times 10^{-6} \\
0.812 \times 10^{-6} \\
\underline{11.1 \quad \times 10^{-6}} \\
13.152 \times 10^{-6} = 13.2 \times 10^{-6} = \underline{1.32 \times 10^{-5}}
\end{array}
$$

EXAMPLE 2-6

Carry out the following operations. Express the answer to the proper number of significant figures.

(a) $(8.25 \times 10^{-5}) \times (5.442 \times 10^{-3})$
(b) $(4.68 \times 10^{16}) \div (9.1 \times 10^{-5})$

SOLUTION

(a) In the first step, group the coefficients and the powers of 10. Carry out each step. In multiplication, exponents are added.

$$(8.25 \times 5.442) \times (10^{-5} \times 10^{-3}) = 44.9 \times 10^{-8}$$
$$= \underline{4.49 \times 10^{-7}} \text{ (three significant figures)}$$

(b) Group the coefficients and the powers of 10. In division, exponents are subtracted.

$$\frac{4.68 \times 10^{16}}{9.1 \times 10^{-5}} = \frac{4.68}{9.1} \times \frac{10^{16}}{10^{-5}}$$
$$= 0.51 \times 10^{16-(-5)}$$
$$= 0.51 \times 10^{21}$$
$$= \underline{5.1 \times 10^{20}} \text{ (two significant figures)}$$

The use of scientific notation can help remove the ambiguity of numbers containing zero that may or may not be significant. For example, the number 9000 can have from one to four significant figures as written. The following expressions, however, are clear as to the number of significant figures:

$$9 \times 10^3 \text{ has one significant figure}$$
$$9.0 \times 10^3 \text{ has two significant figures}$$
$$9.00 \times 10^3 \text{ has three significant figures}$$
$$9.000 \times 10^3 \text{ has four significant figures}$$

■
REFER TO APPENDIX C FOR FURTHER REVIEW.

The preceding discussion is supplemented in Appendix C. This appendix includes a diagnostic test and a discussion of the mathematical manipulation of numbers expressed in scientific notation. You are urged to take the short test at this time and work through the appendix if practice is indicated by the results of the test.

■
SEE PROBLEMS 2-14 THROUGH 2-24.

2-4 MEASURE-MENTS OF MASS, LENGTH, AND VOLUME

If you were skillful (or lucky) enough to catch a trophy fish, you would certainly describe this fish by means of its length and mass. Even more detail could be given by including its girth (circumference), which would give the listener an idea of its volume. In the United States, you would most likely use the English system and report numbers such as 2 ft 4 in. and 6 lb 3 oz for the fish. Note that we often use two units (e.g., ft and in.) for one measurement. This awkward situation arises because of a lack of systematic relationships between units. That is, units are not related by the same number. For example, for length we have relationships of 12 in. = 1 ft, 3 ft = 1 yd, and 1750 yd =

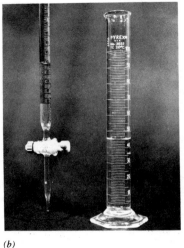

(a)   (b)   (c)

Figure 2-3   LENGTH, VOLUME, MASS. These properties of a quantity of matter are measured with common laboratory equipment. (a) Length: metric rulers. (b) Volume: a buret and a graduate cylinder. (c) Mass: an electric balance.

TABLE 2-1   Metric or SI Units

Measurement	Unit	Symbol
Mass	gram	g
Length	meter	m
Volume	liter	L
Time	second	s
Temperature	degrees Celsius	°C
	kelvin	K
Quantity	mole	mol
Energy	joule	J
Pressure	pascal	Pa

1 mile. Our monetary system is an exception because it is based on the decimal system. Therefore, only one unit (dollars) is needed to show a typical student's dismal financial condition (e.g., $11.98). The English system that the United States almost adopted required the use of two units (e.g., 2 pounds, 3 shillings).

The rest of the world and the sciences use the metric system of measurement for length, volume, and mass (see Figure 2-3). This system, like our monetary system, uses decimal fractions. Units in the metric system are conveniently related by multiples of 10. The United States still plans to adopt the metric system, but many years will pass before a complete switch is made. Metric units also form the basis of the SI system after the French *Système International* (International System). The basic SI or metric units that concern us are listed in Table 2-1. There are many other SI units that designate quantities that are not used in this text. Most SI units have very precisely defined standards based on certain precisely known properties of matter and light. Although English units were formerly based on a king's anatomy, in modern times many have been redefined more precisely based on a corresponding metric unit.

Use of SI or metric units is simplified by their exact relationships by powers of 10. This is illustrated in Table 2-3 by use of the more common prefixes listed in Table 2-2.

TABLE 2-2   Prefixes Used in the Metric System

Prefix	Symbol	Relation to Basic Unit	Prefix	Symbol	Relation to Basic Unit
tera-	T	10^{12}	deci-	d	10^{-1}
giga-	G	10^{9}	centi-	c	10^{-2}
mega-	M	10^{6}	milli-	m	10^{-3}
kilo-	k	10^{3}	micro-	μ[a]	10^{-6}
hecto-	h	10^{2}	nano-	n	10^{-9}
deca-	da	10^{1}	pico-	p	10^{-12}

[a] A Greek letter, mu.

TABLE 2-3   Relationships among Metric Units Using Common Prefixes

Mass Unit	Symbol	Relation to Basic Unit	Volume Unit	Symbol	Relation to Basic Unit	Length Unit	Symbol	Relation to Basic Unit
kilogram	kg	10^3 g	kiloliter	kL	10^3 L	kilometer	km	10^3 m
decigram	dg	10^{-1} g	deciliter	dL	10^{-1} L	decimeter	dm	10^{-1} m
centigram	cg	10^{-2} g	centiliter	cL	10^{-2} L	centimeter	cm	10^{-2} m
milligram	mg	10^{-3} g	milliliter	mL	10^{-3} L	millimeter	mm	10^{-3} m
microgram	μg	10^{-6} g						

In the metric system there is also an exact relationship between length and volume. Thus, one liter is defined as the volume occupied by one cubic decimeter (i.e., 1 L = 1 dm³). One milliliter is the volume occupied by one cubic centimeter (i.e., 1 mL = 1 cm³ = 1 cc). The units milliliter (mL) and cubic centimeter (cm³ or cc) can be used interchangeably when expressing volume. (See Figure 2-4.)

There is often confusion between the terms "mass" and "weight." **Mass** *is the quantity of matter that a sample contains.* It is the same for the sample anywhere in the universe. **Weight** *is a measure of the attraction of gravity for the sample.* An astronaut has the same mass on the moon as on the earth. An astronaut who weighs 170 lb on earth, however, weighs only about 29 lb on the moon. In earth orbit, where the effect of gravity is counteracted, the astronaut is "weightless." Obviously, the astronaut lost weight on the moon,

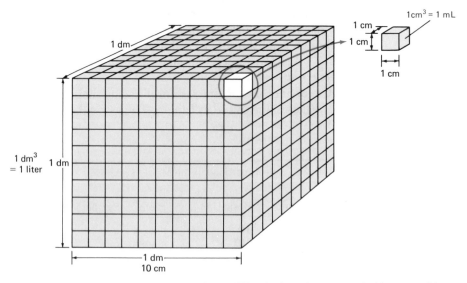

Figure 2-4   VOLUME AND LENGTH. One milliliter is the volume occupied by one cubic centimeter.

Figure 2-5 METRIC UNITS. The use of metric units in this country is becoming increasingly evident.

but matter (mass) was not lost. However, because the attraction of gravity is essentially the same anywhere on the surface of the earth, weight is used as a measure of mass. As a result, the terms "mass" and "weight" are often used interchangeably.

Several relationships between the metric and English systems are listed in Table 2-4. (See also Figure 2-5.) The relationships *within* systems (e.g., 12 in. = 1 ft, 10^3 m = 1 km) are always defined and exact. The relationships *between* systems, however, can be expressed to various degrees of precision. All of the relationships shown are known to more than the three significant figures given. Although there are exact, defined relationships between the two systems, the definitions usually include many more significant figures than our needs in chemistry require. For example, one yard is defined by the U.S. Bureau of Standards as exactly equal to 0.91440183 meter.

TABLE 2-4 The Relationship between English and Metric Units

	English	Metric Equivalent
Length	1.00 in.	2.54 cm
	1.00 mile	1.61 km
Mass	1.00 lb	454 g
	2.20 lb	1.00 kg
Volume	1.06 qt	1.00 L
	1.00 gal	3.78 L
Time	1.00 sec	1.00 sec

**2-5
CONVERSION
OF UNITS BY
THE FACTOR-
LABEL
METHOD**

It is useful to have a feeling for the relationships between English and metric units (see Figure 2-6), but in fact, it is necessary to make calculations that convert precise measurements in one system to the other. A system known as the factor-label method, or dimensional analysis, has been shown to be the most helpful in making these conversions and many others that we will encounter in later chapters. A brief introduction of the method is given in this chapter; supplemental explanations and exercises are presented in Appendix D.

The **factor-label method** *converts from one unit to another by the use of conversion factors.* **A conversion factor** *is an equality or equivalency relationship between units or quantities expressed in fractional form.* For example, in the metric system we have the exact relationship

$$10^3 \text{ m} = 1 \text{ km}$$

This can be expressed in fractional or factor form as

$$\frac{1 \text{ km}}{10^3 \text{ m}}$$

or the reciprocal

$$\frac{10^3 \text{ m}}{1 \text{ km}} \quad \left(\text{or simply } \frac{10^3 \text{ m}}{\text{km}} \right)$$

This is read as "one kilometer per 10^3 meters" or "10^3 meters per one kilometer." The latter factor is usually simplified to "10^3 meters per kilometer." When just a unit (no number) is read or written in the denominator, the number is assumed to be "one."

Conversion factors are also constructed from the equalities between the English and metric units. For example, the equality

$$1.00 \text{ in.} = 2.54 \text{ cm}$$

can be expressed in factor form as

$$\frac{1.00 \text{ in.}}{2.54 \text{ cm}} \quad \text{or} \quad \frac{2.54 \text{ cm}}{1.00 \text{ in.}}$$

Such factors are sometimes referred to as **unit factors,** since *they relate a quantity to "one" of another.* As such, the factors are often written in a somewhat simplified form:

$$\frac{1 \text{ in.}}{2.54 \text{ cm}} \quad \text{or} \quad \frac{2.54 \text{ cm}}{\text{in.}}$$

Unit factors are written as such in this text. *It should be understood, however, that "one" (as written in the numerator or implied in the denominator) is known exactly.*

These factors can be used to convert a measurement in one unit to the other. In the factor-label method all units of the numbers are maintained in the

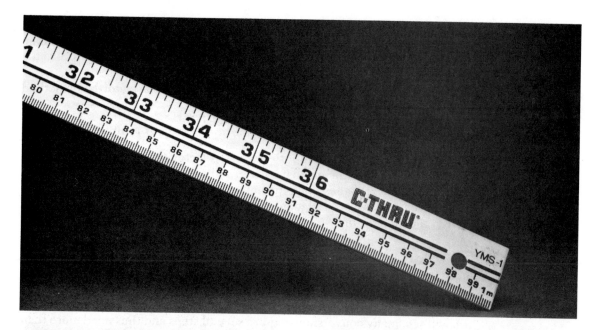

Figure 2-6 COMPARISON OF METRIC AND ENGLISH UNITS.

calculation. Like the numbers themselves, the units are multiplied, divided, and canceled in the course of the calculation. This adds a little time to the calculation, but the payoff is a better understanding of the conversion process as well as correct answers. Students who consistently use this method are often amazed at the orderliness and logic of chemistry. In more complex conversions, where several conversion factors are required, the student can plan the calculation in a stepwise and orderly manner from what is given to what is requested.

The general procedure of a one-step conversion by the factor-label method is

$$(\text{what's given}) \times (\text{conversion factor}) = (\text{what's requested})$$

In most calculations, the proper conversion factor that converts "given" to "requested" has the unit of what is given in the denominator (the old unit) and the unit of what is requested in the numerator (the new unit). In this way, the old unit *cancels*, as would an identical numerical quantity. This leaves the new unit in the numerator.

$$\left(\begin{array}{c}\text{given}\\\text{quantity}\end{array}\right)\text{old unit} \times \left(\begin{array}{c}\text{conversion}\\\text{factor}\\\text{quantity}\end{array}\right)\frac{\text{new unit}}{\text{old unit}} = \left(\begin{array}{c}\text{requested}\\\text{quantity}\end{array}\right)\text{new unit}$$

The key, then, is to select the conversion factor that does the right job. As we will see, many conversions require several steps and thus need more than one conversion factor to convert from what's given to what's requested. Proper organization and planning help avoid difficulties with these problems. The following examples illustrate the use of the factor-label method in one-step conversion between units.

EXAMPLE 2-7

Convert 0.468 m to (a) kilometers and (b) millimeters.

(a)

1 Given:0.468 m.　Requested: ___?___ km.

2 Procedure: Use a conversion factor that cancels m and leaves km in the numerator. In shorthand our procedure is

$$\text{m} \longrightarrow \text{km}$$

3 Relationship: 10^3 m $= 1$ km (from Table 2-3).

4 Conversion factor: Of the two possible conversion factors that originate from the above relationship, we choose the one with km (requested) in the numerator and m (given) in the denominator. This is 1 km/10^3 m.

5 Solution:

$$0.468 \text{ m} \times \frac{1 \text{ km}}{10^3 \text{ m}} = 0.468 \times 10^{-3} \text{ km} = \underline{4.68 \times 10^{-4} \text{ km}}$$

(b)

1   Given: 0.468 m.   Requested: __?__ mm.
2   Procedure: m → mm.
3   Relationship: 1 mm $= 10^{-3}$ m.
4   Conversion factor: 1 mm/10^{-3} m.
5   Solution:

$$0.468 \, \cancel{m} \times \frac{1 \text{ mm}}{10^{-3} \, \cancel{m}} = 0.468 \times 10^3 \text{ mm} = \underline{\underline{468 \text{ mm}}}$$

EXAMPLE 2-8

Convert 825 cm to inches.

1   Given: 825 cm.   Requested: __?__ in.
2   Procedure: cm → in.
3   Relationship: 1.00 in. = 2.54 cm (from Table 2-4).
4   Conversion factor: 1 in./2.54 cm.
5   Solution:

$$825 \, \cancel{cm} \times \frac{1 \text{ in.}}{2.54 \, \cancel{cm}} = \underline{\underline{325 \text{ in.}}}$$

EXAMPLE 2-9

Convert 6.85 qt to liters.

1   Given: 6.85 qt.   Requested: __?__ L.
2   Procedure: qt → L.
3   Relationship: 1.00 L = 1.06 qt.
4   Factor: 1 L/1.06 qt.
5   Solution:

$$6.85 \, \cancel{qt} \times \frac{1 \text{ L}}{1.06 \, \cancel{qt}} = \underline{\underline{6.46 \text{ L}}}$$

What would have happened if we had multiplied instead of divided? In that case, the units would have served as a red flag indicating that a mistake had been made. The units of the answer would have been qt²/L, which is obviously not correct.

$$6.85 \text{ qt} \times \frac{1.06 \text{ qt}}{1 \text{ L}} = 7.26 \, \frac{\text{qt}^2}{\text{L}} \quad ??????$$

The one-step conversions that have been worked so far are analogous to a direct, nonstop airline flight between your home city and your destination. The problems that follow are analogous to the situation in which a nonstop flight is not available and you have to make the flight in several steps before reaching your destination. Each step in the flight is a separate journey, but it gets you closer to your ultimate destination. For multistep conversions you must plan a step-by-step path from your origin (what's given) to your destination (what's requested). Each step along the path toward the answer requires one conversion factor.

EXAMPLE 2-10

Convert 4978 mg to kilograms.

1   Given 4978 mg.   Requested: ___?___ kg.

2   Procedure: In Table 2-3, note that one relationship between milligrams and kilograms is not directly available. However, we can take a two-step journey by (a) converting milligrams to grams and then (b) converting grams to kilograms. In shorthand the procedure is

$$\text{mg} \xrightarrow{(a)} \text{g} \xrightarrow{(b)} \text{kg}$$

3   Relationships: $1 \text{ mg} = 10^{-3} \text{ g}$; $1 \text{ kg} = 10^3 \text{ g}$.

4   Conversion factors: Treat each step as a distinct operation or problem. In step (a) we need g in the numerator and mg in the denominator. In step (b) we need kg in the numerator and g in the denominator.

(a) 10^{-3} g/mg   (b) 1 kg/10^3 g

5   Solution

$$4978 \text{ mg} \times \underset{\text{mg}}{\overset{(a)}{\frac{10^{-3} \text{ g}}{}}} \times \underset{10^3 \text{ g}}{\overset{(b)}{\frac{1 \text{ kg}}{}}} = 4978 \times 10^{-6} \text{ kg} = \underline{\underline{4.978 \times 10^{-3} \text{ kg}}}$$

EXAMPLE 2-11

Convert 9.85 L to gallons.

1   Given: 9.85 L.   Requested: ___?___ gal.

2   Procedure

$$\text{L} \xrightarrow{(a)} \text{qt} \xrightarrow{(b)} \text{gal}$$

3   Relationships: $1.06 \text{ qt} = 1.00 \text{ L}$; $4 \text{ qt} = 1 \text{ gal}$

4 Factors

$$\text{(a)} \ \frac{1.06 \ \text{qt}}{L} \qquad \text{(b)} \ \frac{1 \ \text{gal}}{4 \ \text{qt}}$$

5 Solution

$$\overset{\text{(a)}}{} \qquad \overset{\text{(b)}}{}$$

$$9.85 \ \cancel{L} \times \frac{1.06 \ \cancel{\text{qt}}}{\cancel{L}} \times \frac{1 \ \text{gal}^*}{4 \ \cancel{\text{qt}}} = \underline{\underline{2.61 \ \text{gal}}}$$

---

* *An exact number does not limit the number of significant figures in the answer.*

EXAMPLE 2-12

Convert 55 miles/hr to meters per minute.

1 Given: 55 miles/hr.   Requested: __?__ m/min.

2 Procedure: Note in this problem that both the numerator and denominator must be converted to other units. The first two steps convert the numerator and the third the denominator.

$$\text{Numerator:} \quad \text{miles} \xrightarrow{\text{(a)}} \text{km} \xrightarrow{\text{(b)}} \text{m}$$

$$\text{Denominator:} \quad \text{hr} \xrightarrow{\text{(c)}} \text{min}$$

3 Relationships: $1.00 \ \text{mile} = 1.61 \ \text{km}$; $10^3 \ \text{m} = 1 \ \text{km}^*$, $60 \ \text{min} = 1 \ \text{hr.}^*$

4 Conversion factors

$$\text{(a)} \ \frac{1.61 \ \text{km}}{\text{mile}} \qquad \text{(b)} \ \frac{10^3 \ \text{m}}{\text{km}}$$

To convert a denominator to another unit, note that the conversion factor must be set up with what's given in the *numerator* and what's requested in the *denominator*.

$$\text{(c)} \ \frac{1 \ \text{hr}}{60 \ \text{min}}$$

5 Solution

$$\overset{\text{(a)}}{} \qquad \overset{\text{(b)}}{} \qquad \overset{\text{(c)}}{}$$

$$55 \ \frac{\text{miles}}{\text{hr}} \times \frac{1.61 \ \cancel{\text{km}}}{\cancel{\text{mile}}} \times \frac{10^3 \ \text{m}}{\cancel{\text{km}}} \times \frac{1 \ \cancel{\text{hr}}}{60 \ \text{min}} = \underline{\underline{1.5 \times 10^3 \ \text{m/min}}}$$

---

* *Exact numbers.*

The general procedure for working conversion problems can be summarized as follows:

1 Write down what is given and what is requested.

2 Outline a procedure in shorthand for a sequence of calculations from what is given to what is requested. To do this evaluate the appropriate relationships available. Sometimes it is easier to start with what is requested and work backward to what is given. Don't forget that common sense is your best tool.

3 Write down the relationship for each step in the conversion.

4 Following the shorthand procedure from step 2, write the relationships in factor form for each step. Remember the unit for what is wanted in each step should be in the numerator and the unit for what is given, in the denominator. (An exception is a problem requiring a change in the denominator, such as miles per hour to miles per minute.)

5 Put the equation together. Make sure that the proper units cancel and that you are left with only the unit or units requested. Carefully do the math, making sure to express your answer to the proper number of significant figures.

■
SEE PROBLEMS
2-25 THROUGH
2-54.

■
REFER TO
APPENDIX D
FOR ADDITIONAL
DISCUSSION ON
PROBLEM
SOLVING BY
THE FACTOR-
LABEL METHOD.

If these examples and the problems at the end of this chapter cause you any difficulties, you are strongly urged to work through Appendix D. There you will find an extensive supplement including solved problems and exercises employing familiar English units, imaginary units, and actual chemical units. It is designed to increase your proficiency in the mechanics of the factor-label method so that it can serve as your "backup system" for obtaining correct

Figure 2-7 CELSIUS TEMPERATURE. Thanks to the evening news, Americans are becoming more familiar with the Celsius scale.

answers. (Your primary system should eventually be a "feeling" for the type and general magnitude of the answer you are seeking.)

Another important property of a substance is its temperature. **Temperature** *is a measure of the intensity of heat of a substance.* **A thermometer** *is a device that measures temperature.* The thermometer scale with which we are most familiar is probably the Fahrenheit scale (°F), but the Celsius scale (°C) is used in most of the rest of the world and in science. As you've probably noticed on the TV news, most weather reports now give the temperature readings in both scales as this country slowly switches to use of the Celsius scale. (See Figure 2-7.)

We learned in Chapter 1 that the boiling and melting points of pure water (at sea level pressure) are constant and definite properties. We can take advantage of this fact to compare the two temperature scales and establish a relationship between them. In Figure 2-8 the temperature of an ice and water mixture is shown to be exactly 0 °C. This temperature was originally established by definition. This corresponds to exactly 32 °F on the Fahrenheit thermometer. The boiling point of pure water is exactly 100 °C, which corresponds to 212 °F.

On the Celsius scale there are 100 equal divisions between these two temperatures, whereas on the Fahrenheit scale there are $212 - 32 = 180$

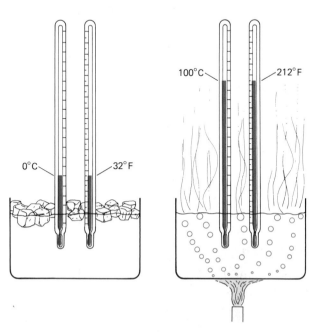

Figure 2-8 THE TEMPERATURE SCALES. The freezing and boiling points of water are used to calibrate the temperature scales.

equal divisions between the two temperatures. Thus, we have the following relation between scale divisions:

$$100 \text{ C div.} = 180 \text{ F div.}$$

This relationship can be used to construct conversion factors between an equivalent number of Celsius and Fahrenheit degrees.

$$\frac{1.8 \text{ °F}}{1 \text{ °C}} \text{ and } \frac{1 \text{ °C}}{1.8 \text{ °F}}$$

To convert the Celsius temperature [$t(C)$] to the Fahrenheit temperature [$t(F)$]:

1   Multiply the Celsius temperature by the proper conversion factor (1.8 °F/1 °C) to convert to the equivalent number of Fahrenheit degrees.

2   Add 32 °F to this number so that both scales start at the same point (the freezing point of water).

$$t(F) = \left[t(C) \times \frac{1.8 \text{ °F}}{1 \text{ °C}} \right] + 32 \text{ °F} = [t(C) \times 1.8] + 32$$

To convert the Fahrenheit temperature to the Celsius temperature:

1   Subtract 32 °F from the Fahrenheit temperature so that both scales start at zero.

2   Multiply that number by the proper conversion factor (1 °C/1.8 °F) to convert to the equivalent number of Celsius degrees.

$$t(C) = [t(F) - 32 \text{ °F}] \times \frac{1 \text{ °C}}{1.8 \text{ °F}} = \frac{[t(F) - 32]}{1.8}$$

EXAMPLE 2-13

A person with a cold has a fever of 102 °F. What would be the reading on a Celsius thermometer?

$$t(C) = \frac{[t(F) - 32]}{1.8} = \frac{(102 - 32)}{1.8} = \underline{\underline{39 \text{ °C}}}$$

EXAMPLE 2-14

On a cold winter day the temperature is −10.0 °C. What is the reading on the Fahrenheit scale?

$$t(F) = [t(C) \times 1.8] + 32$$
$$t(F) = (1.8 \times -10.0) + 32.0 = -18.0 + 32.0 = \underline{14.0 \text{ °F}}$$

TABLE 2-5 Density (at 20 °C)

Substance (Liquid)	Density (g/mL)	Substance (Solid)	Density (g/mL)
Ethyl alcohol	0.790	Aluminum	2.70
Gasoline (a mixture)	~0.67 (variable)	Gold	19.3
Carbon tetrachloride	1.60	Ice	0.92 (0 °C)
Kerosene	0.82	Lead	11.3
Water	1.00	Lithium	0.53
Mercury	13.6	Magnesium	1.74
		Table salt	2.16

The SI temperature unit is called **the kelvin (K)**. The zero on the Kelvin scale is theoretically the lowest possible temperature (the temperature at which the heat energy is zero). This corresponds to -273 °C (or more precisely -273.15 °C). Since the magnitude of a kelvin and a Celsius degree unit is the same, we have the following simple relationship between the two scales, where T represents the number of kelvins and $t(C)$ represents the Celsius degrees.

$$T = [t(C) + 273]$$

■
SEE PROBLEMS 2-55 THROUGH 2-64.

Thus, the freezing point of water is 0 °C or 273 K and the boiling point is 100 °C or 373 K. We will use the Kelvin scale more in later chapters.

2-7 DENSITY AND SPECIFIC GRAVITY

Both the mass and the volume of a substance are variable properties that depend on the amount of the substance present. For example, we all know that Styrofoam is considered "light" and lead "heavy"; however, a truckload of Styrofoam would certainly be heavy so we must compare equal volumes if our relative terms are to be used, for identification. *The ratio of the mass (usually in grams) to the volume (usually in milliliters or liters) is a physical property known as* **density** *and is the same for any amount of homogeneous matter under the same conditions of pressure and temperature.* Since pure substances (elements and compounds) are homogeneous, density can be an important tool in identification. The densities of several liquids and solids are listed in Table 2-5. Because 1 mL is the same as 1 cm³ (cc), density is sometimes expressed as g/cc or g/cm³. The densities of gases are discussed in Chapter 10.

The following examples illustate how density is calculated and how it is used to identify pure substances.

EXAMPLE 2-15

A young woman was interested in purchasing a sample of pure gold having a mass of 8.99 g. Being wise, she wished to confirm that it was actually gold before she paid for it. With a quick test using a graduated

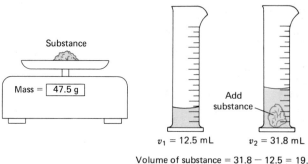

Substance

Mass = | 47.5 g |

Add
substance

$v_1 = 12.5$ mL $v_2 = 31.8$ mL

Volume of substance = 31.8 − 12.5 = 19.3 mL

$$\text{Density} = \frac{47.5 \text{ g}}{19.3 \text{ mL}} = 2.46 \text{ g/mL}$$

Figure 2-9   DENSITY. To determine the density, both the mass and the volume of a substance are measured.

cylinder like that shown in Figure 2-9, she found that the "gold" had a volume of 0.796 mL. Was the substance gold?

SOLUTION

From the volume and the mass, the density can be calculated and compared with that of pure gold.

$$\text{density} = \frac{8.99 \text{ g}}{0.796 \text{ mL}} = 11.3 \text{ g/mL}$$

The sample was *not* gold. It apparently was lead that had been dipped in gold paint.

EXAMPLE 2-16

A sample of a pure substance was found to have a mass of 47.5 g. As shown in Figure 2-9, a quantity of water has a volume of 12.5 mL. When the substance is added to the water the volume reads 31.8 mL. The difference in volume is the volume of the substance. What is the density?

SOLUTION

Refer to Figure 2-9.

In the previous examples, density was calculated from a given mass and volume. Density itself is used as a conversion factor to:

1   Convert a given mass to an equivalent volume.
2   Convert a given volume to an equivalent mass.

EXAMPLE 2-17

Using Table 2-5, determine the volume occupied by 485 g of table salt.

1 Given: 485 g of salt. Requested: __?__ mL of salt.
2 Procedure: g → mL.
3 Relationship: 1.00 mL = 2.16 g.
4 Conversion factor: 1 mL/2.16 g.
5 Solution

$$485 \text{ g} \times \frac{1 \text{ mL}}{2.16 \text{ g}} = \underline{\underline{225 \text{ mL}}}$$

EXAMPLE 2-18

What is the mass of 1.52 L of kerosene?

1 Given: 1.52 L. Requested: __?__ g of kerosene.
2 Procedure

$$L \xrightarrow{\text{(a)}} mL \xrightarrow{\text{(b)}} g$$

3 Relationships: 1 mL = 10^{-3} L, 0.82 g = 1.0 mL.
4 Conversion factors: (a) 1 mL/10^{-3} L; (b) 0.82 g/mL.
5 Solution

$$1.52 \text{ L} \times \frac{1 \text{ mL}}{10^{-3} \text{ L}} \times \frac{0.82 \text{ g}}{\text{mL}} = \underline{\underline{1.2 \times 10^3 \text{ g}}}$$

Assuming that there is no reaction or dissolving, *a substance with a density lower than that of a certain liquid floats or is* **buoyant** *in that liquid.* In the case of water, anything with a density less than 1.00 g/mL floats. Note that gasoline and ice float on water, but most solids and carbon tetrachloride sink.

Liquids can be mixed in such proportions to provide a range of densities. Gemologists use this principle to determine the authenticity of gemstones. For example, liquids can be mixed whereby an emerald (density 2.70 g/mL) floats in one mixture but sinks in another. Fakes do not have the same density as the authentic stone and can be discovered as shown in Figure 2-10.

The density of a liquid can be determined with a device called a **hydrometer.** The hydrometer tube, as shown in Figure 2-11, is exactly balanced with weights in the bottom so that its level in pure water is exactly at the 1.00 g/mL mark. When immersed in liquids of other densities, it becomes more or less buoyant. The scale is calibrated (the divisions previously determined and checked) in such a manner that the density of the liquid is read directly from

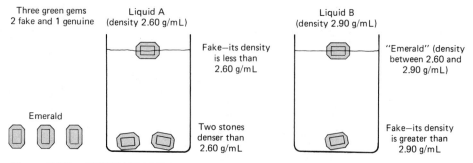

Figure 2-10   BUOYANCY OF AN EMERALD. A stone sinks in a liquid that is less dense than the stone but floats in a more dense liquid.

the scale. A hydrometer is used to measure the density of the acid in a car battery because this indicates its charge condition.

"Specific gravity" is also a term that relates mass to volume. **Specific gravity** *is a comparison of the density of a substance to that of water at 4 °C.* Since the density of water at 4 °C is 1.00 g/mL, specific gravity in effect is simply the density expressed without units.

■
SEE PROBLEMS 2-65 THROUGH 2-82.

$$\text{density of Hg} = 13.6 \text{ g/mL}$$
$$\text{density of water} = 1.00 \text{ g/mL}$$

$$\text{specific gravity} = \frac{\text{density of mercury}}{\text{density of water}} = \frac{13.6 \text{ g/mL}}{1.00 \text{ g/mL}} = \underline{\underline{13.6}}$$

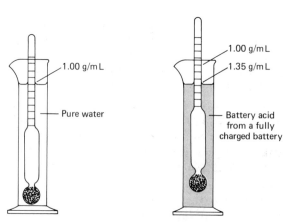

Figure 2-11   THE HYDROMETER. This device is used to measure the density of a liquid.

2-8 SPECIFIC HEAT

Most of us have probably noted how long it takes a pot of water to come to a boil. Yet a heavy iron skillet seems to heat up almost instantly. This phenomenon relates to a property of matter known as the specific heat capacity or more simply as specific heat. **Specific heat** *is defined as the amount of heat required to raise the temperature of one gram of a substance one degree Celsius (or Kelvin).* Specific heat, like density, is a physical property that is the same for any amount of a pure substance. Therefore, it can be used as a means of identification. The amount of heat energy is reported in either of two units used extensively in science: the *calorie* and the *joule.* The calorie was originally defined as the amount of heat required to raise the temperature of one gram of water from 14.5 to 15.5 °C. The calorie is now defined in terms of its equivalent SI unit, the joule. The modern definition of the calorie is

$$1 \text{ cal} = 4.184 \text{ joules (J)} \quad \text{(exactly)}$$

The nutritional calorie (1 Cal) is actually one kilocalorie (10^3 cal). The number of Calories contained in a portion of food such as a piece of pie is the amount of heat that is liberated when the portion is dried and burned. Although both calories and joules are used in chemistry, the joule is becoming the unit of choice.

From the definition of the calorie one can see that the specific heat of water is therefore 1.00 cal/(g · °C) or 4.184 J/(g · °C). (The °C in the denominator of the unit represents a temperature change and not an actual temperature reading.) Specific heats of several pure substances are listed in Table 2-6.

Note that the specific heat of water is comparatively quite large. This fortunate property of water helps even out the temperature of our planet. For example, the Gulf Stream is a river of warm water within the Atlantic Ocean that transports a large amount of heat energy from the Gulf of Mexico to England and Scandinavia. If it were not for this warm current and the ability of water to hold heat energy, these countries would be covered by ice, as is Greenland. Climate is more moderate near any ocean or large lake. The water can slowly absorb large amounts of heat energy in the summer, thus keeping the surroundings comparatively cool. In the winter, the water releases the stored heat energy and keeps the surroundings comparatively warm.

TABLE 2-6   Specific Heats

Substance	Specific Heat [cal/(g · °C)]	Specific Heat [J/(g · °C)]
Water	1.00	4.18
Ice	0.492	2.06
Aluminum (Al)	0.214	0.895
Gold (Au)	0.031	0.13
Copper (Cu)	0.092	0.38
Zinc (Zn)	0.093	0.39
Iron (Fe)	0.106	0.444

EXAMPLE 2-19

If 150 cal of heat energy is added to 50.0 g of water at 25 °C, what is the final temperature of the water?

$$\frac{150 \text{ cal}}{50.0 \text{ g}} \times \frac{1 \text{ g} \cdot {}^\circ\text{C}}{1.00 \text{ cal}} = 3 \text{ °C rise} \qquad t(\text{C}) = 25 + 3 = \underline{\underline{28 \text{ °C}}}$$

EXAMPLE 2-20

It takes 62.8 J to raise a 125-g quantity of silver 0.714 °C. What is the heat capacity of silver?

$$\frac{62.8 \text{ J}}{125 \text{ g} \cdot 0.714 \text{ °C}} = \underline{\underline{0.704 \text{ J/(g} \cdot {}^\circ\text{C)}}}$$

EXAMPLE 2-21

A 255-g quantity of gold is heated from 28 to 100 °C. How many joules were added to the sample?

Rise in temperature = 100 − 28 = 72 °C

$$255 \text{ g} \times 72 \text{ °C} \times \frac{0.13 \text{ J}}{\text{g} \cdot {}^\circ\text{C}} = \underline{\underline{2400 \text{ J}}}$$

SEE PROBLEMS 2-83 THROUGH 2-96.

A calculation similar to those in the preceding examples would tell us why an iron skillet heats up so much faster than water. If we put the same amount of heat into the same masses of iron and water, the temperature of the iron would rise more than 9 °C for every 1 °C rise for the water. Iron is also a good conductor of heat so the heat is rapidly distributed.

2-9 CHAPTER REVIEW

Chemistry, as a physical science, is vitally concerned with quantitative physical properties. These properties result from a measurement that expresses a numerical value as well as the appropriate unit. The numerical value has two qualities: precision and accuracy.

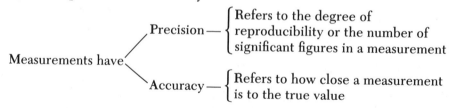

Rules were given for handling zero as a significant figure and for rounding off. We then discussed how to express the answers to mathematical operations:

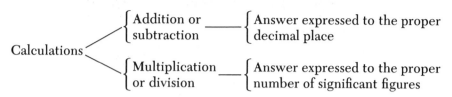

Scientific notation was then discussed as a convenient way of handling the very large and the very small numbers encountered in chemistry.

The metric or SI system of measurement was introduced specifically with regard to length, volume, and mass. The decimal relationships between units of the metric system are convenient compared with the unsystematic English system. Relationships within a system are exact numbers, while relationships between systems can be expressed to various degrees of precision.

The factor-label method was introduced as a convenient tool to convert a measurement to other units within the same system or to another system. The relationships listed in Tables 2-3 and 2-4 were used as conversion factors to convert what is "given" to what is "requested." The careful use of units helps guide us to the proper mathematical manipulation.

Temperature is a property of matter that measures heat intensity. Thermometers measure temperature. Three temperature scales are common: Fahrenheit (°F), principally used in the United States and Britain; Celsius (°C), used in most of the world and science; and Kelvin (K), used almost entirely in science as the SI unit. Relationships between scales were given.

Density (or specific gravity) and specific heat are two properties of homogeneous matter that are independent of the mass or dimensions of the sample. Thus, these properties can be used as identifying characteristics of a pure substance. Density relates the mass of a sample to its volume. Substances that are less dense than a liquid are buoyant, or float, in the liquid. A hydrometer

Property	Principal Unit	Comment
Length	meter (m)	m^2 refers to area
Volume	liter (L)	m^3 refers to volume, $1\ dm^3 = 1\ L$
Mass	gram (g)	In the metric system, 1 mL (1 cm^3) of H_2O has a mass of 1 g
Temperature	°C	$T = [t(C) + 273]$ K; zero on the Kelvin scale is the lowest possible temperature
Density	g/mL	This property can be used to identify pure substances and as a conversion factor between volume and mass
Specific heat	J/(g · °C)	This property can be used to identify pure substances and as a conversion factor among heat, mass, and temperature change

uses the principle of buoyancy to determine the density or specific gravity of a liquid. Specific heat is a property that relates the amount of heat in calories or joules to a mass and a change of temperature.

The six properties of matter that were discussed are summarized on the preceding page.

■ EXERCISES

Throughout the text, more difficult problems are marked with an asterisk.

SIGNIFICANT FIGURES

2-1   Which of the following measurements is the most precise?
   (a) 75.2 gal   (c) 75.22 gal
   (b) 74.212 gal   (d) 75 gal

2-2   How can a measurement be precise but not accurate?

2-3   The actual length of a certain plank is 26.782 in. Which of the following measurements is the most precise and which is the most accurate?
   (a) 26.5 in.   (c) 26.202 in.
   (b) 26.8 in.   (d) 26.98 in.

2-4   How many significant figures are in each of the following measurements?
   (a) 7030 g   (e) 4002 m
   (b) 4.0 kg   (f) 0.060 hr
   (c) 4.01 lb   (g) 8200 km
   (d) 0.01 ft   (h) 0.00705 ton

2-5   How many significant figures are in each of the following measurements?
   (a) 0.045 in.   (e) 7.060 qt
   (b) 405 ft   (f) 2.0010 yd
   (c) 0.340 cm   (g) 0.0080 in.
   (d) 21.0 m   (h) 2200 lb

SIGNIFICANT FIGURES IN MATHEMATICAL OPERATIONS

2-6   Round off each of the following numbers to three significant figures.
   (a) 15.9994   (e) 87,550
   (b) 1.0080   (f) 0.027225
   (c) 0.6654   (g) 301.4
   (d) 4885

2-7   Round off each of the following numbers to two significant figures.
   (a) 115   (e) 55.6
   (b) 27.678   (f) 0.0396
   (c) 37500   (g) 1,557,000
   (d) 0.47322   (h) 321

2-8   Carry out each of the following operations. Assume that the numbers represent measurements so that the answer should be expressed to the proper decimal place.
   (a) $14.72 + 0.611 + 173$
   (b) $0.062 + 11.38 + 1.4578$
   (c) $1600 - 4 + 700$
   (d) $47 + 0.91 - 0.286$
   (e) $0.125 + 0.71$

2-9   Carry out each of the following operations. Assume that the numbers represent measurements so that the answer should be expressed to the proper decimal place.
   (a) $0.013 + 0.7217 + 0.04$
   (b) $15.3 + 1.12 - 3.377$
   (c) $35.48 - 4 + 0.04$
   (d) $337 + 0.8 - 12.0$

2-10   A container holds 32.8 qt of water. The following portions of water are then added to the container: 0.12 qt, 3.7 qt, and 1.266 qt. What is the new volume of water?

2-11   A container holds 3760 lb (three significant figures) of sand. The following portions are added to the container: 1.8 lb, 32 lb, and 13.55 lb. What is the final mass of the sand?

2-12   Carry out the following calculations. Express your answer to the proper number of significant figures. Specify units.

(a) 40.0 cm × 3.0 cm
(b) 179 ft × 2.20 lb
(c) 4.386 cm² ÷ 2 cm
(d) (14.65 in. × 0.32 in.) ÷ 2.00 in.

2-13 Carry out the following calculations. Express your answer to the proper number of significant figures. Specify units.
(a) 243 m² ÷ 0.05 m
(b) 3.0 ft × 472 lb
(c) 0.0575 in. × 21.0 in.
(d) (1.84 yd × 42.8 yd) ÷ 0.8 yd

SCIENTIFIC NOTATION

2-14 Express the following numbers in scientific notation with one digit to the left of the decimal point in the coefficient.
(a) 157   (e) 0.0349
(b) 0.157   (f) 32,000
(c) 0.0300   (g) 32 billion
(d) 40,000,000   (h) 0.000771
    (two signifi-   (i) 2340
    cant figures)

2-15 Express the following numbers in scientific notation with one digit to the left of the decimal point in the coefficient.
(a) 423,000   (f) 82,000,000
(b) 433.8   (three signifi-
(c) 0.0020   cant figures)
(d) 880   (g) 75 trillion
(e) 0.00008   (h) 0.00000106

2-16 Express the following numbers in common notation.
(a) 4.76×10^{-4} (d) 0.489×10^{6}
(b) 6.55×10^{3} (e) 475×10^{-2}
(c) 788×10^{-5} (f) 0.0034×10^{-3}

2-17 Change the following numbers to standard scientific notation (one digit to the left of the decimal point).
(a) 489×10^{-6} (d) 571×10^{-4}
(b) 0.456×10^{-4} (e) 4975×10^{5}
(c) 0.0078×10^{6} (f) 0.030×10^{-2}

2-18 Change the following numbers to standard scientific notation.
(a) 0.078×10^{-8}
(b) 72000×10^{-5}
(c) 3450×10^{16}
(d) 280.0×10^{8}
(e) 0.000690×10^{-10}
(f) 0.0023×10^{6}

2-19 Carry out the following operations. Assume that the numbers represent measurements so that the answer should be expressed to the proper decimal place.
(a) $(1.82 \times 10^{-4}) + (0.037 \times 10^{-4}) + (14.11 \times 10^{-4})$
(b) $(13.7 \times 10^{6}) - (2.31 \times 10^{6}) + (116.28 \times 10^{5})$
(c) $(0.61 \times 10^{-6}) + (0.11 \times 10^{-4}) + (0.0232 \times 10^{-3})$
(d) $(372 \times 10^{12}) + (1200 \times 10^{10}) - (0.18 \times 10^{15})$

2-20 Carry out the following operations. Assume that the numbers represent measurements so that the answer should be expressed to the proper decimal place.
(a) $(1.42 \times 10^{-10}) + (0.17 \times 10^{-10}) - (0.009 \times 10^{-10})$
(b) $(146 \times 10^{8}) + (0.723 \times 10^{10}) + (11 \times 10^{8})$
(c) $(1.48 \times 10^{-7}) + (2911 \times 10^{-9}) + (0.6318 \times 10^{-6})$
(d) $(299 \times 10^{10}) + (823 \times 10^{8}) + (0.75 \times 10^{11})$

2-21 Carry out the following operations. Assume that the numbers represent measurements so that the answer should be expressed to the proper number of significant figures.
(a) $(149 \times 10^{6}) \times (0.21 \times 10^{3})$
(b) $(0.371 \times 10^{14}) \div (2 \times 10^{4})$
(c) $(6 \times 10^{6}) \times (6 \times 10^{6})$
(d) $(0.1186 \times 10^{6}) \times (12 \times 10^{-5})$
(e) $(18.21 \times 10^{-10}) \div (0.0712 \times 10^{6})$

2-22 Carry out the following operations. Assume that the numbers represent measurements so that the answer should be expressed to the proper number of significant figures.
(a) $(76.0 \times 10^{7}) \times (0.6 \times 10^{8})$
(b) $(7 \times 10^{-5}) \times (7.0 \times 10^{-5})$
(c) $(0.786 \times 10^{-7}) \div (0.47 \times 10^{7})$
(d) $(3798 \times 10^{18}) \div (0.00301 \times 10^{12})$
(e) $(0.06000 \times 10^{18}) \times (84921 \times 10^{-9})$

2-23 Using scientific notation, express the number 87,000,000 to (a) one significant figure, (b) two significant figures, and (c) three significant figures.

2-24 Using scientific notation, express the number 23,600 to (a) one significant figure, (b) two significant figures, (c) three significant figures, and (d) four significant figures.

LENGTH, VOLUME AND MASS IN THE METRIC SYSTEM

2-25 Complete the following table.

	mm	cm	m	km
(Example)	108	10.8	0.108	1.08×10^{-4}
(a)	7.2×10^3	____	____	_____
(b)	_____	____	56.4	_____
(c)	_____	____	____	0.250

2-26 Complete the following table.

	mg	g	kg
(a)	8.9×10^3	____	____
(b)	_____	25.7	____
(c)	_____	____	1.25

2-27 Complete the following table.

	mL	L	kL
(a)	____	____	6.8
(b)	____	0.786	____
(c)	4,452	____	____

CONVERSIONS BETWEEN THE ENGLISH AND METRIC SYSTEMS

2-28 What of the following are "exact" relationships?
(a) 12 = 1 dozen
(b) 1 gal = 3.78 L
(c) 3 ft = 1 yd
(d) 1.06 qt = 1 L
(e) 10^3 m = 1 km
(f) 454 g = 1 lb

2-29 Write a relationship in factor form that would be used in making the following conversions.
(a) in. to ft   (d) L to qt
(b) in. to cm   (e) pt to L
(c) miles to ft   (two factors)

2-30 Write a relationship in factor form that would be used in making the following conversions.
(a) qt to gallons   (d) ft to km (two
(b) kg to lb   factors)
(c) gal to L

2-31 Complete the following table.

	miles	ft	m	km
(a)	____	____	7.8×10^3	____
(b)	0.450	_____	_____	____
(c)	____	8.98×10^3	_____	____
(d)	____	_____	_____	6.78

2-32 Complete the following table.

	gal	qt	L
(a)	6.78	____	_____
(b)	____	670	_____
(c)	____	____	7.68×10^3

2-33 Complete the following table.

	lb	g	kg
(a)	____	____	0.780
(b)	____	985	____
(c)	16.0	____	____

2-34 If a person has a mass of 122 lb, what is her mass in kilograms?

2-35 A punter on a professional football team averaged 28.0 m per kick. What is his average in yards? Should he be kept on the team?

2-36 If a student drinks a 12-oz (0.375 qt) can of soda, what volume did she drink in liters?

2-37 A prospective basketball player is 6 ft $10\frac{1}{2}$ in. tall and weighs 212 lb. What is his height in meters and his weight in kilograms?

2-38 Gasoline is sold by the liter in Europe. How many gallons does a 55.0-L gas tank hold?

2-39 If the length of a football field is changed to 100 m from 100 yd, will the field be longer or shorter than the current field? How many yards would a "first and ten" be on the metric field?

2-40 Bourbon used to be sold by the "fifth" (one-fifth of a gallon). A bottle now contains 750 mL. Which is greater?

2-41 If the speed limit is 65.0 miles/hr, what is the speed limit in kilometers per hour?

2-42 Mount Everest is 29,028 ft high. How high is this in kilometers?

2-43 If gold costs $415/oz, what is the cost of 1.00 kg of gold? (Assume 16 oz per pound.)

2-44 It is 525 miles from St. Louis to Detroit. How far is this in kilometers?

2-45 Gasoline sold for $0.899 per gallon in 1987. What was the cost per liter? What did it cost to fill an 80.0-L tank?

2-46 Using the price of gas from the preceding problem, how much does it cost to drive 551 miles if your car averages 21.0 miles/gal? How much does it cost to drive 482 km?

2-47 Using information from the two preceding problems, how many kilometers can one drive for $45.00?

2-48 An aspirin contains 0.324 g (5.00 grains) of aspirin. How many pounds of aspirin are in a 500-aspirin bottle?

2-49 A hamburger in Canada sells for $2.55 (Canadian dollars). That seems expensive until we realize that the exchange rate is $1.26 Canadian per one U.S. dollar. What is the cost in U.S. dollars?

2-50 A certain type of nail costs $0.95/lb. If there are 145 nails per pound, how many nails can you purchase for $2.50?

2-51 Another type of nail costs $0.92/lb, and there are 185 nails per pound. What is the cost of 5670 nails?

*2-52 At a speed of 35 miles/hr, how many centimeters do you travel per second?

*2-53 The planet Jupiter is about 4.0×10^8 miles from Earth. If radio signals travel at the speed of light, which is 3.0×10^{10} cm/sec, how long does it take a radio command from Earth to reach a Voyager spacecraft passing Jupiter?

2-54 If grapes sell for $1.15/lb and there are 255 grapes per pound, how many grapes can you buy for $5.15?

TEMPERATURE

2-55 The temperature of the water around a nuclear reactor core is about 300 °C. What is this temperature in degrees Fahrenheit?

2-56 The temperature on a comfortable day is 76 °F. What is this temperature in degrees Celsius?

2-57 The lowest possible temperature is −273 °C and is called absolute zero. What is absolute zero on the Fahrenheit scale?

2-58 Mercury thermometers cannot be used in cold arctic climates because mercury freezes at −39 °C. What is this temperature in degrees Fahrenheit?

2-59 The coldest temperature recorded on earth was −110 °F. What is this temperature in degrees Celsius?

2-60 A hot day in the midwest is 35.0 °C. What is this in degrees Fahrenheit?

2-61 Convert the following temperatures to the Kelvin scale.
(a) 47 °C   (d) −12 °C
(b) 23 °C   (e) 65 °F
(c) −73 °C   (f) −20 °F

2-62 Make the following temperature conversions.
(a) 37 °C to K   (d) 127 K to °F
(b) 135 °C to K   (e) 100 °F to K
(c) 205 K to °C   (f) −25 °C to K

2-63 Convert the following Kelvin temperatures to degrees Celsius.
(a) 175 K   (d) 225 K
(b) 295 K   (e) 873 K
(c) 300 K

*2-64 At what temperature are the Celsius and Fahrenheit scales numerically equal?

DENSITY

2-65 A handful of sand has a mass of 208 g and displaces a volume of 80.0 mL. What is its density?

2-66 A 125-g quantity of iron has a volume of 15.8 mL. What is its density?

2-67 The volume of liquid in a graduate cylinder is 14.00 mL. When a certain solid is added, the volume reads 92.45 mL. The mass of the solid is 136.5 g. What might the solid be? (Refer to Table 2-5.)

2-68 A certain liquid has a volume of 0.657 L and a mass of 1064 g. What might the liquid be? (Refer to Table 2-5.)

2-69 What is the volume in milliliters occupied by 285 g of mercury? (See Table 2-5.)

2-70 What is the mass of 671 mL of table salt? (See Table 2-5.)

2-71 What is the mass of 1.00 L of gasoline (See Table 2-5.)

2-72 What is the mass of 1.50 L of gold? (See Table 2-5.)

2-73 What is the volume in milliliters occupied by 1.00 kg of carbon tetrachloride?

2-74 Pumice is a volcanic rock that contains many trapped air bubbles. A 155-g sample is found to have a volume of 163 mL. What is the density of pumice? What is the volume of a 4.56-kg sample? Will pumice float or sink in water? In ethyl alcohol?

2-75 The density of diamond is 3.51 g/mL. What is the volume of the Hope diamond if it has a mass of 44.0 carats (1 carat = 0.200 g)?

2-76 A small box is filled with liquid mercury. The box is 3.00 cm wide, 8.50 cm long, and 6.00 cm high. What is the mass of the mercury in the box? (1.00 mL = 1.00 cm³.)

2-77 Which has a greater volume, 1 kg of lead or 1 kg of gold?

2-78 Which has the greater mass, 1 L of gasoline or 1 L of water?

2-79 What is the mass of 1 gal of carbon tetrachloride in grams? In pounds?

*2-80 Calculate the density of water in pounds per cubic foot?

*2-81 In certain stars, matter is tremendously compressed. In some cases the density is as high as 2.0×10^7 g/mL. A tablespoon full of this matter is about 4.5 mL. What is this mass in pounds?

2-82 What is the difference between density and specific gravity?

SPECIFIC HEAT

2-83 It took 73.2 J of heat to raise the temperature of 10.0 g of a substance 8.58 °C. What is the specific heat of the substance?

2-84 When 365 g of a certain pure metal cooled from 100 to 95 °C, it liberated 56.6 cal. Identify the metal from among those listed in Table 2-6.

2-85 A 10.0-g sample of a metal requires 22.4 J of heat to raise the temperature from 37.0 to 39.5 °C. Identify the metal from those listed in Table 2-6.

2-86 If 150 cal of heat energy is added to 50.0 g of copper at 25 °C, what is the final temperature of the copper? Compare this temperature rise with that of an equivalent amount of water in Example 2-19.

2-87 How many joules are evolved if 43.5 g of aluminum is cooled by 13.0 Celsius degrees?

2-88 What mass of iron is needed to absorb 16.0 cal if the temperature of the sample rises from 25 to 58 °C?

2-89 When 486 g of zinc absorbs 265 J of heat energy, what is the rise in temperature of the metal? (See Table 2-6.)

2-90 Given 12.0-g samples of iron, gold, and water, calculate the temperature rise that would occur when 50.0 J of heat is added to each?

2-91 A 10.0-g sample of water cools 2.00 °C. What mass of aluminum is required to undergo the same temperature change?

2-92 A can of diet soda contains 1.00 Cal (1.00 kcal) of heat energy. If this energy were transferred to 50.0 g of water at 25 °C, what would be the final temperature?

2-93 If 50.0 g of aluminum at 100.0 °C is allowed to cool to 35.0 °C, how many joules are evolved?

*2-94 In the preceding problem assume that the hot aluminum was added to water originally at 30.0 °C. If all of the heat lost by the aluminum was gained by the water, what mass of water was present if the final temperature was 35.0 °C?

*2-95 If 50.0 g of water at 75.0 °C is added to 75.0 g of water at 42.0 °C, what is the final temperature?

*2-96 If 100.0 g of a metal at 100.0 °C is added to 100 g of water at 25.0 °C, the final temperature is 31.3 °C. What is the specific heat of the metal? Identify the metal from Table 2-6.

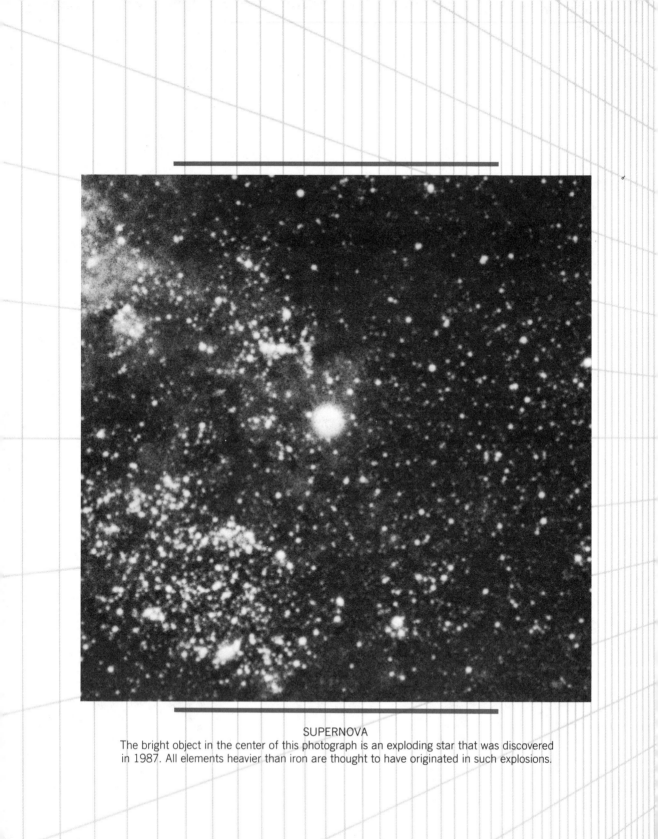

SUPERNOVA
The bright object in the center of this photograph is an exploding star that was discovered in 1987. All elements heavier than iron are thought to have originated in such explosions.

THE ATOM, THE
■
STRUCTURE OF
■
MATTER, AND
■
NUCLEAR
■
REACTIONS
■

PURPOSE OF CHAPTER

In Chapter 3 we look closely at the atom, the basic building block of the elements: we study its composition and how it exists in elements and compounds. We also describe the changes that the nucleus of the atom can undergo.

OBJECTIVES FOR CHAPTER

After completion of this chapter, you should be able to:

1   List the names and symbols of some common elements. (3-1)

2   Describe the conclusions of Dalton's atomic theory. (3-2)

3   Describe the nuclear atom, including the name, location, mass (in amu), and electrical charge of the three primary particles in the atom. (3-3)

4   Give the atomic number and mass number of a specified isotope. (3-4)

5   Determine the number of protons, neutrons, and electrons from the symbol for a specified isotope. (3-4)

3

6 Define atomic mass and describe how it differs from mass number for an isotope. (3-4)

7 Calculate the atomic weight of an element given the percent distribution and atomic masses of its isotopes. (3-4)

8 Describe the function of a covalent bond in a molecule. (3-5)

9 List the elements and the number of atoms of that element in a compound from the formula. (3-5)

10 Write definitions for the terms "ion," "cation," "anion," and "electrostatic force." (3-6)

11 Distinguish between molecular and ionic compounds. (3-6)

12 Determine the number of protons and electrons present in a specified ion. (3-6)

(OPTIONAL: NUCLEAR REACTIONS)

13 Explain what is illustrated by a nuclear equation. (3-7)

14 Define radioactivity and the three common kinds of radiation. (3-7)

15 Complete a nuclear equation given appropriate data. (3-7)

16 Describe the concepts of half-life and radioactive decay series. (3-8)

17 Describe how the three major types of radiation affect nearby matter. (3-9)

18 Explain how nuclear reactions take place and how elements are artificially synthesized. (3-10)

19 Describe nuclear fission and how a chain reaction occurs. (3-11)

20 Explain how fission is controlled in a nuclear reactor. (3-11)

21 Explain what is meant by nuclear fusion. (3-12)

In late February of 1987, a young scientist on a desolate mountaintop in Chile looked out into the night sky through a telescope. He expected a routine study of the skies but what he saw at first glance looked like a bright new star. Only this wasn't a normal star burning steadily like our sun. It was a star in its death throes undergoing a catastrophic explosion known as a supernova. It was the brightest supernova seen on the earth in almost 400 years. While this dramatic event signaled the end of this star, it also meant the creation of vast quantities of heavy elements.

For billions of years, the stars have been churning, boiling, and eventually undergoing catastrophic explosions. In these great infernos throughout space the creation of all of the elements from the simplest element, hydrogen, has been occurring. Blasted by the force of exploding suns, these elements and some simple compounds of the elements have drifted through cold, dark space as bits of dust and gases. About four and a half billion years ago a large cloud of this gas and dust condensed, forming our sun and the planets. On the third planet from the sun, which we call earth, there arose living systems composed of these elements. We and all of nature as we know it are, in fact, composed of "stardust."

In the past 100 years, the miracle of the human mind and spirit has unraveled step-by-step some of the mysteries of the composition of these elements as well as their origin. Can a sample of an element be divided into infinitely smaller pieces or is there a basic unit that cannot be divided further? If there is a basic unit, is it composed of even more fundamental units? What feature makes one element different from another? What is the basic composition of compounds? These are questions that we are now ready to answer and are the subjects of this chapter. But first, we have more to say about the elements from which all matter is constructed.

3-1 THE NAMES AND SYMBOLS OF THE ELEMENTS

Of the 108 known elements only 90 are known to occur in nature. The others have a transitory existence because they spontaneously change into other elements by means of radioactive decay. All elements above element 93 were not discovered in the earth but were synthesized in laboratories or nuclear reactors. Of all the elements found in nature only 10 constitute over 99% of the crust of the earth (the outer portion including the oceans and the atmosphere). About 93% of the mass of our bodies is composed of only three elements: carbon, hydrogen, and oxygen. (See Table 3-1.)

The names of the elements come from many sources. Some elements derive their names from the Greek, Latin, or German word for a color [e.g., bismuth (white mass), iridium (rainbow), rubidium (deep red), and chlorine (greenish yellow)]. Some derive their names from the locality where they were discovered (e.g., germanium, francium, and californium). Four elements (yttrium, erbium, terbium, and ytterbium) are named after a town in Sweden (Ytterby). Other elements are named after noted scientists (e.g., einsteinium, fermium, and curium) and still others after mythological figures [e.g., plutonium, uranium, titanium, mercury, and promethium (the fire-giver)]. Many of the oldest known elements have names that have obscure origins. In any case, the names of all of the elements are listed inside the front cover. The names of elements 104 through 109 are considered temporary until appropriate names are assigned. Permanent names can be assigned only by the International Union of Pure and Applied Chemistry (IUPAC).

An element can be designated by a symbol as shorthand for its full name. *In most cases the first one or two letters of the name are used as the element's symbol.* When an element has a two-letter symbol, the first is capitalized but the second is not. The symbols of the elements are listed along with the names

TABLE 3-1   Distribution of the Elements in the Earth's Crust[a] and Human Body

Earth's Crust				Human Body	
Element	Weight Percent	Element	Weight Percent	Element	Weight Percent
Oxygen	49.1	Magnesium	1.9	Oxygen	64.6
Silicon	26.1	Hydrogen	0.88	Carbon	18.0
Aluminum	7.5	Titanium	0.58	Hydrogen	10.0
Iron	4.7	Chlorine	0.19	Nitrogen	3.1
Calcium	3.4	Carbon	0.09	Calcium	1.9
Sodium	2.6	Sulfur	0.06	Phosphorus	1.1
Potassium	2.4	All others	0.50	Chlorine	0.4
				All others	0.9

[a] Includes the oceans and atmosphere.

TABLE 3-2   Some Common Elements

Element	Symbol	Element	Symbol
Aluminum	Al	Iodine	I
Bromine	Br	Magnesium	Mg
Calcium	Ca	Nickel	Ni
Carbon	C	Nitrogen	N
Chlorine	Cl	Oxygen	O
Chromium	Cr	Phosphorus	P
Fluorine	F	Silicon	Si
Helium	He	Sulfur	S
Hydrogen	H	Zinc	Zn

inside the front cover. In a chart of the elements called the periodic table (which is shown inside the back cover), the symbols of the elements are listed in addition to other important information that is discussed in this chapter. Some common elements whose symbols derive from their English names are shown in Table 3-2.

In addition to these elements there are eleven elements whose symbols do not seem to relate to their names. These elements (see Table 3-3) derive their symbols from their original Latin names (except for wolfram, which is German).

The names of compounds are usually derived from the names of the elements of which they are composed. The names of most of the compounds referred to in this text contain two words where the second word ends in *ide, ite,* or *ate.* Thus, such names as carbon dioxide, sodium sulfite, and iron(III) nitrate (pronounced "iron three nitrate") all refer to specific compounds. There are a few names for compounds that have three words, such as sodium hydrogen carbonate. A number of compounds also have common names, such as water, ammonia, lye, and methane. We will talk more about the names of compounds in Chapter 7.

SEE PROBLEMS
3-1 THROUGH
3-5.

TABLE 3-3   Elements with Symbols from Earlier Names

Element	Symbol	Former Name
Antimony	Sb	Stibium
Copper	Cu	Cuprum
Gold	Au	Aurum
Iron	Fe	Ferrum
Lead	Pb	Plumbum
Mercury	Hg	Hydragyrum
Potassium	K	Kalium
Silver	Ag	Argentum
Sodium	Na	Natrium
Tin	Sn	Stannum
Tungsten	W	Wolfram

3-2 THE COMPOSITION OF THE ELEMENTS

Only 200 years ago, the most knowledgeable scientists had a view of matter very different from our current understanding. They thought that if you took a sample of an element such as copper it could (theoretically) be divided into infinitely smaller pieces without changing its nature. In other words, they thought that matter was continuous. We now have a completely different image of matter. If we were able to greatly magnify a small copper wire, we would see that it is composed of closely packed spheres. (See Figure 3-1.) These small spheres are called atoms. *An atom is the smallest particle of an element that can exist and enter into chemical reactions.* Atoms are so small, however, that only the most sophisticated microscope (called a scanning tunneling microscope) in use today can produce fuzzy images of the larger atoms.

The phenomenon of the copper wire can be likened to the appearance of a brick wall: it looks completely featureless and continuous from a distance, but up close it becomes obvious that the wall is constructed of closely packed basic units (bricks). In the wire, the atoms of copper are firmly packed together, as would be any spheres of the same size such as marbles. The small size of the atom is difficult to comprehend. Since the diameter of a typical atom is on the order of 10^{-8} cm, it would take 100 million atoms in a line to extend for 1 cm (less than half an inch).

Our acceptance of the atom as the basic particle of an element did not await the invention of the powerful tunneling microscopes that produce the images of some large atoms. In 1803 an English scientist named John Dalton (1766 – 1844) proposed what is now known as the atomic theory. The major conclusions of this theory are as follows.

1   Matter is composed of small indivisible particles called atoms.
2   Atoms of the same element are identical and have the same properties.
3   Chemical compounds are composed of atoms of different elements combined in small whole-number ratios.
4   Chemical reactions are merely the rearrangement of atoms into different combinations.

This theory, with some adjustments, serves as the basis of our modern understanding of the nature of matter. In fact, it was a restatement of a long dormant

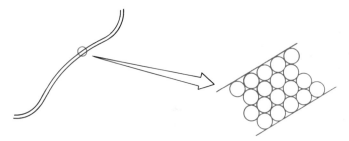

Figure 3-1   ATOMS OF COPPER. Most solid elements exist in nature as closely packed atoms.

concept proposed by a Greek philosopher, Democritus, more than 2000 years earlier. Dalton's theory, like other modern scientific theories, correlated a large amount of quantitative experimental observations, particularly the law of conservation of mass (see p. 18). The theory of Democritus was simply a product of his mind and not of experiment and observation.

The first question that we attempt to answer concerns the atom itself. Of special interest is how the atoms of one element differ from those of another element.

3-3 THE COMPOSITION OF THE ATOM

From earth with the naked eye, the moon looks like a solid, round sphere. Through a telescope, however, the moon is seen to be complex, with mountains, valleys, and large craters. The best image of an atom currently possible also shows the atom as a featureless sphere. If we could peer within the atom we would find that like the moon through a telescope, the atom is more complex than the simple image indicates. What follows is a simplification of the accepted view of the composition of the atom. How this model evolved is an intriguing story beginning in the late 1800s and continuing to the present day. The ingenious experiments and observations of such scientists as Thomson, Rutherford, Becquerel, Curie, and Roentgen, among others, contributed to this view of the atom, known as the nuclear model of the atom.

Before we peer into the confines of the atom we should note that it had been known for centuries that matter is electrical in nature. This electrical nature of matter has been a topic of intense curiosity. We know that even Ben Franklin experimented with lightning and was fortunate enough to live through it. It seemed that certain substances could take on an electrostatic charge. **Electrostatic charges** *can be either positive or negative.* Interaction between two charges produces what is known as an electrostatic force. **Electrostatic forces** *refer to the forces of attraction between opposite charges and the forces of repulsion between like charges.* (See Figure 3-2.)

If we were to journey within the confines of the atom we would first encounter a very small particle called an *electron.* The electron was the first subatomic particle in the atom to be identified. In 1897, J. J. Thomson charac-

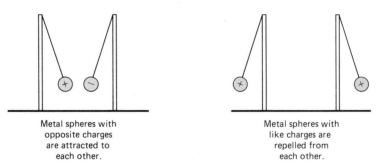

Metal spheres with opposite charges are attracted to each other.

Metal spheres with like charges are repelled from each other.

Figure 3-2   ELECTROSTATIC FORCES. Opposite charges attract; like charges repel.

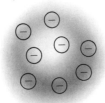

Figure 3-3   THOMSON'S MODEL. Electrons were seen as tiny particles evenly distributed in a positively charged medium.

terized the electron by proving that it had a negative electrical charge (assigned a value of −1) and was common to the atoms of all elements. Apparently, matter was electrical in nature because it contained electrically charged particles. From this discovery, Thomson suggested a model of the atom something like a ball of cotton. (See Figure 3-3.) The cotton would represent a spread out positive charge, with the tiny electrons distributed like seeds in the cotton exactly balancing the positive charge.

In 1911 Ernest Rutherford in England conducted experiments meant to support the model of Thomson, but instead he was forced to come to a radically different conclusion about the atom. When he bombarded a thin foil of gold with alpha particles (positively charged helium atoms) he expected that all of these very small particles would pass right through the atoms of the metal, like bullets through a stick of butter. That is, none of the alpha particles should be affected by the metal atoms. Instead, he found that some of the particles were deflected and even a very few were deflected right back to the source. (See Figure 3-4.) After repeated experiments he reluctantly con-

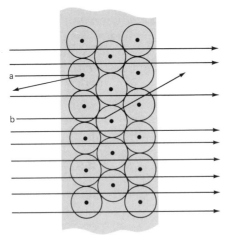

Figure 3-4   RUTHERFORD'S EXPERIMENT. To account for the deflection of some of the positively charged alpha particles, Rutherford proposed that the atom is mostly empty space with the positive charge located in a small, dense core. When an alpha particle encountered a nucleus (a and b) electrostatic repulsions would either deflect it or repel it back.

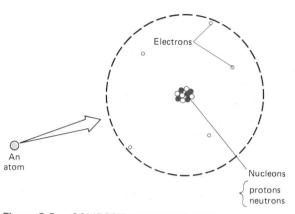

Figure 3-5 COMPOSITION OF THE ATOM. The atom is composed of electrons, neutrons, and protons.

cluded that Thomson's model was in error. Rutherford realized that there could be only one explanation for the results of his experiments and this required that all of the positive charge and almost all of the mass be centered in a small, densely packed core of the atom. This core was called the *nucleus*. The large volume of the atom is mostly empty space occupied by the very small electrons. The nucleus is so small compared with the total volume of a typical atom that if it were expanded to the size of a softball, the radius of the atom would extend for about one mile. Later experiments would show that the nucleus is composed of particles called nucleons. There are two types of nucleons: protons, which have a positive charge (assigned a value of $+1$, equal and opposite to that of an electron), and neutrons, which do not carry a charge. (See Figure 3-5.) The data on the three particles in the atom are summarized in Table 3-4. The proton and neutron have roughly the same mass, which is about "1 amu" *(atomic mass unit)*. This unit of mass is a convenient unit for the masses of individual atoms.

For some time it was thought that the atom was composed of just these three particles. As experimental procedures became more elaborate and sophisticated, however, the picture became more complex. It now appears that the three particles are themselves composed of various combinations of even more fundamental particles called quarks. Fortunately, the three-particle model of the atom still meets the needs of the chemist.

TABLE 3-4 Atomic Particles

Name	Symbol	Electrical Charge	Mass (amu)	Mass (g)
Electron	e	-1	0.000549	9.110×10^{-28}
Proton	p	$+1$	1.00728	1.673×10^{-24}
Neutron	n	0	1.00867	1.675×10^{-24}

3-4 ATOMIC NUMBER, MASS NUMBER, AND ATOMIC WEIGHT

A typical atom of the element copper is made up of a nucleus containing a total of 63 nucleons, of which 29 are protons and 34 are neutrons. The neutral atom also contains 29 electrons. *The number of protons in the nucleus* (which is equal to the total positive charge) *is referred to as the atom's* **atomic number.** *The total number of nucleons is called the* **mass number.** Therefore, this particular copper atom has an atomic number of 29 and a mass number of 63. If we were to examine a number of copper atoms, we would soon come upon one that is different from the one we just described. This atom of copper also has an atomic number of 29 but a mass number of 65, so it has 36 neutrons. *Atoms having the same atomic number but different mass numbers are known as* **isotopes.** (Obviously, Dalton wasn't completely right; all of the atoms of an element are not necessarily identical.)

If we looked at the atoms of other elements, we would find the same phenomenon—almost all elements exist in nature as a mixture of isotopes. What we would also notice is that different elements have different atomic numbers. *Indeed, it is the atomic number that distinguishes one element from another.*

Isotopes of elements are identified by the mass number in the upper left-hand corner of the symbol of the element. Sometimes, the atomic number of the element is represented in the lower left-hand corner. This is strictly a convenience, however, since the atomic number determines the identity of the element. For example, if the atomic number is 28, the element is nickel; if it is 30, the element is zinc. The two isotopes of copper are shown below.

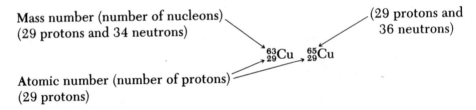

Mass number (number of nucleons)
(29 protons and 34 neutrons)

(29 protons and
36 neutrons)

$$^{63}_{29}\text{Cu} \quad ^{65}_{29}\text{Cu}$$

Atomic number (number of protons)
(29 protons)

EXAMPLE 3-1

How many protons, neutrons, and electrons are present in $^{90}_{38}\text{Sr}$?

$$\text{atomic number} = \text{number of protons} = \underline{38}$$

$$\text{number of neutrons} = \text{mass number} - \text{number of protons}$$
$$90 - 38 = \underline{52}$$

In a neutral atom the number of protons and the number of electrons are the same.

$$\text{number of electrons} = \underline{38}$$

EXAMPLE 3-2

How many protons, neutrons, and electrons are in the three isotopes of oxygen, $^{16}_{8}O$, $^{17}_{8}O$, and $^{18}_{8}O$?

$^{16}_{8}O$ Protons = atomic number = $\underline{8}$

Neutrons = mass number − atomic number = $16 - 8 = \underline{\underline{8}}$

Electrons = protons = $\underline{8}$

$^{17}_{8}O$ Protons = $\underline{\underline{8}}$

Neutrons = $17 - 8 = \underline{\underline{9}}$

Electrons = $\underline{\underline{8}}$

$^{18}_{8}O$ Protons = $\underline{\underline{8}}$

Neutrons = $18 - 8 = \underline{10}$

Electrons = $\underline{\underline{8}}$

Note that the only difference among the isotopes is the number of neutrons in the nucleus.

■
*SEE PROBLEMS
3-6 THROUGH
3-14.*

Because the proton and the neutron each have a mass of about 1 amu, the mass number of an isotope is a convenient but imprecise measure of its comparative mass. A more precise measure of the mass of one isotope relative to another is the atomic mass. *The **atomic mass** of an isotope is determined by comparison with a standard, ^{12}C, which is defined as having a mass of exactly 12 amu.* By comparison with ^{12}C, the atomic mass of ^{10}B is 10.013 amu and that of ^{11}B is 11.009 amu. Since boron, as well as most other naturally occurring elements, are found in nature as a mixture of isotopes, the atomic weight* of the element reflects this mixture. *The **atomic weight** of an element as found in nature is the average of the atomic masses of all isotopes present.* The average is related, however, to the percentage of each isotope present, as illustrated by the following examples.

Inside the back cover you will find an arrangement of all of the elements in what is called the *periodic table*. The symbol of each element is shown in a box. Also included in the box is the atomic number and the atomic weight of the element. (See Figure 3-6.) Note that the elements are listed in order of in-

---

* Atomic mass is actually the correct term. However, long usage and tradition require that we still use "atomic weight" to indicate the average masses of isotopes of an element. We will continue to use "atomic mass" for individual atoms or isotopes.

EXAMPLE 3-3

In nature the element boron is present as 19.9% $^{10}_{5}B$ and 80.1% $^{11}_{5}B$. If the atomic mass of $^{10}_{5}B$ is 10.013 amu and that of $^{11}_{5}B$ is 11.009 amu, what is the atomic weight of boron?

PROCEDURE

Find the contribution of each isotope toward the atomic weight by multiplying the percent in decimal fraction form by the mass of the isotope.

$$^{10}_{5}B \quad 0.199 \times 10.013 \text{ amu} = \quad 1.99 \text{ amu}$$
$$^{11}_{5}B \quad 0.801 \times 11.009 \text{ amu} = \quad \underline{8.82 \text{ amu}}$$
$$\underline{10.81 \text{ amu}}$$

EXAMPLE 3-4

Naturally occurring lead is composed of four isotopes: 1.40% $^{204}_{82}Pb$ (203.97 amu), 24.10% $^{206}_{82}Pb$ (205.97 amu), 22.10% $^{207}_{82}Pb$ (206.98 amu), and 52.40% $^{208}_{82}Pb$ (207.98 amu). What is the atomic weight of lead?

$$^{204}_{82}Pb \quad 0.0140 \times 203.97 \text{ amu} = \quad 2.86 \text{ amu}$$
$$^{206}_{82}Pb \quad 0.2410 \times 205.97 \text{ amu} = \quad 49.64 \text{ amu}$$
$$^{207}_{82}Pb \quad 0.2210 \times 206.98 \text{ amu} = \quad 45.74 \text{ amu}$$
$$^{208}_{82}Pb \quad 0.5240 \times 207.98 \text{ amu} = \underline{109.0 \quad \text{amu}}$$
$$\underline{207.2 \quad \text{amu}}$$

creasing atomic number. The elements with whole-number atomic weights in parentheses are synthetic elements and are not found in nature. These elements undergo nuclear disintegration, as discussed in a later section of this chapter. The atomic weight listed for these elements represents the mass number of the most common isotope.

SEE PROBLEMS 3-15 THROUGH 3-25.

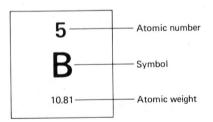

Figure 3-6   ATOMIC WEIGHT AND ATOMIC NUMBER FROM THE PERIODIC TABLE.

**3-5
MOLECULES,
COMPOUNDS,
AND
FORMULAS**

The copper wire that was carefully examined in Section 3-2 was composed of individual atoms packed closely together. Most, but not all, elements exist in nature as simple congregations of individual atoms. In some elements, however, the basic units are composed of molecules. *A molecule is a group of two or more atoms where the atoms are held together by a force called a covalent chemical bond.*

Iodine is an example of an element whose atoms are joined by covalent bonds to form molecules. If we could magnify a small sample of solid iodine, the image would show that the atoms are bonded to each other in pairs. (See Figure 3-7.) Two spherical atoms in each molecule are firmly joined together and even overlap to a small extent. Atoms of adjoining molecules just touch, however, as did the individual atoms of copper. The nature of the covalent bond is discussed in more detail in Chapter 6.

Examples of other elements that exist as diatomic (two-atom) molecules under normal conditions, such as are found on the surface of the earth, are oxygen, nitrogen, hydrogen, fluorine, chlorine, and bromine. A form of phosphorus consists of molecules composed of four atoms, and a form of sulfur consists of molecules composed of eight atoms.

The molecules of an element are composed of atoms of the same type. Of more significance to the chemist is the fact that atoms of different elements also combine to form molecules. *A molecular compound is composed of molecules made up of atoms of two or more different elements.* If it were possible to magnify a droplet of water so as to visualize its basic units we would see that it is an example of a molecular compound. Each molecule of water is composed of two atoms of hydrogen joined to one atom of oxygen by covalent bonds. (See Figure 3-8.)

A compound is represented by using the symbols for the elements of which it is composed. This is called the **formula** *of the compound or element.* The familiar formula for water is therefore

$$H_2O$$

This is, of course, pronounced "H-two-O." Note that the "2" is written as a subscript, indicating that the molecule has two hydrogen atoms. When there is only one atom of a certain element present (e.g., oxygen), a subscript of "1" is assumed.

Each chemical compound has a unique formula or arrangement of atoms in its molecules. For example, another compound whose molecules are composed of

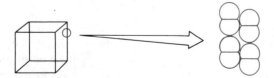

Figure 3-7 THE MOLECULES OF IODINE. Some elements exist in nature as two or more atoms bound together.

Figure 3-8   A WATER MOLECULE. A water molecule is composed of two hydrogen atoms bound to one oxygen atom.

hydrogen and oxygen is called hydrogen peroxide. Since the molecules of hydrogen peroxide are composed of two hydrogen atoms and two oxygen atoms, its formula is H_2O_2 and its properties are distinctly different from those of water (H_2O). The formulas of several other well-known molecular compounds are $C_{12}H_{22}O_{11}$ (table sugar — sucrose), $C_9H_8O_4$ (aspirin), NH_3 (ammonia), and CH_4 (methane gas). Elements that exist as molecules have formulas such as I_2 (iodine) and P_4 (phosphorus).

It is possible for two compounds to have the same formula. For example, ethyl alcohol and dimethyl ether both have the formula C_2H_6O. The two compounds are unique, however, because of the different arrangement of atoms within the molecules, as shown below. (The dashes between atoms represent covalent bonds.)

ethyl alcohol   dimethyl ether

$$\begin{array}{ccc} & H \quad H & \\ & | \quad\; | & \\ H- & C-C & -O-H \\ & | \quad\; | & \\ & H \quad H & \end{array} \qquad \begin{array}{ccc} H & & H \\ | & & | \\ H-C & -O- & C-H \\ | & & | \\ H & & H \end{array}$$

Formulas such as these that show the order and arrangement of atoms in a molecule are known as **structural formulas.**

3-6 IONS AND IONIC COMPOUNDS

Ordinary table salt is composed primarily of the compound sodium chloride, which has the formula NaCl. This type of compound is fundamentally different from the molecular compounds discussed previously. In this compound, the sodium and chlorine atoms have a net electrical charge. *Atoms or groups of atoms that have a net electrical charge are known as* **ions.** *Positive ions are called* **cations** *and negative ions are known as* **anions.** In NaCl, the sodium is present as a cation with a single positive charge and the chlorine is present as an anion with a charge equal and opposite that of the sodium cation. This is illustrated with a + or − as a superscript to the right of the symbol of the element.

$$Na^+Cl^-$$

Since the sodium cations and chlorine anions are oppositely charged, the ions are held together by electrostatic forces of attraction. (See Figure 3-2.)

Compounds consisting of ions are known as **ionic compounds.** *The electrostatic forces holding the ions together are known as* **ionic bonds.** In Figure 3-9 the ions in NaCl are shown as they would appear if magnification were possible. Note that each cation is not attached to just one anion. In fact, each ion is

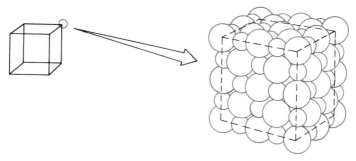

Figure 3-9   THE IONS IN SODIUM CHLORIDE. An ionic compound exists as an
arrangement of charged particles.

surrounded by six oppositely charged ions. This situation is different from that
of the molecular compounds discussed previously, in which atoms actually
overlapped to form discrete groups of molecules. The formula of an ionic
compound, such as NaCl, represents the simplest ratio of ions present and this
group of ions is known as a **formula unit.**

Formulas of certain ionic compounds may at first appear more complex than
necessary. For example, barium nitrite has the formula $Ba(NO_2)_2$ rather than
the simpler formula BaN_2O_4. This ionic compound consists of Ba^{2+} cations and
NO_2^- anions. The two oxygen atoms and one nitrogen atom in the anion are
bound together by covalent bonds and the three atoms act as a single unit with
an electrical charge of -1. The parentheses around the NO_2 indicate that
there are two NO_2^- ions present for each Ba^{2+} ion. The NO_2^- anion is known as
a **polyatomic** *(more than one atom)* **ion.** Note that an ion (e.g., NO_2^-) does not
exist alone in nature as does a neutral molecule (e.g., H_2O). Anions and cations
always exist together in such a ratio that the total positive charge is balanced
by the total negative charge. If only one polyatomic ion is present in a formula,
parentheses are not used. For example, $CaCO_3$ (limestone) contains one Ca^{2+}
cation and one CO_3^{2-} (carbonate) anion per formula unit. Other familiar examples of ionic compounds are K_2O (potash), $NaHCO_3$ (baking soda), and NaF
[sodium fluoride (in toothpaste)].

An ion has a net electrical charge because it does not have an equal number
of protons in its nucleus (or nuclei for polyatomic ions) and electrons. Cations
have fewer electrons than the neutral atom (one less electron for each positive
charge) and anions have more electrons than the neutral atom (one extra
electron for each negative charge). Thus, the Na^+ cation has 11 protons in its
nucleus but only 10 electrons ($+11 - 10 = +1$) and the Cl^- anion has 17
protons in its nucleus and 18 electrons ($+17 - 18 = -1$).

In Figure 3-10, the discussion in Sections 3-5 and 3-6 has been summarized.
Under normal conditions found on the earth, matter is composed of either
atoms, molecules, or ions. The formula of a compound represents either a
discrete molecular unit in the case of molecules or a ratio of ions in the case of
ionic compounds. Eventually (Chapter 6), we will discuss why some elements

■
*SEE PROBLEMS
3-26 THROUGH
3-37.*

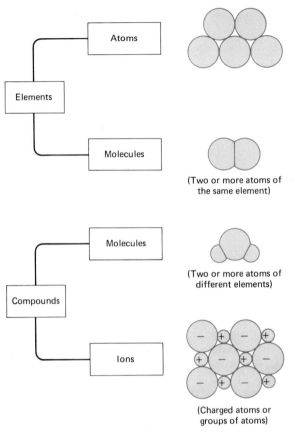

Figure 3-10   BASIC STRUCTURES OF ELEMENTS AND COMPOUNDS.

are composed of molecules and why one compound is molecular whereas another is ionic.

(*Note:* The material in the remainder of this chapter concerns a more indepth study of the nucleus of the atom. It is not directly relevant to the chapters that follow. It may be covered now as a logical extension of the nature of the nucleus, omitted completely, or covered later as desired.)

3-7 NATURAL RADIO-ACTIVITY

What was to eventually have such monumental effects on the human race all began quite innocently in 1896. A French scientist by the name of Henri Becquerel was investigating how sunlight interacted with a compound of uranium. He suspected that the sunlight prompted the compound to give off the mysterious "X-rays" discovered the previous year by Wilhelm Roentgen. One cloudy day he was unable to do an experiment but he left the uranium compound on top of an unexposed photographic plate. Several days later, he developed the plate anyway and discovered that the part of the photographic plate on which the compound sat was exposed, even though the plate was covered. Obviously, the compound emitted some form of radiation even though it was not exposed to sunlight. Eventually, it was discovered what was happening. The major isotope of uranium (^{238}U) spontaneously (without warning or stimulation) emits the nucleus of a helium atom. The loss of the four nucleons of helium from the uranium nucleus leaves the nucleus of a different element. Ironically, the very process that the alchemists of the Middle Ages tried to promote occurred quite naturally. One element could spontaneously change into another.

The radiation process discovered by Becquerel can be illustrated conveniently by means of a nuclear equation. *A **nuclear equation** shows the initial nucleus to the left of the arrow and the product nuclei to the right of the arrow*, as follows.

$$^{238}_{92}\text{U} \longrightarrow {}^{234}_{90}\text{Th} + {}^{4}_{2}\text{He}$$

Note that the loss of four nucleons from the original (parent) nucleus leaves the remaining (daughter) nucleus with 234 nucleons ($238 - 4 = 234$); the loss of two protons leaves 90 protons in the daughter nucleus ($92 - 2 = 90$). Because a nucleus with 90 protons is an isotope of thorium, one element has indeed changed into another.

*The helium nucleus, when emitted from another nucleus, is called an **alpha (α) particle.** (See Figure 3-11.) The process of emitting energy or particles from a nucleus is known as **radioactive decay** or simply **radioactivity.** The particles or "rays" that are emitted are called **radiation.** The radioactive isotopes that undergo decay are sometimes referred to as **radionuclides.***

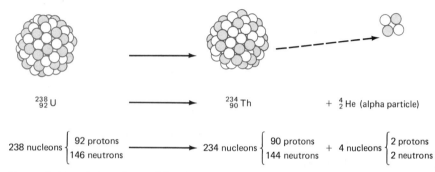

$^{238}_{92}$U $\longrightarrow$ $^{234}_{90}$Th $+$ $^{4}_{2}$He (alpha particle)

238 nucleons $\begin{cases} 92 \text{ protons} \\ 146 \text{ neutrons} \end{cases}$ $\longrightarrow$ 234 nucleons $\begin{cases} 90 \text{ protons} \\ 144 \text{ neutrons} \end{cases}$ $+$ 4 nucleons $\begin{cases} 2 \text{ protons} \\ 2 \text{ neutrons} \end{cases}$

Figure 3-11   ALPHA RADIATION. An alpha particle is a helium nucleus emitted from a larger nucleus.

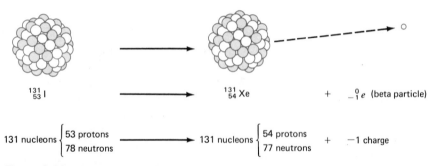

$$^{131}_{53}\text{I} \longrightarrow {}^{131}_{54}\text{Xe} \quad + \quad {}^{0}_{-1}e \text{ (beta particle)}$$

$$131 \text{ nucleons} \begin{cases} 53 \text{ protons} \\ 78 \text{ neutrons} \end{cases} \longrightarrow 131 \text{ nucleons} \begin{cases} 54 \text{ protons} \\ 77 \text{ neutrons} \end{cases} + \quad -1 \text{ charge}$$

Figure 3-12   BETA RADIATION. A beta particle is an electron emitted from a nucleus.

A second form of radioactive decay involves the emission of an electron from the nucleus. *An electron emitted* from a nucleus *is known as a* **beta (β) particle.** (See Figure 3-12.) The following equation illustrates beta particle emission:

$$^{131}_{53}\text{I} \longrightarrow {}^{131}_{54}\text{Xe} + {}^{0}_{-1}e$$

Note that the loss of a beta particle from a nucleus has the effect of changing a neutron in the nucleus into a proton, which increases the atomic number by one but leaves the mass number (number of nucleons) unchanged.

A third type of radioactive decay involves the emission of a high-energy form of light called a **gamma (γ) ray.** Like all light, gamma rays travel at 3.0×10^{10} cm/sec (186,000 miles/sec). Light as a form of energy is discussed in the next chapter. This type of radiation may occur alone or in combination with alpha or beta radiation as discussed above. The following equation illustrates gamma ray emission:

$$^{60}_{27}\text{Co}^* \longrightarrow {}^{60}_{27}\text{Co} + \gamma$$

(The asterisk indicates that the nucleus is in a high-energy state.)

Note that gamma radiation by itself does not result in a change in the nucleus. Excess energy contained in a nucleus has simply been released, much as burning coal releases chemical energy.

The three major types of radiation are summarized in Table 3-5.

TABLE 3-5	Types of Radiation		
Radiation	Mass Number	Charge	Identity
Alpha (α)	4	+2	Helium nucleus
Beta (β)	0	−1	Electron
Gamma (γ)	0	0	Light

EXAMPLE 3-5

Complete the following nuclear equations:

(a) $^{218}_{84}\text{Po} \rightarrow$ _____ $+ \, ^{4}_{2}\text{He}$

(b) $^{210}_{81}\text{Tl} \rightarrow \, ^{210}_{82}\text{Pb} +$ _____

PROCEDURE

For the missing isotope or particle:

1 Find the number of nucleons; it is the same on both sides of the equation.
2 Find the charge or atomic number; it is also the same on both sides of the equation.
3 If what's missing is the isotope of an element, find the symbol of the element that matches the atomic number from the list of elements inside the front cover.

SOLUTION

(a) Nucleons: $218 = x + 4$, so $x = 214$.
    Atomic number: $84 = y + 2$, so $y = 82$.

From the inside cover, the element having an atomic number of 82 is Pb. The isotope is

$$\underline{^{214}_{82}\text{Pb}}$$

(b) Nucleons: $210 = 210 + x$, so $x = 0$.
    Atomic number: $81 = 82 + y$, so $y = -1$.

An electron or beta particle has negligible mass (compared with a nucleon) and a charge of -1.

$$\underline{^{0}_{-1}e}$$

■
SEE PROBLEMS 3-38 THROUGH 3-41.

3-8 RATES OF DECAY OF RADIOACTIVE ISOTOPES

The isotopes of all elements heavier than $^{209}_{83}\text{Bi}$ are radioactive. If this is true, how can elements such as $^{238}_{92}\text{U}$ occur naturally in the earth? The answer is that many of these isotopes have very long "half-lives." *The half-life ($t_{1/2}$) of a radioactive isotope is the time required for one-half of a given sample to decay.* Each radioactive isotope has a specific and constant half-life. For example, $^{238}_{92}\text{U}$ has a half-life of 4.5×10^9 years. Since that is roughly the age of the earth, about one-half of the $^{238}_{92}\text{U}$ originally present when the earth formed from hot

TABLE 3-6   Half-Lives

Isotope	$t_{1/2}$	Mode of Decay	Product
$^{238}_{92}U$	4.5×10^9 years	α, γ	$^{234}_{90}Th$
$^{234}_{90}Th$	24.1 days	β	$^{234}_{91}Pa$
$^{226}_{88}Ra$	1620 years	α	$^{222}_{86}Rn$
$^{14}_{6}C$	5760 years	β	$^{14}_{7}N$
$^{131}_{53}I$	8.0 days	β, γ	$^{131}_{54}Xe$
$^{218}_{85}At$	1.3 sec	α	$^{214}_{83}Bi$

gases and dust is still present. Half of what is now present will be gone in another 4.5 billion years. The half-lives and modes of decay of some radioactive isotopes are listed in Table 3-6.

Note in Table 3-6 that when $^{238}_{92}U$ decays, it forms an isotope ($^{234}_{90}Th$) that also decays (very rapidly compared with $^{238}_{92}U$). The $^{234}_{91}Pa$ formed from the decay of $^{234}_{90}Th$ also decays and so forth, until finally the stable isotope $^{206}_{82}Pb$ is formed. This is known as the $^{238}_{92}U$ radioactive decay series. *A radioactive decay series starts with a naturally occurring radioactive isotope with a half-life near the age of the earth* (if it was very much shorter, there wouldn't be any left). *The series ends with a stable isotope.* There are at least two other naturally occurring decay series: the $^{235}_{92}U$ series, which ends with $^{207}_{82}Pb$, and the $^{232}_{90}Th$ series, which ends with $^{208}_{82}Pb$. As a result of the $^{238}_{92}U$ decay series, where uranium is found in rocks we also find other radioactive isotopes as well as lead. In fact, by examining the ratio of $^{238}_{92}U/^{206}_{82}Pb$ in a rock, its age can be determined. For example, a rock from the moon showed about half of the original $^{238}_{92}U$ had decayed to $^{206}_{82}Pb$. This meant that the rock was about 4.5×10^9 years old.

EXAMPLE 3-6

If we started with 4.0 mg of $^{14}_{6}C$, how long would it take until only 0.50 mg remained?

SOLUTION

$$\text{After 5760 years, } \tfrac{1}{2} \times 4.0 \text{ mg} = 2.0 \text{ mg remaining}$$
$$\text{After another 5760 years, } \tfrac{1}{2} \times 2.0 \text{ mg} = 1.0 \text{ mg remaining}$$
$$\text{After another } \frac{5760 \text{ years}}{17,280 \text{ years}}, \tfrac{1}{2} \times 1.0 \text{ mg} = 0.50 \text{ mg remaining}$$

Therefore, in 17,280 years, 0.50 mg remains.

3-9 THE EFFECTS OF RADIATION

We mentioned in Chapter 1 that energy-rich compounds undergo chemical reactions to produce energy-poor compounds, releasing the chemical energy as heat. Radioactive isotopes can be considered in a similar manner. When the nucleus of an isotope is radioactive, it is considered to be an energy-rich nucleus. When it decays, the radiation carries away this excess energy from the nucleus by means of a high-energy particle (alpha or beta) or high-energy light (gamma). This nuclear energy carried away by the particles or rays may also be converted to heat energy. In fact, much of the heat generated in the interior of the earth and other planets is a direct result of natural radiation from unstable elements. The presence of radioactive elements in wastes from nuclear reactors makes these materials extremely hot; they must be cooled continuously to avoid melting.

Radiation also affects matter when the high-energy particles or rays penetrate. *The radiation changes neutral molecular compounds into ions in a process known as* **ionization.** For example, if high-energy particles collide with an H_2O molecule, an electron may be removed from the molecule, leaving an H_2O^+ ion behind. The properties of the ion are distinctly different from those of the neutral molecule. If the molecule in question is a large complex molecule that is part of a cell of a living system, the ionization causes damage or even eventual destruction of the cell. As shown in Figure 3-13, an alpha particle causes the most ionization and is the most destructive along its path. However, these particles do not penetrate matter to any extent and can be stopped, even by a piece of paper. The danger of alpha emitters (such as uranium and plutonium) is that they can be ingested through food or inhaled into the lungs where these heavy elements tend to accumulate in bones. There, in intimate contact with the blood-producing cells of the bone marrow, they slowly do damage. Ultimately, the radiation can cause certain cells to change into abnormal cells that reproduce rapidly. These are the dreaded cancer cells such as found with leukemia.

Beta radiation is less ionizing than alpha radiation but is more penetrating. Still, a couple of inches of solid or liquid matter absorbs beta radiation. This type of radiation can cause damage to surface tissue such as skin and eyes but

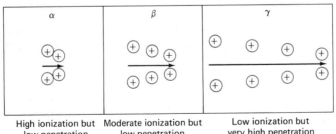

High ionization but   Moderate ionization but   Low ionization but
low penetration   low penetration   very high penetration

Figure 3-13   IONIZATION AND PENETRATION. Gamma radiation is the most damaging of the three types of radiation.

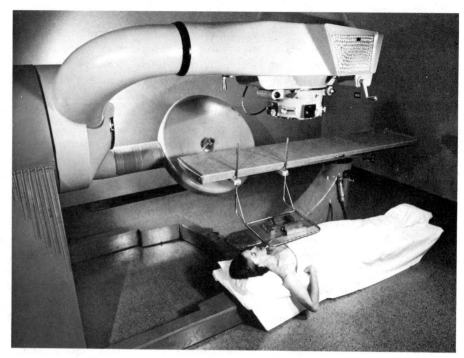

Figure 3-14 ^{60}Co RADIATION TREATMENT. A beam of gamma rays can be used to destroy cancerous tissue.

does not reach internal organs unless ingested. The most damaging radiation is gamma radiation. Although much less ionizing along its path than the others, it has tremendous penetrating power. Several feet of concrete or blocks of lead are needed for protection from gamma radiation.

The damage done to cells by gamma radiation is cumulative. That means small doses over a long period of time can be as harmful as one large dose at once. That is why the dosage of radiation absorbed by those who work in radioactive environments such as a nuclear research laboratory must be carefully and continually monitored. If one receives too much radiation over a specified period, that person must be removed from additional exposure for a length of time.

Isotopes that emit gamma radiation (e.g., $^{60}_{27}$Co) can be useful as well as destructive because of their penetrating rays. The gamma radiation of $^{60}_{27}$Co can be narrowly focused like a beam of light to destroy cancer cells in localized tumors located deep within the body. (See Figure 3-14.) Healthy cells are destroyed as well in the process, so this form of cancer treatment has many undesirable side effects. The purpose is to destroy or damage more cancer cells than normal cells. The normal cells can recover faster than the damaged abnormal cells.

■
SEE PROBLEMS
3-42 THROUGH
3-47.

3-10 NUCLEAR REACTIONS

Transmutation *of an element is the conversion of that element into another.* We have discussed how this occurs naturally by the spontaneous emission of an alpha or beta particle from a nucleus. Transmutation can also occur by artificial or synthetic means. The first example of transmutation was given by Ernest Rutherford in 1919. (Rutherford had earlier proposed the nuclear model of the atom.) Rutherford found that $^{14}_{7}N$ atoms could be bombarded with alpha particles, causing a nuclear reaction that produced $^{17}_{8}O$ and a proton. This nuclear reaction is illustrated by the nuclear equation

$$^{14}_{7}N + ^{4}_{2}He \longrightarrow ^{17}_{8}O + ^{1}_{1}H$$

Note that the total number of nucleons is conserved by the reaction [14 + 4 (on the left) = 17 + 1 (on the right)]. The total charge is also conserved during the reaction [7 + 2 (on the left) = 8 + 1 (on the right)]. Both the number of nucleons and the total charge must be balanced (the same on both sides of the equation) in a nuclear reaction.

Since the time of Rutherford, thousands of isotopes have been prepared by nuclear reactions. Two other examples are

$$^{209}_{83}Bi + ^{2}_{1}H \longrightarrow ^{210}_{84}Po + ^{1}_{0}n$$
$$^{27}_{13}Al + ^{4}_{2}He \longrightarrow ^{30}_{15}P + ^{1}_{0}n$$

($^{2}_{1}H$ is a deuterium nucleus or a deuteron; $^{1}_{0}n$ is a neutron.)

Because the nucleus has a positive charge and many of the bombarding nuclei (except neutrons) also have a positive charge (e.g., $^{4}_{2}He$, $^{1}_{1}H$, $^{2}_{1}H$), the particles must have a high energy (velocity) to overcome the natural repulsion between these two like charges. From the collision of the nucleus and the particle, a high-energy nucleus is formed. As a result, another particle (usually a neutron) is then emitted by the nucleus to carry away the excess energy from the collision. This is analogous to a head-on collision of two cars, where fenders, doors, and bumpers come flying out after impact. This is illustrated in Figure 3-15.

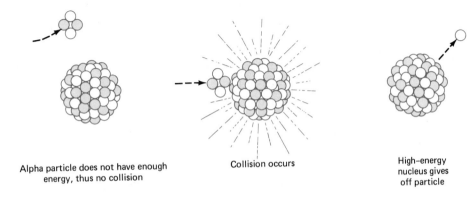

Alpha particle does not have enough energy, thus no collision

Collision occurs

High-energy nucleus gives off particle

Figure 3-15 COLLISION OF AN ALPHA PARTICLE WITH A NUCLEUS. Alpha particles must have sufficient energy to overcome the repulsion of the target nucleus for a collision to take place.

Figure 3-16   FERMI NATIONAL ACCELERATOR LABORATORY. This aerial view of the accelerator gives an indication of its huge size.

The invention of particle accelerators, which speed nuclei and subatomic particles to high velocities and energies, has opened the door to vast possibilities for artificial nuclear reactions. The largest accelerators also allow studies of collisions so powerful that the colliding nuclei are blasted apart. From these experiments, scientists have obtained a glimpse at the most basic composition of matter. These accelerators can be large and expensive. Completed in the early 1970s, the Fermi National Accelerator in Batavia, Illinois, has a circumference of about 6 miles and cost $245 million. (See Figure 3-16.) A huge accelerator now in the advanced planning stage will have a circumference of 52 miles and cost U.S. federal agencies more than $4.5 billion.

One interesting development from particle accelerators has been the preparation of new, heavy elements. In fact, all elements between plutonium (atomic number 94) and the last element made (atomic number 109, which has not yet been named) are synthetic elements. The following nuclear reactions illustrate the formation of heavy elements:

$$^{238}_{92}\text{U} + ^{12}_{6}\text{C} \longrightarrow ^{244}_{98}\text{Cf} + 6^{1}_{0}n$$
$$^{238}_{92}\text{U} + ^{16}_{8}\text{O} \longrightarrow ^{250}_{100}\text{Fm} + 4^{1}_{0}n$$

EXAMPLE 3-7

Complete the following nuclear equations:

(a) ${}^{9}_{4}\text{Be} + {}^{2}_{1}\text{H} \rightarrow$ _____ $+ {}^{1}_{0}n$

(b) ${}^{252}_{98}\text{Cf} + {}^{10}_{5}\text{B} \rightarrow$ _____ $+ 5{}^{1}_{0}n$

PROCEDURE

Same as Example 3-5.
(a) Nucleons: $9 + 2 = x + 1$, so $x = 10$.
    Atomic number: $4 + 1 = y + 0$, so $y = 5$.

From the list of elements, we find that the element is boron (atomic number 5). The isotope is

$$\underline{\underline{{}^{10}_{5}\text{B}}}$$

(b) Nucleons: $252 + 10 = x + (5 \times 1)$, so $x = 257$.
    Atomic number: $98 + 5 = y + (5 \times 0)$, so $y = 103$.

From the list of elements, the isotope is

$$\underline{\underline{{}^{257}_{103}\text{Lr}}}$$

Many nuclear reactions form isotopes that are themselves radioactive. For example, ${}^{60}\text{Co}$ is prepared as follows:

$$\text{}^{59}_{27}\text{Co} + {}^{1}_{0}n \longrightarrow {}^{60}_{27}\text{Co}^{\circ}$$

This type of procedure (called *neutron activation*) is used to prepare many radioactive isotopes whose naturally occurring isotopes are stable. There has been a great deal of application of these synthetic radioactive isotopes in the field of medicine, mainly in the area of diagnosis of diseases. Each year, more than 10 million diagnostic tests are run that involve the use of radioactive isotopes.

3-11 NUCLEAR FISSION

In 1938 a discovery was made that would eventually change the course of the world for better and worse. It started with some very nonspectacular experiments by two German scientists named Otto Hahn and Fritz Strassman. They certainly had no intention of launching the human race into the atomic age. They were attempting to make an element with an atomic number higher than that of uranium by bombarding uranium with neutrons. The extra neutron would hopefully produce a uranium isotope that would undergo beta decay. What they hoped to accomplish is illustrated by the equation

$$\text{}^{238}_{92}\text{U} + {}^{1}_{0}n \longrightarrow {}^{239}_{92}\text{U} \longrightarrow {}^{239}_{93}\text{X} + {}^{0}_{-1}e$$

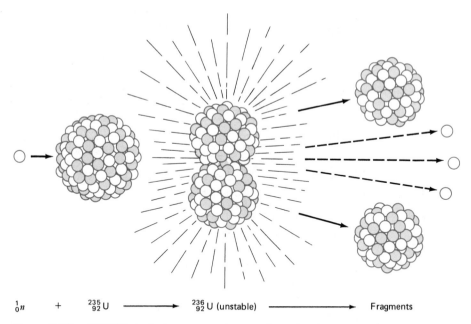

$$_0^1 n \quad + \quad _{92}^{235}U \quad \longrightarrow \quad _{92}^{236}U \text{ (unstable)} \quad \longrightarrow \quad \text{Fragments}$$

Figure 3-17 FISSION. In fission, one neutron produces two fragments of the original atom and an average of three neutrons.

Indeed, this reaction seemed to occur, but of more importance was the discovery by these scientists of isotopes present that were about half of the mass of the uranium atoms (e.g., $_{56}^{139}Ba$, $_{38}^{90}Sr$). The German scientists were able to show that the smaller atoms resulted from the reaction of ^{235}U and not the more abundant ^{238}U. (Naturally occurring uranium is 99.3% ^{238}U and only 0.7% ^{235}U.) The following equation represents a typical nuclear reaction that occurs when ^{235}U absorbs a neutron.

$$_{92}^{235}U + _0^1 n \longrightarrow _{56}^{139}Ba + _{36}^{94}Kr + 3_0^1 n$$

This type of nuclear reaction is called fission. **Fission** *is the splitting of a large nucleus into two smaller nuclei of similiar size.* (See Figure 3-17.)

Two points about this reaction had monumental consequences for the world, and scientists in Europe and America were quick to grasp their meaning in a world about to go to war.

1 Comparison of the masses of the product nuclei with that of the original nucleus indicated that a significant amount of mass was lost in the reaction. According to Einstein's equation (see Section 1-8), the mass lost must be converted to a tremendous amount of energy. Fission of a few kilograms of ^{235}U could produce energy equivalent to tens of thousands of tons of the conventional explosive, TNT.

2 What made the rapid fission of a large sample of ^{235}U feasible was the potential for a chain reaction. *A **nuclear chain reaction** is a reaction that is*

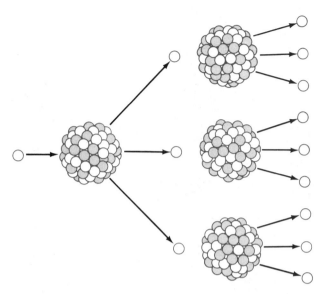

Figure 3-18 CHAIN REACTION. The fission of one nucleus produces neutrons, which induces fission in other nuclei.

self-sustaining. The reaction generates the means to trigger additional reactions. Note in Figure 3-17 that the reaction of one original neutron caused three to be released. These three neutrons cause the release of nine neutrons and so forth, as illustrated in Figure 3-18. If a certain densely packed "critical mass" of ^{235}U is present, the whole mass of uranium can undergo fission in an instant with a quick release of the energy in the form of radiation.

Unfortunately, the world thus entered the nuclear age in pursuit of a bomb. After a massive but secret effort, the first nuclear bomb was exploded over Alamogordo Flats in New Mexico on July 16, 1945. This bomb and the bomb exploded over Nagasaki, Japan, were made of ^{239}Pu, which is a synthetic fissionable isotope. The bomb exploded over Hiroshima, Japan, was made of ^{235}U.

This enormously destructive device, however, can be tamed. The chain reaction can be controlled by absorbing excess neutrons with cadmium bars. A typical nuclear reactor is illustrated in Figure 3-19. In a reactor core the uranium in the form of pellets is encased in long rods called fuel elements. Cadmium bars are raised and lowered among the fuel elements to control the rate of fission by absorbing neutrons. If the cadmium bars are lowered all the way, the fission process can be halted altogether. In normal operation, the bars are raised just enough that the fission reaction occurs at the desired level. Energy released by the fission and the decay of radioactive products formed from the fission is used to heat water, which circulates among the fuel elements. This water, which is called the primary coolant, becomes very hot

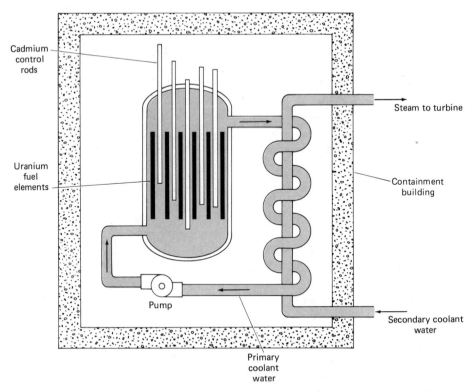

Figure 3-19   A NUCLEAR REACTOR. Nuclear energy is converted to heat energy, the heat energy to mechanical energy, and the mechanical energy to electrical energy.

(about 300 °C) without boiling because it is under high pressure. The water heats a second source of water, changing it to steam. The steam from the secondary coolant is cycled outside of the containment building, where it is used to turn a turbine that generates electricity.

Perhaps the greatest advantage of nuclear energy is that it does not pollute the air with compounds (called oxides) of sulfur and carbon as do conventional power plants. Sulfur oxides have been implicated as a major cause of "acid rain." It is also feared that large amounts of carbon dioxide will eventually lead to significant changes in the weather as a result of the "greenhouse effect." Another advantage is that nuclear power does not deplete the limited supply of fossil fuels (oil, coal, and natural gas). Until the tragedy in Chernobyl, Russia, in April 1986, there had been no loss of human life from the use of nuclear energy to generate electricity. In 1983, 9% of the world's electrical power was from nuclear energy. In fact, nuclear power is now the primary source of electricity in France. Proponents of nuclear power feel that if the proper systems are used with adequate safeguards and backup systems a catastrophic accident should never occur. In support of this position, a reactor known as the Integral Fast Reactor (IFR) has been successfully tested and is claimed to be inherently safe. That is, even if "all the plugs are pulled" and the reactor is ignored completely, it shuts itself down and will not overheat or explode.

The disadvantages are also obvious. The catastrophic explosion and subsequent fire of reactor 4 in Chernobyl in 1986 dispersed more dangerous radiation into the environment than all of the atmospheric atomic bomb tests put together. The surroundings for 30 miles around are still uninhabitable. The reactor itself has been entombed in concrete and must remain so for hundreds of years. Although there were relatively few deaths immediately (considering the force of the explosion), estimates range into the thousands for those whose lives will be shortened because of this accident. In the United States, the accident at Three Mile Island in Pennsylvania in March 1979 also led to partial melting of the reactor core, although there was no explosion and little loss of radioactivity to the environment. Still, the cleanup will take several more years and cost well over $1 billion.

In both of these accidents a core "meltdown" did not occur. A meltdown would occur if the temperature of the reactor core exceeded 1130 °C, the melting point of uranium. Theoretically, the mass of molten uranium, together with all of the highly radioactive decay products, could accumulate on the floor of the reactor, melt through the many feet of protective concrete, and eventually reach ground water. If this happened, vast amounts of deadly radioactive wastes could be spread through the ground water into streams and lakes.

The type of accident that occurred in Russia is unlikely in most of the world's commercial reactors. Unlike the Russian reactors, these reactors are surrounded by an extremely strong building made of reinforced concrete and called a containment building. (See Figure 3-19.) These buildings should contain an explosion such as occurred in Russia. Also, most reactors do not contain graphite, which is highly combustible and led to the prolonged fire in Chernobyl that spread so much radioactivity into the atmosphere. There is one U.S. reactor similar to the Russian reactors, in Hanford, Washington, but it has been shut down indefinitely.

A growing problem in the use of nuclear energy involves the used or spent fuel. Originally, the used fuel was to be processed at designated centers where the unused fuel could be separated and reused and the radioactive wastes disposed. However, problems in transportation of this radioactive material and the danger of theft of the ^{239}Pu that is produced in a reactor have hindered solution of the problem. (^{239}Pu is also fissionable and could be used in a nuclear bomb.) Currently, spent fuel elements, which must be replaced every 4 years, are being stored at the reactor sites. These fuel rods are not only highly radioactive but remain extremely hot and must be continually cooled. It is estimated that they will remain dangerously radioactive for at least 1000 years. This is a problem that must soon be resolved.

Because of the accidents, new regulations regarding safeguards have made the cost of a new nuclear power plant considerably higher than that of a

Figúre 3-20   A NUCLEAR POWER PLANT. In these plants the nuclear energy of uranium is converted into electrical energy.

conventional power plant using fossil fuels. As a result, no new nuclear plants are planned in the United States in the immediate future. (See Figure 3-20.)

3-12 NUCLEAR FUSION

Soon after the process of nuclear fission was demonstrated, an even more powerful type of nuclear reaction was shown to be theoretically possible. *This reaction involved the* **fusion** *or bringing together of two small nuclei.* An example of a fusion reaction is

$$\ _1^3\text{H} + \ _1^2\text{H} \longrightarrow \ _2^4\text{He} + \ _0^1 n + \text{energy}$$

($_1^3$H is called tritium; it is a radioactive isotope of hydrogen.)

As in the fission process, a significant amount of mass is converted into energy. Fusion energy is the origin of almost all of our energy, since it powers the sun. Millions of tons of matter are converted to energy in the sun every second. Because of its large mass, however, the sun contains enough hydrogen to "burn" for additional billions of years.

The principle of fusion was first demonstrated on this planet with a tremendously destructive device called the hydrogen bomb. This bomb can be more than 1000 times more powerful than the atomic bomb, which uses the fission process.

Fission can be controlled, but what about fusion? This is a big and very important question at present. Research into this area is currently of high priority in much of the industrial world, especially in the Soviet Union and the United States. Technically, controlling fusion for the generation of power is an extremely difficult problem. Temperatures on the order of 100 million degrees Celsius are needed just to start the fusion process. This temperature is higher than that of the interior of the sun. No known materials can withstand these temperatures, so alternative containment procedures are being examined. So far, the research efforts of several nations have not reached the "breakeven" point, where as much energy is released from the fusion process as is put into it to get it started.

The advantages of controlled fusion power are impressive.

1 It should be relatively clean. Few radioactive products are formed.
2 Fuel is inexhaustable. The oceans of the world contain enough deuterium, one of the reactants, to provide the world's energy needs for a trillion years. On the other hand, there is a very limited supply of fossil fuels and uranium.
3 There is no possibility of the reaction going out of control and causing a meltdown. Fusion will occur in power plants in short bursts of energy that can be easily stopped in case of mechanical problems.

The disadvantage is, of course, the expense of research and development. Billions of dollars have been spent and billions more will have to be spent before the technical problems are solved. It is hoped that before the end of this century, experimental plants will be in operation that will prove the feasibility of fusion as a source of power. Unfortunately, there is little choice in this matter as there are few alternatives for power in the next century.

■
*SEE PROBLEMS
3-48 THROUGH
3-52.*

**3-13 CHAPTER
REVIEW**

It has been less than 200 years since John Dalton first explained a large body of experimental data with the atomic theory of matter. It quickly became accepted that there is a basic particle of an element called an atom that is unique and characteristic of that element. Since that time we have discovered that the atom itself has structure and is composed of three more basic particles called electrons, protons, and neutrons. The relative charges and masses of these three particles are summarized in Table 3-4.

Experimental results indicated that protons and neutrons (called nucleons) exist in a small, dense core of the atom called the nucleus. Electrons occupy the vast expanse of the atom outside the nucleus which is otherwise empty space. The number of protons in an atom is known as its atomic number (symbolized by Z below), and it is this that distinguishes the atoms of one element from those of another. A neutral atom has the same number of electrons as protons. The total number of nucleons in an atom is known as its mass number (symbolized by A below). Atoms of the same element may have different mass numbers and are known as isotopes of that element. This information for a particular isotope is illustrated as follows.

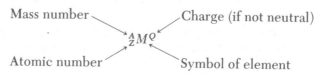

Mass number — Charge (if not neutral)

$$_Z^A M^Q$$

Atomic number — Symbol of element

As protons and neutrons do not have exactly the same mass, the mass number is not an exact measure of the comparative masses of isotopes. A more precise measure of mass is the atomic mass. This is obtained by comparing the mass of the particular isotope with the mass of ^{12}C, which is defined as the exact standard. The atomic weight of an element found in nature is the weighted average of all of the naturally occurring isotopes.

Pure matter is composed of either elements or compounds. Elements exist in nature mostly as aggregates of individual atoms but may also be present as molecules. A molecule is a combination of two or more atoms chemically combined by covalent bonds. Compounds can be composed of molecules or ions. The molecules of compounds are made up of the atoms of different elements. Ions are atoms or groups of atoms that have an electrical charge as a result of an imbalance in the numbers of electrons and protons in the nucleus (or nuclei of polyatomic ions). Ionic compounds are always composed of enough positive ions (cations) and negative ions (anions) to balance the charge. The formulas of molecular compounds or molecular elements give the number of atoms of each element chemically combined in the molecule. The formula of an ionic compound represents the type and number of atoms in a formula unit, which is the smallest whole-number ratio of cations and anions present. This information is summarized as follows.

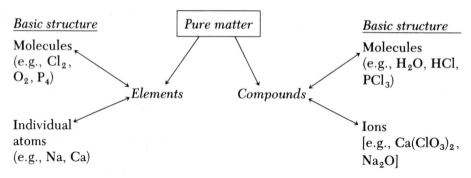

Basic structure

Molecules
(e.g., Cl_2, O_2, P_4)

Individual atoms
(e.g., Na, Ca)

Pure matter

Elements   *Compounds*

Basic structure

Molecules
(e.g., H_2O, HCl, PCl_3)

Ions
[e.g., $Ca(ClO_3)_2$, Na_2O]

Nuclear Reactions

The nuclei of all elements with atomic number greater than 83 (bismuth) are unstable. That is, they disintegrate or decay to the nuclei of other elements and emit radiation. There are three major types of radiation: (1) alpha particles, which are helium nuclei; (2) beta particles, which are electrons; and (3) gamma rays, which are high-energy forms of light.

Radioactive isotopes decay at widely different rates. A measure of the rate of decay is called the half-life. Half-lives vary from billions of years for isotopes such as ^{238}U to fractions of a second for others.

Radiation affects living tissue. When radiation passes through matter, it causes ionization that damages or kills cells. Alpha and beta radiation do not cause damage (except to skin tissue or unless ingested) because of low penetration. Gamma radiation is extremely penetrating and thus can cause considerable damage from exposure.

Nuclear changes can be brought about by artificial means in which nuclei are combined with the help of particle accelerators. Many new heavy elements as well as medically useful isotopes have been produced in this manner.

Fission is the breaking of certain heavy isotopes into fragments. Such a process can occur through a chain reaction and liberate a large amount of energy. The energy originates from the conversion of mass according to Einstein's law. Fission is controllable and is the energy source in nuclear power reactors. Fusion is the combination of light nuclei to form heavier ones. Again, a loss of mass accompanies the reaction, producing energy. Fusion is the process that occurs in the hydrogen bomb and the sun. Efforts are underway to control fusion so as to generate power.

The different types of reactions that nuclei undergo that have been discussed are summarized as follows:

$^{240}_{96}Cm \rightarrow {}^{236}_{94}Pu + {}^4_2He$ Alpha radiation

$^{71}_{30}Zn \rightarrow {}^{71}_{31}Ga + {}^0_{-1}e$ Beta radiation

$^{253}_{99}Es + {}^4_2He \rightarrow {}^{256}_{101}Md + {}^1_0n$ Synthetic element from nuclear reaction

$^{235}_{92}U + {}^1_0n \rightarrow {}^{94}_{38}Sr + {}^{139}_{54}Xe + 3{}^1_0n$ Fission

$^2_1H + {}^3_1H \rightarrow {}^4_2He + {}^1_0n$ Fusion

EXERCISES

NAMES AND SYMBOLS OF THE ELEMENTS

3-1 Without reference to a table of elements, write the symbols of the following elements.
(a) bromine   (d) tin
(b) oxygen   (e) sodium
(c) lead   (f) sulfur

3-2 The following elements all have symbols that begin with the letter C: cadmium, calcium, californium, carbon, cerium, cesium, chlorine, chromium, cobalt, copper, and curium. The symbols are C, Ca, Cd, Ce, Cf, Cl, Cm, Co, Cr, Cs, and Cu. Match the symbol with the element and then check with the table of elements inside the front cover.

3-3 Using the table inside the front cover of the book, write the symbols for the following elements.
(a) barium   (d) platinum
(b) neon   (e) manganese
(c) cesium   (f) tungsten

3-4 Without reference to the table, give the elements corresponding to the following symbols.
(a) S   (d) N
(b) K   (e) Mg
(c) Fe   (f) Al

3-5 Using the table, give the elements corresponding to the following symbols.
(a) B   (e) Co
(b) Bi   (f) Hg
(c) Ge   (g) Be
(d) U   (h) As

COMPOSITION OF THE ATOM

3-6 Which of the following were not part of Dalton's atomic theory?
(a) Atoms are the basic building blocks of nature.
(b) Atoms are composed of electrons, neutrons, and protons.
(c) Atoms are reshuffled in chemical reactions.

(d) The atoms of an element are identical.

(e) Different isotopes can exist for the same element.

3-7 Which of the following describes a neutron?

(a) +1 charge, mass number 1 amu

(b) +1 charge, mass number 0 amu

(c) 0 charge, mass number 1 amu

(d) −1 charge, mass number 0 amu

3-8 Which of the following describes an electron?

(a) +1 charge, mass number 1 amu

(b) +1 charge, mass number 0 amu

(c) −1 charge, mass number 1 amu

(d) −1 charge, mass number 0 amu

3-9 The elements O, N, Si, and Ca are among several that are composed primarily of one isotope. Using the Table of Atomic Weights inside the front cover, write the atomic number and mass number of the principal isotopes of these elements.

3-10 Give the mass numbers and atomic numbers of the following isotopes.

(a) ^{193}Au   (c) ^{118}I

(b) ^{132}Te   (d) ^{39}Cl

3-11 Give the numbers of protons, neutrons, and electrons in each of the following isotopes.

(a) $^{45}_{21}$Sc   (c) $^{223}_{87}$Fr

(b) $^{232}_{90}$Th   (d) $^{90}_{38}$Sr

3-12 Three isotopes of uranium are $^{234}_{92}$U, $^{235}_{92}$U, and $^{238}_{92}$U. How many protons, neutrons, and electrons are in each isotope?

3-13 Using the table of elements inside the front cover, complete the following table for neutral isotopes.

Isotope Name	Isotope Symbol	Atomic Number	Mass Number	Subatomic Particles		
				Protons	Neutrons	Electrons
Molybdenum-96	$^{96}_{42}$Mo	42	96	42	54	42
(a) _____	$^{?}_{?}$Ag	_____	_____	_____	61	_____
(b) _____	_____	14	_____	_____	14	_____
(c) _____	_____	_____	39	_____	20	_____
(d) Cerium-140	_____	_____	_____	_____	_____	_____
(e) _____	_____	_____	_____	26	30	_____
(f) _____	_____	50	110	_____	_____	_____
(g) _____	$^{118}_{?}$I	_____	_____	_____	_____	_____
(h) mercury-____	_____	_____	_____	_____	116	_____

3-14 Using the table of elements inside the front cover, complete the following table for neutral isotopes.

Isotope Name	Isotope Symbol	Atomic Number	Mass Number	Subatomic Particles		
				Protons	Neutrons	Electrons
(a) Tungsten-____	_____	_____	184	_____	_____	_____
(b) _____	_____	_____	_____	_____	12	11
(c) _____	$^{200}_{?}$Au	_____	_____	_____	_____	_____
(d) _____	$^{?}_{?}$Pm	_____	_____	_____	87	_____
(e) _____	_____	_____	109	46	_____	_____
(f) _____	_____	_____	48	_____	_____	23
(g) _____	_____	21	_____	_____	29	_____

ATOMIC WEIGHT

3-15   A certain element has a mass 5.81 times that of ^{12}C. What is the atomic weight of the element? What is the element?

3-16   The atomic weight of a certain element is about $3\frac{1}{3}$ times that of ^{12}C. Give the atomic weight, the name, and the symbol of the element.

3-17   If the atomic mass of ^{12}C is defined as exactly 8 instead of 12, what would be the atomic weight of the following elements to three significant figures? Assume that the elements have the same weights relative to each other as before. That is, hydrogen still has a mass about one-twelfth that of carbon.
(a) H   (b) N   (c) Na   (d) Ca

3-18   Assume that the atomic mass of ^{12}C is defined as exactly 10 and that the atomic weight of an element is 43.3 amu on this basis. What is the element?

3-19   Assume that the atomic mass of ^{12}C is defined as exactly 20 instead of 12 and that the atomic weight of an element is 212.7 amu on this basis. What is the element?

3-20   Bromine is composed of 50.5% ^{79}Br and 49.5% ^{81}Br. The atomic mass of ^{79}Br is 78.92 amu and that of ^{81}Br, 80.92 amu. What is the atomic weight of bromine?

3-21   Silicon occurs in nature as a mixture of three isotopes: ^{28}Si (27.98 amu), ^{29}Si (28.98 amu), and ^{30}Si (29.97 amu). The mixture is 92.21% ^{28}Si, 4.70% ^{29}Si, and 3.09% ^{30}Si. Calculate the atomic weight of naturally occurring silicon.

3-22   Naturally occurring Cu is 69.09% ^{63}Cu (62.96 amu). The only other isotope present is ^{65}Cu (64.96 amu). What is the atomic weight of copper?

*3-23   Chlorine occurs in nature as a mixture of ^{35}Cl and ^{37}Cl. If the atomic mass of ^{35}Cl is approximately 35.0 amu and that of ^{37}Cl is 37.0 amu and the atomic weight of the mixture as it occurs in nature is 35.5 amu, what is the proportion of the two isotopes?

*3-24   The atomic weight of the element gallium is 69.72 amu. If it is composed of two isotopes, ^{69}Ga (68.926 amu) and ^{71}Ga (70.925 amu), what is the percent of ^{69}Ga present?

3-25   Using the periodic table, determine the atomic number and the atomic weight of each of the following elements.
(a) Re   (b) Co   (c) Br   (d) Si

FORMULAS, MOLECULES, AND IONS

3-26   Which of the following is the formula of a diatomic element and which is that of a diatomic compound?
(a) NO_2 (d) $(NH_4)_2S$
(b) CO (e) N_2
(c) K_2O (f) CO_2

3-27   Determine the number of atoms of each element in the formulas of the following compounds.
(a) $C_6H_4Cl_2$
(b) C_2H_5OH (ethyl alcohol)
(c) $CuSO_4 \cdot 9H_2O$ (H_2O's are parts of a single molecule)
(d) $C_9H_8O_4$ (aspirin)
(e) $Al_2(SO_4)_3$
(f) $(NH_4)_2CO_3$

3-28   What is the total number of atoms in each molecule or formula unit for the compounds listed in Problem 3-27?

3-29   How many carbon atoms are in each molecule or formula unit of the following compounds?
(a) C_8H_{18} (octane in gasoline)
(b) $NaC_7H_4O_3NS$ (saccharin)
(c) $Fe(C_2O_4)_2$
(d) $Al_2(CO_3)_3$

3-30   Write the formulas of the following compounds.
(a) sulfur dioxide (one sulfur and two oxygens)
(b) carbon dioxide
(c) sulfuric acid (two hydrogens, one sulfur, and four oxygens)
(d) calcium perchlorate (one calcium and two ClO_4^- ions)
(e) ammonium phosphate (three NH_4^+ ions and one PO_4^{3-} ion)
(f) acetylene (two carbons and two hydrogens)

3-31   Write the formulas of the following compounds.

(a) phosphorus trichloride (one phosphorus and three chlorines)

(b) naphthalene (ten carbons and eight hydrogens)

(c) dibromine trioxide (two bromines)

(d) calcium hypochlorite (one Ca^{2+} ion and two ClO^- ions)

(e) magnesium phosphate (three Mg^{2+} ions and two $PO_4{}^{3-}$ ions)

3-32 Neon gas is composed of individual atoms, and oxygen gas is composed of O_2 molecules. Describe how these two elements may appear on the atomic or molecular level. In the gaseous state the particles are comparatively far apart.

3-33 The compound HF contains covalent bonds, and the compound KF contains ionic bonds. Describe how the basic particles of these two compounds may appear.

3-34 How many total protons and electrons are present in each of the following ions?

(a) K^+ (d) $NO_2{}^-$

(b) Br^- (e) Al^{3+}

(c) S^{2-} (f) $NH_4{}^+$

3-35 How many total protons and electrons are present in each of the following ions?

(a) Sr^{2+} (d) NO^+

(b) P^{3-} (e) $SO_3{}^{2-}$

(c) V^{3+}

3-36 Write the symbol of the element, mass number, atomic number, and electrical charge given the following information.

(a) An isotope of Sr contains 36 electrons and 52 neutrons.

(b) An isotope contains 24 protons, 28 neutrons, and 21 electrons.

(c) An isotope contains 36 electrons and 45 neutrons and has a -2 charge.

(d) An isotope of nitrogen contains 7 neutrons and 10 electrons.

(e) An isotope contains 54 electrons and 139 nucleons and has a $+3$ charge.

3-37 Write the symbol of the element, mass number, atomic number, and electrical charge given the following information.

(a) An isotope of Sn contains 68 neutrons and 48 electrons.

(b) An isotope contains 204 nucleons and 78 electrons and has a $+3$ charge.

(c) An isotope contains 45 neutrons and 36 electrons and has a -1 charge.

(d) An isotope of aluminum has 14 neutrons and a $+3$ charge.

NUCLEAR RADIATION

3-38 Complete the following nuclear equations involving alpha and beta particles.

(a) $^{214}_{83}Bi \rightarrow {}^{214}_{84}Po + \underline{\qquad}$

(b) $^{90}_{37}Rb \rightarrow \underline{\qquad} + {}^{0}_{-1}e$

(c) $^{235}_{92}U \rightarrow \underline{\qquad} + {}^{4}_{2}He$

(d) $\underline{\qquad} \rightarrow {}^{41}_{21}Sc + {}^{0}_{-1}e$

(e) $\underline{\qquad} \rightarrow {}^{210}_{82}Pb + {}^{0}_{-1}e$

3-39 Complete the following nuclear equations involving alpha and beta particles.

(a) $^{239}_{93}Np \rightarrow \underline{\qquad} + {}^{0}_{-1}e$

(b) $^{226}_{88}Ra \rightarrow {}^{222}_{86}Rn + \underline{\qquad}$

(c) $\underline{\qquad} \rightarrow {}^{235}_{92}U + {}^{4}_{2}He$

(d) $^{32}_{15}P \rightarrow {}^{32}_{16}S + \underline{\qquad}$

3-40 From the following information, write a nuclear equation that includes all isotopes and particles.

(a) $^{230}_{90}Th$ decays to form $^{226}_{88}Ra$.

(b) $^{214}_{84}Po$ emits an alpha particle.

(c) $^{210}_{84}Po$ emits a beta particle.

(d) An isotope emits an alpha particle and forms $^{235}_{92}U$.

(e) $^{14}_{6}C$ decays to form $^{14}_{7}N$.

*3-41 The decay series of $^{238}_{92}U$ to $^{206}_{82}Pb$ involves alpha and beta emissions in the following sequence: $\alpha, \beta, \beta, \alpha, \alpha, \alpha, \alpha, \alpha, \beta, \alpha, \beta, \beta, \beta, \alpha$. Identify all isotopes formed in the series.

NUCLEAR DECAY AND HALF-LIFE

3-42 The half-life of a certain isotope is 10 years. If we start with a 10.0-g sample of the isotope, how much is left after 20 years?

3-43 Start with 12.0 g of a certain radioactive isotope. After 11 years only 3.0 g is left. What is the half-life of the isotope?

3-44 The radioactive isotope $^{90}_{38}Sr$ can accumulate in bones, where it replaces calcium. It emits a high-energy beta particle, which eventually can cause cancer.

(a) What is the product of the decay of $^{90}_{38}Sr$?

(b) How long would it take for a 0.10-mg sample of $^{90}_{38}Sr$ to decay to where only 2.5×10^{-2} mg was left? (The half-life of $^{90}_{38}Sr$ is 25 years.)

3-45 The radioactive isotope $^{131}_{53}I$ accumulates in the thyroid gland. On the one hand, this can be useful in detecting diseases of the thyroid and even in treating cancer at that location. On the other hand, exposure to excessive amounts of this isotope, such as from a nuclear power plant, can *cause* cancer of the thyroid. $^{131}_{53}I$ emits a beta particle with a half-life of 8.0 days. What is the product of the decay of $^{131}_{53}I$? If one started with 8.0×10^{-6} g of $^{131}_{53}I$, how much would be left after 32 days?

3-46 The isotope $^{14}_{6}C$ is used to date fossils of formerly living systems such as prehistoric animal bones. The radioactivity due to $^{14}_{6}C$ of a sample of the fossil is compared with the radioactivity of currently living systems. The longer something has been dead, the lower will be the radioactivity due to $^{14}_{6}C$ of the sample. If the radioactivity of a sample of bone from a mammoth is one-fourth of the radioactivity of the current level, how old is the fossil? ($t_{1/2} = 5760$ years.)

3-47 Of the three types of radiation, which is the most ionizing? Which is the most penetrating? Which is the most damaging to living cells?

NUCLEAR REACTIONS

3-48 When $^{235}_{92}U$ and $^{239}_{94}Pu$ undergo fission, a variety of reactions actually take place.

Complete the following.

(a) $^{235}_{92}U + ^{1}_{0}n \rightarrow$ _____ $+ ^{146}_{58}Ce + 3^{1}_{0}n$

(b) $^{235}_{92}U + ^{1}_{0}n \rightarrow ^{90}_{37}Rb + ^{142}_{55}Cs +$ _____

(c) $^{239}_{94}Pu + ^{1}_{0}n \rightarrow ^{141}_{56}Ba +$ _____ $+ 2^{1}_{0}n$

3-49 Complete the following nuclear reactions.

(a) $^{35}_{17}Cl + ^{1}_{0}n \rightarrow$ _____ $+ ^{1}_{1}H$

(b) $^{27}_{13}Al +$ _____ $\rightarrow ^{25}_{12}Mg + ^{4}_{2}He$

(c) $^{27}_{13}Al + ^{4}_{2}He \rightarrow$ _____ $+ ^{1}_{0}n$

°(d) $^{238}_{92}U + 15^{1}_{0}n \rightarrow ^{253}_{100}Es +$ _____

(e) $^{244}_{96}Cm + ^{12}_{6}C \rightarrow$ _____ $+ 2^{1}_{0}n$

(f) _____ $+ ^{2}_{1}H \rightarrow ^{238}_{93}Np + ^{1}_{0}n$

(g) $^{242}_{94}Pu + ^{22}_{10}Ne \rightarrow ^{260}_{104}$ _____ $+$ _____

3-50 Complete the following nuclear reactions.

(a) $^{249}_{96}Cm +$ _____ $\rightarrow ^{260}_{103}Lr + 4^{1}_{0}n$

(b) $^{15}_{7}N + ^{1}_{1}H \rightarrow$ _____ $+ ^{4}_{2}He$

(c) $^{27}_{13}Al +$ _____ $\rightarrow ^{25}_{12}Mg + ^{4}_{2}He$

(d) $^{249}_{98}Cf +$ _____ $\rightarrow ^{263}_{106}$ _____ $+ 4^{1}_{0}n$

3-51 What is the difference between fission and fusion? What is the source of energy for these two processes?

3-52 The fissionable isotope $^{239}_{94}Pu$ is made from the abundant isotope of uranium, $^{238}_{92}U$, in nuclear reactors. When $^{238}_{92}U$ absorbs a neutron from the fission process, $^{239}_{94}Pu$ eventually forms. This is the principle of the "breeder reactor," although $^{239}_{94}Pu$ is formed in all reactors. Complete the following reaction.

$^{238}_{92}U + ^{1}_{0}n \rightarrow$ _____ $+ ^{0}_{-1}e$

$\rightarrow ^{239}_{94}Pu +$ _____

REVIEW TEST ON CHAPTERS 1–3*

The following multiple-choice questions have one correct answer.

1. A substance completely fills any size container. The substance is a
   (a) gas (b) solid (c) liquid

2. A substance has a distinct melting point and density. It cannot be decomposed into chemically simpler substances. The substance is a
   (a) heterogeneous mixture
   (b) solution
   (c) compound
   (d) element

3. A sample of a certain liquid has uniform properties, but the temperature at which the liquid boils slowly changes as it boils away. The liquid sample is a
   (a) heterogeneous mixture
   (b) solution
   (c) compound
   (d) element

4. Which of the following is a chemical property of iodine?
   (a) It melts at 114 °C.
   (b) It has a density of 4.94 g/mL.
   (c) When heated, it forms a purple vapor.
   (d) It forms a compound with sodium.

5. A charged car battery has _____ energy which can be converted into _____ energy.
   (a) chemical, electrical
   (b) heat, mechanical
   (c) light, electrical
   (d) mechanical, chemical

6. Which of the following relates to the precision of a measurement?
   (a) the number of significant figures
   (b) the location of the decimal point
   (c) the closeness to the actual value
   (d) the number of zeros that are significant

7. Which of the following is not an exact relationship?
   (a) 1 mile = 5280 ft   (c) 10^3 m = 1 km
   (b) 0.62 mile = 1 km   (d) 4 qt = 1 gal

8. How many significant figures are in the measurement 0.0880 m?
   (a) 5 (b) 4 (c) 3 (d) 2

9. The number 785×10^{-8} expressed in standard scientific notation is
   (a) 7.85×10^{-10} (c) 7.85×10^{-6}
   (b) 78.5×10^{-7} (d) 0.785×10^{-6}

10. One kiloliter is the same as
    (a) 10^6 mL   (c) 10^3 mL
    (b) 10^{-3} L   (d) 10 L

11. The temperature of a sample of matter relates to
    (a) its density
    (b) its specific heat
    (c) its physical state
    (d) the intensity of heat

12. Which of the following is the lowest possible temperature?
    (a) 0 K   (c) 0 °F
    (b) 0 °C   (d) -273 K

13. Which of the following has the highest density?
    (a) a truckload of feathers
    (b) an ounce of gold
    (c) a cup of water
    (d) a gallon of gasoline

14. Which of the following is the symbol for silicon?
    (a) Se (b) S (c) Si (d) Sc

15. Lead dioxide is the name of
    (a) two elements   (c) a compound
    (b) a mixture   (d) a solution

16. Which of the following has a charge of +1 and a mass of 1 amu?
    (a) a neutron   (c) a proton
    (b) an electron   (d) a helium nucleus

17. Isotopes have
    (a) the same mass number and the same atomic number
    (b) the same mass number but different atomic numbers
    (c) different mass numbers and different atomic numbers

---

* Answers to all questions and problems in Review Tests are given in Appendix I.

(d) different mass numbers but the same atomic number

18. Which of the following describes an isotope with a mass number of 99 that contains 56 neutrons in its nucleus?
    (a) $^{99}_{56}Ba$
    (b) $^{43}_{56}Ba$
    (c) $^{99}_{43}Tc$
    (d) $^{56}_{43}Tc$
    (e) $^{155}_{99}Es$

19. Which of the following isotopes is used as the standard for atomic weight?
    (a) ^{12}C (b) ^{16}O (c) ^{13}C (d) ^{1}H

20. Which of the following is a diatomic molecule of a compound?
    (a) NO   (b) NH_4Cl (c) H_2 (d) CO_2

21. Which of the following is not a basic particle of an element?
    (a) an atom   (b) a molecule   (c) an ion

22. Which of the following is a polyatomic cation?
    (a) CO   (b) K^+ (c) SO_4^{2-} (d) NH_4^+

23. What would be the electrical charge on a sulfur atom containing 18 electrons?
    (a) 2−   (b) 1−   (c) 0   (d) 2+

24. One formula unit of the compound $Al_2(CO_3)_3$ contains
    (a) nine oxygen atoms
    (b) three oxygen atoms
    (c) six aluminum atoms
    (d) six carbon atoms

25. What would be the formula of an ionic compound made up of Ba^{2+} ions and Cl^- ions?
    (a) Ba_2Cl
    (b) $BaCl_2$
    (c) BaCl
    (d) $BaCl_3$

PROBLEMS

Carry out the following calculations.

1. Round off the following numbers to three significant figures.
   (a) 173.8   (b) 0.0023158   (c) 18,420

Carry out the following calculations. Express your answer to the proper decimal place or number of significant figures.

2. 1.09 cm + 0.078 cm + 16.0021 cm
3. 4.2 in. × 16.33 in.

4. 398 g ÷ 31 mL
5. 0.892×10^4 cm × 0.0022×10^6 cm
6. 7784×10^{-6} kg ÷ 0.56 kg/mL

Express the answers to the following problems in standard scientific notation. Assume that the numbers are measurements so that the answer should be expressed to the proper decimal place or number of significant figures.

7. $(92.8 \times 10^6) \times (1.3 \times 10^4)$
8. $(0.4887 \times 10^{-4}) \div (0.03 \times 10^6)$
9. $(4.0 \times 10^6)^2$

Make the following conversions.

10. 189 cm to in.
11. 43 qt to L
12. 197 miles/hr to km/hr
13. 6.74 lb/ft to g/cm

Solve the following problems.

14. A cube of metal measures 2.2 cm on a side. It has a mass of 47.68 g. What is its density?
15. What is the volume of a 678-g sample of a substance that has a density of 11.3 g/mL?
16. What is the temperature in degrees Celsius if the reading is 112°F?
17. If the temperature on the Celsius scale increases by 27 degrees, what is the equivalent increase on the Fahrenheit scale?
18. The temperature of a certain liquid increased by 12.0 °C if 69.0 joules of heat is added to a 20.0-g sample. What is the specific heat of the liquid?
19. The atomic weight of an element is 10.15 times that of the defined standard. What is the atomic weight of the element, and what is the name and symbol of the element?
20. The atoms of a certain element are distributed between two isotopes: 65.0% of the atoms have a mass number of 116.0, and the remainder have a mass number of 112.0. Using mass number as an approximation of atomic mass, calculate the atomic weight of the element to tenths of an atomic mass unit.

FIREWORKS
The various colors that are created in firework displays are a result of the presence of certain elements.

THE
ARRANGEMENT
OF THE
ELECTRONS
IN THE
ATOM

PURPOSE OF CHAPTER

In Chapter 4 we describe the arrangement of electrons in the atom so as to understand the basis of the periodic properties of the elements, their atomic spectra, and eventually chemical bonds between elements.

OBJECTIVES FOR CHAPTER

After completion of this chapter, you should be able to:

1   Describe how to locate elements with similar chemical properties in the periodic table. (4-1)

2   Distinguish between the properties of a continuous and a discrete spectrum. (4-2)

3   Describe the relationship between the wavelength of light and energy. (4-2)

4   List the important assumptions of Bohr's model for the hydrogen atom. (4-3)

5   Give the principal quantum number designation, the letter designation, and the capacity of the first five shells. (4-4)

4

6 Give the four sublevels found in known elements and their order of energy. (4-5)

7 Show the distribution of electrons in the sublevels of a specified shell. (4-5)

8 Demonstrate the Aufbau principle by listing the normal order of filling of sublevels. (4-6)

9 Write the electron configuration for the atoms of a specified element. (4-6)

10 Relate electron configuration of the elements to the periodic table. (4-6)

11 Use only the periodic table and noble gas core shorthand to give the expected electron configuration of a specified element. (4-6)

12 Describe what is meant by an orbital. (4-7)

13 Give the number of orbitals in each type of sublevel. (4-7)

14 Describe the shapes of s, p, and d orbitals. (4-8)

15 Demonstrate the Pauli exclusion principle and Hund's rule by writing box diagrams of the outer electrons for a specified element. (4-9)

An element is composed of atoms. Atoms of an element join with atoms of other elements in various ratios to form compounds that may be molecular or ionic. At this point it would seem that chemistry becomes hopelessly complex, with each element having its own chemistry unrelated to that of any other element. Certainly, each element does have its particular characteristics but there is considerable order, making the situation quite tolerable. Fortunately, groups of elements share many common characteristics. The fact that chemical and physical similarities occur in a regular pattern among the elements is summarized by the periodic law. *The* **periodic law** *of the elements states that when elements are arranged by atomic number, their properties vary periodically.* This phenomenon has been known for well over a century, but until the past 60 years there was no understanding of why it was so.

As it turns out, explanation of the periodic law came as a consequence of a model of the atom advanced to explain some other phenomenon. This puzzle concerned the colors or pattern of light (called the emission spectrum) emitted by the atoms of a hot, gaseous element. The spectrum of each element is unique much as fingerprints are unique to each individual. The key to understanding atomic emission spectra seemed to lie in the nature of the electrons in the atom. We haven't said much about electrons yet. In the previous chapter, they were left free to roam outside the nucleus in the vast expanse of the atom that was otherwise empty space.

A model of the atom that emphasized the electrons was first advanced in 1913 by a student of Rutherford named Niels Bohr. His ingenious model of the electrons in an atom had the immediate effect of solving the puzzle of the emission spectrum of hydrogen. Of more consequence to our purposes, however, is that Bohr's model led eventually to a new improved model that made the existence of chemical similarities among certain elements reasonable and predictable.

In this chapter, we begin an important journey that takes us from a discussion of the arrangement of electrons to a logical understanding of the nature of the chemical bonds that hold the atoms of elements together. The latter is not discussed until Chapter 6. However, let us not put the cart before the horse, or, in this case, the answers before the questions. Our first order of business, then, is to appreciate the two questions (the periodic law and emission spectra) that had hounded scientists for logical explanations.

4-1 THE PERIODIC TABLE OF THE ELEMENTS

Long before there was any understanding of the composition of the atom, it was known that if the elements were arranged in order of increasing atomic weight, elements with similar characteristics would appear in a regular manner. In fact, for the lighter elements, it was observed that every eighth element seemed to be related by having similar properties. Although elements are now arranged in order of increasing atomic number rather than weight, this periodic nature of the elements serves as the cornerstone of chemists' ability to correlate and simplify a large body of experimental observations. *In an ar-*

Figure 4-1   METALS IN WATER. Sodium is a metal that reacts violently with water. The copper, silver, and gold in coins, however, are completely unreactive in water. Here, sodium (left) reacts vigorously with water while the coins (right) are unaffected.

rangement of the elements known as the **periodic table,** *elements with periodically related properties are ordered in vertical columns.* **A** periodic table is shown inside the back cover.

An example of this vertical relationship, which you will recognize, is found in the vertical column labeled "11 (**IB**)." Vertical columns are known as *groups* and are discussed in greater detail in the next chapter. The familiar metals in this group are copper, silver, and gold. They are used in coins and jewelry because of their low chemical reactivity as well as their lustrous beauty and comparative rarity. (See Figure 4-1.) These elements can be used for this purpose because they do not rust as does iron or become covered with a dull coating as does aluminum. On the other hand, lithium, sodium, and potassium are also metals, but they are very reactive. (See Figure 4-1.) They easily form compounds with the oxygen in the air and dissolve in water (explosively in the case of Na and K). Note that these closely related elements are all members of group 1 (**IA**). The chemical relationships among the elements have been known and studied for more than 150 years, and the periodic table (with, of course, fewer elements) has been displayed for a little more than a century.

Elements in the groups or vertical columns appear in a somewhat repeating or periodic relationship. For example, the numbers of elements *between* the elements of group 1 (**IA**) are 2, 8, 8, 18, 18, and 32. This is one example of the many periodic properties of nature. The sun and the moon rise and set in a regular, periodic manner. Seasons of the year, tides in the oceans, and final exams all occur in predictable and repeating cycles. Thus, elements not only exist in families, but do so in a periodic manner. An important goal of this chapter is to establish the reason for this *periodicity* of elements.

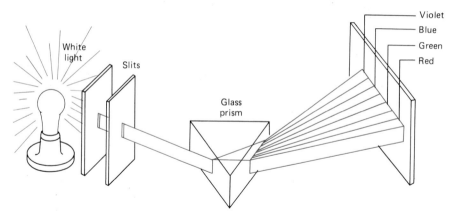

Figure 4-2 SPECTRUM OF INCANDESCENT LIGHT. Incandescent light is composed of all colors of the rainbow.

The information implied in this table is very important to the chemist. It is worthy of discussion in some detail. We do this in Chapter 5 after we answer the question "Why are elements periodic?"

4-2 THE EMISSION SPECTRA OF THE ELEMENTS

When any substance is heated to a high enough temperature, it glows. The tungsten filament in an incandescent light bulb glows as a result of the heat generated by the flow of electricity through the filament. Let's take a close look at the white light that comes from the bulb. *When a beam of the light is allowed to pass through a glass prism, the light separates into a rainbow of colors called its* **spectrum.** *Since one color blends gradually into another, this is known as a* **continuous spectrum.** (See Figure 4-2.) Raindrops in the air also serve as prisms after a rainstorm, breaking white sunlight into the rainbow.

Light is a form of energy, like electricity and heat. Light, however, dissipates in all directions with a wave nature. This property of light is analogous to the waves formed when a pebble lands in a calm lake. Light waves, however, travel at a speed of 3.0×10^{10} cm/sec (186,000 miles/sec). One of the characteristics of a wave is its **wavelength.** *This is the distance between two peaks on a wave,* as shown below. Wavelength is designated by the Greek letter λ (lambda).

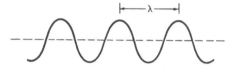

Each specific color of light in the rainbow has a specific λ. In the rainbow, red has the longest λ and violet has the shortest. When light has a somewhat longer λ than red, it is invisible to the eye and is called **infrared** (beneath red)

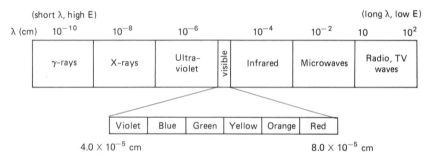

Figure 4-3   SPECTRUM OF LIGHT. Light is classified according to its wavelength.

light. When light has a somewhat shorter λ than violet, it is also invisible and is called **ultraviolet** (beyond violet) light.

The energy of light is inversely proportional to its wavelength (see **Appendix B** for a discussion of proportionalities). That is,

$$E \propto \frac{1}{\lambda} \qquad E = \frac{hc}{\lambda}$$

REVIEW
APPENDIX B-3
ON PROPORTION-
ALITIES.

where h is a constant of proportionality and is known as Planck's constant, and c is also a constant and is equal to the speed of light.

Note that the larger the value of λ, the smaller the energy, and vice versa. Therefore, red light, with the longer λ, has less energy than violet light. Indeed, ultraviolet light has enough energy to damage living organisms. X-rays and gamma rays have extremely small λ values and are accordingly extremely powerful. (Medical X-rays, however, are given in small doses to minimize the damage.) In Figure 4-3 these and other forms of light are shown in terms of relative wavelengths.

As shown in Figure 4-2, white light from a glowing tungsten filament is a combination of all wavelengths and energies of light in the visible range. When an element is heated to such an extent that it breaks into individual gaseous atoms, the spectrum produced is different from the continuous spectrum of the rainbow. In this case, only definite or discrete colors are produced. Recall that if the light produced has a specific color (λ), it has a specific energy (E). The spectrum of a gaseous atomic element (known as its *atomic emission spectrum*) is particular and characteristic of that element. You may have observed this phenomenon in the chemicals that creat colors when added to the logs in a fireplace or to the powder in fireworks. Strontium compounds create a red color, barium green, copper blue, and sodium yellow. *Since only certain colors are produced by the atoms of hot, gaseous elements, their atomic emission spectra are referred to as* **discrete** *or* **line spectra**.

Hydrogen, the simplest of the elements, has the most orderly atomic spectrum, as shown in Figure 4-4. In a sense, this made the hydrogen spectrum the most puzzling to the scientists.

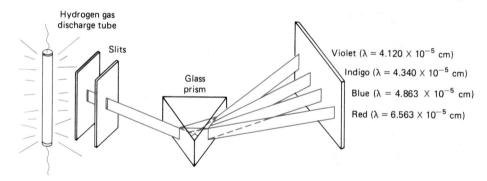

Figure 4-4   ATOMIC SPECTRUM OF HYDROGEN IN THE VISIBLE RANGE. The atomic spectrum of an element is composed of discrete colors.

The fact that light, as a form of energy, is emitted by atomic hydrogen in a discrete and yet orderly manner was also a matter that required an explanation.

4-3 A MODEL FOR THE ELECTRONS IN THE ATOM

By 1913 the nuclear model of the atom proposed by Rutherford was generally accepted by scientists. In that year, Bohr set out to explain the origin of the atomic spectrum of hydrogen by expanding on this model. Bohr's contribution would provide not only a reasonable picture of what forces hold the atom together but also the mathematical relations corresponding to the known experimental facts about the hydrogen atom.

First, Bohr proposed that the electrons revolve around the nucleus in fixed, circular trajectories called orbits. (See Figure 4-5.) The attractive forces be-

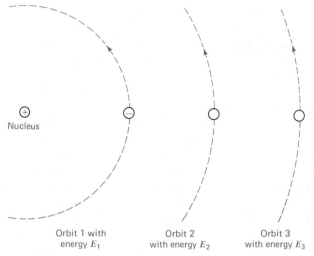

Orbit 1 with energy E_1 Orbit 2 with energy E_2 Orbit 3 with energy E_3

Figure 4-5   BOHR'S MODEL OF THE HYDROGEN ATOM. In this model the electron can exist only in definite or discrete energy states.

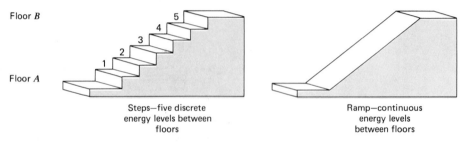

Floor *B*

Floor *A*

Steps—five discrete
energy levels between
floors

Ramp—continuous
energy levels
between floors

Figure 4-6   DISCRETE VERSUS CONTINUOUS ENERGY LEVELS. The steps represent discrete energy levels; the ramp, continuous energy levels.

tween the negative electron and the positive nucleus would be exactly balanced by the centripetal force* of the orbiting electron. This was a controversial and bold statement by Bohr, because classical physics predicted that this was not possible — the electron would collapse into the nucleus. Bohr neatly sidestepped this problem by postulating (suggesting without proof as a necessary condition) that classical physics does not apply in the small dimensions of the atom.

Second, Bohr postulated that the orbits in which electrons reside can be located only at certain, definite distances from the nucleus. An electron cannot exist anywhere between these available orbits unless it is in the process of jumping from one to another. Electrons in such a stationary orbit would have a definite or discrete energy. *Since only discrete energy states are available to the electron in an atom, the electronic energy is said to be* **quantized** *according to Bohr's theory.*

The discrete energy levels in the atom can be compared with the stairway in your home. In Figure 4-6 you can see that between floor A and floor B there are five steps that are available energy levels. Since you cannot stand (with both feet together) between steps, you can see that only these five energy states are possible between floors. A *stairway is thus quantized.* On the other hand, a ramp would be continuous, as all positions or energy levels between floors are possible.

When the electron in the hydrogen atom is in the lowest energy level (which is also closest to the nucleus), the atom is in its **ground state.** Just as a definite amount of energy is required for a person to climb one step, an electron may absorb a discrete amount of energy and "jump" to a higher energy level. The electron then has potential energy and the atom is said to be in an **excited state.** When an electron is in an excited energy state, it can fall back to a lower level; but the electron must then give up the same amount of energy that it previously absorbed. The energy that is emitted by this process comes off in the form of light. Since the orbits have discrete energy, the difference in energy

---

* *The force on a rotating object that tends to move it radially outward. For example, when a discus thrower lets go of the discus, it continues in a straight line from the point of release.*

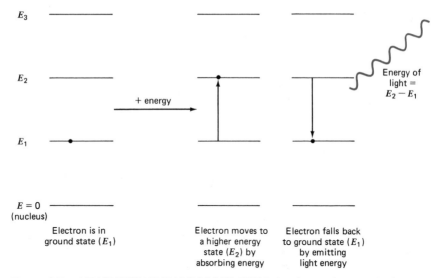

E_3

E_2

+ energy

Energy of
light =
$E_2 - E_1$

E_1

$E = 0$
(nucleus)

Electron is in
ground state (E_1)

Electron moves to
a higher energy
state (E_2) by
absorbing energy

Electron falls back
to ground state (E_1)
by emitting
light energy

Figure 4-7 LIGHT FROM THE HYDROGEN ATOM. An electron in an excited state emits energy in the form of light when it drops to a lower energy level.

between the levels is fixed. This in turn means that the energy of the light is discrete, with a discrete wavelength and hence a specific color. (See Figure 4-7.)

Bohr did significantly more than just present a model to explain discrete spectra. From the mathematical relationships suggested by his model, he was able to calculate the wavelengths of the light emitted by the hydrogen atom. His theory was proven when the calculated values corresponded with the values measured from actual experiments (one series of lines is shown in Figure 4-4).

Because of the beautiful simplicity of Bohr's model of the atom, modern scientists still use his picture in certain situations. However, our theories have now gone well beyond this model. For one thing, Bohr's model was not correct for atoms with more than one electron. That essentially limits the theory to hydrogen. Also, new experiments and new knowledge of the nature of matter have forced us away from viewing electrons as particles orbiting at definite distances from the nucleus with definite speeds and energies. Instead, the electron is now viewed as having a wave nature as well as a particle nature. It is thus described mathematically as a particle–wave (whatever that is) with *probabilities* of having certain energies and locations. This description is known as the **wave mechanical model.** Despite the loss of Bohr's easily visualized particle model, the use of the more complicated wave mechanical model has been rewarded by excellent results over a long period of time. Although mathematical solutions to the wave mechanical model are extremely complex, the results can be easily grasped and applied to an understanding of the periodic nature of the elements.

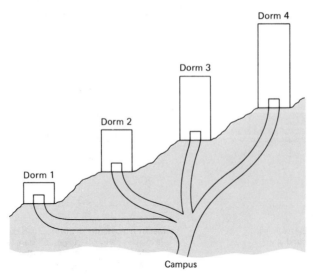

Figure 4-8   ENERGY LEVELS OF DORMS. The dorms on the hill represent different energy levels.

4-4 SHELLS   Bohr's model implied that *the electrons in all atoms* (assuming that the atoms of other elements are like hydrogen) *existed in* **principal energy levels** called **shells.** (This conclusion is consistent with the wave mechanical model.) Naturally, the shell closest to the nucleus is lowest in energy and each successive shell from the nucleus is higher in energy. Electrons occupy available shells of the lowest energy or those closest to the nucleus. In this respect, shells in an atom are analogous to dormitories near a campus. A dormitory is a defined region of space where students reside. In a series of dorms on a hill, obviously the one lowest on the hill and closest to the campus requires the least amount of effort to reach and is thus the lowest in energy. (See Figure 4-8.) Also analogous to electrons in shells, a definite amount of energy is required for a student to move uphill from dorm 1 to dorm 2.

The shells in an atom are labeled in two ways. (See Table 4-1.) Historically, the letters K, L, M, N, and so on, were used, with the K shell being closest to the nucleus (and the lowest in energy), the L shell next, and so on. The modern way is to label the shells with a number (n). *This number is called the* **principal**

TABLE 4-1   Shells

	K	L	M	N	O
Letter designation	K	L	M	N	O
Quantum number designation	1	2	3	4	5
Capacity ($2n^2$)	2	8	18	32	50

quantum number and describes the energy of the shell. The $n = 1$ shell corresponds to the K shell, $n = 2$ to the L shell, and so on. Both models of the atom agree that each shell holds $2n^2$ electrons. Obviously, the larger the principal quantum number, the greater the electron capacity. This result is reasonable, because the farther a shell is from the nucleus the greater the volume of space available to the electrons.

■
SEE PROBLEMS
4-1 THROUGH
4-8.

4-5 SUBLEVELS

Bohr's model described the electron only in terms of the principal quantum number. However, the wave mechanical model tells us that there is more to describing the arrangement of electrons than just the shell. To understand this, let's look again at the analogy of the dorms. If the careful student wants to live in the dorm of lowest energy, he or she would naturally look to the one lowest on the hill with space available. But another point should be raised as to the energy required to get to a certain room: "On what floor is the available room?" Since you can count on the elevators being full, on the top floor, or out of order, it is not far-fetched to assume that the student will have to use the stairs. In our example, the first dorm has only one floor, the second has two, the third has three, and so on. (This unusual situation is for the sake of the analogy.) Thus, not only do the students need to know how far up the hill they must walk but how many flights of stairs must be climbed once they are in the dorm. Each floor within a dorm is at a different energy level. (See Figure 4-9.)

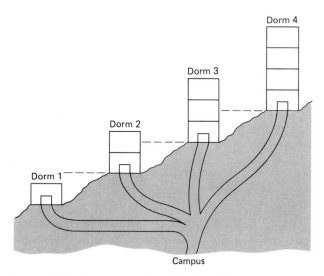

Figure 4-9 ENERGY LEVELS WITHIN A DORM. The floors in the dorms represent different energy levels within each dorm.

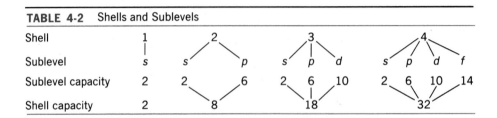

TABLE 4-2   Shells and Sublevels

Shell	1	2		3			4			
Sublevel	s	s	p	s	p	d	s	p	d	f
Sublevel capacity	2	2	6	2	6	10	2	6	10	14
Shell capacity	2	8		18			32			

Just as each dorm has floors, *a shell or principal energy level has* **sublevels.** The sublevels found in the currently known elements are labeled in order of increasing energy:

$$s < p < d < f$$

For each shell, the number of sublevels is equal to the value of its n quantum number. Thus, analogous to the dorms, the first shell ($n = 1$) has one sublevel (an s), the $n = 2$ shell has two (an s and a p), the $n = 3$ shell has three (an s, a p, and a d), and the $n = 4$ shell has four (an s, a p, a d, and an f). Each type of sublevel has a different electron capacity. The s sublevel holds 2 electrons, the p holds 6 electrons, the d holds 10 electrons, and the f holds 14 electrons. This information has been summarized for the first four shells in Table 4-2. Note how the total capacity of a shell is distributed among the sublevels present within the shell. The information in Table 4-2 should be committed to memory before proceeding.

In Figure 4-10, the first four shells and their sublevels are shown in order of increasing energy, analogous to the dorms shown in Figure 4-9.

Figure 4-10   ENERGY LEVELS IN THE ATOM. Each successive shell contains one additional sublevel.

4-6 ELECTRON CONFIGURATION OF THE ELEMENTS

At this point, we observe that *each electron in an atom can be assigned to a specific sublevel within a specific shell*. This is known as the element's **electron configuration**. If we proceed through the elements and list the electron configuration of each, the basis of the periodic law will become clear. Naturally, we start with the simplest atom, which is the one-electron atom of hydrogen. We proceed from there to the two-electron atom of helium and so forth. As we do this, we follow certain rules and guidelines, two of which are:

1 *A shell is designated by the principal quantum number n, a sublevel by the appropriate letter, and the number of electrons in that particular sublevel by a superscript number.*

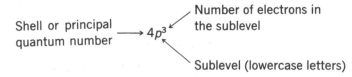

2 *Electrons fill the lowest available energy levels first (shell and sublevel).* This is known as the **Aufbau principle.** It is analogous to the student who wants the available room of lowest energy. Obviously, the most desirable room is on the lowest available floor in the lowest available dorm on the hill.

The first element, H, has one electron, which goes into the first shell, which has only an *s* sublevel. The electron configuration for the one electron is

$$1s^1$$

The next element, He, has two electrons. Since the first shell has room for two electrons, the electron configuration is

$$1s^2$$

Next comes Li with three electrons. Two fill the first shell, but the third must move on to the second shell. The second shell has two sublevels, the *s* and *p*, but as you know the *s* is the lower in energy. The electron configuration of Li is

$$1s^2 2s^1$$

The next seven elements complete the filling of the second shell:

Be	$1s^2 2s^2$	C	$1s^2 2s^2 2p^2$	O	$1s^2 2s^2 2p^4$	Ne	$1s^2 2s^2 2p^6$
B	$1s^2 2s^2 2p^1$	N	$1s^2 2s^2 2p^3$	F	$1s^2 2s^2 2p^5$		

Note in the periodic table that He and Ne are in a vertical column, indicating that they are members of the same family or group of elements. This particular family is known as the noble gases. As a whole, the noble gases are distinguished by having little or no tendency to enter into chemical reactions to form chemical bonds. In fact, they are sometimes referred to as the inert gases. The electron configurations of He and Ne indicate that they do have something in common: they both have "filled shells." Perhaps that is the reason for the particular behavior of the noble gases. This idea can be tested later as we look at the electron configurations of heavier elements. As we proceed, we can adopt a shorthand way of writing that saves us from repeating all of the filled inner sublevels. A noble gas in brackets will be equivalent to all of the electrons of that noble gas (e.g., $[Ne] = 1s^2 2s^2 2p^6$). The electron configurations of the next eight elements are:

Na	$[Ne]3s^1$	P	$[Ne]3s^2 3p^3$
Mg	$[Ne]3s^2$	S	$[Ne]3s^2 3p^4$
Al	$[Ne]3s^2 3p^1$	Cl	$[Ne]3s^2 3p^5$
Si	$[Ne]3s^2 3p^2$	Ar	$[Ne]3s^2 3p^6$

Now note in the periodic table that Ar is also a noble gas even though it does not have a completely filled shell. Our first observation about the electron configurations of noble gases must obviously be discarded or modified. An alternative suggestion is that perhaps filled outer s and p sublevels are characteristic of a noble gas (except for He, which doesn't have a $1p$ sublevel). Also note that Li and Na have something in common, that is, one electron beyond a noble gas. Since Li and Na are also in the same group, it appears at this point that similar electron configurations relate to the periodic relationship of the elements.

The next element after Ar presents a problem. Normally we might expect K, element 19, to have one electron in a $3d$ sublevel. In fact its nineteenth electron is in the $4s$ sublevel. How could this be possible? For the answer, let's go back to the analogy shown in Figure 4-9. Note that although dorm 4 is higher up the hill than dorm 3, all of the floors in dorm 4 are not necessarily higher than those in dorm 3. In fact, the top floor of dorm 3 is higher than the ground floor of dorm 4. The sharp student would thus choose the ground floor of dorm 4 in preference to the top floor of dorm 3. In the case of elements a similar phenomenon is true. For the neutral atom the $4s$ sublevel is lower in energy than the $3d$ sublevel and fills first. After the $4s$ sublevel, the $3d$ fills, followed by the $4p$. In Figure 4-10, note that the $4s$ sublevel is lower in energy than the $3d$ sublevel.

K	$[Ar]4s^1$
Ca	$[Ar]4s^2$
Sc	$[Ar]4s^2 3d^1$
	.
	.
	.
Zn	$[Ar]4s^2 3d^{10}$
Ga	$[Ar]4s^2 3d^{10} 4p^1$
	.
	.
	.
Kr	$[Ar]4s^2 3d^{10} 4p^6$

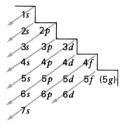

Figure 4-11 ORDER OF FILLING OF SUBLEVELS. The arrows indicate the order of filling.

Our observations on periodicity seem valid. Li, Na, and K, which are all in the same group, have an ns^1 electron configuration (n is a principal quantum number). Kr, a noble gas, has filled outer $4s$ and $4p$ sublevels, which was suggested earlier to be characteristic of this family of elements. *Apparently, vertical columns of elements have the same sublevel electron configuration but different shells.*

At this point, a scheme may be helpful to remember the order of filling of the sublevels. As you go to shells of higher energy, the shells get larger and much more mixing of sublevels occurs. A way to remember is illustrated in Figure 4-11. Write in horizontal columns all of the sublevels that exist starting with the $1s$. Under $1s$ put $2s$, followed by $2p$, and so on. Draw a stairstep down on the right. Now draw a diagonal arrow through each corner of the stairstep. The top arrow points to the first sublevel filled. The second arrow points to the second sublevel, the third arrow points to the $2p$ followed by the $3s$, and so on.

EXAMPLE 4-1

Write the electron configuration for iron.

SOLUTION

From the table inside the front cover, we find that iron has an atomic number of 26, which means the neutral atom has 26 electrons. Write the sublevels in the order of filling plus a running summation of the total number of electrons involved.

Sublevel	Number of Electrons in Sublevel	Total Number of Electrons
$1s$	2	2
$2s$	2	4
$2p$	6	10
$3s$	2	12
$3p$	6	18
$4s$	2	20
$3d$	10	30 (over 26)

> Iron therefore has six electrons past the filled $4s$ sublevel, but it does not completely fill the $3d$ sublevel. The electron configuration for iron is written out as
>
> $$1s^2 2s^2 2p^6 3s^2 3p^6 4s^2 3d^6$$
>
> or, if we use noble gas shorthand,
>
> $$[\text{Ar}]4s^2 3d^6$$

After some practice, the easiest way to predict the electron configuration of the elements is by use of the periodic table. In this endeavor the fact that vertical columns usually have the same electron sublevel configuration can be used to our advantage. In Figure 4-12 the sublevel configuration is listed above the group and the shell (n) is listed in the box for the element. With practice, one can quickly give the electron configuration of any element with only a periodic table for reference. It should be pointed out that certain elements do deviate from the filling predicted by Figure 4-11, especially those elements where d and f sublevels are partially filled. Two well-known exceptions are Cr (atomic number 24) and Cu (atomic number 29). Cr has the configuration $[\text{Ar}]4s^1 3d^5$ rather than the expected $[\text{Ar}]4s^2 3d^4$ and copper has the configuration $[\text{Ar}]4s^1 3d^{10}$ rather than $[\text{Ar}]4s^2 3d^9$. In addition to these exceptions there are numerous deviations of the configuration predicted from the periodic table (Figure 4-12) where the $5d$, $6d$, $4f$, and $5f$ sublevels are

s^1
1

s^2
1

s^1	s^2	d^1	d^2	d^3	d^{5*}	d^5	d^6	d^7	d^8	d^{10*}	d^{10}	s^2p^1	s^2p^2	s^2p^3	s^2p^4	s^2p^5	s^2p^6
2	2											2	2	2	2	2	2
3	3	d^1	d^2	d^3	d^{5*}	d^5	d^6	d^7	d^8	d^{10*}	d^{10}	3	3	3	3	3	3
4	4	3	3	3	3	3	3	3	3	3	3	4	4	4	4	4	4
5	5	4	4	4	4	4	4	4	4	4	4	5	5	5	5	5	5
6	6	5	5	5	5	5	5	5	5	5	5	6	6	6	6	6	6
7	7	6	6	6								7	7	7	7	7	7
8	8																

f^1	f^2	f^3	f^4	f^5	f^6	f^7	f^8	f^9	f^{10}	f^{11}	f^{12}	f^{13}	f^{14}
4	4	4	4	4	4	4	4	4	4	4	4	4	4
5	5	5	5	5	5	5	5	5	5	5	5	5	5

*This has an ns^1 configuration rather than ns^2.

Figure 4-12   ELECTRON CONFIGURATION AND THE PERIODIC TABLE. The electron configuration of an element can be determined from its position in the periodic table. (The value of "n" is shown in the boxes.)

partially filled. Although these exceptions may be important in the discussion of the chemistry of these particular elements, it is not important at this point to know all of the exceptions to the normal order of filling.

It should be noted that the order of filling predicted by Figure 4-11 for the element La (atomic number 57) is in conflict with that shown by the periodic table. Figure 4-11 predicts a configuration of $[Xe]6s^24f^1$ and the table, $[Xe]6s^25d^1$. The periodic table is correct. Lanthanum and the 14 elements that follow generally have one electron in the $5d$ sublevel. A similar situation exists for Ac (atomic number 89) and the 14 elements that follow it.

EXAMPLE 4-2

Write the electron configurations for (a) vanadium and (b) lead.

SOLUTION

(a) Inside the front cover of the book, note that vanadium (V) has atomic number 23. Locate V in the periodic table shown in Figure 4-12. V has the electron configuration

$$1s^22s^22p^63s^23p^64s^23d^3$$

or, starting with the previous noble gas

$$[Ar]4s^23d^3$$

(b) Lead (Pb) has atomic number 82. Locate Pb in the periodic table shown in Figure 4-12. Using the noble gas shorthand, note that the following sublevels come after Xe in lead:

$$[Xe]6s^24f^{14}5d^{10}6p^2$$

EXAMPLE 4-3

What element has the electron configuration $[Kr]5s^24d^{10}5p^5$?

SOLUTION

In Figure 4-12, locate s^2p^5 in the upper right. Atomic number 53 has the given electron configuration. Atomic number 53 is the element <u>iodine</u>.

■
SEE PROBLEMS
4-9 THROUGH
4-35.

4-7 ORBITALS   The wave mechanical model goes farther than just describing the electron in an atom in terms of shell and sublevel. Electrons can be assigned to orbitals. An **orbital** *is the region or volume of space where there is the highest probability of*

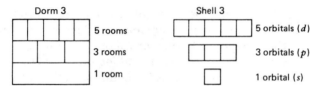

Figure 4-13   DORM ROOMS AND ORBITALS. The rooms on each floor of a dorm represent different regions of space at the same energy level, analogous to the orbitals of a sublevel.

finding a particular electron. Let us return once again to the analogy of the dorms. If one were to attempt to locate a student in a dorm, more information would be needed than just the particular dorm and the floor. One would have to know the location of the student's room. A room in a dorm is analogous to an orbital in an atom. It is the region of highest probability of finding the student (room) or the electron (orbital). Also, like most dorm rooms, each orbital has a capacity of two. (Unlike a dorm room, however, there is no way that more than two electrons can squeeze into one orbital.) An *s* sublevel is like a one-room floor, so it is all one orbital holding up to two electrons. The *p* sublevel is made up of three separate orbitals holding a total of six electrons. A *d* sublevel is composed of five orbitals, holding two each for a maximum capacity of 10 electrons, and finally the *f* sublevel is composed of seven orbitals holding a total of 14 electrons. (See Figure 4-13.)

The energy of an electron is determined only by the shell and sublevel. Under normal conditions all orbitals within a certain sublevel (e.g., the 4*p*) have the same energy. In the dorm analogy, once you have reached the floor, it doesn't matter whether you go to the room on the right, left, or straight ahead. The same amount of energy is required.

4-8 THE SHAPES OF ORBITALS

Each type of orbital (*s*, *p*, *d*, or *f*) is distinguished by its shape. By shape we refer to the dimensions of the region of highest probability of finding the particular electron or electrons. Thus, when we represent a sphere we consider that the electron will have at least a 90% probability of being within that volume. Despite Bohr's model, scientists no longer think of the electron as orbiting the nucleus in some designated path within the orbital.

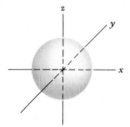

Figure 4-14   *s* ORBITAL. An *s* orbital has a spherical shape.

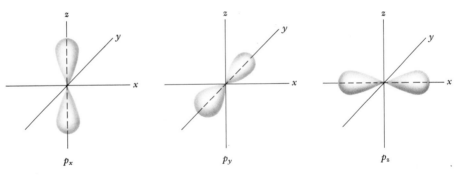

Figure 4-15 p ORBITALS. Each p orbital has two lobes.

An s orbital has a perfectly spherical shape. Electrons in s orbitals are free to occupy any position within the sphere. (See Figure 4-14.)

The three different p orbitals, on the other hand, are roughly shaped like a weird baseball bat with two fat ends called "lobes." If the axis through one of the p orbitals is defined as the x axis, the orbital is designed as the p_x orbital. The other two p orbitals are at 90° angles to the p_x orbital and to each other. They are referred to as p_y and p_z orbitals. (See Figure 4-15.)

The d orbitals are more complex. Four of the orbitals consist of four lobes each. The fifth consists of two lobes along the z axis and a torus (a doughnut shape) in the xy plane. (See Figure 4-16.) No matter how many lobes an orbital has, however, it still holds no more than two electrons.

The f orbitals are even more complex. However, since they are rarely involved in bonding, their shapes are not usually important to the chemist.

An important topic in the study of chemistry concerns the geometry or shape of molecules. The geometry of molecules is related to the shape of the orbitals of the electrons involved in bonding.

4-9 BOX DIAGRAMS OF THE ELECTRONS

The electron configuration of an element represents the location of electrons in shells and sublevels. However, it does not indicate the details of electron distribution among the orbitals of a particular sublevel. We can gain some additional insight into the elements by representing the individual orbitals of a sublevel. To represent an orbital in an atom we employ a box (☐) for each separate orbital. Each electron is represented as an arrow pointing either up or down (↑ or ↓). An electron in an atom behaves as a spinning particle just as the earth spins on its axis as it rotates around the sun. When two electrons occupy the same orbital, the two electrons have opposite or "paired" spins. This phenomenon is represented by the two arrows pointing in opposite directions. This is called the **Pauli exclusion principle,** *which states that no two electrons can reside in the same region of space (orbital) with the same spin* (represented

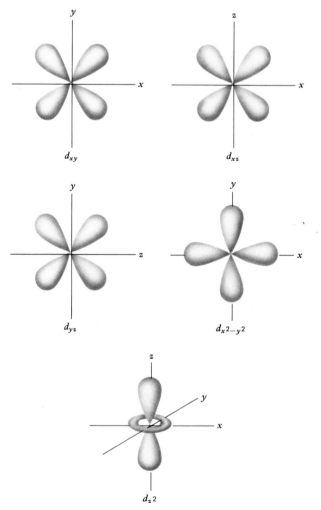

Figure 4-16   *d* ORBITALS. Except for the d_{z^2} the *d* orbitals have four lobes.

by the direction of the arrow). Thus, two electrons in one orbital are represented as

$$\boxed{\uparrow\downarrow}$$

The box diagrams of the first four elements are

		1*s*			1*s*	2*s*
H		$\uparrow$		**Li**	$\uparrow\downarrow$	$\uparrow$
He		$\uparrow\downarrow$		**Be**	$\uparrow\downarrow$	$\uparrow\downarrow$

The box diagrams of the next six elements are

	1s	2s	2p				1s	2s	2p		
B	↑↓	↑↓	↑			O	↑↓	↑↓	↑↓	↑	↑
C	↑↓	↑↓	↑	↑		F	↑↓	↑↓	↑↓	↑↓	↑
N	↑↓	↑↓	↑	↑	↑	Ne	↑↓	↑↓	↑↓	↑↓	↑↓

Note that the element carbon has two unpaired electrons in two different $2p$ orbitals. This follows **Hund's rule,** *which states that electrons occupy separate orbitals of the same energy (same sublevel) with parallel spins.* This result is understandable, since electrons, having the same charge, would prefer to get as far away from each other as possible within the same sublevel. In the dorm analogy, a similar situation develops between two students of the same charge (sex). Given two unoccupied rooms in the same building on the same floor, the two students will probably prefer private rooms. Pairing occurs when no more empty rooms are available.

A similar phenomenon occurs with elements that have a d sublevel that is partially occupied. For example, Mn (atomic number 25) has the electron configuration $[Ar]4s^23d^5$. All five electrons in the $3d$ sublevel occupy separate orbitals. Thus, there are five unpaired electrons in Mn as illustrated by the box diagram

	4s	3d				
Mn [Ar]	↑↓	↑	↑	↑	↑	↑

EXAMPLE 4-4

Show the box diagrams for the electrons beyond the noble gas core for the elements S, Ni, Cr, and Pr.

SOLUTION

First determine the electron configuration for the element and then show the box diagram for the outer orbitals beyond the noble gas core.

		3s	3p		
S $[Ne]3s^23p^4$	[Ne]	↑↓	↑↓	↑	↑

		4s	3d				
Ni $[Ar]4s^23d^8$	[Ar]	↑↓	↑↓	↑↓	↑↓	↑	↑

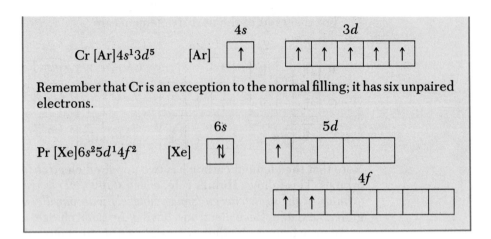

Cr $[Ar]4s^13d^5$ [Ar]

Remember that Cr is an exception to the normal filling; it has six unpaired electrons.

Pr $[Xe]6s^25d^14f^2$ [Xe]

■
SEE PROBLEMS
4-36 THROUGH
4-46.

4-10 CHAPTER REVIEW

In this chapter, we turned our attention to the nature of the electrons in the atoms of elements. The discrete colors or spectra emitted by the atoms of hot, gaseous elements constituted the phenomenon that led eventually to a model of the atom that emphasized electrons. In 1913 Niels Bohr suggested a model in which electrons existed in discrete energy levels or shells, and transitions between these discrete levels accounted for the discrete spectrum of hydrogen (and, presumably, the spectra of other elements). Since the energy of light is related to its wavelength, discrete changes in energy produce light of a discrete wavelength (discrete color if the wavelength is in the visible part of the spectrum). Of more importance to us is that Bohr's model led to a more sophisticated model of the atom called the wave mechanical model. The results of the latter model provided an explanation of the periodic law. The periodic law tells us that the properties of elements are related in a regular periodic manner.

Although Bohr's model is still useful, in the wave mechanical model the electron is viewed as having a certain probability of a location in space rather than a definite location as suggested by Bohr. The results of the wave mechanical picture indicate that an electron in an atom is described by four criteria: (1) shell or principal energy level, (2) sublevel, (3) orbital, and (4) electron spin.

The shell and sublevel describe the energy of an electron. Shells are designated in order of increasing energy by the quantum number n.

$$n = 1 < n = 2 < n = 3 < n = 4, \text{ etc.}$$

The four sublevels found in known elements are designated by letters in order of increasing energy

$$s < p < d < f$$

The number of electrons each shell can hold is $2n^2$, and the number of sublevels in each shell is equal to n. Electrons fill these sublevels according to the Aufbau principle—the sublevel lowest in energy fills first. The electron configuration of an element tells us the electron population of the various sublevels for the atoms of an element. The order of filling of sublevels becomes somewhat complex as we go to higher energy shells. The best way to remember the order of filling is by reference to the periodic table. Vertical columns have the same sublevel configuration but have consecutively larger values of n. This then is the reason for these chemical relationships and periodicity. *Families of elements lie in vertical columns and have the same population in their sublevels beyond the previous noble gas.*

These concepts give us a reasonable understanding of the theoretical basis of the periodic table as well as the emission spectra of the elements. By looking a little deeper into the results of wave mechanics, however, more characteristics of the elements fall into place. The model also tells us that each sublevel in an atom has one or more orbitals. Orbitals are regions of space where there is the greatest probability of finding the electron. Each separate orbital can hold up to two electrons. An s sublevel consists of just one spherically shaped orbital. The p sublevel consists of three propeller-shaped orbitals oriented at angles of $90°$ from each other. The d sublevel consists of five orbitals, and the f sublevel consists of seven orbitals. The information on shells, sublevels, and orbitals is summarized for the first four shells as follows. Each orbital is represented as a box.

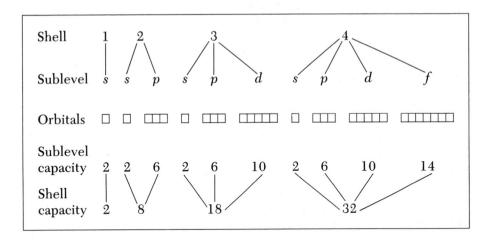

The final property concerning the electron in an atom is its spin. The Pauli exclusion principle states that electrons in the same orbital must have opposite or paired spins. Hund's rule states that electrons occupy separate orbitals of the same energy with parallel spins rather than pair up in the same orbital. All of this can be summarized by the box diagram for the outer electrons of the element As (atomic number 33).

As $[Ar]4s^2 3d^{10} 4p^3$

- **EXERCISES**

4-1 Referring to the periodic table, determine which of the following elements is expected to be chemically similar to calcium?

(a) K   (c) Cl   (e) Al
(b) Sr   (d) Mg

4-2 The elements fluorine, chlorine, and bromine all exist as diatomic molecules. Explain this behavior in terms of chemical families.

4-3 Ultraviolet light causes sunburns, but visible light does not. Explain.

4-4 The $n = 8$ and the $n = 9$ shells are very close in energy. Using the Bohr model, describe how the wavelengths of light compare as an electron falls from these two shells to the $n = 1$ shell.

4-5 The $n = 3$ shell is of considerably greater energy than the $n = 2$ shell. Using the Bohr model, describe how the wavelengths of light compare as an electron falls from these two shells to the $n = 1$ shell.

4-6 What is the total capacity of the fourth shell from the nucleus?

4-7 What is the total capacity of the sixth shell from the nucleus?

4-8 An electron in the lithium atom is in the third energy level. Is the atom in the ground or excited state? Can the atom emit light, and if so, how?

SUBLEVELS AND ELECTRON CONFIGURATIONS

4-9 Which of the following sublevels do not exist? $6s, 3p, 2d, 5f, 1p, 4d, 3f$

4-10 What is the electron capacity of each sublevel in the $n = 4$ shell?

4-11 What is the electron capacity of each sublevel in the $n = 3$ and the $n = 2$ shells?

4-12 Explain how a sublevel in one shell can be lower in energy than a sublevel in a lower shell.

4-13 Theoretically, there should be a fifth sublevel in the $n = 5$ shell. This would be called the g sublevel. What is the electron capacity of the g sublevel?

4-14 Which of the following sublevels fills first?

(a) $6s$ or $6p$ (c) $4p$ or $4f$
(b) $5d$ or $5p$ (d) $4f$ or $4d$

4-15 Which of the following sublevels is lower in energy?

(a) $6s$ or $5p$ (d) $4d$ or $5p$
(b) $6s$ or $4f$ (e) $4f$ or $6d$
(c) $5s$ or $4d$ (f) $3d$ or $4p$

4-16 Write the following sublevels in order of increasing energy: $4s, 5p, 4p, 5s, 4f, 4d, 6s$.

4-17 Write the total electron configuration for each of the following elements.

(a) Mg (b) Ge (c) Pd (d) Si

4-18 Write the electron configuration that is implied by [Ar].

4-19 Using the noble gas shorthand, write electron configurations for the following elements.

(a) S (b) Zn (c) Pu (d) I

4-20 Using the noble gas shorthand, write electron configurations for the following elements.

(a) Sn (b) Ni (c) Cl (d) Au

4-21 What elements have the following electron configurations? (Use only the periodic table.)

(a) $1s^2 2s^2 2p^5$
(b) $[Ar]4s^2 3d^{10} 4p^1$
(c) $[Xe]6s^2$
(d) $[Xe]6s^2 5d^1 4f^7$
(e) $[Ar]4s^1 3d^{10}$ (exception to rules)

4-22 What elements have the following electron configurations? (Use only the periodic table.)

(a) $[He]2s^2 2p^3$
(b) $[Kr]5s^2 4d^2$
(c) $[Ar]4s^2 3d^{10} 4p^6$
(d) $[Rn]7s^2 6d^1$
(e) $[Xe]6s^2 5d^{10} 4f^{14} 6p^1$
(f) $[Ar]4s^2 3d^{10}$

4-23 If any of the elements in Problem 4-21 belong to a numerical group [e.g., group 1 (IA)] in the periodic table, indicate the group.

4-24 Write the number designation of the two groups that may have five electrons beyond a noble gas configuration.

4-25 Write the number designation of a

group that has two electrons beyond a noble gas configuration. Write the number designation of a group with 12 electrons beyond a noble gas configuration.

4-26 Which two configurations belong to the same group in the periodic table?
(a) $[Kr]5s^2$
(b) $[Kr]5s^24d^{10}5p^2$
(c) $[Xe]6s^25d^14f^2$
(d) $[Ar]4s^23d^2$
(e) $[Ne]3s^23p^2$

4-27 Which two configurations belong to the same group in the periodic table?
(a) $[Kr]5s^24d^{10}$
(b) $[Ar]4s^13d^{10}$
(c) $[Ne]3s^1$
(d) $[Xe]6s^2$
(e) $[Xe]6s^24f^{14}5d^{10}$

4-28 What group in the periodic table has the following general electron configuration?
(a) ns^2np^2 (n is the principal quantum number)
(b) ns^2np^6
(c) $ns^1(n-1)d^{10}$ (e.g., $4s^13d^{10}$)
(d) $ns^2(n-1)d^1(n-2)f^2$ (two elements)

4-29 Write the symbol of the element that corresponds to the following.
(a) the first element with a p electron
(b) the first element with a filled $3p$ sublevel
(c) three elements with one electron in the $4s$ sublevel
(d) the first element with one p electron that also has a filled d sublevel
(e) the element after Xe that has two electrons in a d sublevel

4-30 Write the symbol of the element that corresponds to the following.
(a) the first element with a half-filled p sublevel
(b) the element with three electrons in a $4d$ sublevel
(c) the first two elements with a filled $3d$ sublevel
(d) the element with three electrons in a $5p$ sublevel

4-31 Write the general electron configuration for the following groups.

(a) 2 (IIA)   (c) 16 (VIA)
(b) 12 (IIB)   (d) 4 (IVB)

4-32 How does He differ from the other elements in group 18 (0)?

4-33 Which of the following elements fits the electron configuration $ns^2(n-1)d^{10}np^4$?
(a) Cr   (b) Te   (c) S   (d) O   (e) Si

4-34 What would be the atomic number of the element with one electron in the $5g$ sublevel?

*4-35 Element number 114 has not yet been made, but scientists believe that it will form a comparatively stable isotope. What would be its electron configuration and in what group would it be?

ORBITALS AND BOX DIAGRAMS

4-36 How many orbitals are in the following sublevels?
(a) $2p$ (d) $3s$
(b) $4d$ (e) $5f$
(c) $2d$

4-37 How many orbitals are in the $n = 3$ shell?

4-38 How many orbitals would be in the g sublevel? (See Problem 4-13.)

4-39 What is meant by the "shape" of an orbital?

4-40 Describe the shape of the $4p$ orbitals.

4-41 Describe the shape of the $3s$ orbital.

4-42 Which of the following box diagrams is excluded by the Aufbau principle? Which by the Pauli exclusion principle? Which by Hund's rule? Which is correct? Explain why a principle or rule may be violated.

4-43 Write the box diagrams for electrons beyond the noble gas core for the following elements.
(a) S   (b) V   (c) Br   (d) Pm

4-44 Write the box diagrams for electrons beyond the previous noble gas core for the following elements.
(a) As (b) Ar (c) Tc (d) Tl

4-45 How many unpaired electrons are in the atoms of elements in groups 12 (IIB), 5 (VB), 16 (VIA), 17 (VIIA); an atom with electron configuration given by $[ns^1 (n-1)d^5]$; and Pm (atomic number 61)?

4-46 Three groups in the periodic table have all of their electrons paired. What are the groups?

NEON LIGHTS
Argon as well as neon are gases used in "neon" lights. The chemical similarity
of these two elements is inferred from their position in the periodic table.

THE PERIODIC
■
NATURE OF THE
■
ELEMENTS
■

PURPOSE OF CHAPTER

In Chapter 5 we describe the origin of the periodic table, some important periodic trends, and how the properties of elements reflect these trends.

OBJECTIVES FOR CHAPTER

After completion of this chapter, you should be able to:

1 Give a brief description of the origin of the periodic table and how it was first constructed. (Introduction)

2 Locate on the periodic table the elements in the first seven periods. (5-1)

3 Give the characteristics of the electron configurations of the four general categories of elements. (5-1)

4 Locate on the periodic table those elements existing as gases, solids, and liquids at 25 °C. (5-2)

5 Describe and locate on the periodic table metals, nonmetals, and metalloids. (5-2)

6 Give the general trends of the atomic radii of the representative elements. (5-3)

7 Describe the general trends in the first ionization energies of the representative elements. (5-4)

8 Predict which representative element metals may be able to form $+2$ or $+3$ cations. (5-4)

9 Describe the general trends in the electron affinities for the representative elements. (5-5)

10 Predict which representative element nonmetals may be able to form -2 or -3 anions. (5-5)

11 Locate on the periodic table the seven representative element groups and describe some unique properties of elements within each group. (5-6)

12 Give a unique property of the noble gases and of the transition elements. (5-7)

The existence of related elements has been known for around 200 years, but it wasn't until 1869 that the forerunner of the modern periodic table was first introduced. In that year, two scientists, Dimitri Mendeleev of Russia and Lothar Meyer of Germany, independently presented a table of the elements arranged in order of increasing atomic weight (the concept of atomic number was still unknown). When the elements were ordered in this way, it was found that for the most part, elements in families appeared in a regular or periodic manner. Mendeleev, however, is usually credited with solving two problems with this ordering. First, Mendeleev found that if elements were ordered by increasing atomic weight and elements with similar chemical properties were placed in vertical columns, it was necessary to leave some spaces blank! For example, a space was left under silicon and above tin for a yet undiscovered element, which Mendeleev called "eka-silicon." Later, an element, germanium, was discovered with the properties predicted by the location of the blank space. The second problem involved some misfits if the elements were ordered according to atomic weight. For example, tellurium is heavier than iodine, but Mendeleev realized that iodine clearly belonged under bromine and tellurium under selenium, and not vice versa. Mendeleev simply reversed the order, suggesting that perhaps the atomic weights were in error. We now know, of course, that the elements in the periodic table should be listed in order of increasing atomic number instead of atomic weight. We also know that the basis for the periodic nature of the elements lies in their outer electron configurations, as discussed in Chapter 4.

Nevertheless, there is considerably more to understanding the periodicity of elements than just noting electron configurations. For example, even though two elements have similar outer sublevel electron configurations they may still have radically different properties (e.g., nitrogen gas and bismuth metal). This is possible because several other periodic trends also have important implications with respect to properties. In this chapter, we examine these trends so as to appreciate the basis of a rather broad classification of elements into metals and nonmetals. We also prepare the foundation for a discussion of the chemical bond in the next chapter. In the last part of this chapter, we examine some unique properties of individual elements as well as general properties of families of elements.

5-1 PERIODS AND GROUPS IN THE PERIODIC TABLE

In the next four sections, we examine some important trends in properties among the elements of the periodic table. As we will see, both horizontal and vertical trends within the table can be significant. *Horizontal rows of elements are referred to as* **periods.** Each period is composed of all of the elements between noble gases. The periods and the sublevels that compose each are listed in Table 5-1.

TABLE 5-1 Periods

Period	Number of Electrons	Sublevels	Noble Gas at End of Period
1	2	1s	He
2	8	2s, 2p	Ne
3	8	3s, 3p	Ar
4	18	4s, 3d, 4p	Kr
5	18	5s, 4d, 5p	Xe
6	32	6s, 5d, 4f, 6p	Rn
7	20	7s, 6d, 5f	?

Vertical columns of elements are referred to as **groups** *and sometimes as* **families.** As we found in the previous chapter (see Figure 4-12), elements in groups have the same type of sublevel configuration of electrons.

Each group in the periodic table is designated by a number. Traditionally, groups are labeled with Roman numerals followed by an "A" or a "B". Recently, IUPAC has suggested that the groups be numbered 1 through 18. Since the IUPAC convention is not yet in general use, we use both conventions. In this text, a group is labeled by its IUPAC number followed by the traditional number in parentheses [e.g., group 13 (IIIA)].

The elements can be classified into four general categories.

1   The **representative elements** (also known as **main group elements**) are metals and nonmetals in which *s* and *p* sublevels are being filled. These elements are in groups 1, 2, and 13 – 17 (IA – VIIA).

2   The **noble** or **rare gases** [group 18 (0)] are nonmetals at the end of each period. Except for **He**, all have filled outer *s* and *p* sublevels.

3   The **transition elements** are metals in which the *d* sublevel is being filled. These elements are in groups 3 – 12 (IB – VIII).

4   The **inner transition elements** are metals in which the *f* sublevel is being filled. The elements in which the 4*f* sublevel is being filled are known as the **lanthanides** or **rare earths** (atomic numbers 58 – 71) and the elements in which the 5*f* sublevel is being filled are known as the **actinides** (atomic numbers 90 – 103). There is very little chemical similarity *between* the lanthanides and actinides, but there is a strong similarity *among* the lanthanides and *among* the actinides.

SEE PROBLEMS
5-1 THROUGH
5-9.

After discussing some additional information concerning the periodic trends in the elements we will return to a more detailed discussion of some of these groups later in the chapter.

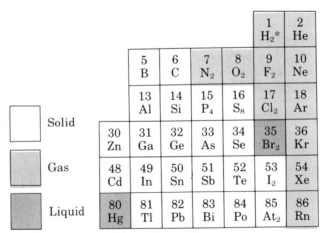

*H$_2$ indicates diatomic molecules at 25 °C.

Figure 5-1   PHYSICAL STATES OF THE ELEMENTS.

5-2 THE PHYSICAL STATES OF THE ELEMENTS

At room temperature (25 °C), various elements exist in all three physical states of matter. However, reference to the periodic table indicates that there is some order in the distribution of elements in the three states. The 11 gaseous elements are all found at the far upper right of the periodic table. A periodic table emphasizing physical state is shown in Figure 5-1. The gases are shown in color shade. The two liquids among the elements are shown in dark shade; these are bromine (atomic number 35) and mercury (atomic number 80). All other elements are solids at room temperature, including all of the elements not shown in Figure 5-1. The temperature that we specify is important, however, because slightly above room temperature (29 °C), two solids, gallium and cesium, melt to become liquids.

Also shown in Figure 5-1 are the formulas of the elements that exist at room temperature as multiatom molecules. Note that these elements are also found on the right-hand side of the table.

One of the most important periodic classifications of the elements, at least to the chemist, is that of *metals* and *nonmetals*. Metals are usually hard and lustrous. They are malleable (can be pounded into sheets) and ductile (can be drawn into wires) and can conduct electricity and heat. Nonmetals, on the other hand, are generally gases or soft solids. They are not good conductors of electricity or heat. Unfortunately, it is sometimes difficult to categorize elements as metals or nonmetals strictly on the basis of physical properties. For example, many metals are soft and one, mercury, is even a liquid. A form of carbon, a nonmetal, conducts electricity, whereas another form of carbon is the hardest substance known. Actually, the difference in chemical properties between metals and nonmetals is more significant and important for our purposes than physical differences. The properties that lead to the fundamental differences in chemistry will become evident shortly.

In any case, the location of the metals and nonmetals is easily determined from the periodic table. In Figure 5-2, the heavy stairstep line separates the

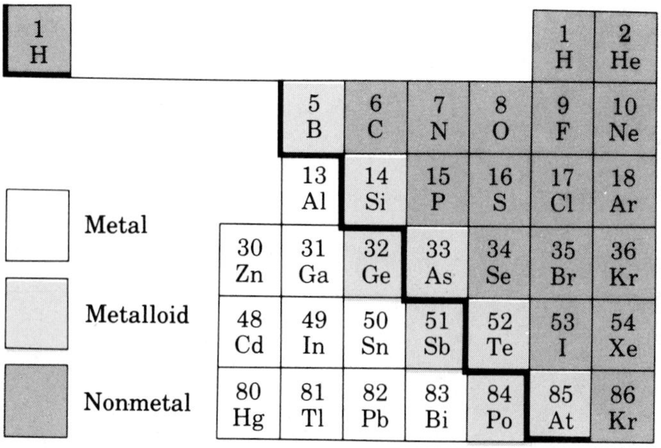

Figure 5-2 METALS, NONMETALS, AND METALLOIDS.

metals (light) from the nonmetals (dark), which are located on the upper right side of the table. Elements on the borderline have properties that are intermediate between the two classes. For example, silicon is a nonmetal but has some metallic properties. Germanium, which is directly under silicon, is a metal with some nonmetallic properties. Most of these borderline elements are sometimes referred to as *metalloids* (color shade). As you can see, most known elements (79%) are metals including all of those not shown in Figure 5-2.

■
SEE PROBLEMS
5-10 THROUGH
5-17.

5-3 ATOMIC RADIUS

The radius of an atom is the distance from the nucleus to the outermost electrons. This distance is in fact neither clearly defined nor easily measured. As mentioned in Chapter 4, the modern theory of the atom discourages a model of the atom in which an electron orbits the nucleus at a fixed distance. Instead, the electrons have a certain probability of being at a range of distances from the nucleus. Therefore, the radius does not actually have a fixed value. Besides the theoretical problem, there is a more practical problem. It is possible to determine the distance between the centers of two atoms in a solid, but it is difficult to decide where one atom starts and the other ends. Nevertheless, consistent values for the radii of neutral atoms have been compiled and are listed for the first three rows of representative elements in Figure 5-3. Two units are usually used for such small distances, the nanometer (nm, 10^{-9}

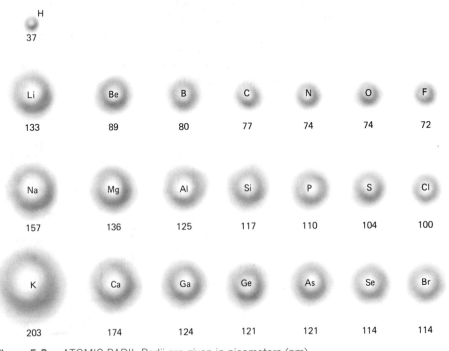

Figure 5-3   ATOMIC RADII. Radii are given in picometers (pm).

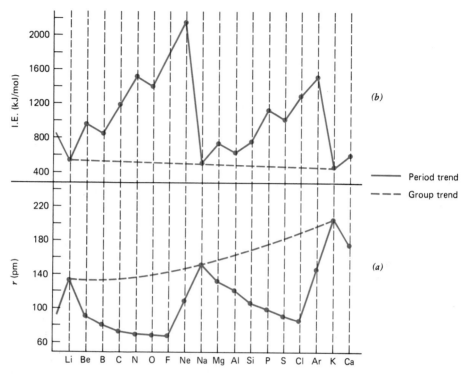

Figure 5-4 TRENDS IN ATOMIC RADII AND IONIZATION ENERGIES. Generally, as atomic radii *(a)* decrease across a period, ionization energies *(b)* increase.

meter) and the picometer (pm, 10^{-12} meter). We will use picometers since the numbers are easier to express (e.g., 37 pm rather than 0.037 nm).

It is more evident from a graph such as shown in Figure 5-4 that there are indeed significant periodic trends in the radii of the atoms. In curve *a* note that there is a general decrease in the radii across a period and an increase in radii down a group. It seems reasonable that the radii down a group would increase since each successive element has more subatomic particles and more mass. However, the same is true across a period, yet, for the most part, each successive element is more compact. A closer look at the location of the electrons may make these trends more understandable. Two factors are important. The first concerns the shell of the outermost electrons and the second concerns the net positive charge felt by the outermost electrons.

First, let us examine why the radii of atoms increase down a group. Electrons that lie between the nucleus and the outermost electrons are said to be in an inner shell. Each of these electrons essentially cancels the effect of one positive charge from the nucleus. For example, consider the atoms of the elements in group 1 (IA), all of which have only one electron in the outermost shell. If all of the inner electrons cancel one positive charge, the outer electron in each case feels the attraction from a net positive charge of +1. For example,

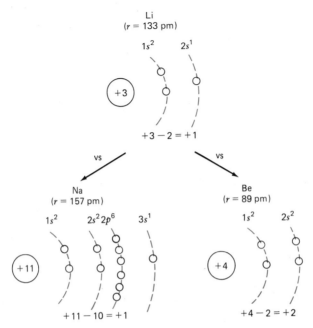

Figure 5-5   RADII OF LITHIUM, SODIUM, AND BERYLLIUM. The outer electrons in lithium and sodium feel about the same net positive charge; the outer electrons in beryllium feel a greater net positive charge.

consider the elements lithium and sodium. The net positive charge felt by the $2s$ electron in lithium is

$$+3 \text{ (nuclear charge)} - 2 \text{ (inner electrons)} = +1 \text{ (net charge)}$$

And the net positive charge felt by the $3s$ electron in sodium is

$$+11 \text{ (nuclear charge)} - 10 \text{ (inner electrons)} = +1 \text{ (net charge)}$$

Although the atoms of each successive element feel essentially the same net charge, the outermost electron in each successive element is in a shell farther from the nucleus. Thus, the atoms are progressively larger. (See Figure 5-5.)

Now visualize, if you will, a number of electrons in an atom that are in the same shell. Since they are all essentially the same distance from the nucleus, the negatively charged electrons all feel the attraction of the positively charged nucleus more or less equally. In this case, electrons in the same shell do not effectively cancel a positive charge of the nucleus from the other electrons in the same shell. Proceeding across a period, note that each successive element has one additional positive charge in the nucleus and one additional electron in the outer shell. However, since each additional electron goes into the same shell as in the preceding element, all of the outer electrons of each succeeding element feel an increasing net positive force of attraction from the nucleus. The increasing force of attraction felt by the outer electrons of each succeeding element accounts for the general contraction of the radii of the atoms across a period. For example, as mentioned above the outer electron of lithium would feel a net positive charge of $+1$. However, the outer two electrons of the next element, beryllium, would feel a net positive charge of $+2$. The higher net charge in the latter case accounts for the smaller size of beryllium atoms. (See Figure 5-5.) It should be pointed out that the model used in this discussion is somewhat simple compared with the actual situation. More exact calculations of the net charge felt by the outer electrons give a value of $+1.3$ for the charge felt by the outer electron in Li and $+1.8$ for the two outer electrons in Be.

**5-4
IONIZATION
ENERGY**

Perhaps the most significant periodic trend for the purpose of understanding chemistry is that of ionization energy. **Ionization energy** *(I.E.) is defined as the energy required to remove an electron from a gaseous atom to form a gaseous ion.* Since the outermost electron is generally the least firmly attached, it is the first to go.

$$M(g) \longrightarrow M^+(g) + e^-$$

This is always an endothermic process (requires energy), since an electron is held in the atom by its attraction to the nucleus. This attractive force must be overcome to separate the electron from the atom, leaving the rest of the atom as a positive ion. *The amount of energy that this process requires depends on how strongly the outermost electron is attracted to the nucleus.* Fortunately, the same reasoning that explains the trends in atomic radii also explains the trends in ionization energy. The ionization energies for the second and third periods are shown in Table 5-2. The energy unit abbreviated kJ/mol stands for kilojoules per mole, which is energy per a certain defined quantity of atoms.

Across a period the I.E. generally gets larger, which means increased difficulty in removing an electron. This corresponds to what was discussed in Section 5-3. Since the outer electrons are more strongly attracted to the nucleus, the atoms of each successive element are smaller and the outermost electron is harder to remove.

Table 5-2   Ionization Energy of Some Elements

Element	I.E. (kJ/mol)	Element	I.E. (kJ/mol)
Li	520	Na	496
Be	900	Mg	738
B	801	Al	578
C	1086	Si	786
N	1402	P	1102
O	1314	S	1000
F	1681	Cl	1251
Ne	2081	Ar	1520

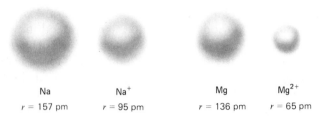

Na	Na$^+$	Mg	Mg^{2+}
r = 157 pm	r = 95 pm	r = 136 pm	r = 65 pm

Figure 5-6   IONIC RADII. A positive ion is always smaller than the neutral atom.

Since the smaller the atom the higher its ionization energy, it is reasonable to expect the reverse to be true. Indeed, the larger size of each succeeding atom down a group accounts for the successively lower ionization energies.

Again, the periodic trends in ionization energy may appear more meaningful when the information is presented in graphical form. The information on ionization energies is included in Figure 5-4 (curve *b*). Although the units are different on the ordinate axis for the two curves they have been shown together to emphasize their related periodic trends.

The ease with which an atom forms a positive ion by losing an electron is a fundamental difference between a metal and a nonmetal. Note that metallic properties increase to the left and down in the periodic table. This also corresponds to the trend toward lower I.E. Therefore, metals form cations more easily than nonmetals. Nonmetal behavior, to the right and up in the periodic table, is indicated by a comparative difficulty in forming cations. This fundamental trend affects the difference in bonding behavior between metals and nonmetals, which will be discussed in the next chapter.

The second I.E. for an ion involves removal of an electron from a +1 ion to form a +2 ion:

$$M^+(g) \longrightarrow M^{2+}(g) + e^-$$

In a similar manner, the third I.E. forms a +3 ion and so forth. In all cases it becomes increasingly difficult to remove each succeeding electron. When an electron is removed from an atom, the resulting ion is smaller than the parent atom (see Figure 5-6), because the remaining electrons feel a larger share of the nuclear charge. Since the remaining electrons are held more tightly, the ion is smaller and the next electron is harder to remove. The trends in consecutive ionization energies for Na through Al are shown in Table 5-3.

Note that the second I.E. for Na, the third I.E. for Mg, and the fourth I.E. for Al are all very large compared with the preceding number. For example, it

TABLE 5-3	Ionization Energies (kJ/mol)			
Element	First I.E.	Second I.E.	Third I.E.	Fourth I.E.
Na	496	4565	6912	9,540
Mg	738	1450	7732	10,550
Al	577	1816	2744	11,580

takes 2188 kJ (1450 + 738) to remove the first two electrons from Mg but about three times as much energy to remove the third electron (7732 kJ). In the case of Al, the first three electrons can be removed (to form Al^{3+}) at a cost of 5137 kJ, but the fourth would require an input of energy of 11,580 kJ (to form Al^{4+}).

There is a good reason for these sudden jumps in energy required to remove a particular electron. It depends on whether the electron is removed from an outer sublevel or a filled inner sublevel. Note that for Na, the first electron lost comes from the outer $3s$ sublevel but the second must be removed from the inner filled $2p$ sublevel. Removal of the first two electrons from Mg to form a Mg^{2+} ion is easy compared with formation of a Mg^{3+} ion, which also would require removal of one electron from the inner $2p$ sublevel. The same argument holds for the difficulty in formation of an Al^{4+} ion. As we will discuss in greater detail in the next chapter, cations can form in certain chemical compounds. The cations that do form, however, are those that can form with only a modest expense in energy. From the trends discussed in this section, it is possible to make some observations as to what species form cations with comparative ease. First, only metals lose electrons easily and, second, only electrons that come from the outer electron sublevels can be removed at a modest expense in energy. Thus +2 or +3 metal ions form only if the original metal atom had two or three outer sublevel electrons, respectively.

EXAMPLE 5-1

Tell whether the following cations are likely to form and state the reason for your answer: (a) B^+ (b) K^+ (c) Sr^{3+} (d) Tl^{3+}.

ANSWER

(a) No. Nonmetals do not form monatomic cations.

(b) Yes. K is a metal, and the one electron removed is the outer $4s$ electron.

(c) No. Sr is a metal, but the third electron would have to come from an inner $2p$ sublevel. Elements in group 2 (IIA) have only two electrons beyond the noble gas core.

(d) Yes. Tl is a metal, and the three electrons removed all come from the outer $6s$ and $6p$ sublevels.

5-5 ELECTRON AFFINITY

The final periodic trend that we will discuss at this time concerns a phenomenon that relates to the tendency of an atom of an element to accept an additional electron. This is known as **electron affinity** *(E.A.) and is defined as the energy released when a gaseous atom adds an electron to form a gaseous ion.*

$$X(g) + e^- \longrightarrow X^-(g)$$

Table 5-4   Electron Affinity[a]

Element	E.A. (kJ/mol)	Element	E.A. (kJ/mol)	Element	E.A. (kJ/mol)
B	−30				
C	−121	Si	−140		
N	+9	P	−75		
O	−140	S	−200	Se	−160
F	−334	Cl	−349	Br	−325

[a] *A negative value means that the process is exothermic.*

The electron affinities of some nonmetals are given in Table 5-4. The trends are somewhat uneven but, generally, the formation of a negative ion becomes more favorable to the right in the periodic table. The *general* trend is to lower electron affinities down a group although we do find that the third period (Si to Cl) is actually higher than the second. Thus, a high electron affinity (ease in formation of an anion) is generally a nonmetal property just as a low ionization energy (ease in formation of a cation) is a metallic property. Although formation of a nonmetal cation in chemical reactions is energetically unlikely, formation of a nonmetal anion is favorable.

Just as we find that metals lose one or more electrons to form cations in compounds, we find that representative element nonmetals can gain one or more electrons to form anions. For the atoms of an element to gain more than one electron, however, there must be room in the outer electron sublevel for the additional electrons. For example, oxygen ($[He]2s^2 2p^4$) is a nonmetal with two vacancies in its outer $2p$ sublevel. It can therefore add two electrons to form a −2 anion. If oxygen were to add a third electron, it would have to be placed in the $3s$ sublevel, and this would incur a large expense in energy. On the other hand, nitrogen has three vacancies in its outer sublevel and can add three electrons to form a −3 ion in certain compounds. In summary, a monatomic anion is feasible if it forms from a representative element nonmetal and, in the case of −2 and −3 anions, if the electrons can be added to the outer p sublevel.

We have already seen that cations are smaller than their parent atoms. (See Figure 5-6.) Conversely, anions are considerably *larger* than their parent atoms since the nuclear charge felt by the outer electrons is reduced somewhat because it is spread out over more electrons. (See Figure 5-7.)

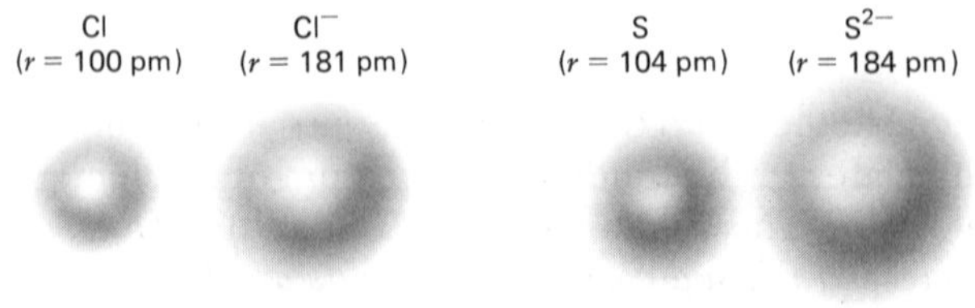

SEE PROBLEMS 5-18 THROUGH 5-39.

| Cl ($r = 100$ pm) | Cl⁻ ($r = 181$ pm) | S ($r = 104$ pm) | S²⁻ ($r = 184$ pm) |

Figure 5-7   IONIC RADII. A negative ion is always larger than the neutral atom.

EXAMPLE 5-2

Tell whether formation of the following anions is likely and state the reason for your answer: (a) Ba^- (b) Br^- (c) S^{3-} (d) N^{3-}.

ANSWER

(a)   No. Metals do not form monatomic anions.

(b)   Yes. Br is a nonmetal and group 17 (VIIA) elements (ns^2np^5) all have one vacancy in the outer sublevel.

(c)   No. Sulfur is a nonmetal but group 16 (VIA) elements (ns^2np^4) have only two vacancies in the p sublevel.

(d)   Yes. Nitrogen is a nonmetal and is a group 15 (VA) element (ns^2np^3), which has three vacancies in the p sublevel.

5-6 THE REPRESENTATIVE ELEMENTS

Much of the chemistry discussed in this text centers on the elements in which the outer s and p sublevels are filling. As mentioned, these are known as representative or main group elements. As we will see, the chemical properties vary not only from group to group but also from top to bottom within each group. Since the chemistry of these elements is so important, it is worthwhile to highlight each group to look at its unique characteristics.

Group 1 (IA)

Of all the groups in the periodic table, the members of this group have the most uniform properties from top to bottom. These elements, starting with Li, are known as the **alkali metals.** The most dominant feature of their chemistry

Figure 5-8   GROUP 1 (IA), THE ALKALI METALS. Sodium is a typical alkali metal that must be stored under mineral oil because of its chemical reactivity.

involves the comparatively easy loss of the one outer electron to form a + 1 ion. The easy loss of an electron means that these elements are highly reactive. Thus, in nature they are always found in a chemically combined state. The pure metals are difficult to handle safely since they react rapidly with the oxygen in air or with water. In fact, some of these metals react explosively with water. They are all soft, shiny metals and must be stored under oil or kerosene to prevent exposure to air and water. If an alkali metal is designated by the symbol "M" and a member of group 17 (VIIA) is designated by the symbol "X", then all combinations of these two groups of elements have the general formula MX (e.g., LiCl, KBr, CsI, and NaF). (See Figure 5-8.)

Group 2 (IIA)

Members of this group are known as the **alkaline earth metals.** These elements are also somewhat uniform in properties in that they all form the + 2 ion in compounds by loss of the two outer electrons. The loss of the electrons is not as easy as in the preceding group, however, so they are not as chemically reactive. For example, the heavier alkaline earths react with water but do so very slowly. They also react with the oxygen in air but form an oxide coating that protects the metal from further reaction. (See Figure 5-9.) Beryllium chemistry is somewhat different from that of the rest of the group. Magnesium, when

Figure 5-9   GROUP 2 (IIA), THE ALKALINE EARTH METALS. Magnesium is a typical alkaline earth metal. It is not as reactive as the alkali metals and can be stored in the open air. Shown here is a piece of magnesium ribbon.

Figure 5-10 GROUP 13 (IIIA). In this group boron (left) is a nonmetal but aluminum (right) and other members of the group are metals.

mixed with aluminum, forms a strong, lightweight **alloy** *(a homogeneous mixture of metals)*. It is also used in flashbulbs because of the bright light it emits when ignited. With group 17 (VIIA) elements, this group forms compounds with the general formula MX_2 (e.g., $CaCl_2$, MgF_2, and BaI_2). With oxygen, the general formula is MO.

Group 13 (IIIA)   In this group, a significant difference exists between the chemical properties of the nonmetal at the top, boron, and those of the metals below. Thus, boron is usually discussed separately as a typical nonmetal. Aluminum and the other metals may lose their three outer electrons to form $+3$ ions, but other types of compounds form as well. Aluminum is, of course, the most important metal in this group because, as a metal, it combines low density and high strength. It is a reactive metal but forms a coating of oxide when exposed to air that protects it from further reaction. Aluminum is the most common metal in the earth's crust (the outer layers including the atmosphere). Compounds of aluminum are components of most clays. (See Figure 5-10.)

Group 14 (IVA)   There is a wide range of properties within this group, with two nonmetals at the top, carbon and silicon, and three metals at the bottom, germanium, tin, and lead. Carbon is the most amazing of all elements as its compounds form the basis of life as we know it. *The millions of known compounds of carbon are studied in a special branch of science called* **organic** *chemistry. The complex organic compounds involved in life processes are studied in another but closely allied branch of science called* **biochemistry.** Chapter 16 in this text, is devoted to an introduction to organic chemistry. Pure carbon is found in nature in two forms: graphite and diamond. *Different forms of a pure element in the same physical state are called* **allotropes.** (See Figure 5-11.)

  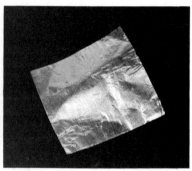

Figure 5-11   GROUP 14 (IVA). Carbon is found in two crystalline allotropes, graphite and diamond (left and center). Tin is a typical metal (right).

Silicon and germanium can both be classified as metalloids and as such have properties intermediate between those of metals and nonmetals. For example, both conduct a limited amount of electricity and thus are known as semiconductors. Artificial intelligence (i.e., computers) is based on the use of these materials as semiconductors. Silicon, a major component of sand, is also a plentiful element in the earth's crust. Tin and lead have properties typical of metals.

Group 15 (VA)

This group exhibits the most dramatic changes in properties from the gaseous nonmetal at the top, nitrogen, to the solid metal at the bottom, bismuth. Elemental nitrogen (N_2) is the major component of the atmosphere (about 80%). Nitrogen is also an important constituent of living organisms. Phosphorus is a nonmetallic solid that exists as P_4 molecules in a common allotrope called "white phosphorus." It is very reactive in air, burning violently with the evolution of large amounts of heat. For this reason, it has an unfortunate application in weapons. Arsenic and antimony are metalloids, but bismuth is definitely metallic. (See Figure 5-12.)

Group 16 (VIA)

Like the first two groups, this group also has a name — the **chalcogens**. Actually, this name is not used to a large extent because there is a wide variety of properties among the elements of this group. At the top, oxygen has a unique chemistry and for this reason is usually discussed separately from the other elements in the group (like most other elements in the second period). Oxygen is the most abundant element in the earth's crust. The atmosphere itself is composed of more than 19% elemental oxygen. It exists primarily as O_2 in the air but another important and lifesaving allotrope of oxygen, ozone (O_3), is present in the upper atmosphere (the stratosphere). There it absorbs the powerful ultraviolet rays from the sun that would otherwise penetrate to the surface and damage living organisms. In the lower atmosphere, ozone is a toxic, pungent pollutant. Oxygen is a very reactive element and forms com-

Figure 5-12   GROUP 15 (VA). Phosphorus (left) is a reactive nonmetal and must be stored under water. Bismuth (right) is a typical metal.

pounds with every other element except the noble gases helium, neon, and argon. Sulfur and selenium are also nonmetals and form many important compounds. Tellurium and polonium are both metalloids. The chemistry of polonium has not been studied extensively. It is a highly radioactive element that rapidly decays to other elements, making comprehensive studies difficult. (See Figure 5-13.)

Figure 5-13   GROUP 16 (VIA). The two famous members of this group are oxygen, which is a gas, and sulfur, which is a nonmetallic solid.

Figure 5-14   GROUP 17 (VIIA), THE HALOGENS. At room temperature chlorine (left) is a gas, bromine (center) is a liquid, and iodine (right) is a solid. The elements of this group are very reactive chemically. (Left, Time/Life Books, Inc.)

Group 17 (VIIA)   Once again we have a group with fairly similar properties from top to bottom, so it has a commonly used name, the **halogens.** All of these elements exist as diatomic molecules in the elemental state and all are chemically reactive. In fact, fluorine is the most reactive nonmetal and, like oxygen, forms compounds with every other element except helium, neon, and argon. Fluorine is so reactive in elemental form that it must be stored in special containers because it reacts with most common containers. Astatine is a metalloid that has not been studied extensively because of its radioactive instability. All three physical states are represented in this group: fluorine and chlorine are gases, bromine is a liquid, and iodine is a solid at room temperature. (See Figure 5-14.)

Although hydrogen is not a halogen, it is sometimes included in group 17 (VIIA) because it precedes a noble gas. Like the halogens, hydrogen exists as a diatomic gas (H_2) and forms a −1 anion. Unlike the halogens, however, hydrogen is not a particularly reactive element.

5-7 THE NOBLE GASES AND TRANSITION METALS

Group 18 (0)   Elements of this group are known as the **noble** or **rare gases.** At one time this group was also referred to as the inert gases because it was thought that these elements could not form chemical bonds and existed only as individual atoms. This is still true for He, Ne, and Ar. These elements find use where noncombustible gases are needed. In 1962, a scientist proved that Xe forms compounds with the two most reactive elements, fluorine and oxygen. Later, it was discovered that Rn and Kr show similar behavior by forming bonds with oxygen and fluorine. Despite the existence of these compounds the noble gases as a group are still considered the most unreactive elements.

Transition Elements

Many of our most familiar metals are transition elements [groups 3 – 12 (IB – VIII)] such as chromium, iron, silver, and gold. For the most part, transition metals are much less reactive than representative metals. Also, the chemistry of transition elements is more complex than that of the representative elements, making it harder to generalize. For example, manganese forms the following compounds with oxygen: MnO, Mn_3O_4, Mn_2O_3, MnO_2, and Mn_2O_7. A discussion of the chemistry of transition elements is beyond the scope of this text.

■
SEE PROBLEMS 5-40 THROUGH 5-51.

5-8 CHAPTER REVIEW

Perhaps the greatest timesaving device for the chemist, with the possible exception of the computer, is the periodic table. A glance at this table hanging on the wall can provide a great deal of information about an element just from its location in the table. The main concern of this chapter has been to emphasize a number of periodic relationships within the periods (horizontal rows between noble gases) and groups (vertical columns). In addition to the location of gaseous and solid elements, we discussed the periodic relationships of atomic radii, ionization energies, and electron affinities. The latter two trends relate directly to two important classes of elements: metals and nonmetals. The major difference between these two classes of most importance to the chemist relates to ion formation. Metals lose electrons comparatively easily to form cations, and representative element nonmetals can gain electrons to form anions. The information on trends and consequences is summarized as follows.

	Physical State	Periodic Trends	Consequences
Metals	All solids except for one liquid	1. Larger radii 2. Smaller first ionization energy 3. Smaller electron affinity	Can form cations (M^+) relatively easy; can also form M^{2+} and M^{3+} if electrons are available outside noble gas core
Nonmetals	1 liquid, 11 gases, 10 solids	1. Smaller radii 2. Larger first ionization energy 3. Larger electron affinity	Can form anions (X^-); can also form X^{2-} and X^{3-} if there is room in the outer p sublevel

We then took a brief look at the seven groups of representative elements, noting the trends from top to bottom within a particular group. Finally, the noble gases and transition metals were briefly mentioned. The locations of the four main categories of elements, as well as the named groups of representative elements, are shown below and are followed by a summary of the properties of the groups.

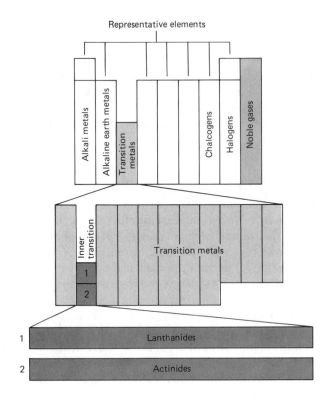

Group	Outer Sublevel Configuration	Name	Comments
1 (IA)	ns^1	Alkali metals	Many common properties from top to bottom; very reactive
2 (IIA)	ns^2	Alkaline earths	Less reactive metals and larger variation in properties from top to bottom than in group 1 (IA)
13 (IIIA)	ns^2np^1	—	Nonmetal at the top, others are metals; large variation in properties
14 (IVA)	ns^2np^2	—	Large variation in properties; one nonmetal, two metalloids, and two metals
15 (VA)	ns^2np^3	—	Wide range in properties; two nonmetals, two metalloids, and one metal

Group	Outer Sublevel Configuration	Name	Comments
16 (VIA)	ns^2np^4	Chalcogens	Some similarities among nonmetals at the top
17 (VIIA)	ns^2np^5	Halogens	All nonmetals with many similar properties from top to bottom
18 (0)	ns^2np^6	Noble gases	Nonmetal gases at the end of each period
3–12 (IB–VIIIB)	$ns^2(n-1)d^x$	Transition elements	Many familiar metals, but complex chemistry
—	$6s^25d^14f^x$	Lanthanides	Horizontal rather than vertical similarities
—	$7s^26d^15f^x$	Actinides	

▪ EXERCISES

PERIODS AND GROUPS

5-1 What is the electron configuration of the noble gas at the end of the third period?

5-2 What is the electron configuration of the noble gas at the end of the fourth period?

5-3 How many elements would be in the seventh period if it were complete?

5-4 How many elements would be in a complete eighth period? (*Hint:* Consider the 5g sublevel.)

5-5 Classify the following elements into one of the four main categories of elements.
(a) Fe (c) Pm (e) Xe (g) In
(b) Te (d) La (f) H

5-6 Classify the following elements into one of the four main categories of elements.
(a) Se (c) Ni (e) Zn (g) Er
(b) U (d) Sr (f) I

5-7 Which of the following elements are transition metals?
(a) In (c) Ca (e) Pd (g) Ag
(b) Ti (d) Xe (f) Tl

5-8 Classify the following electron configurations into one of the four main categories of elements.
(a) $[Ar]4s^23d^2$
(b) $[Kr]5s^24d^{10}5p^5$
(c) $[Ar]4s^23d^{10}4p^6$
(d) $[Rn]7s^26d^15f^3$

5-9 Classify the following electron configurations into one of the four main categories of elements.
(a) $[Ne]3s^23p^6$
(b) $[Ne]3s^23p^5$
(c) $[Xe]6s^24f^{14}5d^2$
(d) $[Xe]6s^25d^14f^7$

PHYSICAL STATES OF THE ELEMENTS

5-10 Referring to Figure 5-1, tell which of the following are gases at room temperature.
(a) Ne (c) B (e) Br (g) Na
(b) S (d) Cl (f) N

5-11 Referring to Figure 5-1, tell which of the following elements exist as diatomic molecules under normal conditions.
(a) N (c) Ar (e) H (g) Xe
(b) C (d) F (f) B (h) Hg

5-12 Referring to Figure 5-2, tell which of the following elements are metals.
(a) Ru (c) Hf (e) Ar (g) Se
(b) Sn (d) Te (f) B (h) W

5-13 Which, if any, of the elements in Problem 5-12 can be classified as a metalloid?

5-14 Identify the following elements using the periodic table and the information in Figures 5-1 and 5-2.
(a) a nonmetal, monatomic gas in the third period

(b) a transition metal that is a liquid

(c) a diatomic gas in group 15 (VA)

(d) a metalloid in the fourth period that has three electrons in its outer sublevel

(e) the second metal in the second period

(f) the only member of a group that is a metal

5-15 Identify the following elements using the periodic table and the information in Figures 5-1 and 5-2.

(a) a nonmetal, diatomic liquid

(b) the last element in the third period

(c) a metalloid that is more metal than nonmetal and has only two electrons in an outer p sublevel

(d) a nonmetal, diatomic solid

(e) a metal with one p electron and no d electrons

*5-16 Assume that elements with atomic number greater than 109 will eventually be made. If so, what would be the atomic number and group number of the next nonmetal?

*5-17 What is the atomic number of the next metalloid that would appear in the periodic table after element number 109?

PERIODIC TRENDS

5-18 In each of the following pairs of elements, which has the larger radius?

(a) As or Se   (c) Sr or Ba

(b) Ru or Rh   (d) F or I

5-19 In each of the following pairs of elements, which has the larger radius?

(a) Tl or Pb   (c) Pr or Ce

(b) Sc or Y   (d) P or Br

5-20 Four elements have the following radii: 117 pm, 122 pm, 129 pm, and 134 pm. The elements are V, Cr, Nb, and Mo. Which element has a radius of 117 pm and which has a radius of 134 pm?

5-21 Four elements have the following radii: 180 pm, 154 pm, 144 pm, and 141 pm. The elements are In, Sn, Tl, and Pb. Which element has a radius of 141 pm and which has a radius of 180 pm?

*5-22 Zirconium and hafnium are in the same group and have almost the same radius despite the general trends. As a result the two elements have almost identical chemical and physical properties. The fact that these elements have almost the same radius is due to the *lanthanide contraction*. With the knowledge that atoms get progressively smaller as a sublevel is being filled, can you explain this phenomenon? (*Hint:* Follow all of the expected trends between the two elements.)

5-23 Which of the following elements has the higher ionization energy?

(a) Ti or V   (d) Fe or Os

(b) P or Cl   (e) B or Br

(c) Mg or Sr

5-24 Four elements have the following first ionization energies (in kJ/mol): 762, 709, 579, and 558. The elements are Ga, Ge, In, and Sn. Which element has an ionization energy of 558 kJ/mol and which element has an ionization energy of 762 kJ/mol?

5-25 Four elements have the following first ionization energies (in kJ/mol): 869, 941, 1010, and 1140. The elements are Se, Br, Te, and I. Which element has an ionization energy of 869 kJ/mol and which element has an ionization energy of 1140 kJ/mol?

5-26 The first four ionization energies for Ga are 578.8, 1979, 2963, and 6200 kJ/mol. How much energy is required to form each of the following ions: Ga^+, Ga^{2+}, Ga^{3+}, and Ga^{4+}? Why does the formation of Ga^{4+} require a comparatively large amount of energy?

5-27 The first five ionization energies for carbon are 1086, 2353, 4620, 6223, and 37,830 kJ/mol. How much energy is required to form the following ions: C^+, C^{2+}, C^{3+}, C^{4+}, and C^{5+}? In fact even C^+ does not form in compounds. Compare the energy required to form this ion with that needed to form some metal ions and explain.

5-28 Which of the following monatomic cations are likely to form? If a certain ion requires a large amount of energy to form, give the reason.

(a) Cs^+ (c) Ne^+ (e) Sc^{3+}
(b) Rb^{2+} (d) Tl^{4+} (f) Te^{2+}

5-29 Which of the following monatomic cations are likely to form? If a certain ion requires a large amount of energy to form, give the reason.
(a) In^{3+} (c) Ca^{2+} (e) B^{3+}
(b) I^+ (d) K^+

5-30 Which of the following monatomic anions are likely to form? If a certain ion is unlikely, explain.
(a) F^{2-} (c) Be^- (e) Ar^-
(b) Se^{2-} (d) N^{3-}

5-31 Which of the following monatomic anions are likely to form? If a certain ion is unlikely, explain.
(a) I^- (c) K^- (e) Sb^{3-}
(b) Bi^{2-} (d) Se^{2-}

5-32 Which of the following atoms would most easily form a cation?
(a) B   (b) Al   (c) Si   (d) C

5-33 Which of the following atoms would not be able to form a $+2$ cation?
(a) Sr   (b) Li   (c) B   (d) Ba

5-34 Which of the following atoms would most likely form an anion?
(a) Be   (b) Al   (c) Ga   (d) I

5-35 Which of the following atoms would not be likely to form a -2 anion?
(a) Se   (b) Po   (c) Cl   (d) H

5-36 Noble gases form neither anions nor cations. Why?

5-37 Arrange the following ions and atoms in order of increasing radii.
(a) Mg   (c) S   (e) K^+
(b) S^{2-} (d) Mg^{2+} (f) Se^{2-}

5-38 Arrange the following ions and atoms in order of increasing radii.
(a) Br   (c) K   (e) Br^-
(b) K^+ (d) I^- (f) Ca^{2+}

*5-39 On the planet Zerk, the periodic table of elements is slightly different from ours. On Zerk, there are only two p orbitals, so a p sublevel holds only four electrons. There are only four d orbitals, so a d sublevel holds only eight electrons. Everything else is the same as on earth, such as the order of filling ($1s$, $2s$, etc.) and the characteristics of noble gases, metals, and nonmetals. Construct a Zerkian periodic table using numbers for elements up to element number 50. Then answer these questions.
(a) How many elements are in the second period? In the fourth period?
(b) What are the atomic numbers of the noble gases at the ends of the third and fourth periods?
(c) What is the atomic number of the first inner transition element?
(d) Which element is more likely to be a metal: element number 5 or element number 11; element number 17 or element number 27?
(e) Which element has the larger radius: element number 12 or element number 13; element number 6 or element number 12?
(f) Which element has a higher ionization energy: (1) element number 7 or element number 13; (2) element number 7 or element number 5; (3) element number 7 or element number 9?
(g) Which ions are reasonable?
(1) 16^{2+} (3) 7^+ (5) 17^{4+}
(2) 9^{2+} (4) 13^- (6) 15^+
                            (7) 1^-

THE REPRESENTATIVE ELEMENT GROUPS

5-40 Why is hydrogen sometimes classified with the alkali metals? How is hydrogen different from the alkalis? Why is hydrogen sometimes classified with the halogens?

5-41 Which of the following elements are halogens?
(a) O_2 (c) I_2 (e) Li   (g) Br_2
(b) P_4 (d) N_2 (f) H_2

5-42 Which of the following elements are alkaline earth metals?
(a) Sr   (c) B   (e) Na
(b) C   (d) Be   (f) K

5-43 Which of the following elements is the most chemically reactive?
(a) O_2 (c) Na   (e) Xe
(b) Cl_2 (d) F_2

5-44 Which of the following elements does not form chemical bonds?
(a) Au   (c) Ne   (e) Xe
(b) F_2 (d) O_2 (f) Li

5-45 Which of the following is an allotrope of oxygen?
(a) O   (c) H_2O (e) O_4
(b) O^{2-} (d) O_3

5-46 Which of the following is an allotrope of carbon?
(a) diamond   (c) bronze
(b) ozone   (d) carbon monoxide

5-47 Which of the following elements is the most reactive metal?
(a) Au   (c) Mg   (e) Pb
(b) Rb   (d) Sn

5-48 Which of the following metals reacts rapidly with water?
(a) Be   (c) Fe   (e) Pb
(b) Ag   (d) K

5-49 Which of the following electron configurations corresponds to a halogen?
(a) $[Ne]3s^2 3p^4$
(b) $[Kr]5s^2 4d^5$
(c) $[Xe]6s^2 5d^1 4f^7$
(d) $[Kr]5s^2 4d^{10} 5p^5$

5-50 Which of the following electron configurations corresponds to an alkaline earth?
(a) $[Xe]6s^2$ (c) $1s^2$
(b) $[Ar]4s^2 3d^{10}$ (d) $[Kr]5s^2 4d^2$

*5-51 It has been argued that the elements Lu and Lr should be placed in group 3 (IIIB) rather than with the lanthanides and actinides because of a better fit of their chemical properties to the other group 3 (IIIB) elements, Sc and Y. Do Lu and Lr have the proper electron configurations for group 3 (IIIB)? What is the one difference in the electron configurations of La and Lu?

WATER—SOFT, COOL, AND REFRESHING
The properties of water can be attributed to the fact that it is a
molecular compound—one of the two basic types of compounds.

THE NATURE OF
THE CHEMICAL
BOND

PURPOSE OF CHAPTER

In Chapter 6 we consider the basis for the formation of bonds between elements. We describe the two types of bonds and how to conveniently represent this information.

OBJECTIVES FOR CHAPTER

After completion of this chapter, you should be able to:

1   Write Lewis dot symbols for the atoms of any representative element. (6-2)

2   Determine the charge acquired by a specified atom of a representative element by application of the octet rule. (6-1, 6-3)

3   Write the formula of binary ionic compounds formed by representative elements. (6-4)

4   Demonstrate the octet rule by writing the Lewis structures of some simple binary molecular compounds. (6-1, 6-5)

5   Distinguish among single, double, and triple covalent bonds. (6-6)

6

6 Determine the number of valence electrons present in a polyatomic ion. (6-7)

7 Write Lewis structures for a variety of molecules and ions. (6-8)

8 Write equivalent resonance structures for specified molecules or ions. (6-9)

9 Give the general periodic trends in electronegativity. (6-10)

10 Determine the direction of a bond dipole from the electronegativities of the elements. (6-10)

11 Predict ionic bonding from differences in electronegativities. (6-10)

12 Predict polarity of molecules from a knowledge of the molecular geometry. (6-11)

Two of the compounds most essential to life and health are water (H_2O) and table salt (NaCl). Although both of these compounds are composed of only two elements, it is obvious from their physical state that they are fundamentally very different types of compounds. We mentioned in Chapter 3 that there are two fundamental types of compounds, molecular and ionic. We will find that water is a typical molecular compound and salt is a typical ionic compound. What is the reason for the difference? Why is a metal–nonmetal bond (as in NaCl) so different from a bond between two nonmetals (as in H_2O)? This difference is easily understandable and even predictable after the basis for bond formation is examined. These then are the questions that we are now prepared to address concerning the bonding between elements:

1 Why do certain elements combine to form ionic compounds and others combine to form molecular compounds?

2 What determines the formula of a compound? (That is, Why is the formula for water H_2O and not H_3O or HO_4?)

3 Why do certain elements, namely the noble gases, have little or no tendency to form any chemical bonds?

We will find that to a large extent, the electron configuration of the atoms of an element determines the type and number of bonds that the element forms. To be amply prepared for this discussion the following topics and sections should be fresh in your mind.

1 Nature of ionic and molecular compounds (Sections 3-5 and 3-6)
2 Electron configuration of the elements (Section 4-6)
3 Relationship of electron configuration to the periodic table (Section 4-6)
4 Formation of cations (Section 5-4) and formation of anions (Section 5-5)

6-1 BOND FORMATION AND THE REPRE- SENTATIVE ELEMENTS

In a way, people are much like the atoms of which they are made. Both people and atoms are generally social creatures, forming bonds with others because of a mutual need for a more stable arrangement. Unlike most people, however, there is a family or group of elements that have little or no tendency to join with others to form bonds. As mentioned previously, these are the noble gas elements, which all exist as individual gaseous atoms under normal conditions.

Elements that form chemical bonds do so through some alteration of their atoms' electron configuration. Since noble gases form few chemical bonds, it seems reasonable to assume that there is a special stability connected with their electron configuration as it is. As we will find in this chapter, *representative elements (and a few transition elements) form many bonds in such a manner that the electron configuration of their atoms is altered to become like that of a noble gas element.*

In Chapter 4 we showed that noble gas elements are characterized by filled outer s and p sublevels (i.e., ns^2np^6). Since this is a total of eight electrons (except for He), it is referred to as an **octet** of electrons. The types of bonds and formulas of many compounds of the representative elements can be understood and even predicted by reference to the octet rule. The **octet rule** *states that atoms of the representative elements form bonds so as to have access to eight outer electrons.* The outer s and p electrons in the atoms of an element are referred to as the **valence electrons.** Representative elements that border helium in the periodic table are influenced by a "duet" rule. These elements (H, Li, and Be) can alter their electron configuration to be like helium, which has only two electrons.

An octet of valence electrons can be obtained in three ways.

1 A metal may lose one to three electrons to form a cation with the electron configuration of the previous noble gas.
2 A nonmetal may gain one to three electrons to form an anion with the electron configuration of the next noble gas.
3 Atoms (usually two nonmetals) may share electrons with other atoms to obtain access to the number of electrons in the next noble gas.

We have already discussed the relative ease of formation of cations and anions in Chapter 5. The formation of cations and the formation of anions thus

TABLE 6-1   Lewis Dot Symbols

1 (IA)	2 (IIA)	13 (IIIA)	14 (IVA)	15 (VA)	16 (VIA)	17 (VIIA)	18 (O)
H·							He
Li·	Be·	B·	·C·	·N·	·O:	:F:	:Ne:
Na·	Mg·	Al·	·Si·	·P·	·S:	:Cl:	:Ar:
K·	Ca·	Ga·	·Ge·	·As·	·Se:	:Br:	:Kr:

complement each other to form ionic compounds. Case 3 produces molecular compounds in which atoms are held together by covalent bonds.

6-2 LEWIS DOT SYMBOLS

The bonding of the representative elements focuses mainly on the valence electrons, which are the electrons in the outer *s* and *p* sublevels. **Lewis dot symbols** * *of these elements represent the valence electrons as dots around the symbol of the element.* Since bonding of these elements involves access to eight electrons (four pairs), the electrons are represented as one or two dots on the four sides of the element's symbol. Although the valence electrons come from two different sublevels (*s* and *p*), only the total number of these electrons is important for discussions of bonding. Thus, it is convenient to place one electron on each side of the symbol first [groups 1 (IA) through 14 (IVA)] and then represent pairs of electrons [groups 15 (VA) through 18 (O)]. † In this manner, the Lewis dot symbols of the first four periods are shown in Table 6-1. Note that the last digit (or the Roman numeral) of the group number also represents the number of dots or outer electrons for a neutral atom of a representative element.

■

SEE PROBLEMS 6-1 THROUGH 6-4.

6-3 THE FORMATION OF IONS

In the previous chapter (Section 5-4), we studied the ionization energies of the elements. We found that the ionization energy was lowest for removal of the valence electrons of a metal. On the other hand, removal of electrons from nonmetals or from inner sublevels of a metal was found to require a comparatively large amount of energy and thus is unfavorable. Significant to us, however, is that loss of the valence electrons from a representative element metal

---

* *Named after the American chemist G. N. Lewis (1875–1946), who developed this theory of bonding.*
† *This procedure of representing the* s *electrons as unpaired for bonding purposes has a basis in fact. For example, although we learned in Chapter 4 that carbon has two paired electrons (2s²) and two unpaired electrons (in 2p orbitals), it forms bonds as if all four valence electrons were in equivalent or what are referred to as "hybridized" orbitals.*

leaves a cation with the electron configuratin of the previous noble gas. Since the resulting cation then has an octet of electrons (unless the noble gas is He), it is especially stable. For example, the loss of the one outer electron ($3s^1$) of sodium leaves a cation (Na^+) with the same octet of electrons as the previous noble gas, neon. This can be illustrated by the use of Lewis dot symbols:

$$\dot{Na} \longrightarrow Na^+ + e^-$$
$$[Ne]3s^1 \qquad [Ne]$$

All of the atoms of the group 1 (IA) *metals* can lose one electron to form a $+1$ ion with the electron configuration of the preceding noble gas.

The element magnesium can lose two electrons, to leave an ion with the electron configuration of Ne:

$$\dot{Mg}\cdot \longrightarrow Mg^{2+} + 2e^-$$
$$[Ne]3s^2 \qquad [Ne]$$

All other metals in group 2 (IIA) can lose two electrons to form a $+2$ ion with the configuration of the preceding noble gas.

Group 13 (IIIA) *metals* can lose three electrons to form a $+3$ ion*:

$$\dot{\underset{\cdot}{Al}}\cdot \longrightarrow Al^{3+} + 3e^-$$
$$[Ne]3s^2 3p^1 \qquad [Ne]$$

In this group, boron is *not* a metal and does not form a $+3$ ion. Its bonds are covalent in nature and are discussed later.

Four electrons would have to be removed from a group 14 (IVA) atom to attain a noble gas configuration. As mentioned in the previous chapter, each successive electron is harder to remove. Therefore, metal ions with a $+4$ charge are not generally considered to exist in ionic compounds because of the large amount of energy required for their formation. The elements Ge, Sn, and Pb bond either by electron sharing or by forming a $+2$ ion that does not follow the octet rule.

In summary, *metals of the representative elements may lose up to three electrons to form ions with a noble gas electron configuration.*

Positive ions do not form independently, as the lost electrons must go somewhere. The electrons are added to atoms that form negative ions. In Chapter 5 we learned that this process is more favorable for nonmetals.

Atoms of group 17 (VIIA) elements (including H) are all one electron short of a noble gas configuration. If they add one electron they will then have an octet of electrons or the same electron configuration of the next noble gas. The

---

* Ions such as Tl^{3+}, Ga^{3+}, and Zn^{2+} have a filled d sublevel in addition to a noble gas configuration. This is sometimes referred to as a pseudo-noble gas configuration. The filled d sublevel does not seem to affect the stability of these ions. In this text we do not distinguish between noble gas and pseudo-noble gas electron configurations. Transition metals also form positive ions, but for the most part these ions do not relate to a noble gas configuration. Some of these ions are discussed in Chapter 7.

resulting ion will then have a -1 charge. (Remember that hydrogen, an exception to the octet rule, needs only a total of two electrons to achieve the electron configuration of He.)

$$e^- + H\cdot \longrightarrow H\colon^-$$
$$1s^1 \qquad\qquad 1s^2 = [He]$$

$$e^- + \quad \colon\!\ddot{C}l\colon \longrightarrow \colon\!\ddot{C}l\colon^-$$
$$[Ne]3s^2 3p^5 \qquad [Ne]3s^2 3p^6 = [Ar]$$

Group 16 (VIA) nonmetals can attain a noble gas configuration by adding two electrons to form a -2 ion:

$$2e^- + \quad \cdot\ddot{O}\colon \longrightarrow \colon\!\ddot{O}\colon^{2-}$$
$$[He]2s^2 2p^4 \qquad [He]2s^2 2p^6 = [Ne]$$

In group 15 (VA), nitrogen and phosphorus can add three electrons to form -3 ions:

$$3e^- + \quad \cdot\ddot{N}\cdot \longrightarrow \colon\!\ddot{N}\colon^{3-}$$
$$[He]2s^2 2p^3 \qquad [He]2s^2 2p^6 = [Ne]$$

For the most part, group 14 (IVA) nonmetals bond by electron sharing instead of ion formation. Although there is some evidence for a C^{4-} ion, formation of such a highly charged ion is an energetically unfavorable process.

In summary, *nonmetals may add up to three electrons to form ions with a noble gas electron configuration.*

■
SEE PROBLEMS
6-5 THROUGH
6-16.

6-4

**FORMULAS OF
BINARY IONIC
COMPOUNDS**

At this point, we have established that metals can attain an octet of electrons by formation of cations, and nonmetals can attain an octet of electrons by formation of anions. Since one type of element loses electrons comparatively easily and the other type has a tendency to gain electrons, we have the basis for compatible combinations between atoms from the two basic classes of elements. This is obvious when a small piece of metallic sodium is placed in a bottle containing chlorine gas. The sodium ignites and a coating of white sodium chloride is formed. (See Figure 6-1.) The chemical combination of these two elements produces a **binary** *(two-element)* compound. The fundamental process in this chemical reaction is the transfer of an electron from a metallic sodium atom to a nonmetallic chlorine atom:

$$Na\overset{\frown}{+}\cdot\ddot{C}l\colon \longrightarrow Na^+\colon\!\ddot{C}l\colon^-$$

$$\text{Formula} = NaCl\,*$$

---

* *The metal is always written first.*

Figure 6-1 REACTION OF SODIUM WITH CHLORINE.

The transfer of the one electron results in two ions with equal but opposite charges. Both ions have noble gas electron configurations. When lithium combines with oxygen, however, two lithium atoms are required to satisfy the need of oxygen for two electrons. The anion formed has a -2 charge from the added electrons. Note that the resulting compound is electrically neutral, as all compounds must be. The two $+1$ ions balance the -2 charge on the oxygen [i.e., $2(+1) - 2 = 0$]. The chemical formula for the compound formed is Li_2O.

$$Li \cdot \qquad \qquad $$
$$+ \ \cdot \ddot{O} : \longrightarrow 2(Li^+) : \ddot{\ddot{O}} :^{2-}$$
$$Li \cdot \qquad \qquad Formula = Li_2O$$

When calcium combines with bromine, the opposite situation exists. Two Br atoms are needed to take up the two electrons from one Ca.

$$\qquad \qquad \cdot \ddot{Br} :$$
$$Ca \cdot \qquad \longrightarrow \ Ca^{2+} 2 \left(: \ddot{Br} :^- \right)$$
$$\qquad \qquad \cdot \ddot{Br} : \qquad Formula = CaBr_2$$

When aluminum combines with oxygen, it is somewhat more complex to follow the transfer of the three electrons of aluminum onto the oxygens, which need two each. In this case, two Al's and three O's are needed to achieve a charge balance of zero and to satisfy the octet rule.

$$\text{Al}\quad\ddot{\text{O}}:$$
$$\ddot{\text{O}}: \longrightarrow 2(\text{Al}^{3+})3(:\ddot{\text{O}}:^{2-})$$
$$\text{Al}\quad\ddot{\text{O}}:$$
$$\text{Formula} = \text{Al}_2\text{O}_3$$

To predict the formulas of binary ionic compounds, only the group numbers of the two elements are needed to give the charges on the respective ions. First, write the appropriate ions, including the charges, adjacent to each other (IA, 1+; IIA, 2+; IIIA, 3+; VA, 3−; VIA, 2−; VIIA, 1−). The numerical value of the charge becomes the subscript of the other element as shown by the arrows. Note that this method predicts the formula of the compound containing calcium and oxygen to be Ca_2O_2. Formulas of ionic compounds, however, should be expressed with the simplest whole numbers for the subscripts. The formula is expressed as CaO.

$$\text{Na}\overset{1+}{\longleftrightarrow}\text{N}\overset{3-}{} = \text{Na}_3\text{N}$$

$$\text{Ga}\overset{3+}{\longleftrightarrow}\text{S}\overset{2-}{} = \text{Ga}_2\text{S}_3$$

$$\text{Ca}\overset{2+}{\longleftrightarrow}\text{O}\overset{2-}{} = \text{Ca}_2\text{O}_2 = \text{CaO} \qquad \text{(write the simplest formula)}$$

EXAMPLE 6-1

What is the formula of the ionic compound formed between (a) aluminum and fluorine and (b) barium and sulfur?

SOLUTION

(a) Aluminum is in group 3 (IIIA) and fluorine is in group 17 (VIIA), and they have the dot symbols

$$\cdot\dot{\text{Al}}\cdot \qquad \cdot\ddot{\text{F}}:$$

To have a noble gas configuration (an octet) the Al, a metal, must lose all three outer electrons to form a +3 ion. Three fluorine atoms are needed to add one electron each to form three −1 ions. Note that each fluorine can add only one electron, which gives the F^- ion an octet. The compound formed is

$$\text{Al}^{3+}3\left(:\ddot{\text{F}}:^-\right) = \underline{\underline{\text{AlF}_3}}$$

We could also determine the formula from the charges of the respective ions formed. Al becomes Al^{3+} and F becomes F^{1-}.

$$\text{Al}\overset{3+}{\longleftrightarrow}\text{F}\overset{1-}{} = \underline{\underline{\text{AlF}_3}}$$

(b) Barium is in group 2 (IIA) and sulfur is in group 16 (VIA), and they have the dot symbols

$$\text{Ba·} \quad \ddot{\underset{..}{S}}\text{:}$$

One Ba atom gives up two electrons and one S atom takes up two electrons, forming the compound

$$\text{Ba}^{2+}\text{:}\ddot{\underset{..}{S}}\text{:}^{2-} = \underline{\text{BaS}}$$

We could also determine the formula from the charges of the respective ions formed. Ba becomes Ba^{2+} and S becomes S^{2-}

$$\text{Ba}\overset{\textcircled{2}+}{\underset{\longleftrightarrow}{}}\text{S}\overset{\textcircled{2}}{}{}^{-} = \text{Ba}_2\text{S}_2 = \underline{\underline{\text{BaS}}}$$

Besides single atoms, two or more atoms can also bond together by covalent bonds with the whole unit behaving as an ion (e.g., CO_3^{2-}). These are called **polyatomic ions** and are discussed later in this chapter.

As mentioned in Chapter 3, ionic compounds do not consist of discrete molecular units. That is, in a crystal of a compound such as NaCl one Na^+ ion is not attached to any one Cl^- ion. As shown in Figure 6-2, each Na^+ ion is surrounded by six Cl^- ions and each Cl^- ion is surrounded by six Na^+ ions in a lattice. (**A lattice** *is a particular three-dimensional array of particles.*) To understand the relative sizes of the ions in Figure 6-2 recall from Chapter 5 that cations are smaller than their parent atoms and that anions are larger than their parent atoms. Thus, anions are considerably larger than cations. *The electrostatic forces of attraction between adjacent oppositely charged ions are known as* **ionic bonds** and hold the crystal together. A common physical characteristic of ionic compounds is that they are all hard, high-melting-point solids. Thus, like NaCl, all ionic compounds are found in the solid state at room temperature.

■
SEE PROBLEMS
6-17 THROUGH
6-21.

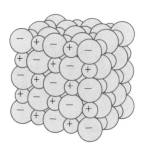 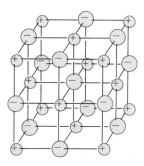

Figure 6-2 AN IONIC SOLID. Each cation (small spheres) is surrounded by six anions. Each anion (large spheres) is surrounded by six cations.

6-5 THE COVALENT BOND

The formation of ions and ionic bonds adequately explains the bonding in table salt (NaCl) and many other similar compounds. The atoms in molecular compounds (e.g., H_2O), however, are not held together in this manner. In most of these cases, a metal is not involved so there are no atoms present that easily give up electrons to form ions. The only alternative left for atoms of these elements is to attain a noble gas configuration or octet by the sharing of electrons. *A shared pair of electrons between two atoms is known as a **covalent bond**.* Molecular compounds, which contain only covalent bonds, tend to be gases, liquids, or low-melting-point solids under laboratory conditions.

There is a simple analogy to the formation of an octet of electrons by sharing. For two people to each have access to $8, ideally both should have separate savings accounts containing $8. However, if they only have $7 each there is still a way for both to claim access to $8. They would have to open separate accounts containing $6 and then contribute $1 each (for a total of $2) to a joint account. Thus, each person would claim access to a total of $8. (See Figure 6-3.)

Likewise, two fluorine atoms each have seven valence electrons. If each atom were to hold six electrons to itself and share one each (for a total of two electrons) both fluorine atoms would have access to eight electrons, thus satisfying the octet rule:

$$:\ddot{F}\cdot \quad \curvearrowright \quad \cdot\ddot{F}: \longrightarrow \qquad :\ddot{F}{\textstyle\colon}\ddot{F}:$$

Shared pair of electrons
(one from each F)

We are now ready to extend the concept of Lewis dot representations to covalent molecules. **A Lewis structure** *for a molecule shows the order and arrangement of atoms in a molecule (the structural formula) as well as all of the valence electrons for the atoms involved.* There are several variations of how Lewis structures are represented for molecules. A pair of electrons is sometimes shown as a pair of dots (:) or as a dash (—). In this text, we use a pair of dots to represent unshared pairs (also called *lone pairs*) of electrons on an atom and a dash to represent a pair of electrons shared between atoms. In this way

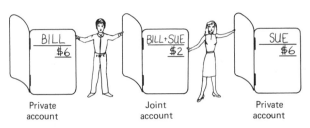

Figure 6-3   INCREASE YOUR MONEY BY SHARING. The money in a joint account is analogous to electrons in a covalent bond.

the two different environments of electrons (shared and unshared) can be distinguished.

The Lewis structure of F_2* is illustrated as follows.

Total of 14 outer electrons (7 from each F)

:F — F: — Three lone pairs on each F

Two shared electrons in a covalent bond

Similarly, all of the halogens exist as diatomic molecules, like F_2. Hydrogen, which forms the simplest of all molecules, also exists as a diatomic gas with one covalent bond between atoms:

$$H—H$$

Note again that hydrogen does not follow the octet rule. Its outer sublevel holds only two electrons, which is the electron configuration of He.

How does the covalent bond hold the two hydrogen atoms together? Taken alone, the two nuclei (represented as H_2^{2+} in Figure 6-4) would repel each other because of their like charges ($\rightarrow$ represents a repulsive force). The presence of one electron (with a negative charge) between the two hydrogen nuclei (represented as H_2^+) changes the situation. The nuclei are now held together (at a certain distance) by the two nucleus–electron attractions. These two attractions are stronger than the nucleus–nucleus repulsion. ($\overline{\text{ooo}}$ represents an attractive force in Figure 6-4). If there are two electrons between the two hydrogen atoms, as in the H_2 molecule, the nuclei are held together even tighter by the *four* nucleus–electron attractions even though the second electron adds an electron–electron repulsion.

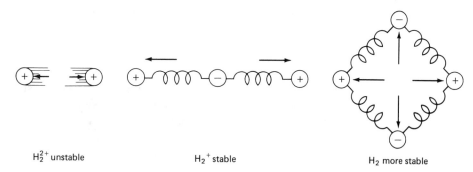

H_2^{2+} unstable H_2^+ stable H_2 more stable

Figure 6-4 COVALENT BOND. The presence of electrons between the two nuclei holds the atoms together.

---

* An F atom has one unpaired electron in a 2p orbital. Formation of a covalent bond pairs the electrons in the two F atoms so that the F_2 molecule has all electrons paired. Although most atoms of the representative elements have unpaired electrons, most molecules or ions formed from these elements do not have unpaired electrons.

We are now ready to consider the Lewis structures of the molecules of some simple binary compounds. This endeavor along with observance of the octet rule will have an immediate benefit. First, we will be able to justify the formulas of some simple binary molecular compounds just as we were able to understand the formulas of simple binary ionic compounds. Second, we can even make a few predictions about expected formulas.

Perhaps the simplest molecule of a compound is formed between hydrogen and any halogen [group 17 (VIIA)]. For example, consider the molecule HF. A shared pair of electrons (one from each atom) gives both atoms access to the same number of electrons as a noble gas.

$$H\cdot \longrightarrow \cdot \ddot{F}\colon \longrightarrow H\!-\!\ddot{F}\colon$$

Now let us consider a compound formed between hydrogen and a chalcogen [group 16 (VIA)] such as oxygen. Since the oxygen needs to gain access to two more electrons it must form a bond with two hydrogens to satisfy the octet rule. The simplest compound between hydrogen and oxygen has the formula H_2O:

$$
\begin{array}{c}
H\odot \\
\searrow \\
\cdot\ddot{O}\colon \longrightarrow H\!-\!\ddot{O}\colon \\
\nearrow | \\
H\odot H
\end{array}
$$

Likewise, the octet rule allows us to justify the formulas of the simplest compounds that hydrogen forms with nitrogen and carbon. (See also some hydrogen compounds of third-period elements in Figure 6.5.)

$$
\begin{array}{ccc}
H\cdot & & H \\
\searrow & & | \\
H\cdot \longrightarrow \cdot\ddot{N}\colon \longrightarrow H\!-\!\ddot{N}\colon & & \\
\nearrow & & | \\
H\cdot & & H
\end{array}
$$

Ammonia (NH_3)   Methane (CH_4)

Now let us make a prediction. What is the simplest compound formed between phosphorus and fluorine? The dot symbol for phosphorus indicates that it needs three more electrons to attain an octet. The dot symbol for fluorine indicates that it needs one more electron. The compound requires three fluorines, each providing one electron for the phosphorus:

$$
\colon\!\ddot{F}\cdot \longrightarrow \cdot\ddot{P}\cdot \longleftarrow \cdot\ddot{F}\colon \longrightarrow \colon\!\ddot{F}\!-\!\ddot{P}\!-\!\ddot{F}\colon
$$

$$
\colon\!\ddot{F}\colon \qquad\qquad\qquad \colon\!\ddot{F}\colon
$$

Simplest formula $=$ PF_3*

---

* In binary molecular compounds, the element closest to the metal–nonmetal border is written first. NH_3 and CH_4 are exceptions to this rule.

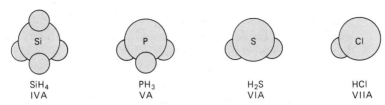

Figure 6-5 FORMULAS OF HYDROGEN COMPOUNDS. The formulas of some simple hydrogen compounds can be predicted from the octet rule.

■
*SEE PROBLEMS
6-22 THROUGH
6-24.*

Eventually, some guidelines will be presented that will make the writing of Lewis structures quite systematic. But first, we will consider the multiple covalent bond and the bonding in polyatomic ions.

6-6 THE MULTIPLE COVALENT BOND

In Figure 6-3 an analogy to the covalent bond illustrated how two people could each claim access to \$8 with only a total of \$14 between them. What if they only had \$12 between them? In this case they could still claim \$8 each if they shared \$4 in the joint account and kept \$4 each in private accounts.

Two atoms may also share four electrons (two pairs) to achieve an octet of valence electrons. *The sharing of two pairs of electrons (between the same two atoms) is known as a* **double bond.** An example of a molecule containing a double bond is illustrated by the following Lewis structure.

$$\overset{\cdot\cdot}{\underset{\ddot{O}\diagup\quad\diagdown\ddot{O}}{S}}$$

Note that all atoms in the molecule have access to eight electrons.*

The N_2 molecule has a total of 10 outer electrons available to achieve octets. In this molecule, each N holds only one pair of electrons to itself and shares six electrons (three from each N). *When three pairs of electrons are shared between atoms, the bond is known as a* **triple bond.** The Lewis structure for N_2 is

$$:N\equiv N:$$

---

* *It would seem that O_2, with 12 valence electrons, would make an excellent example of the simplest molecule with a double bond:*

$$\overset{\cdot\cdot}{O}=\overset{\cdot\cdot}{O}$$

This Lewis structure implies that all of the electrons in O_2 are paired. However, experiments show that O_2 has two unpaired electrons. Although writing Lewis structures works very well in explaining the bonding in most simple molecules, you should keep in mind that a Lewis structure is simply the representation of a theory. For O_2 the theory doesn't work perfectly. Thus, O_2 is usually not represented by a Lewis structure. Other theories on bonding work well in explaining the bonding in this molecule but are not discussed here.

Only C, N, and O among the representative elements are ever involved in triple bonds. Triple and quadruple bonds (eight shared electrons) also exist between certain transition metals.

6-7 POLYATOMIC IONS

As we mentioned earlier, *two or more atoms can bond together by covalent bonds to form what is known as a* **polyatomic ion.** The charge on the ion is not located on any particular atom in the ion but belongs to the entire structure. A simple example of a polyatomic ion is the hypochlorite ion, ClO^-. The existence of the -1 charge tells us that the ion contains one electron *in addition* to those from neutral Cl and O atoms. The total number of outer electrons involved in the ClO^- ion is calculated as follows:

$$\begin{aligned} \text{From a neutral Cl} &= 7 \\ \text{From a neutral O} &= 6 \\ \text{Additional electron indicated by charge} &= \underline{1} \\ \text{Total number of electrons} &= 14 \end{aligned}$$

Two atoms bonded together with 14 electrons have a Lewis structure like F_2 with a single covalent bond:

$$\left[:\overset{..}{\underset{..}{Cl}}-\overset{..}{\underset{..}{O}}: \right]^-$$

■
SEE PROBLEMS
6-25 AND 6-26.
Note that the brackets indicate that the total ion has a -1 charge. The extra electron has not been specifically identified, since electrons are all identical and the charge belongs to the ion as a whole.

6-8 WRITING LEWIS STRUCTURES

Perhaps one of the most important fundamental skills to be developed in basic chemistry is that of writing Lewis structures. As we have seen, this ability not only allows us to rationalize certain formulas but also to *predict* formulas of compounds between particular elements. In more advanced courses in chemistry, the Lewis structure provides significant information as to what type of chemical behavior a molecule may display. The Lewis structure (including unshared electrons) can also be used to predict the geometry of the molecule. From geometry even more information becomes available with respect to the types of chemical reactions a compound undergoes. To the skilled chemist a Lewis representation of a molecule is like a detailed personal resume.

The process of writing correct Lewis structures is quite straightforward when certain guidelines or rules are systematically applied. These rules, of course, require considerable practice in their application. The rules for writing the Lewis structure, given a formula of a compound, are as follows:

1   Check to see whether any ions are involved in the compounds. Write any ions present.
   (a) Metal–nonmetal binary compounds are mostly ionic.
   (b) If group 1 (IA) or group 2 (IIA) metals (except Be) are part of the formula, ions are present. For example, KClO is K^+ClO^- because K is a group 1 *(IA) element*

and forms only a +1 ion. If K is +1, the ClO must be −1 to have a neutral compound. Likewise, $Ba(NO_3)_2$ contains ions because Ba is a group 2 *(IIA)* *element and forms only a +2 ion.* To maintain neutrality, each NO_3 ion must have a −1 charge. The ions are represented as Ba^{2+} $2(NO_3^-)$.

(c) Compounds composed only of nonmetals contain only covalent bonds.

2 For a molecule, add all of the outer (valence) electrons of the neutral atoms. For an ion, add (if negative) or subtract (if positive) the number of electrons indicated by the charge.

3 Write the symbols of the atoms of the molecule or ion in a skeletal arrangement.

(a) A hydrogen atom can form only one covalent bond and therefore bonds to only one atom at a time. They are situated on the periphery of the molecule.

(b) The atoms in molecules and polyatomic ions tend to be arranged in a symmetrical pattern around a central atom. The central atom is generally a nonmetal other than oxygen or hydrogen. Oxygens do not usually bond to each other. Thus, SO_3 has an S surrounded by three O's,

O

S

O O

rather than such structures as

In most cases, the first atom in a formula is the central atom, and the other atoms are bound to it.

4 Put a dash representing a shared pair of electrons between adjacent atoms that have covalent bonds (not between ions). Subtract the electrons used for this (two for each bond) from the total calculated in step 2.

5 Distribute the remaining electrons among the atoms so that no atom has more than eight electrons.

6 Check all atoms for an octet (except H). If an atom has access to fewer than eight electrons, put an electron pair from an adjacent atom into a double bond. Each double bond increases by two the number of electrons available to the atom needing electrons. Remember that you cannot satisfy an octet for an atom by adding any electrons at this point.

An alternate method combines steps 4, 5, and 6. In this method, a count of electrons is used to determine the number of multiple bonds present and then electrons and bonds are added to the skeletal structure accordingly. This has been conveniently summarized as the "6N + 2 rule,"* where N stands for the number of atoms other than hydrogen in the formula. If the number of valence electrons in the formula equals 6N + 2, then only single bonds are present. If

---

*Zandler, M. E., and Talaty, E. R., J. Chem. Educ. **61**, 124 (1984).

the number of valence electrons is two less than $6N + 2$, then one double bond is present. (It could also mean that a ring structure is present, but these are discussed only briefly, in Chapter 16.) If the number of valence electrons is four less than $6N + 2$, then one triple bond or two double bonds are present. Consider the following molecules:

(a) PF_3 $N = 4$ $6(4) + 2 = 26$
Valence electrons: $5[P] + (3 \times 7)[F] = 26$
Therefore, only single bonds are present.

(b) CO_3^{2-} $N = 4$ $6(4) + 2 = 26$
Valence electrons: $4[C] + (3 \times 6)[O] + 2[charge] = 24$ $26 - 24 = 2$
Therefore, one double bond is present in this ion.

(c) C_6H_{10} $N = 6$ $6(6) + 2 = 38$
Valence electrons: $(6 \times 4)[C] + (10 \times 1)[H] = 34$ $38 - 34 = 4$
Therefore, two double bonds or one triple bond is present in this molecule.

In this text we use the "$6N + 2$ rule" as a check to confirm the Lewis structure determined by applying steps 1 through 6.

EXAMPLE 6-2

Write the Lewis structure for NCl_3.

1   This is a binary compound between two nonmetals. Therefore, it is not ionic.

2   The total number of electrons available for bonding is

$$
\begin{array}{ll}
N & 1 \times 5 = 5 \\
Cl & 3 \times 7 = \underline{21} \\
& \text{Total} = 26
\end{array}
$$

3   The skeletal arrangement is

$$Cl \quad N \quad Cl$$
$$Cl$$

4   Use six electrons to form bonds:

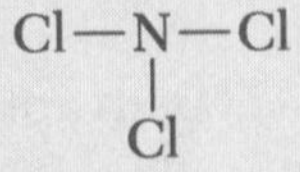

$$Cl{-}N{-}Cl$$
$$|$$
$$Cl$$

5   Distribute the remaining 20 electrons ($26 - 6 = 20$):

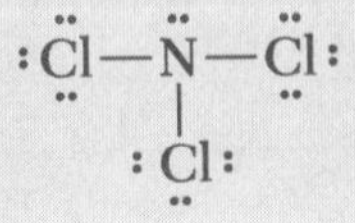

$$:\!\ddot{C}l{-}\ddot{N}{-}\ddot{C}l\!:$$
$$|$$
$$:\!\ddot{C}l\!:$$

6 Check to make sure that all atoms satisfy the octet rule:

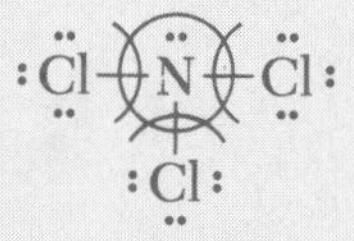

$$:\ddot{C}l\!-\!N\!-\!\ddot{C}l:$$
$$:\ddot{C}l:$$

The $6N + 2$ rule confirms this structure.

For NCl_3 $N = 4$ $6(4) + 2 = 26$

From step 2 there are 26 valence electrons, so there are only single bonds.

EXAMPLE 6-3

Write the Lewis structure for K_2S.

This is a binary compound between a group 1 (IA) metal and a group 16 (VIA) nonmetal. It is therefore ionic. Since K is a $+1$ ion, the S is a -2 ion. The Lewis structure is simply that of an ionic compound (no dashes between ions).

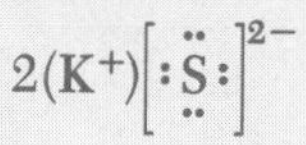

$$2(K^+)\left[:\ddot{S}:\right]^{2-}$$

EXAMPLE 6-4

Write the Lewis structure for N_2O_4.

1 This is not ionic.

2 The total number of outer electrons available is

$$\begin{array}{ll} N & 2 \times 5 = 10 \\ O & 4 \times 6 = \underline{24} \\ & \text{Total} = 34 \end{array}$$

3 The symmetrical skeletal structure is

$$\begin{array}{cc} O & O \\ N & N \\ O & O \end{array}$$

4 Add 10 electrons for the five bonds:

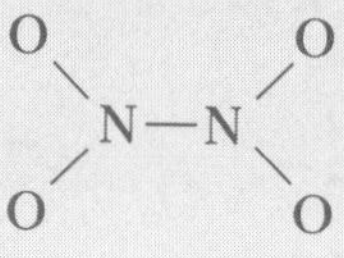

$$\begin{array}{ccc} O & & O \\ & \diagdown \quad \diagup & \\ & N\!-\!N & \\ & \diagup \quad \diagdown & \\ O & & O \end{array}$$

5 Add the remaining 24 electrons ($34 - 10 = 24$):

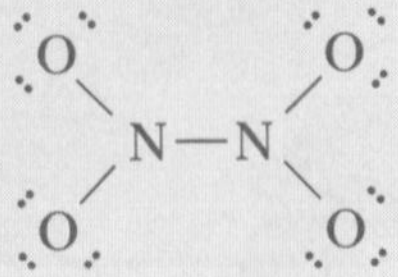

6 Check for octets. The oxygens are all right, but both N's need access to one more pair of electrons each. Therefore, make one double bond for each N using a lone pair of electrons from an adjacent O.

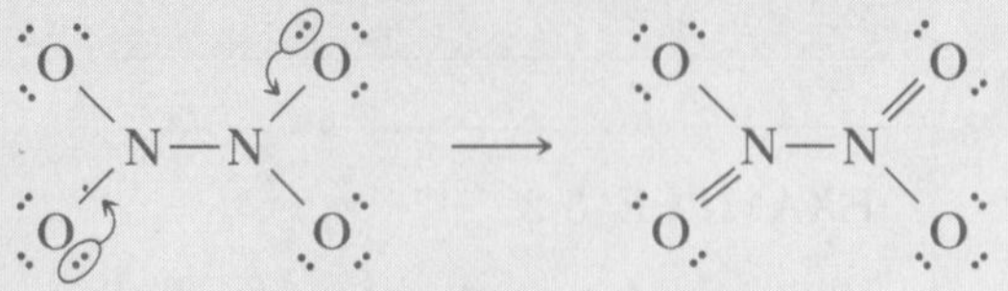

The $6N + 2$ rule is consistent with this structure.

For N_2O_4 $N = 6$ $6(6) + 2 = 38$

Valence electrons $= 34$ $38 - 34 = 4$ (two double bonds)

EXAMPLE 6-5

Write the Lewis structure for $CaCO_3$.

1 This is an ionic compound composed of Ca^{2+} and CO_3^{2-} ions [since you know that Ca is in group 2 (IIA), it must have a $+2$ charge; therefore the polyatomic anion must be -2]. A Lewis structure can be written for CO_3^{2-}.

2 For the CO_3^{2-} ion the total number of outer electrons available is:

$$
\begin{array}{ll}
\text{C} & 1 \times 4 = 4 \\
\text{O} & 3 \times 6 = 18 \\
\text{From charge} = \underline{2} \\
\text{Total} = 24
\end{array}
$$

3, 4 The skeletal structure with bonds is

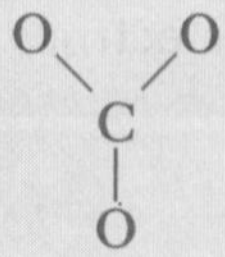

5 Add the remaining 18 electrons ($24 - 6 = 18$):

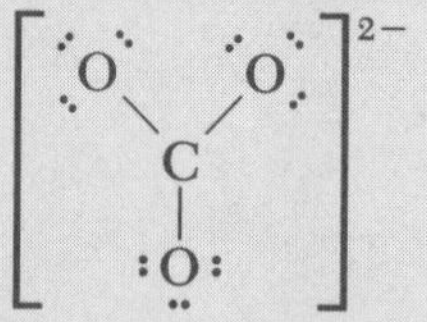

$$\left[\overset{..}{\underset{..}{:}O:} \quad \overset{..}{\underset{..}{:}O:} \right]^{2-}$$
$$C$$
$$:\overset{}{\underset{..}{O}}:$$

6 The C needs two more electrons, so one double bond is added using one lone pair from one oxygen:

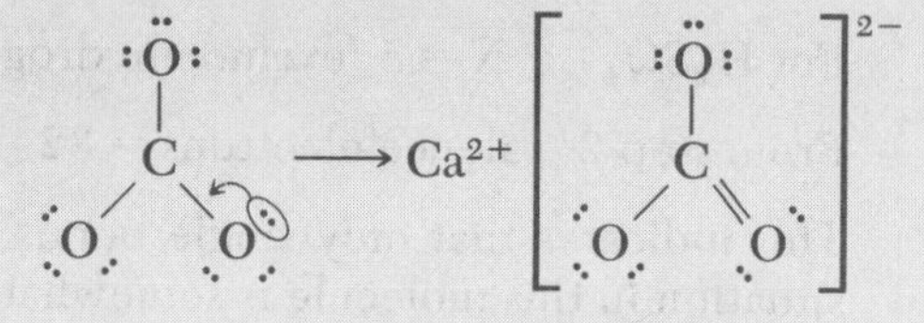

$$:\overset{..}{\underset{}{O}}:$$
$$C \longrightarrow Ca^{2+} \left[:\overset{..}{\underset{}{O}}: \atop C \right]^{2-}$$

EXAMPLE 6-6

Write the Lewis structure for H_2SO_4.

1 All three atoms are nonmetals, which means that all bonds are covalent.

2 The total number of outer electrons available is

$$
\begin{array}{lrl}
\text{H} & 2 \times 1 = & 2 \\
\text{S} & 1 \times 6 = & 6 \\
\text{O} & 4 \times 6 = & \underline{24} \\
& \text{Total} = & 32
\end{array}
$$

3 In most molecules containing H and O the H is bound to an O and the O to some other atom, which in this case is S. The skeletal structure is

$$
\begin{array}{ccccc}
& & \text{O} & & \\
\text{H} & \text{O} & \text{S} & \text{O} & \text{H} \\
& & \text{O} & &
\end{array}
$$

4 Add 12 electrons for the six bonds:

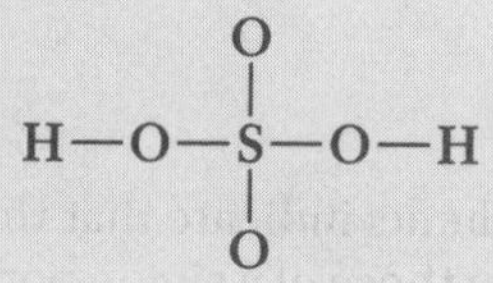

$$
\begin{array}{c}
\text{O} \\
| \\
\text{H—O—S—O—H} \\
| \\
\text{O}
\end{array}
$$

5 Add the remaining 20 electrons ($32 - 12 = 20$):

$$\overset{\displaystyle :\overset{\cdot\cdot}{\underset{}{O}}:}{\underset{\displaystyle :\overset{}{\underset{\cdot\cdot}{O}}:}{H-\overset{\cdot\cdot}{\underset{\cdot\cdot}{O}}-S-\overset{\cdot\cdot}{\underset{\cdot\cdot}{O}}-H}}$$

6 All octets are satisfied.

The $6N + 2$ rule is consistent with this structure.

For H_2SO_4 $N = 5$ (exclude hydrogens) $6(5) + 2 = 32$

From step 2, valence electrons $= 32$

This indicates that only single bonds are present, although the actual situation in this molecule is somewhat more complex.

EXAMPLE 6-7

Write the Lewis structure for NO.

    1 It is not ionic.

    2 There are $5 + 6 = 11$ outer electrons.

3, 4, 5 With a bond and the other electrons the structure is

$$\cdot N - \overset{\cdot\cdot}{\underset{\cdot\cdot}{O}} :$$

    6 As written, the N has access to five electrons. By making one double bond, the nitrogen can be brought up to seven electrons. However, there is no way that an odd number of electrons can be arranged so that all atoms have access to the even number of eight electrons. In this case one of the atoms does not have an octet:

$$\cdot N = \overset{\cdot\cdot}{\underset{\cdot\cdot}{O}} :$$

SEE PROBLEMS
6-27 THROUGH
6-38.

**6-9
RESONANCE
HYBRIDS**

The Lewis structure for SO_2 is

$$\underset{\overset{\cdot\cdot}{\underset{\cdot\cdot}{O}} \quad \quad \overset{\cdot\cdot}{\underset{\cdot\cdot}{O}}}{\overset{\overset{\cdot\cdot}{S}}{\diagup \quad \diagdown}}$$

Experiments indicate that the length and strength of a double bond are different from those of a single bond between the same two elements. However, the structure above by itself implies something that isn't true according to available evidence. It implies that one S—O bond is different from the other. In fact, both S—O bonds are known to be identical and have length and strength between those of a single and a double bond. Since there is no way to illustrate this fact with one Lewis structure, it must be done with two. The two structures below (connected by a double-headed arrow) are known as **resonance structures.** The actual structure is a **hybrid** of the resonance structures.

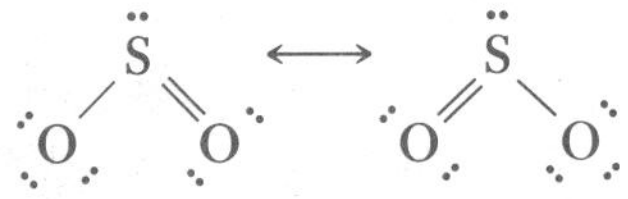

Resonance structures exist for compounds when equally correct Lewis structures can be written without changing the basic skeletal geometry or changing the position of any atoms.

EXAMPLE 6-8

Write equivalent resonance structures for the CO_3^{2-} ion.

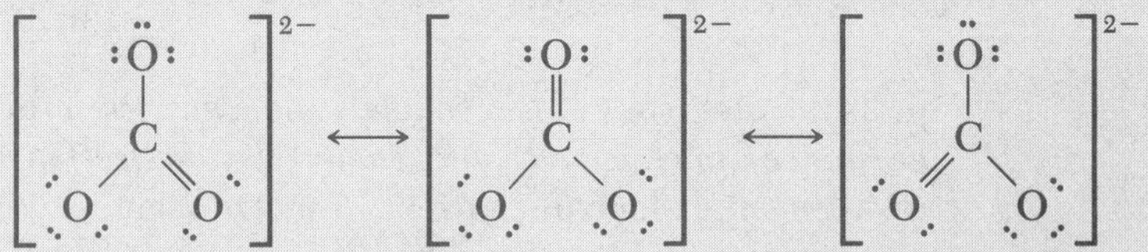

The three resonance structures indicate that the C—O bond has one-third double-bond properties and two-thirds single-bond properties.

The word "resonance" is an unfortunate choice. It implies that the S—O bond is changing (resonating) between a double and a single bond and that what is seen is actually just an average of the two extremes. This is not the case; the S—O bond actually *exists* full-time as part double and part single. In this sense a resonance hybrid is like the hybrid tomato that you may grow in the garden or buy in the grocery store. The tomato has a sweet taste and large size but not because it changes rapidly back and forth between a large tomato and a small, sweet tomato. It exists full-time with the properties of its two parents.

SEE PROBLEMS 6-39 THROUGH 6-43.

6-10 ELECTRO-NEGATIVITY AND POLARITY

At this point, it would seem that only two situations are possible when the atoms of two elements combine: (1) a complete transfer of electrons to form an ionic bond and (2) an equal sharing of electrons to form a covalent bond. The situation is not so simple. In fact, the electrons in a covalent bond are rarely shared equally between atoms of different elements. To some extent one atom

Decreases ↓   Increases →

1	2	3	4	5	6	7	8	9	10	11	12	13	14	15	16	17
1 H 2.1																
3 Li 1.0	4 Be 1.5											5 B 2.0	6 C 2.5	7 N 3.0	8 O 3.5	9 F 4.0
11 Na 0.9	12 Mg 1.2											13 Al 1.5	14 Si 1.8	15 P 2.1	16 S 2.5	17 Cl 3.0
19 K 0.8	20 Ca 1.0	21 Sc 1.3	22 Ti 1.5	23 V 1.6	24 Cr 1.6	25 Mn 1.5	26 Fe 1.8	27 Co 1.8	28 Ni 1.8	29 Cu 1.9	30 Zn 1.6	31 Ga 1.6	32 Ge 1.8	33 As 2.0	34 Se 2.4	35 Br 2.8
37 Rb 0.8	38 Sr 1.0	39 Y 1.2	40 Zr 1.4	41 Nb 1.6	42 Mo 1.8	43 Tc 1.5	44 Ru 2.2	45 Rh 2.2	46 Pd 2.2	47 Ag 2.4	48 Cd 1.7	49 In 1.7	50 Sn 1.8	51 Sb 1.9	52 Te 2.1	53 I 2.5
55 Cs 0.7	56 Ba 0.9	57 La 1.1	72 Hf 1.3	73 Ta 1.5	74 W 1.7	75 Re 1.9	76 Os 2.2	77 Ir 2.2	78 Pt 2.2	79 Au 2.4	80 Hg 1.9	81 Ti 1.8	82 Pb 1.8	83 Bi 1.9	84 Po 2.0	85 At 2.2
87 Fr 0.7	88 Ra 0.9	89 Ac 1.1														

Figure 6-6 ELECTRONEGATIVITY.

pulls the electrons closer to itself and away from the other atom. *The ability of an atom of an element to attract electrons to itself in a covalent bond is known as the element's* **electronegativity.** Electronegativity is a periodic property that increases up and to the right in the periodic table. Thus, a comparatively large value for electronegativity is a property characteristic of a nonmetal. The electronegativity values in Figure 6-6 were calculated by Linus Pauling (winner of two Nobel Prizes and of current vitamin C fame). Although more refined values are now available, the actual values are not as important as how the electronegativity of one element compares with that of another.

Electronegativity is significant because it indicates how the electrons are distributed in a covalent bond. If one atom is more electronegative than another it attracts a greater share of electrons in the bond and thus acquires a partial negative charge (symbolized by δ^-). The other atom is left with an equal but opposite positive charge (symbolized by δ^+). The distribution of charge in a molecule determines to a large extent both the physical and the chemical properties of the compound. For example, in Chapter 11, we will see how the charge distribution in the water molecule gives it so many of the familiar properties that we take for granted.

A covalent bond that has a partial separation of charge because of unequal sharing of electrons is known as a **polar covalent bond.** A polar bond has a negative end and a positive end and is said to contain a **dipole** (two poles).*

---

* The earth is polar (contains a dipole), with a north and south magnetic pole.

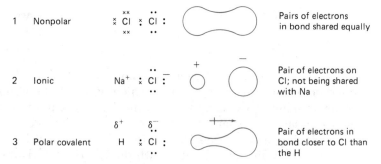

Figure 6-7   NONPOLAR, IONIC, AND POLAR COVALENT BONDS.

The dipole of a bond is represented by an arrow pointing from the positive to the negative end ($\longmapsto$).

$$\overset{\delta^+ \quad \delta^-}{X-Y}\quad\text{—Representation of polarity}$$

$$\longmapsto\quad\text{—Representation of bond dipole}$$

Note from Figure 6-6 that fluorine is the most electronegative element. This means that fluorine has the strongest attraction for electrons in a bond and will always be the negative end of a dipole. Oxygen is the second most electronegative element and will be the negative end of a dipole in all bonds except those with fluorine.

The greater the difference in electronegativity between two atoms, the greater the polarity (the bond has a larger dipole). As a matter of fact, if the difference between two elements is around 1.9 to 2.0 or greater, the bond is generally ionic, meaning that one atom has gained complete control of the electron pair in the bond to acquire a full negative charge.†

In summary, when two atoms compete for a pair of electrons in a bond, three things can happen:

1 Both atoms share the electrons equally.

2 One atom takes the pair of electrons completely to itself, forming an ionic bond.

3 The two atoms share the electrons but not equally.

The molecule Cl_2 illustrates case 1, as shown in Figure 6-7. Since both atoms obviously have the same electronegativity, the electron pair is shared equally between the two atoms. Each Cl has access to its three lone pairs of electrons plus exactly one-half of the electron pair in the bond.

---

† *This is not a flawless way for predicting whether a bond is ionic or covalent. However, it does point out that in many cases the borderline between ionic and covalent character is not necessarily sharp.*

Case 1:   *Each Cl*   Unshared electrons $= 6$

Portion of shared electrons $(\frac{1}{2} \times 2) = \underline{1}$

Total electrons $= 7$

The seven electrons leave each Cl [group 17 (**VIIA**)] exactly neutral. Thus, a bond between atoms of the same element is said to be **nonpolar**. Bonds between elements of the same or nearly the same electronegativity (e.g., C and S) are essentially nonpolar.

The ionic compound NaCl illustrates case 2. As shown in Figure 6-7 the pair of electrons is not shared but is with the Cl. The Cl has three lone pairs of electrons plus the two electrons in question.

Case 2:   *Each Cl*   Unshared electrons $= 6$

Portion of shared electrons $(1 \times 2) = \underline{2}$

Total electrons $= 8$

Since seven valence electrons leave a Cl atom neutral, eight electrons give Cl a charge of -1. The Na has an equal but opposite charge of $+1$. Note that there is a large difference in electronegativity ($3.0 - 0.9 = 2.1$) between the Cl and the Na, leading to a prediction of ionic character for the bond.

Between these extremes lie the many bonds formed between two atoms that share electrons but not equally. In HCl (case 3), the Cl has a greater electronegativity than H but not enough to form an ionic bond. The Cl has three lone pairs of electrons plus more than one-half of the electrons in the bond. This gives the Cl atom a partial negative charge and leaves the H atom with a partial positive charge. (See Figure 6-7).

Case 3:   *Each Cl*   Unshared electrons $=\quad 6$

Portion of shared electrons $(>\frac{1}{2} \times 2) = \underline{\,>1\,}$

Total electrons $= >7$

EXAMPLE 6-9

Referring to Figure 6-6, rank the following bonds in order of increasing polarity. Write the atom with the positive end of the dipole first. Indicate if any of the bonds are predicted to be ionic on the basis of electronegativity differences.

Ba—Br,   C—N,   Be—F,   B—H,   Be—Cl

SOLUTION

Calculate the difference in electronegativity between the elements.

Ba—Br $2.8 - 0.9 = 1.9$ Ba positive (less electronegative)
C—N $3.0 - 2.5 = 0.5$ C positive
Be—F $4.0 - 1.5 = 2.5$ Be positive

B—H $2.1 - 2.0 = 0.1$ B positive
Be—Cl $3.0 - 1.5 = 1.5$ Be positive

B—H $<$ C—N $<$ Be—Cl $<$ Ba—Br $<$ Be—F

The difference in electronegativity suggests that Ba—Br and Be—F are ionic bonds.

SEE PROBLEMS 6-44 THROUGH 6-49.

6-11 THE POLARITY OF MOLECULES

Almost all bonds formed between atoms of different elements are polar at least to some extent. Nevertheless, not all molecules of compounds are polar. In fact, the polarity of a molecule is determined by *two* factors: (1) the polarity of the bonds in the molecule and (2) the geometry of the molecule.*

The geometry of a molecule is an important factor because a bond dipole is a force that has direction as well as magnitude.† The magnitude of the bond dipole depends on the difference in electronegativity between the two elements. The way in which the directions of the various bond dipoles of the molecule interact with each other, however, determines whether the molecule itself is polar.

In a molecule such as CO_2, both C—O bonds have a dipole as shown below. (Since O is more electronegative than C, the C is the positive end and the O the negative end.)

$$\overset{\longleftarrow\ \ \longmapsto}{\ddot{O}{=}C{=}\ddot{O}}$$

The dipoles are the same size, but their directions are opposite since CO_2 is a linear molecule. *Thus, the bond dipoles cancel, leaving the molecule nonpolar.* On the other hand, the molecule COSe is also linear but the two bond dipoles do not cancel. In fact, they reinforce each other since they both have the same direction. As a result, there is a net molecular dipole. *The **molecular dipole** is the net or resultant effect of all of the individual bond dipoles.*

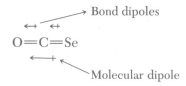

$$\overset{\longleftarrow\ \ \longleftarrow}{O{=}C{=}Se}$$

The relation of individual bond dipoles to the total molecular dipole is analogous to the effect of two people pulling on a block of concrete, as illustrated in Figure 6-8. The force exerted by each person is analogous to a bond

---

* The geometry of simple molecules can be predicted from either the Lewis structure or a knowledge of the shapes of what are called "hybridized" orbitals. This topic is not covered in this text, so the geometry of the sample molecules should be accepted as fact.
† Size and direction together make up what is known as the dipole moment of the bond.

Figure 6-8   FORCES AT 180°. If forces are equal but opposite (180°), they cancel. If the forces are not equal, there is a resultant force.

dipole, and the net effect (the resultant) of their efforts is analogous to the molecular dipole. Two people of equal strength pulling in opposite directions (an angle of 180°) cancel each other's force and there is no net movement.

Other molecules that are nonpolar although their bonds are polar include:

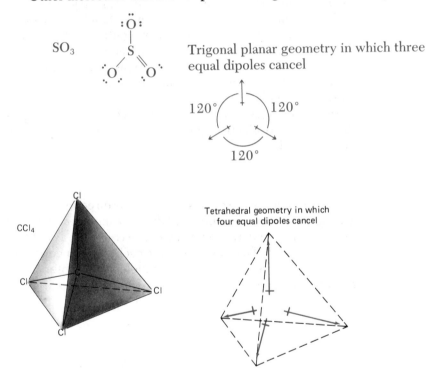

Many molecules exist in which the equal bond dipoles do not cancel. Ordinary water is an important example of this. Although an H_2O molecule has two equal bond dipoles, they are not at an angle of 180° and thus do not cancel. If H_2O were a linear molecule, it would be nonpolar and have properties vastly different from those discussed in Chapter 11.

Bond dipoles at an angle less than 180° constitute a situation analogous to two people pulling on a block of concrete at an angle, as illustrated in Figure 6-9. Both direction and resultant force depend on the strength of the two

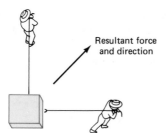

Resultant force
and direction

Figure 6-9   FORCES AT LESS THAN 180°. Both
direction and resultant force depend on the angle
between the two people.

people and the angle between them. (Resultant force and direction of movement also depend on their comparative strengths, if different.) In a similar
manner, the resultant dipole for the H_2O molecule is determined by the size of
each H—O dipole and the angle between the two dipoles:

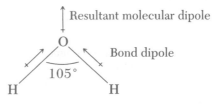

Resultant molecular dipole

Bond dipole

105°

In later chapters the importance of the polarity of molecules will become
evident. Physical state and other physical properties as well as chemical properties of compounds are determined to a large extent by the polarity of the
molecules. We will refer back to this discussion at that time.

■
*SEE PROBLEMS
6-50 THROUGH
6-56.*

**6-12 CHAPTER
REVIEW**

We have now answered a question first posed in Chapter 3. Why do elements
form bonds in two different ways? That is, what determines whether elements
combine to form ionic compounds or molecular compounds? The answer lies
in the electron configuration of the element and its tendency to adjust its
configuration to acquire a certain stability. The adjustment a specific representative element makes in its electron configuration can be predicted as a
result of these three observations:

1   Noble gases do not readily form chemical bonds.
2   Noble gases do not form bonds (are stable) because they have an octet of
outer or valence electrons. This corresponds to filled outer *s* and *p* sublevels.
3   Other atoms of representative elements achieve the same stability of noble
gases by gaining, losing, or sharing electrons with other atoms to attain an
octet of electrons.

To follow the changes in electron configuration of elements as bonds are
formed, Lewis dot symbols were introduced. Dot symbols emphasize only the
valence electrons.

Our first look at bonding concerned binary compounds involving a metal and a nonmetal. In this case, metals attain an octet by losing electrons and nonmetals do so by gaining electrons. Two nonmetals form binary compounds by electron sharing, which is known as covalent bonding and is summarized as follows:

Elements	Type of Bond	Comments
Metal–nonmetal	Ionic	Nonmetal can form −1, −2, or −3 ion. Metal can form +1, +2, or +3 ion.
Nonmetal–nonmetal	Covalent	Single, double, or triple bonds used to form octet.

The construction of Lewis structures of compounds was discussed and the procedural rules were given. The rules are summarized for three compounds as follows:

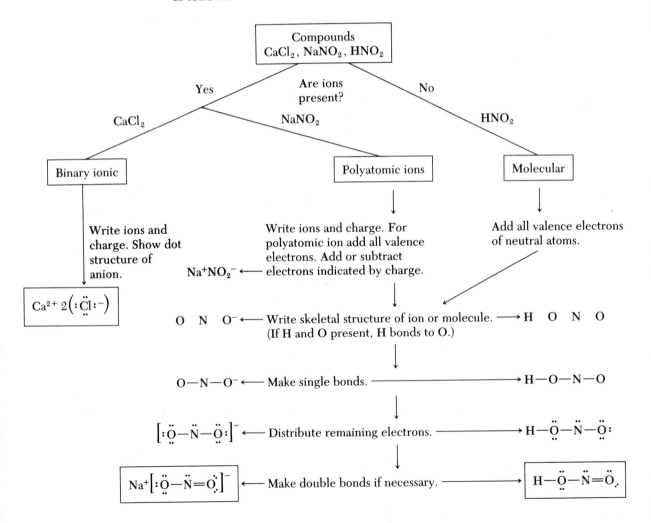

Resonance structures were introduced to illustrate molecules that cannot be represented accurately by a single Lewis structure. They indicate that certain bonds in one molecule are identical and exist as a hybrid of the equivalent resonance structures.

Electronegativity is a periodic property that gives us information about the polarity of covalent bonds. Large differences in electronegativity between two atoms in a bond may indicate ionic nature. Any difference indicates that the bond is polar covalent. Polar bonds contain a positive end and a negative end. The separation of charges is known as a dipole.

Many important physical properties of a compound are determined by the overall or net polarity of the molecule. If the geometry and magnitude of the dipoles are such that the bond dipole forces cancel, the molecule is nonpolar. If there is a resultant molecular dipole, the molecule is polar.

■ EXERCISES

DOT SYMBOLS OF ELEMENTS

6-1 Write Lewis dot symbols for:
   (a) Ca   (c) Sn   (e) Ne
   (b) Sb   (d) I   (f) Bi
   (g)   all group 16 (VIA) elements

6-2 Identify the group from the dot symbols.
   (a) $\cdot \dot{M} \cdot$ (b) $\cdot \ddot{X} \cdot$ (c) $\dot{A} \cdot$

6-3 Why are only outer electrons represented in dot symbols?

6-4 Which of the following dot symbols are incorrect?
   (a) $\dot{Pb} \cdot$ (c) $: \ddot{He} :$ (e) $\cdot \dot{Te} \cdot$
   (b) $\cdot \ddot{Bi} \cdot$ (d) $\dot{Cs}$ (f) $\cdot \dot{Tl} \cdot$

BINARY IONIC COMPOUNDS

6-5 From the periodic table predict which pairs of elements can combine to form ionic bonds.
   (a) H and Cl   (d) Al and F
   (b) S and Sr   (e) B and Cl
   (c) H and K   (f) Xe and F

6-6 From the periodic table predict which pairs of elements can combine to form ionic bonds.
   (a) Ba and I   (d) Sn and Sb
   (b) Cs and Se   (e) P and S
   (c) C and O   (f) Cs and P

6-7 Which ions do not have a noble gas electron configuration?
   (a) Sr^{2+} (d) Te^{2-} (g) Ba^{2+}
   (b) S^- (e) In^+ (h) Tl^{3+}
   (c) Cr^{2+} (f) Pb^{2+}

6-8 The ions Li^+, Be^{2+}, and H^- do not follow the octet rule. Why?

6-9 Write Lewis dot symbols for the following ions.
   (a) K^+ (d) P^{3-} (g) Sc^{3+}
   (b) O^- (e) Ba^+
   (c) I^- (f) Xe^+

6-10 What is the origin of the octet rule? How does the octet rule relate to s and p sublevels and to noble gases?

6-11 Which of the ions listed in Problem 6-9 do not follow the octet rule?

6-12 Write the charge on the following atoms that would give the element a noble gas configuration.
   (a) Mg (b) Ga (c) Br (d) S (e) P

6-13 Write the charge on the following atoms that would give the element a noble gas configuration.
   (a) Rb   (b) Ba   (c) Te   (d) N

6-14 Write six ions that have the same electron configuration as Ne.

6-15 Write five ions that have the same electron configuration as Kr.

6-16 The Tl^{3+} ion does not have the same electron configuration as Xe although it lost its three outermost electrons. Explain.

6-17 Complete the following table with formulas of the ionic compounds that form between the anion and cation shown.

Cation/Anion	Br⁻	S²⁻	N³⁻
Cs⁺	CsBr	——	——
Ba²⁺	——	——	——
In³⁺	——	——	——

6-18 Write the formula of the compound formed between the following non-metals and the metal calcium.
(a) I   (b) O   (c) N   (d) Te   (e) F

6-19 Write the formula of the compound formed between the following metals and the nonmetal sulfur.
(a) Be   (b) Cs   (c) Ga   (d) Sr

6-20 Most transition metal ions cannot be predicted by reference to the octet rule. Determine the charge on the metal cation from the charge on the anion for the following compounds.
(a) Cr_2O_3 (c) MnS   (e) $NiBr_2$
(b) FeF_3 (d) CoO   (f) VN

6-21 Why isn't a formula unit of $BaCl_2$ referred to as a molecule?

LEWIS STRUCTURES OF COMPOUNDS

6-22 From their Lewis dot symbols predict the formula of the simplest compound formed from the combination of the following pairs of elements.
(a) H and Se   (d) Cl and O
(b) H and Ge   (e) N and Cl
(c) Cl and F   (f) C and Br

6-23 From their Lewis dot symbols, predict the formula of the simplest compound formed from the following pairs of elements.
(a) H and I   (c) Si and Br
(b) Se and Br   (d) H and As

6-24 From a consideration of the octet rule, which of the following compounds are impossible?
(a) PH_3 (c) SCl_2 (e) H_3O
(b) Cl_3 (d) NBr_4

6-25 Determine the charge on the following polyatomic anions from the charge on the cation.
(a) K_2SO_4 (d) $Ca_3(PO_4)_2$
(b) $Ca(ClO_3)_2$ (e) CaC_2
(c) $Al_2(SO_3)_3$

6-26 Determine the charge on the following polyatomic anions from the charge on the cation.
(a) $NaClO_2$ (c) $AlAsO_4$
(b) $SrSeO_3$ (d) $Ca(H_2PO_4)_2$

6-27 Write Lewis structures for the following.
(a) C_2H_6 (e) C_2H_6O (There
(b) H_2O_2 are two correct
(c) NF_3 answers; both
(d) SCl_2 have the lone
                             pairs on the
                             oxygen.)

6-28 Write Lewis structures for the following.
(a) N_2H_4 (c) C_3H_8
(b) AsH_3 (d) CH_4O (all
                             unshared elec-
                             trons on 0)

6-29 Write Lewis structures for the following.
(a) CO   (c) KCN
(b) SO_3 (d) H_2SO_3 (H's on
                             different O's)

6-30 Use the "$6N + 2$ rule" to confirm the structures in Problem 6-29.

6-31 Use the $6N + 2$ rule to predict the number of multiple bonds (if any) in the following molecules or ions.
(a) N_2O (d) H_2S
(b) $Ca(NO_2)_2$ (e) CH_2Cl_2
(c) $AsCl_3$ (f) NH_4^+

6-32 Write Lewis structures for the compounds in Problem 6-31.

6-33 Write Lewis structures for the following.
(a) Cl_2O (c) C_2H_4 (e) BF_3
(b) SO_3^{2-} (d) H_2CO (f) NO^+

6-34 Use the $6N + 2$ rule to predict the number of multiple bonds (if any) in the following molecules or ions.
(a) CO_2 (e) HOCN
(b) H_2NOH (f) $SiCl_4$
(c) $BaCl_2$ (g) C_2H_2
(d) NO_3^- (h) O_3

6-35 Write Lewis structures for the molecules or ions in Problem 6-34.

6-36 Write Lewis structures for the following.

(a) Cs_2Se
(b) $CH_3CO_2^-$ (all H's on one C both O's on the other C)
(c) $LiClO_3$
(d) N_2O_3
(e) PBr_3

*6-37 Refer to Problem 5-39. Use the periodic table from the planet Zerk.

(1) What are the simplest formulas of compounds formed between the following elements? (Example: Between 7 and 7 is 7_2.)

(a) 1 and 7   (e) 7 and 13
(b) 1 and 3   (f) 10 and 13
(c) 1 and 5   (g) 6 and 7
(d) 7 and 9   (h) 3 and 6

(2) Write Lewis structures for all of the above. Indicate which are ionic. (Remember that on Zerk there will be something different than an octet rule.)

(3) What would be the Zerkian equivalent of the $6N + 2$ rule.

6-38 Which of the following compounds contains both ionic and covalent bonds?

(a) H_2SO_3 (c) K_2S (e) C_2H_6
(b) K_2SO_4 (d) H_2S (f) $BaCl_2$

RESONANCE STRUCTURES

6-39 Write all equivalent resonance structures (if any) for the following.
(a) SO_3 (b) NO_2^- (c) SO_3^{2-}

6-40 Write all equivalent resonance structures (if any) for the following.
(a) NO_3^- (b) N_2O_4

6-41 Write the equivalent resonance structures for the $H_3BCO_2^{2-}$ anion. The skeletal structure for the ion is

```
    H
              O
H   B   C
              O
    H
```

6-42 What is meant by a resonance hybrid? What is implied about the nature of the C—O bond from the resonance structures in Problem 6-41?

*6-43 A possible Lewis structure for CO_2 involves a triple bond between C and O. Write the two resonance structures involving the triple bond. What is implied about the nature of the CO bond by these two structures? How does this relate to the common Lewis structure for CO_2 involving two double bonds?

ELECTRONEGATIVITY AND POLARITY

6-44 Rank the following elements in order of increasing electronegativity: B, Ba, Be, C, Cl, Cs, F, O.

6-45 For bonds between the following elements, indicate the positive end of the dipole by a δ^+ and the negative end by a δ^-. Also indicate with a dipole arrow the direction of the dipole.

(a) N—H   (d) F—O   (g) C—B
(b) B—H   (e) O—Cl   (h) Cs—N
(c) Li—H   (f) S—Se   (i) C—S

6-46 Rank the bonds in Problem 6-45 in order of increasing polarity.

6-47 Predict whether the following pairs of elements will form an ionic or covalent bond on the basis of difference in electronegativity.

(a) Sc—Br   (c) B—Br
(b) Cu—B   (d) Al—F

6-48 Predict whether the following pairs of elements will form an ionic or covalent bond on the basis of difference in electronegativity.

(a) Al—Cl   (c) K—O
(b) Ca—I   (d) Mn—Te

6-49 Which of the following bonds is nonpolar?

(a) I—F   (c) C—H   (e) B—N
(b) I—I   (d) N—Br

6-50 How can a molecule be nonpolar if it contains polar bonds?

6-51 From the geometry of the following, determine which molecules are polar.

(a) CS_2 S—C—S   linear
(b) COS   S—C—O   linear
(c) SF_2
```
          S
         / \
        F   F
```
bent

(d) CCl_4
```
        Cl
        |
        C
      / | \
    Cl  Cl  Cl
```
tetrahedron

(e) $CHCl_3$ 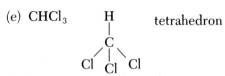   tetrahedron

```
        H
        |
        C
      / | \
    Cl  Cl  Cl
```

(d) NH_2Cl trigonal pyramid

```
         N
       / | \
   H------Cl
        H
```

6-52   From the geometry of the following determine which molecules are polar.

(a) BF_3 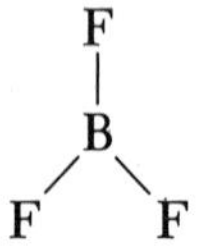   trigonal planar

```
        F
        |
        B
      /   \
    F       F
```

(b) BF_2Cl 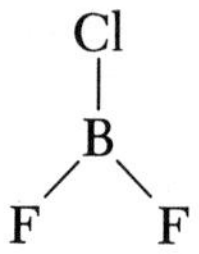   trigonal planar

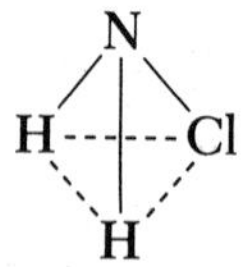

```
        Cl
        |
        B
      /   \
    F       F
```

(c) NH_3 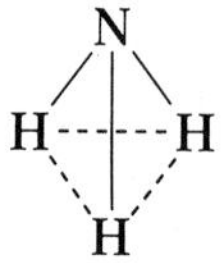   trigonal pyramid

```
         N
       / | \
   H------H
        H
```

6-53   Compare the expected molecular polarities of H_2O and H_2S. Assume that both molecules are bent at the same angle.

6-54   Compare the expected molecular polarities of CH_4 and CH_2F_2. Assume that both molecules are tetrahedral.

6-55   Compare the expected molecular polarities of $CHCl_3$ and CHF_3. Assume that both molecules are tetrahedral.

6-56 CO_2 is a nonpolar molecule, but CO is polar. Explain.

CHEMICAL COMPOUNDS
This beautiful earth—its mountains, trees, and streams—are all composed of various compounds. The names of these compounds range from the simple to the complex.

THE NAMING
OF COMPOUNDS

7

6   Use the Greek prefixes in Table 7-3 to name nonmetal–nonmetal binary compounds and write formulas given their names. (7-4)

7   Name binary acids and oxyacids and write formulas given their names. (7-5)

8   Distinguish between an acid and a salt. (7-5)

Chemistry seems to be taking a bad rap lately. When a "chemical" is mentioned by the media a red flag goes up in our minds and we think of something mysterious and perhaps evil. We hear of the names of many chemicals such as dioxins in the soil, nitrites in our meats, sulfites in vegetables, and sulfuric and nitric acids in the air and in the water. Certainly, these compounds in excess quantity and in the wrong place do cause problems that have to be solved. But how about water, sodium chloride, and amino acids? These are also the names of chemicals but they are essential to life. In fact, everything, including ourselves, is made of chemicals. To put things in perspective, perhaps the first order of business is to understand how compounds are named. *This is what is known as chemical nomenclature.*

One hundred years ago there were not all that many known compounds, so the citizen as well as the chemist did not need much order in nomenclature. Thus many compounds, such as lye, baking soda, ammonia, and water, have *common* names dating from long ago. These names tell little, if anything, about the elements of which they are composed. (The common names of some well-known compounds are given in Problem 7-29.) This, of course, would never do in the naming of the millions of compounds that have now been identified. Obviously, a more orderly approach was needed. Since 1921, a group of chemists who make up an organization called IUPAC (International Union of Pure and Applied Chemistry) have met regularly to discuss the naming of compounds. From this group come *systematic* methods of chemical nomenclature.

This chapter involves the naming of inorganic compounds, which include all compounds except organic compounds. (Organic compounds are composed of carbon, hydrogen, and sometimes other elements.) This is essentially the language of chemistry. Like any language, a good deal of memorization is needed. With some basic ground rules, however, what might otherwise be an impossible task becomes quite reasonable and orderly.

As background to this chapter you should be familiar with the differences between metals and nonmetals and between ionic and covalent compounds as discussed in the preceding chapter.

**7-1
OXIDATION
STATES**

A binary (two-element) compound is conveniently named by reference to its two elements. For example, the name of table salt, sodium chloride, tells us that the compound is composed of sodium and chlorine. Unfortunately, it is not always that simple. For example, iron forms two compounds when combined with chlorine, $FeCl_2$ and $FeCl_3$. Obviously, the name "iron chloride" would be ambiguous. Therefore, before we begin the systematic naming of compounds it is necessary for us to establish a system for distinguishing the iron in $FeCl_2$ from that in $FeCl_3$. This is done by expressing the **oxidation state** (or **oxidation number**) of the iron. The *oxidation state of an element in a compound is the charge the atoms of that element would have if the compound were composed only of monatomic ions.* To determine such a charge the electrons in a covalent bond are assigned to the more electronegative element. This is, in a way, simply a method of electron accounting much like the expenses of running a business are assigned to various divisions. Actually, the assignment of oxidation states in a compound is not complicated since we can take advantage of some generalities and facts that produce the following rules.

1   The oxidation state of an element in its free state is zero. This is true even for polyatomic elements (e.g., Cu, B, O_2, P_4).

2   The oxidation state of a monatomic ion is the same as the charge on that ion (e.g., $Na^+ = +1$ oxidation state, $O^{2-} = -2$, $Al^{3+} = +3$).
    (a) Alkali metals are always $+1$
    (b) Alkaline earth metals are always $+2$.

3   The halogens are in a -1 oxidation state in binary (two-element) compounds, whether ionic or covalent, *when bound to a less electronegative element.*

4   Oxygen in a compound is usually -2. Certain compounds (which are rare) called peroxides or superoxides contain O in a lower negative oxidation state. Oxygen is positive when bound to F.

5   Hydrogen in a compound is usually $+1$. When combined with a less electronegative element, usually a metal, H has a -1 oxidation state (e.g., LiH).

6   The sum of the oxidation states of all of the atoms in a neutral compound is zero. For polyatomic ions, the sum of the oxidation states equals the charge on the ion.

The following examples illustrate the assignment of oxidation states.

EXAMPLE 7-1

What are the oxidation states of the elements in FeO?
Rule 6: Ths oxidation states of the two elements add to zero.

$$(\text{ox. state Fe}) + (\text{ox. state O}) = 0$$

Rule 4: Since O is -2,

$$\text{Fe} + (-2) = 0$$
$$\text{Fe} = \underline{\underline{+2}} \quad \text{(This is an ionic compound.)}$$

EXAMPLE 7-2

What is the oxidation state of the N in N_2O_5?
Rule 6: The oxidation states add to zero, as shown by the equation

$$2(\text{ox. state N}) + 5(\text{ox. state O}) = 0$$

Rule 4: Since O is -2,

$$2N + 5(-2) = 0$$
$$2N = +10$$
$$N = \underline{\underline{+5}} \quad \text{(This is a covalent compound.)}$$

EXAMPLE 7-3

What is the oxidation state of the S in H_2SO_3?
Rule 6: The oxidation states add to zero.

$$2(\text{ox. state H}) + (\text{ox. state S}) + 3(\text{ox. state O}) = 0$$

Rules 4 and 5: H is usually $+1$ and O is usually -2.

$$2(+1) + S + 3(-2) = 0$$
$$S = \underline{\underline{+4}}$$

EXAMPLE 7-4

What is the oxidation state of the As in $AsO_4{}^{3-}$?
Rule 6: The oxidation states add to the charge on the ion.

$$(\text{ox. state As}) + 4(\text{ox. state O}) = -3$$
$$\text{As} + 4(-2) = -3$$
$$\text{As} = \underline{\underline{+5}}$$

■
SEE PROBLEMS
7-1 THROUGH
7-8.

1 (IA)	2 (IIA)	13 (IIIA)	14 (IVA)	15 (VA)	16 (VIA)	17 (VIIA)
						Hydride H^-
Lithium Li^+	Beryllium Be^{2+}		Carbide C^{4-}	Nitride N^{3-}	Oxide O^{2-}	Fluoride F^-
Sodium Na^+	Magnesium Mg^{2+}	Aluminum Al^{3+}		Phosphide P^{3-}	Sulfide S^{2-}	Chloride Cl^-
Potassium K^+	Calcium Ca^{2+}				Selenide Se^{2-}	Bromide Br^-
Rubidium Rb^+	Strontium Sr^{2+}				Telluride Te^{2-}	Iodide I^-
Cesium Cs^+	Barium Ba^{2+}					

Figure 7-1   MONATOMIC IONS OF THE REPRESENTATIVE ELEMENTS.

7-2 NAMING METAL– NONMETAL BINARY COMPOUNDS

We are now ready to begin the task of naming compounds. In the previous section we noted that sodium chloride is a legitimate name but iron chloride is not. This is because certain metals have only one oxidation state in compounds, whereas others have two or more. Let us consider the two cases separately.

Metals with One Oxidation State

Many of the most familiar representative element metals are present in only one oxidation state in their compounds. These are the metals of group 1 (IA), which have a +1 charge; those of group 2 (IIA), which have a +2 charge; and aluminum in group 13 (IIIA), which has a +3 charge. (There are also several transition metals with only one oxidation state, but they are not included in this discussion.) For these metals, the oxidation state is indicated by its position in the periodic table. (See Figure 7-1.)

In both naming and writing the formula for a binary ionic compound, the metal comes first and the nonmetal second. The unchanged English name of the metal is used. (If a metal cation is named alone, the word "ion" is also included to distinguish it from the free metal.) The name of the anion includes only the English root plus *ide*. For example, chlor*ide* is derived from *chlor*ine and ox*ide* from *ox*ygen. Figure 7-1 includes all of the monatomic anions and their names.

EXAMPLE 7-5

Name the following binary ionic compounds: KCl, Li$_2$S, and Mg$_3$N$_2$.

ANSWER

KCl   potassium chloride

Li$_2$S   lithium sulfide

Mg$_3$N$_2$ magnesium nitride

EXAMPLE 7-6

Write the formulas of the following compounds: aluminum fluoride, calcium selenide, and potassium phosphide.

ANSWER

Aluminum is in group 13 (**IIIA**), so it has a +3 charge in compounds. Fluorine is in group 17 (**VIIA**) so it has a −1 charge in compounds.

$$\text{Al}^{(3+)} \quad \text{F}^{(1-)} \qquad = \text{AlF}_3$$

Calcium is in group 2 (**IIA**), so it has a +2 charge in compounds. Selenium is in group 16 (**VIA**), so it has a −2 charge in compounds.

$$\text{Ca}^{(2+)} \quad \text{Se}^{(2-)} \qquad = \text{Ca}_2\text{Se}_2 = \text{CaSe}$$

Potassium is in group 1 (**IA**), so it has a +1 charge in compounds. Phosphorus is in group 15 (**VA**), so it has a −3 charge in compounds.

$$\text{K}^{(1+)} \quad \text{P}^{(3-)} \qquad = \text{K}_3\text{P}$$

Metals with Two or More Oxidation States

The remaining representative element metals and most transition metals have more than one oxidation state. An extreme example is manganese. Oxides of manganese have the formulas MnO, Mn$_3$O$_4$, Mn$_2$O$_3$, MnO$_2$, MnO$_3$, and Mn$_2$O$_7$. To distinguish these oxides of manganese the **Stock method** is used. *In this method the oxidation state of the metal is listed with Roman numerals in parentheses after the name of the metal.* In the **classical method,** *the two most common states are emphasized by using the name of the metal as a root plus* ous *for the lower oxidation state and* ic *for the higher oxidation state.* If the symbol is derived from a Latin name, the Latin root is used. Several metals with variable oxidation states are shown using both methods of nomenclature in Table 7-1.

TABLE 7-1   Metals with Variable Oxidation States

Metal	Stock Method	Classical Method	Metal	Stock Method	Classical Method
thallium	thallium(I)	thallous	lead	lead(II)	plumbous
	thallium(III)	thallic		lead(IV)	plumbic
iron	iron(II)	ferrous	tin	tin(II)	stannous
	iron(III)	ferric		tin(IV)	stannic
chromium	chromium(II)	chromous	copper	copper(I)	cuprous
	chromium(III)	chromic		copper(II)	cupric
cobalt	cobalt(II)	cobaltous	gold	gold(I)	aurous
	cobalt(III)	cobaltic		gold(III)	auric

The Stock method is used almost exclusively in chemistry today. It is more convenient because it is not necessary to know the oxidation state corresponding to each name for a particular metal. This method is used in this text.

As discussed in Chapter 6, most metal–nonmetal binary compounds are ionic. Where the metal is in a $+4$ or higher oxidation state (e.g., $SnBr_4$ and Mn_2O_7), the compounds generally have properties more typical of molecular compounds that are neutral compounds containing only covalent bonds. It should be noted then that an oxidation state for a metal does not necessarily indicate the presence of an ion.

EXAMPLE 7-7

Name the following compounds: $SnBr_4$, CoF_3, Au_2O.

ANSWER

Refer to Section 7-1 to determine the oxidation state of the metal ion.

$SnBr_4$ tin(IV) bromide* or stannic bromide

CoF_3 cobalt(III) fluoride or cobaltic fluoride

Au_2O gold(I) oxide or aurous oxide

---

* Pronounced "tin-four bromide."

EXAMPLE 7-8

Write formulas for lead(IV) oxide, iron(II) chloride, chromium(III) sulfide, and manganese(VII) oxide.

ANSWER

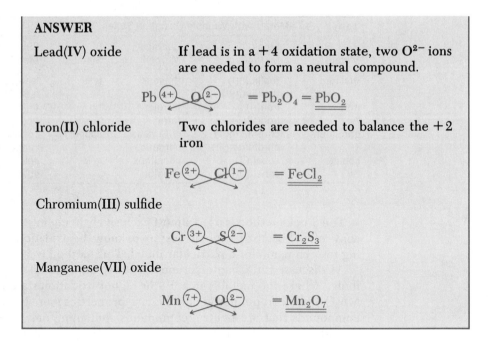

Lead(IV) oxide — If lead is in a $+4$ oxidation state, two O^{2-} ions are needed to form a neutral compound.

$$Pb^{4+} \; O^{2-} \quad = Pb_2O_4 = \underline{\underline{PbO_2}}$$

Iron(II) chloride — Two chlorides are needed to balance the $+2$ iron

$$Fe^{2+} \; Cl^{1-} \quad = \underline{\underline{FeCl_2}}$$

Chromium(III) sulfide

$$Cr^{3+} \; S^{2-} \quad = \underline{\underline{Cr_2S_3}}$$

Manganese(VII) oxide

$$Mn^{7+} \; O^{2-} \quad = \underline{\underline{Mn_2O_7}}$$

■
SEE PROBLEMS
7-9 THROUGH
7-18.

7-3 NAMING COMPOUNDS WITH POLYATOMIC IONS

A polyatomic ion is composed of two or more atoms covalently bound together with a net electrical charge. In ionic compounds, such an ion behaves much as the monatomic ions discussed in the previous section. Table 7-2 contains many of the more common polyatomic anions (and one cation) encountered in chemistry.

In naming and writing the formulas for metal–polyatomic anion compounds, we follow essentially the same rules as before. The metal is named or written first (followed by the oxidation state if necessary), and then the name of the anion is given or written.

TABLE 7-2   Polyatomic Ions

Ion	Name	Ion	Name
$C_2H_3O_2^-$	acetate	HSO_3^-	hydrogen sulfite or bisulfite
NH_4^+	ammonium	OH^-	hydroxide
CO_3^{2-}	carbonate	ClO^-	hypochlorite
ClO_3^-	chlorate	NO_3^-	nitrate
ClO_2^-	chlorite	NO_2^-	nitrite
CrO_4^{2-}	chromate	$C_2O_4^{2-}$	oxalate
CN^-	cyanide	ClO_4^-	perchlorate
$Cr_2O_7^{2-}$	dichromate	MnO_4^-	permanganate
HCO_3^-	hydrogen carbonate or bicarbonate	PO_4^{3-}	phosphate
HSO_4^-	hydrogen sulfate or bisulfate	SO_4^{2-} SO_3^{2-}	sulfate sulfite

There is some systematization possible that will help in learning Table 7-2. In many cases *the anions are composed of oxygen and one other element.* These anions are thus called **oxyanions.** When there are two oxyanions of the same element (e.g., SO_3^{2-} and SO_4^{2-}), they, of course, have different names. The anion with the element other than oxygen in the *higher* oxidation state uses the root of the element plus *ate.* The anion with the element in the *lower* oxidation state uses the root of the element plus *ite.*

$$SO_4^{2-} \quad \text{sulf}\underline{ate} \quad \text{(S is in +6 ox. state)}$$
$$SO_3^{2-} \quad \text{sulf}\underline{ite} \quad \text{(S is in +4 ox. state)}$$

The four oxyanions for Cl are

$$ClO_4^- \quad \underline{per}\text{chlor}\underline{ate} \quad \text{(Cl is in +7 ox. state)}$$
$$ClO_3^- \quad \text{chlor}\underline{ate} \quad \text{(Cl is in +5 ox. state)}$$
$$ClO_2^- \quad \text{chlor}\underline{ite} \quad \text{(Cl is in +3 ox. state)}$$
$$ClO^- \quad \underline{hypo}\text{chlor}\underline{ite} \quad \text{(Cl is in +1 ox. state)}$$

Note that the two oxyanions of intermediate oxidation states are assigned the *ate* (+5) and the *ite* (+3) endings. The highest oxidation state (+7) also received a prefix *per* in addition to the *ate* ending. The lowest oxidation state (+1) receives the prefix *hypo* in addition to the *ite* ending. Oxyanions of Br and I follow the same pattern.

EXAMPLE 7-9

Name the following compounds: K_2CO_3 and $Fe_2(SO_4)_3$.

ANSWER

$$K_2CO_3 \qquad \underline{\text{potassium carbonate}}$$
$$Fe_2(SO_4)_3 \qquad \underline{\text{iron(III) sulfate}}$$

EXAMPLE 7-10

Give the formulas for barium acetate, ammonium sulfate, thallium(III) nitrate, and manganese(III) phosphate.

ANSWER

Barium acetate

Barium is in group 2 (IIA), so it has a +2 charge. Acetate is the $C_2H_3O_2^-$ ion.

$$Ba^{(2+)} \quad C_2H_3O_2^{(1-)} \qquad = Ba(C_2H_3O_2)_2$$

(If more than one polyatomic ion is in the formula, enclose the ion in parentheses.)

Ammonium sulfate   From Table 7-2, ammonium = NH_4^+, sulfate = SO_4^{2-}.

$$NH_4^{(1+)} \quad SO_4^{(2-)} \quad = (NH_4)_2SO_4$$

Thallium(III) nitrate $Tl^{(3+)} \quad NO_3^{(1-)} \quad = Tl(NO_3)_3$

Manganese(III) phosphate $Mn^{(3+)} \quad PO_4^{(3-)} \quad = Mn_3(PO_4)_3 = MnPO_4$

EXAMPLE 7-11

Arsenic forms two oxyanions with the formulas AsO_3^{3-} and AsO_4^{3-}. Name the two anions using the root *arsen*.

ANSWER

AsO_3^{3-} is $\underline{\text{arsenite}}$ (As is in the $+3$ oxidation state).

AsO_4^{3-} is $\underline{\text{arsenate}}$ (As is in the $+5$ oxidation state).

EXAMPLE 7-12

Referring to the nomenclature for Cl, name $NaBrO_4$ and $Ca(IO)_2$.

ANSWER

$NaBrO_4$
The Br is in the $+7$ oxidation state. Thus it is named in an analogous manner to Cl in the $+7$ oxidation state. Since ClO_4^- is the *perchlorate* ion, the BrO_4^- ion is the *perbromate* ion.

$$\underline{\text{sodium perbromate}}$$

$Ca(IO)_2$
The I is in the $+1$ oxidation state. Thus the IO^- ion is analogous to the ClO^- ion.

$$\underline{\text{calcium hypoiodite}}$$

SEE PROBLEMS 7-19 THROUGH 7-31.

TABLE 7-3	Greek Prefixes				
Number	Prefix	Number	Prefix	Number	Prefix
1	mono	5	penta	8	octa
2	di	6	hexa	9	nona
3	tri	7	hepta	10	deca
4	tetra				

7-4 NAMING NONMETAL– NONMETAL BINARY COMPOUNDS

Which element is written first if neither of the elements is a metal? For these compounds, we write the element that is closest to being a metal (the element with the lower electronegativity) first. Generally, it is the element that is closer to the metal–nonmetal border on the periodic table. Thus, we write CO rather than OC and H_2O rather than OH_2. On this basis, we would predict that the formulas of ammonia (NH_3) and methane (CH_4)* would be opposite from the way they are written. The systematic rules are violated, however, for these compounds because that's the way these formulas have always been written. The formulas of these compounds were among the first known and were written long before any attempts were made to systematize nomenclature.

The compounds are also named with the least electronegative element (most metallic) first when its English name is used. The more nonmetallic element is named second, with its root plus *ide* as discussed before. Since nonmetals (except F) have variable oxidation states, some means must be provided to specify particular compounds. The Stock system is *not* preferred in this case, because it can still be ambiguous. For example, both NO_2 and N_2O_4 would have the Stock names of nitrogen(IV) oxide. In these compounds use is made of the Greek prefixes shown in Table 7-3 to indicate numbers of atoms in a compound.

The use of prefixes is illustrated by the names of the oxides of nitrogen listed in Table 7-4.

TABLE 7-4	The Oxides of Nitrogen	
Formula	Name	Oxidation State of N
N_2O	dinitrogen monoxide (sometimes referred to as nitrous oxide)	+1
NO	nitrogen monoxide (sometimes referred to as nitric oxide)	+2
N_2O_3	dinitrogen trioxide	+3
NO_2	nitrogen dioxide	+4
N_2O_4	dinitrogen tetroxide[a]	+4
N_2O_5	dinitrogen pentoxide[a]	+5

[a] The "a" is often omitted from tetra and penta for ease in pronunciation.

■
SEE PROBLEMS
7-32 THROUGH
7-37.

* With few exceptions, in organic compounds containing C, H, and other elements, the C is written first followed by H and then other elements.

TABLE 7-5	Binary Acids		
Anion	Formula of Acid	Compound Name	Acid Name
Cl^-	HCl	hydrogen chloride	hydrochloric acid
F^-	HF	hydrogen fluoride	hydrofluoric acid
I^-	HI	hydrogen iodide	hydroiodic acid
S^{2-}	H_2S	hydrogen sulfide	hydrosulfuric acid

7-5 NAMING ACIDS

Hydrogen compounds of most of the anions listed in Figure 7-1 and Table 7-2 are known as acids. Acids are compounds that are grouped together because of a type of chemical reaction that they all undergo, especially when dissolved in water. *In the pure state these compounds are molecular. That is, the hydrogen is covalently bound to the anion. When present in water solution, however, these compounds undergo ionization to form H^+ ions (also written as H_3O^+, which is called the hydronium ion) plus the anion.* This can be illustrated as shown below, where X^- represents a typical anion with a -1 charge.

$$HX \xrightarrow{\text{H}_2\text{O}} H^+ + X^-$$

The chemical nature of acids is discussed in some detail in Chapter 13. However, since these compounds have special names it is helpful to include their nomenclature at this point.

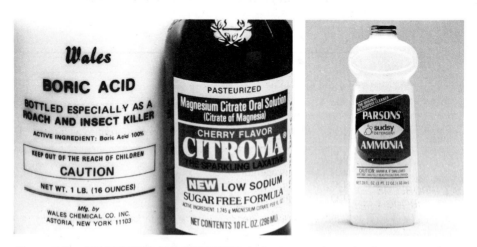

Figure 7-2 COMPOUNDS AND PRODUCTS. Ionic compounds, molecular compounds, and acids are all used in various products.

TABLE 7-6   Oxyacids

Anion	Name of Anion	Formula of Acid	Name of Acid
$C_2H_3O_2^-$	acetate	$HC_2H_3O_2$	acetic acid
CO_3^{2-}	carbonate	H_2CO_3	carbonic acid
NO_3^-	nitrate	HNO_3	nitric acid
PO_4^{3-}	phosphate	H_3PO_4	phosphoric acid
ClO_2^-	chlorite	$HClO_2$	chlorous acid
ClO_4^-	perchlorate	$HClO_4$	perchloric acid
SO_3^{2-}	sulfite	H_2SO_3	sulfurous acid
SO_4^{2-}	sulfate	H_2SO_4	sulfuric acid

Binary acids *are hydrogen compounds of most of the monatomic anions listed in Figure 7-1.* These compounds can exist in the pure molecular state as gases at room temperature. In this form, they are named as any other binary nonmetal–nonmetal compound, as discussed in the previous section. When dissolved in water, however, they are generally referred to by their acid name. The acid name is obtained by adding the prefix *hydro* to the anion root and changing the *ide* ending to *ic* followed by the word *acid*. Both names are illustrated in Table 7-5.

The following hydrogen compounds of anions listed in Figure 7-1 are not generally considered to be binary acids: H_2O, NH_3, CH_4, PH_3, and H_2.

Oxyacids *are hydrogen compounds of most of the oxyanions listed in Table 7-2.* To name an oxyacid, we use the root of the anion to form the name of the acid. If the name of the oxyanion ends in *ate*, it is changed to *ic* followed by the word *acid*. If the name of the anion ends in *ite*, it is changed to *ous* plus the word *acid*. Many hydrogen compounds of oxyanions do not exist in the pure state as do the binary acids. Generally, only the acid name is used in the naming of these compounds. For example, HNO_3 is referred to only as nitric acid and not as hydrogen nitrate. Development of the acid name from the anion name is shown for some anions in Table 7-6.

EXAMPLE 7-13

Name the following acids: H_2Se, $H_2C_2O_4$, and $HClO$.

ANSWER

H_2Se	<u>hydroselenic acid</u>
$H_2C_2O_4$	<u>oxalic acid</u>
$HClO$	<u>hypochlorous acid</u>

Naming Compounds Flow Chart

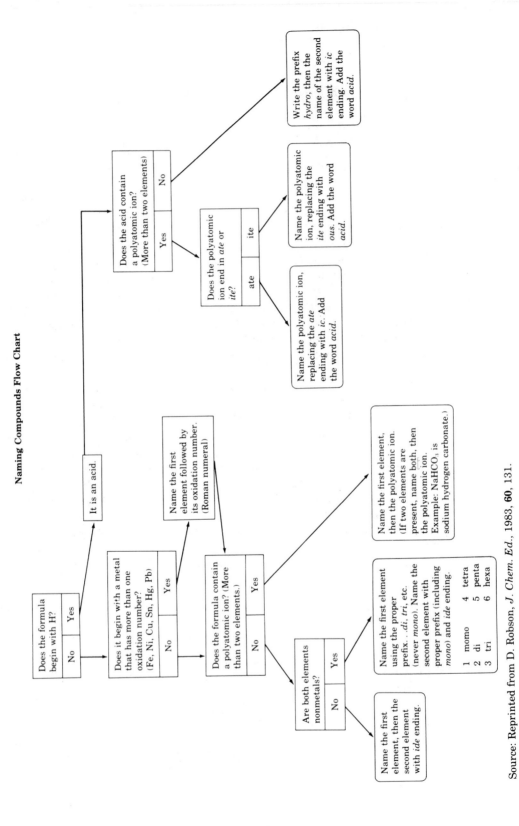

Source: Reprinted from D. Robson, *J. Chem. Ed.*, 1983, **60**, 131. *Source:* Reprinted from D. Robson, J. Chem. Ed., 1983, **60**, 131.

Figure 7-3 FLOW CHART FOR NAMING COMPOUNDS.

> **EXAMPLE 7-14**
>
> Give formulas for the following: permanganic acid, chromic acid, and acetic acid.
>
> **ANSWER**
>
> | | |
> |---|---|
> | permanganic acid | $HMnO_4$ |
> | chromic acid | H_2CrO_4 |
> | acetic acid | $HC_2H_3O_2$ |

■

SEE PROBLEMS
7-38 THROUGH
7-42.

Most of the ionic compounds that have been discussed previously are also known as salts. *A **salt** is an ionic compound formed by the combination of any cation except H^+ with any anion except OH^-.* For example, sodium nitrate is a salt composed of Na^+ and NO_3^- ions and calcium sulfate is a salt composed of Ca^{2+} and SO_4^{2-} ions. Ordinary table salt is composed of Na^+ and Cl^- ions.

7-6 CHAPTER REVIEW

This entire chapter can be summarized and reviewed by taking specific compounds to be named through the flow chart in Figure 7-3.

■ **EXERCISES**

OXIDATION STATES

7-1 Give the oxidation states of the elements in the following compounds.
(a) PbO_2 (d) N_2H_4 (g) Rb_2Se
(b) P_4O_{10} (e) LiH (h) Bi_2S_3
(c) C_2H_2 (f) BCl_3

7-2 Give the oxidation states of the elements in the following compounds.
(a) ClO_2 (c) CO (e) Mn_2O_3
(b) XeF_2 (d) O_2F_2 (f) Bi_2O_5

7-3 Which of the following elements form *only* the $+1$ oxidation state in compounds?
(a) Li (c) Ca (e) K (g) Rb
(b) H (d) Cl (f) Al

7-4 Which of the following elements form *only* the $+2$ oxidation state in compounds?
(a) O (c) Be (e) Sc (g) Ca
(b) B (d) Sr (f) Hg

7-5 What is the only oxidation state of Al in its compounds?

7-6 What is the oxidation state of each of the following?
(a) P in H_3PO_4 (e) S in SF_6
(b) C in $H_2C_2O_4$ (f) N in $CsNO_3$
(c) Cl in ClO_4^- (g) Mn in $KMnO_4$
(d) Cr in $CaCr_2O_7$

7-7 What is the oxidation state of each of the following?
(a) Se in SeO_3^{2-} (d) Cl in $HClO_2$
(b) I in H_5IO_6 (e) N in $(NH_4)_2S$
(c) S in $Al_2(SO_3)_3$

7-8 Nitrogen exists in nine oxidation states. Arrange the following compounds in order of increasing oxidation state of N: K_3N, N_2O_4, N_2, NH_2OH, N_2O, $Ca(NO_3)_2$, N_2H_4, N_2O_3, NO.

METAL–NONMETAL BINARY COMPOUNDS

7-9 Name the following compounds.
(a) LiF   (c) Sr_3N_2 (e) $AlCl_3$
(b) BaTe   (d) BaH_2

7-10 Name the following compounds.
(a) CaI_2 (c) BeSe   (e) RaS
(b) FrF   (d) Mg_3P_2

7-11 Give formulas for the following compounds.
(a) rubidium selenide
(b) strontium hydride
(c) radium oxide
(d) aluminum carbide
(e) beryllium fluoride

7-12 Give formulas for the following compounds.
(a) potassium hydride
(b) cesium sulfide
(c) potassium phosphide
(d) barium telluride

7-13 Name the following compounds using the Stock method.
(a) Bi_2O_5 (c) SnS_2 (e) TiC
(b) SnS   (d) Cu_2Te

7-14 Name the following compounds using the Stock method.
(a) CrI_3 (c) IrO_4 (e) $NiCl_2$
(b) $TiCl_4$ (d) MnH_2

7-15 Give formulas for the following compounds.
(a) copper(I) sulfide
(b) vanadium(III) oxide
(c) gold(I) bromide
(d) nickel(II) carbide
(e) chromium(VI) oxide

7-16 Give formulas for the following compounds.
(a) yttrium(III) hydride
(b) lead(IV) chloride
(c) bismuth(V) fluoride
(d) palladium(II) selenide

7-17 From the magnitude of the oxidation states of the metals predict which of the compounds in Problems 7-13 and 7-15 are primarily covalent.

7-18 From the magnitude of the oxidation states of the metals predict which of the compounds in Problems 7-14 and 7-16 are primarily covalent.

COMPOUNDS WITH POLYATOMIC IONS

7-19 Which of the following is the chlorate ion?
(a) ClO_2^- (c) ClO_3^- (e) ClO_3^+
(b) ClO_4^- (d) Cl_3O^-

7-20 Which of the following ions have a -2 charge?
(a) sulfate   (d) carbonate
(b) nitrite   (e) sulfite
(c) chlorite   (f) phosphate

7-21 What is the name and formula of the most common polyatomic cation?

7-22 Which of the following oxyanions contain four oxygen atoms?
(a) nitrate
(b) permanganate
(c) perchlorate
(d) sulfite
(e) phosphate
(f) oxalate
(g) carbonate

7-23 Name the following compounds. Use the Stock method where appropriate.
(a) $CrSO_4$ (e) $(NH_4)_2CO_3$
(b) $Al_2(SO_3)_3$ (f) NH_4NO_3
(c) $Fe(CN)_2$ (g) $Bi(OH)_3$
(d) $RbHCO_3$

7-24 Name the following compounds. Use the Stock method where appropriate.
(a) $Na_2C_2O_4$ (c) $Fe_2(CO_3)_3$
(b) $CaCrO_4$ (d) $Cu(OH)_2$

7-25 Give formulas for the following compounds.
(a) magnesium permanganate
(b) cobalt(II) cyanide
(c) strontium hydroxide
(d) thallium(I) sulfite
(e) indium(III) bisulfate
(f) iron(III) oxalate
(g) ammonium dichromate
(h) mercury(I) acetate [The mercury(I) ion exists as Hg_2^{2+}.]

7-26 Give formulas for the following compounds.
(a) zirconium(IV) phosphate
(b) sodium cyanide
(c) thallium(I) nitrite
(d) nickel(II) hydroxide
(e) radium hydrogen sulfate

(f) beryllium phosphate
(g) chromium(III) hypochlorite

7-27 Complete the following table.

Cation/anion	HSO$_3^-$	___	___
NH$_4^+$	(formula) ___	___	___
	(name) ___	___	___
___	___	CoTe	___
___	___	___	___
___	___	___	aluminum phosphate

7-28 Complete the following table.

Cation/anion	___	C$_2$O$_4^{2-}$	___
___	(formula) thallium(I) hydroxide	___	___
Sr^{2+}	___	___	___
___	___	___	___
___	___	___	TiN

7-29 Give the systematic name for each of the following.

Common Name	Formula	Systematic Name
(a) table salt	NaCl	
(b) baking soda	NaHCO$_3$	
(c) marble or limestone	CaCO$_3$	
(d) lye	NaOH	
(e) chile saltpeter	NaNO$_3$	
(f) sal ammoniac	NH$_4$Cl	
(g) alumina	Al$_2$O$_3$	
(h) slaked lime	Ca(OH)$_2$	
(i) caustic potash	KOH	

7-30 The perxenate ion has the formula XeO$_6^{4-}$. Write formulas of compounds of perxenate with the following.
(a) calcium   (c) aluminum
(b) potassium

*7-31 Name the following compounds. In these compounds an ion is involved that is not in Table 7-2. However, the name can be determined by reference to other ions of the central element or from ions in Table 7-2 in which the central atom is in the same group.
(a) PH$_4$F   (d) CaSiO$_3$
(b) KBrO   (e) AlPO$_3$
(c) Co(IO$_3$)$_3$ (f) CrMoO$_4$

NONMETAL–NONMETAL BINARY COMPOUNDS

7-32 The following pairs of elements combine to make binary compounds. Which element should be written and named first?
(a) Si and S   (d) Kr and F
(b) F and I   (e) H and F
(c) H and B

7-33 The following pairs of elements combine to make binary compounds. Which element should be written and named first?
(a) S and P   (c) O and F
(b) O and Cl   (d) As and Cl

7-34 Name the following.
(a) CS$_2$ (e) SO$_3$
(b) BF$_3$ (f) Cl$_2$O
(c) P$_4$O$_{10}$ (g) PCl$_5$
(d) Br$_2$O$_3$ (h) SF$_6$

7-35 Name the following.
(a) PF$_3$ (c) ClO$_2$ (e) SeCl$_4$
(b) I$_2$O$_3$ (d) AsF$_5$ (f) SiH$_4$

7-36 Write the formulas for the following.
(a) tetraphosphorus hexoxide
(b) carbon tetrachloride
(c) iodine trifluoride
(d) dichlorine heptoxide
(e) sulfur hexafluoride
(f) xenon dioxide

7-37 Write formulas for the following.
(a) xenon trioxide
(b) sulfur dichloride
(c) dibromine monoxide
(d) carbon disulfide
(e) diboron hexahydride (also known as diborane)

ACIDS

7-38 Name the following acids.
  (a) HCl
  (b) HNO_3
  (c) $HClO$
  (d) $HMnO_4$
  (e) HIO_4
  (f) HBr

7-39 Write formulas for the following acids.
  (a) cyanic acid
  (b) hydroselenic acid
  (c) chlorous acid
  (d) carbonic acid
  (e) hydroiodic acid
  (f) acetic acid

7-40 Write formulas for the following acids.
  (a) oxalic acid
  (b) nitrous acid
  (c) dichromic acid
  (d) phosphoric acid

*7-41 Refer to the ions in Problems 7-30 and 7-31. Write the acid names for the following
  (a) $HBrO$
  (b) HIO_3
  (c) H_3PO_3
  (d) $HMoO_4$
  (e) H_4XeO_6

7-42 Write the formulas and the names of the acids formed from the arsenite $(AsO_3{}^{3-})$ ion and the arsenate $(AsO_4{}^{3-})$ ion.

REVIEW TEST ON CHAPTERS 4–7*

The following multiple choice questions have one correct answer.

1. Which of the following are the sublevels present in the third ($n = 3$) shell?
   (a) s, p
   (b) p, d, f
   (c) s, p, d
   (d) s, p, d, f

2. What is the electron capacity of the $4f$ sublevel?
   (a) 6 (b) 7 (c) 14 (d) 8 (e) 10

3. Which of the following elements has the electron configuration $[Kr]5s^2 4d^2$?
   (a) Sn (b) Zr (c) Ti (d) Pr (e) Sr

4. What is the electron configuration of In (atomic number 49)?
   (a) $[Kr]4s^2 4p^1$
   (b) $[Kr]5s^2 4d^1$
   (c) $[Kr]5s^2 4d^{10} 5p^1$
   (d) $[Kr]4s^2 3d^{10} 4p^1$
   (e) $[Kr]5s^2 4d^6 5p^1$

5. Silicon has two unpaired electrons. This is a result of:
   (a) Hund's rule
   (b) the Pauli exclusion principle
   (c) the Aufbau principle
   (d) Bohr's model

6. How many orbitals are in a d sublevel?
   (a) 5 (b) 3 (c) 10 (d) 14 (e) 2

7. Which of the following is the box diagram for electrons in the outer orbitals of a sulfur atom?

(a) [↑↓ | ↑ | ↑]

(b) [↑↓ | ↓ | ↑]

(c) [↑ | ↑ | | |]

(d) [↑↓]

(e) [↑↓ | ↑↓ |]

8. The following is the shape of what type of orbital?

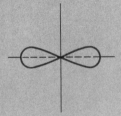

(a) s (b) p (c) d (d) f

9. Which of the following elements is an alkali metal?
   (a) Sc (b) Al (c) Cl (d) K (e) H

10. The element U (atomic number 92) is in which classification of elements?
    (a) representative
    (b) transition
    (c) noble gas
    (d) inner transition

11. Which of the following is an allotrope of carbon?
    (a) ozone
    (b) diamond
    (c) carbon dioxide
    (d) sugar
    (e) quartz

12. Which of the following groups of elements have the general electron configuration $ns^2 np^5$?
    (a) halogens
    (b) alkaline earths
    (c) noble gases
    (d) Group 15 (VA)
    (e) transition elements

13. Which of the following atoms has the smallest radius?
    (a) As (b) Se (c) Sb (d) Te

14. Which of the following atoms has the highest first ionization energy?
    (a) N (b) O (c) P (d) S

15. Which of the following ions requires the most energy to form?
    (a) K^+
    (b) Ca^{3+}
    (c) Al^{3+}
    (d) F^-
    (e) O^{2-}

16. Which of the following ions violates the octet rule?
    (a) I^{2-}
    (b) S^{2-}
    (c) Br^-
    (d) N^{3-}
    (e) Te^{2-}

17. Which of the following pairs of elements would be expected to form an ionic bond?
    (a) Mg, Ca
    (b) B, F
    (c) Mg, Br
    (d) S, O
    (e) C, O

18. Which of the following is the correct formula for the compound formed between beryllium and bromine?
    (a) BeBr
    (b) Br_2Be
    (c) BBr_2
    (d) $BeBr_2$
    (e) Be_2Br

19. Which of the following pairs of elements would be most likely to form a covalent bond?
    (a) Al, F (b) K, Sn (c) As, Br (d) K, H

20. Which of the following is the correct formula for the simplest compound formed between As and Cl?
    (a) Cl_2As (c) $AsCl_3$ (e) $AsCl_4$
    (b) $AsCl_2$ (d) $AsCl$

21. Which of the following compounds contains both ionic and covalent bonds?
    (a) $BaCO_3$ (c) HNO_3 (e) CCl_4
    (b) H_2CO_3 (d) K_3N

22. Using the $6N + 2$ rule, determine the number of double bonds in the N_2O_5 molecules.
    (a) none   (b) one   (c) two   (d) three

23. Which of the following is the most electronegative element?
    (a) B   (b) Na   (c) O   (d) Cl   (e) N

24. Which of the following bonds is the most polar?
    (a) C—O   (c) Be—Cl   (e) N—N
    (b) B—Cl   (d) Be—F

25. Which of the following metals has more than one oxidation state?
    (a) Na   (b) Be   (c) Bi   (d) Al   (e) Ra

26. What is the oxidation state of the N in NO_2?
    (a) $+2$ (c) -4 (e) $+6$
    (b) $+3$ (d) $+4$

27. What is the oxidation state of the S in $Na_2S_2O_3$?
    (a) $+4$ (c) -4 (e) -2
    (b) $+2$ (d) $+1$

28. Which of the following is the formula for the permanganate ion?
    (a) MnO_4 (c) $MnO_3{}^{2-}$ (e) $MnO_4{}^-$
    (b) $MnO_4{}^{2-}$ (d) $MgO_4{}^-$

29. Which of the following is the commonly accepted name for SeO_3?
    (a) selenium(VI) oxide   (c) selenium dioxide
    (b) selenium oxide   (d) selenium trioxide

30. Which of the following is the name for $HClO_3$?
    (a) hydrogen chlorate   (d) chloric acid
    (b) chlorous acid   (e) hydrochloric acid
    (c) perchloric acid

PROBLEMS

1. Answer the following questions about the element aluminum (Al).
   (a) What is its group number and general classification?
   (b) What is its electron configuration?
   (c) Is it a solid, liquid, or gas?
   (d) Is it a metal or nonmetal?
   (e) What is its electrical charge when present as an ion?
   (f) What noble gas has the same electron configuration as its ion?
   (g) What is the simplest formula when it forms a binary compound with (1) sulfur, (2) bromine, and (3) nitrogen?
   (h) What are the names of the three compounds in question (g)?

2. Answer the following questions about the element nitrogen (N).
   (a) What is its group number and general classification?
   (b) What is its electron configuration?
   (c) How many unpaired electrons are in an atom of nitrogen?
   (d) Is it a metal or nonmetal?
   (e) Does it exist as a solid, liquid, or gas under normal conditions?
   (f) What is the formula of the element?
   (g) What is the Lewis structure of the element?
   (h) What is its electrical charge when present as a monatomic ion?
   (i) What noble gas has the same electron configuration as its monatomic ion?
   (j) What is its polarity (negative or positive) when covalently bound to (1) oxygen and (2) boron?
   (k) What is the simplest formula when it forms a binary compound with (1) magnesium, (2) lithium, and (3) fluorine?
   (l) What are the names of the three compounds in question (k)?
   (m) What is the Lewis structure of the compound with fluorine in question (k)?

3. Answer the following with the symbol of the appropriate element.
   (a) a diatomic gas in nature but whose atoms have two unpaired electrons
   (b) a monatomic gas in nature but the atoms do not have an octet of electrons
   (c) the first element with a filled d sublevel
   (d) a nonmetal that forms a -2 ion that has a filled $4d$ sublevel
   (e) a metal that forms a $+3$ ion with a noble gas configuration by loss of a $4d$ electron

(f) a metal with a half-filled p sublevel but no f electrons
(g) the most chemically reactive element
(h) a transition element with 24 electrons beyond the previous noble gas
(i) a solid nonmetal whose atoms can form a -3 ion
(j) a metal that forms a $+2$ ion with the electron configuration of Xe

4. For each of the following compounds write (1) the formula in ionic form if ions are present, (2) the name of the compound, and (3) the Lewis structure.

Formula	Ionic Form	Name	Lewis Structure
NaClO	Na^+ClO^-	sodium hypochlorite	$Na^+\left[:\ddot{Cl}-\ddot{O}:\right]^-$
(a) $MgSO_4$			
(b) HNO_3			
(c) $LiNO_3$			
(d) $Co_2(CO_3)_3$			
(e) Cl_2O			

5. For each of the following compounds write (1) the formula of the compound, (2) the formula in ionic form if ions are present, and (3) the Lewis structure.

Name	Formula	Ionic Form	Lewis Structure
(a) dinitrogen trioxide			
(b) chromium(III) sulfite			
(c) iron(II) hydroxide			
(d) strontium oxalate			
(e) hydroiodic acid			

SAND GRAINS IN A DESERT
The smallness of the atom is difficult for the human mind to comprehend. There are more atoms in ''one mole,'' the counting unit of chemistry, than grains of sand in this picture.

QUANTITATIVE
RELATIONSHIPS:
THE MOLE

PURPOSE OF CHAPTER

In Chapter 8 we introduce the counting unit of chemistry known as the mole and describe how it relates the number of atoms, molecules, or formula units to the mass of elements and compounds.

OBJECTIVES FOR CHAPTER

After completion of this chapter, you should be able to:

1 Describe the unit known as the mole and tell why it is needed in chemistry. (Introduction, 8-1)

2 Determine the molar mass of any element from the periodic table. (8-2)

3 Calculate the mass of an element given the mass of the same number of atoms of a different element. (8-2)

4 Convert between moles, mass, and number of atoms of any element. (8-2)

5 Calculate the formula weight of a compound. (8-3)

8

6 Convert between moles, mass, and number of molecules or formula units of a compound. (8-3)

7 Calculate the percent composition of a compound from its formula. (8-4)

8 Distinguish between an empirical formula and a molecular formula. (8-5)

9 Calculate the empirical formula of a compound from its percent composition or mass composition. (8-5)

10 Use the data from chemical analysis to establish the molecular formula of a compound. (8-5, 8-6)

Elements and compounds; atoms, molecules, and ions; electron configuration and bonding; and finally systematic nomenclature—these are among the many topics of chemistry that have been discussed so far. All of these subjects pertain to the qualitative nature of matter, however. A topic of equal importance is how matter relates in a quantitative manner. For example, in a molecule such as carbon dioxide the chemist is concerned not only with what types of bonds hold the three atoms together but with what portion of the mass of carbon dioxide is distributed among its two elements. Of course, quantitative questions require mathematical calculations. To carry out these calculations in a systematic fashion, we will apply many of the problem-solving techniques discussed in Chapter 2 and in Appendix D. The material in this and subsequent chapters builds steadily but in an orderly, step-by-step manner. Generally, each topic extends the previous one. Thus, it is especially helpful to work at least some of the problems assigned in the margin immediately after each topic is discussed.

We now turn our attention back to an individual atom of an element. In Chapter 3, it was emphasized that an atom is unbelievably small although the most powerful type of microscope in use can produce fuzzy images of the larger atoms. As an example of the smallness of an atom just consider the period at the end of this sentence. You are looking at more than 10^{17} (one hundred quadrillion) atoms of carbon. That is enough to make only a barely visible smudge.

Let us now consider what we learned about the comparative masses of the individual atoms. The atomic mass of a certain isotope is the mass of that isotope relative to the atomic mass of ^{12}C which is defined as exactly 12 amu. The atomic weight of an element is the weighted average of all of the naturally occurring isotopes of that element. The mass unit used for individual atoms, *amu*, is hardly of any practical use to us, however. It is far too small to be measurable by any laboratory instrument. For example, the mass of one carbon atom, which is 12 amu, in grams is

$$12 \text{ amu} = 2.0 \times 10^{-23} \text{ g}$$

Since it would take about 10^{18} atoms of carbon to register on a sensitive laboratory balance, we obviously need to consider vast numbers of atoms at a time to make a dependable measurement. Thus, the chemist needs a counting unit that represents a huge number of particles. Just as the baker has the counting unit "dozen," the chemist has the "mole."

8-1 THE MOLE

Many of the counting units with which we are most familiar represent a number consistent with their use. For example, the baker sells a dozen doughnuts at a time because 12 is a practical number for that purpose. On the other hand, we may buy a ream of typing paper which is 500 sheets. A ream of doughnuts and a dozen sheets of paper are not practical amounts to purchase for most purposes. Likewise, the consideration of a dozen or even a ream of atoms is of little help to the chemist. We would still be dealing with a huge number of those units in a laboratory situation. The unit of amount specially suited to chemistry is known as the **mole** (the SI symbol is "mol"). The value of 1 mol is

$$6.02 \times 10^{23} \text{ objects or particles}$$

This unit is practical for the chemist. For example, one mole of iron nails has a mass of about 56 g, which is a convenient amount in a laboratory situation. *The number of objects in one mole is referred as the* **molar number** *or as* **Avogadro's number** (named in honor of Amedeo Avogadro, 1776–1856, a pioneer investigator of the quantitative aspects of chemistry). It is not an exact and defined number such as 12 in one dozen or 144 in one gross. It is actually known to more significant figures than the three that are shown and used here.

Although Avogadro's number is valuable to the chemist, its size defies description and the ability of the human mind to comprehend. For example, if an atom were the size of a marble and one mole of marbles were spread over the surface of the earth, the earth would be covered by a 50-mile-thick layer. Or, if the marbles were laid end to end and extended into outer space, they would reach past the farthest planets almost to the center of our galaxy. It takes light traveling at 186,000 miles per second over 30,000 years to travel from earth to the center of the galaxy. A new supercomputer can count all of the people in the United States in one-quarter of a second but it would take almost two million years for it to count one mole of people at the same rate.

■
SEE PROBLEMS
8-1 AND 8-2.

8-2 THE MOLAR MASS OF THE ELEMENTS

Obviously, it is out of the question to measure numbers of atoms by counting them one by one. So, we must count another way. In chemistry this is done by weighing. Let us consider an everyday example. Say a builder had 250 lb of £8 nails and wished to acquire an equal number of £12 nails. The builder could spend a few days counting but there is an easier way, using a weighing scale. Since nails are fairly uniform, one could easily find the average mass of a nail by weighing as few as a dozen of that particular size. Assume that the builder established that the average mass of a £8 nail is 0.197 oz and the average mass of a £12 nail is 0.269 oz. The following calculation gives the mass of £12 nails needed.

$$\frac{0.269 \text{ oz (£12)}}{0.197 \text{ oz (£8)}} = \frac{1.365 \text{ oz (£12)}}{\text{oz (£8)}} = \frac{1.365 \text{ lb. (£12)}}{\text{lb (£8)}} \qquad \text{(The mass ratio is the same for any mass units.)}$$

$$\cancel{250 \text{ lb (£8)}} \times \frac{1.365 \text{ lb (£12)}}{\cancel{\text{lb (£8)}}} = \underline{\underline{341 \text{ lb (£12)}}}$$

Likewise, in chemistry the number of atoms in a sample is counted by weighing. The mass of one mole of atoms (6.02×10^{23} atoms) is given in the periodic table and is known as the molar mass. *The* **molar mass** *of an element is the atomic weight expressed in grams.* In this text, the molar mass is usually expressed to three significant figures. Thus, the mass of 1 mole of oxygen atoms is 16.0 g, the mass of 1 mole of helium atoms is 4.00 g, and the mass of 1 mole of uranium atoms is 238 g.

Note that one mole of a certain element implies two things: the molar mass, which is different for each element, and the molar number, which is the same for all elements. (See Figure 8-1.)

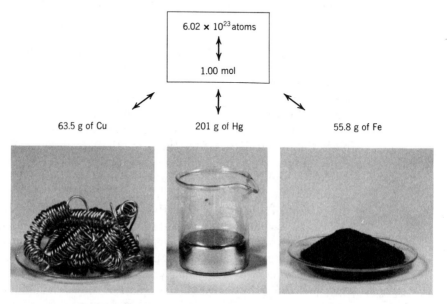

| 6.02×10^{23} atoms |
| $\updownarrow$ |
| 1.00 mol |

63.5 g of Cu   201 g of Hg   55.8 g of Fe

Figure 8-1   MOLES OF ELEMENTS. One mole of Cu, Fe, or Hg contains the same number of atoms.

TABLE 8-1   Mass Relation of C and He

C	He	Number of Atoms of Each Element Present
12.0 amu	4.00 amu	1
24.0 amu	8.00 amu	2
360 amu	120 amu	30
12.0 g	4.00 g	6.02×10^{23}
12.0 lb	4.00 lb	2.73×10^{26}
24.0 ton	8.00 ton	1.09×10^{30}

In the preceding discussion, we mentioned the builder who was confident that he had the same number of nails without knowing the exact number of nails present. Likewise, in most calculations in chemistry it is not necessary to know the actual numbers of atoms present as long as we know the relative numbers present. We will use as an example the elements helium (4.00 amu/ atom) and carbon (12.0 amu/atom). Note in Table 8-1 that whenever helium and carbon are present in a $4:12$ $(1:3)$ mass ratio, regardless of the units of mass, the same number of atoms are present in each case. As in the case of the £8 and £12 nails, once we know the relative masses, we can count by weighing.

EXAMPLE 8-1

The formula of the compound magnesium sulfide (MgS) indicates that there is one atom of Mg for every atom of S. What mass of sulfur is combined with 46.0 lb of magnesium?

PROCEDURE

From the atomic weights in the periodic table, note that there is the same number of magnesium atoms in 24.3 g of magnesium as there are sulfur atoms in 32.1 g of sulfur. There is also the same number of atoms in 24.3 lb of magnesium as there are in 32.1 lb of sulfur. This statement can be converted into two conversion factors, which we can use to convert a mass of one element to an equivalent mass of the other.

$$(1) \quad \frac{24.3 \text{ lb Mg}}{32.1 \text{ lb S}} = \frac{0.757 \text{ lb Mg}}{\text{lb S}} \qquad (2) \quad \frac{32.1 \text{ lb S}}{24.3 \text{ lb Mg}} = \frac{1.32 \text{ lb S}}{\text{lb Mg}}$$

Use factor 2 to convert mass of Mg to an equivalent mass of S.

$$46.0 \text{ lb Mg} \times \frac{1.32 \text{ lb S}}{\text{lb Mg}} = \underline{60.7 \text{ lb S}}$$

There is the same number of atoms in 60.7 lb of sulfur as in 46.0 lb of magnesium.

■ SEE PROBLEMS 8-3 THROUGH 8-10.

For many, the mole is a new and unfamiliar unit. Its huge value sometimes causes undue concern. Actually, the mole is a unit very much like a dozen and many of the calculations that the chemist makes with a "mole of sodium" are much like the calculations that one may make with a "dozen oranges." To set the stage for mole problems perhaps it may help to first work some problems with the everyday "dozen." After this we will work some very similar problems using the less familiar "mole." In the problems that follow, note that the factor-label method of problem solving introduced in Chapter 2 serves us well, especially with the less familiar units.

■
REVIEW THE FACTOR-LABEL METHOD IN CHAPTER 2 AND APPENDIX D.

In these problems, assume that there is such a thing as a standard orange (all oranges are exactly the same). The "dozen mass" (the mass of 12) of those particular oranges is 3.60 lb. The relationship given or implied by this statement are as follows:

$$1 \text{ doz oranges} = 12 \text{ oranges} \quad\quad 1 \text{ doz oranges} = 3.60 \text{ lb}$$

From these relationships, four conversion factors can be written:

$$(1) \ \frac{1 \text{ doz oranges}}{12 \text{ oranges}} \quad (2) \ \frac{12 \text{ oranges}}{\text{doz oranges}}$$

$$(3) \ \frac{3.60 \text{ lb}}{\text{doz oranges}} \quad (4) \ \frac{1 \text{ doz oranges}}{3.60 \text{ lb}}$$

EXAMPLE 8-2a

What is the mass of exactly 3 dozen oranges?

PROCEDURE

Convert dozen to mass. Use a conversion factor that has what you want (lb) in the numerator and what you wish to cancel (doz) in the denominator. This is factor 3. In shorthand, this can be written

$$\text{dozen} \xrightarrow{\text{factor 3}} \text{mass}$$

SOLUTION

$$3.00 \ \cancel{\text{doz oranges}} \times \frac{3.60 \text{ lb}}{\cancel{\text{doz oranges}}} = \underline{10.8 \text{ lb}}$$

EXAMPLE 8-3a

How many dozens of oranges are there in 18.0 lb?

PROCEDURE

This is the reverse of the previous problem, so use the inverse of the previous factor, which is factor 4.

$$\text{mass} \xrightarrow{\text{factor 4}} \text{dozen}$$

SOLUTION

$$18.0 \, \cancel{lb} \times \frac{1 \text{ doz}}{3.60 \, \cancel{lb}} = \underline{5.00 \text{ doz oranges}}$$

EXAMPLE 8-4a

What is the mass of 142 oranges?

PROCEDURE

Since a factor that relates number to mass is not available, use two factors in a two-step process. Note that you can go from number to dozens (factor 1) and from dozens to mass (factor 3).

$$\text{number} \xrightarrow{\text{factor 1}} \text{dozen} \xrightarrow{\text{factor 3}} \text{mass}$$

SOLUTION

$$142 \, \cancel{\text{oranges}} \times \frac{1 \, \cancel{\text{doz oranges}}}{12 \, \cancel{\text{oranges}}} \times \frac{3.60 \text{ lb}}{\cancel{\text{doz oranges}}} = \underline{42.6 \text{ lb}}$$

EXAMPLE 8-5a

How many individual oranges are there in 73.5 lb?

PROCEDURE

This is the reverse of the previous problem, so convert what's given to dozens (factor 4) and then dozens to individual oranges (factor 2).

$$\text{mass} \xrightarrow{\text{factor 4}} \text{dozen} \xrightarrow{\text{factor 2}} \text{number}$$

SOLUTION

$$73.5 \, \cancel{lb} \times \frac{1 \, \cancel{\text{doz oranges}}}{3.60 \, \cancel{lb}} \times \frac{12 \text{ oranges}}{\cancel{\text{doz oranges}}} = \underline{245 \text{ oranges}}$$

■
*REVIEW
SCIENTIFIC
NOTATION IN
APPENDIX C.*

Now, instead of working with dozens and "dozen mass," let's work some very similar problems with moles and molar mass. Review scientific notation in Appendix C if necessary. Note that Example 8-2a is the analog of Example 8-2b, and so forth.

In these problems you will work with sodium atoms instead of oranges. The molar mass (the mass of 6.02×10^{23} atoms) of Na is 23.0 g. The relationships given or implied by this statement are as follows:

$$1 \text{ mol of Na} = 6.02 \times 10^{23} \text{ atoms} \quad 1 \text{ mol of Na} = 23.0 \text{ g}$$

From these relationships, four conversion factors can be written:

(1) $\dfrac{1 \text{ mol Na}}{6.02 \times 10^{23} \text{ atoms Na}}$ (2) $\dfrac{6.02 \times 10^{23} \text{ atoms Na}}{\text{mol Na}}$

(3) $\dfrac{23.0 \text{ g}}{\text{mol Na}}$ (4) $\dfrac{1 \text{ mol Na}}{23.0 \text{ g}}$

EXAMPLE 8-2b

What is the mass of exactly 3 mol of Na?

PROCEDURE

Convert moles to mass. Use a conversion factor that has what you want (g) in the numerator and what you wish to cancel (mol) in the denominator. This is factor 3. In shorthand, this can be written

$$\text{moles} \xrightarrow{\text{factor 3}} \text{mass}$$

SOLUTION

$$3.00 \text{ mol Na} \times \frac{23.0 \text{ g}}{\text{mol Na}} = \underline{\underline{69.0 \text{ g}}}$$

EXAMPLE 8-3b

How many moles are there in 34.5 g of Na?

PROCEDURE

This is the reverse of the previous problem, so use the inverse of the previous factor, which is factor 4.

$$\text{mass} \xrightarrow{\text{factor 4}} \text{moles}$$

SOLUTION

$$34.5 \; g \times \frac{1 \; mol \; Na}{23.0 \; g} = \underline{1.50 \; mol \; Na}$$

EXAMPLE 8-4b

What is the mass of 1.20×10^{24} atoms of Na?

PROCEDURE

Since a factor that relates number to mass is not available, use two factors in a two-step process. Note that you can go from number to moles (factor 1) and from moles to mass (factor 3).

$$\text{number} \xrightarrow{\text{factor 1}} \text{moles} \xrightarrow{\text{factor 3}} \text{mass}$$

SOLUTION

$$1.20 \times 10^{24} \; \text{atoms Na} \times \frac{1 \; mol \; Na}{6.02 \times 10^{23} \; \text{atoms Na}} \times \frac{23.0g}{mol \; Na} = \underline{45.8 \; g}$$

EXAMPLE 8-5b

How many individual atoms are there in 11.5 g of Na?

PROCEDURE

This is the reverse of the previous problem, so convert what's given to moles (factor 4) and then moles to individual atoms (factor 2).

$$\text{mass} \xrightarrow{\text{factor 4}} \text{moles} \xrightarrow{\text{factor 2}} \text{number}$$

SOLUTION

$$11.5 \; g \times \frac{1 \; mol \; Na}{23.0 \; g} \times \frac{6.02 \times 10^{23} \; \text{atoms Na}}{mol \; Na} = \underline{3.01 \times 10^{23} \; \text{atoms Na}}$$

At this point you should pay special attention to the problems at the end of this chapter indicated in the margin. It is essential that you be able to work these types of problems backward and forward. If you master these problems, the rest of this chapter and the next should flow smoothly. For additional practice on problems of this type, refer to Appendix D, especially Examples D-12, D-13, D-17, D-18, and D-19.

■
SEE PROBLEMS
8-11 THROUGH
8-19.

8-3 THE MOLAR MASS OF COMPOUNDS

The periodic table tells us about the relative masses of the atoms of the elements. We are now ready to extend this information to the relative masses of compounds. Just as an element has an atomic weight, a compound has a formula weight. *The **formula weight** of a compound is determined from the number of atoms and the atomic weight of each element indicated by the formula.* Recall that the formula may represent discrete molecular units or ratios of ions in ionic compounds (known as a formula unit). The term *formula weight* refers to either ionic or molecular compounds and will be used in this text. The term *molecular weight* is sometimes used when referring only to molecular compounds. The following examples illustrate the computation of formula weights.

EXAMPLE 8-6

What is the formula weight of CO_2?

Atom	Number of Atoms in Molecule	Atomic Weight	Total Mass of Atom in Molecule
C	1	× 12.0 amu	= 12.0 amu
O	2	× 16.0 amu	= 32.0 amu
			44.0 amu

The formula weight of CO_2 is

<u>44.0 amu</u>

EXAMPLE 8-7

What is the formula weight of $Fe_2(SO_4)_3$?

Atom	Number of Atoms in Molecule	Atomic Weight	Total Mass of Atom in Formula Unit
Fe	2	× 55.8 amu	= 112 amu
S	3	× 32.1 amu	= 96.3 amu
O	12	× 16.0 amu	= 192 amu
			400 amu

■
SEE PROBLEMS 8-20 THROUGH 8-23.

The formula weight of $Fe_2(SO_4)_3$ is

<u>400 amu</u>

Once again, we need to establish a workable amount of molecules for the laboratory. Thus, we can extend the definition of the mole to include compounds. *The mass of one mole (6.02 × 10²³ formula units) is referred to as the* **molar mass** *of the compound and is the formula weight expressed in grams.* As was the case with atoms of elements, the molar masses of various compounds differ, but the number of formula units remains the same. (See Figure 8-2.)

When we extend the concept of the mole to compounds we have expanded the scope of information to a considerable extent. For example, consider one mole of the compound sulfuric acid, which in the pure state is composed of molecules with the formula H_2SO_4. In Table 8-2 the information concerning a quantity of one mole and all of its component atoms is summarized.

All of the information listed in Table 8-2 can be used as relationships to construct useful conversion factors. In the following two examples, some sample calculations illustrate how these relationships answer questions concerned with composition.

Figure 8-2 MOLES OF COMPOUNDS. One mole of $NaHCO_3$, H_2O or NaCl contains the same number of molecules or formula units.

TABLE 8-2   The Composition of One Mole of H_2SO_4

Total		Components	
1.00 mol of molecules		2 mol of H atoms	2.02 g / 1.20 × 10^{24} H atoms
6.02 × 10^{23} molecules	H_2SO_4	1 mol of S atoms	32.1 g / 6.02 × 10^{23} S atoms
98.1 g		4 mol of O atoms	64.0 g / 2.41 × 10^{24} O atoms

EXAMPLE 8-8

How many moles of each atom are present in 0.345 mol of $Al_2(CO_3)_3$? What is the total number of moles of atoms present?

PROCEDURE

In this problem, note that there are 2 Al atoms, 3 C atoms, and 9 O atoms in each formula unit of compound. Therefore, in 1 mol of compound, there are 2 mol of Al, 3 mol of C, and 9 mol of O atoms. This can be expressed with conversion factors as

$$\frac{2 \text{ mol Al}}{\text{mol Al}_2(CO_3)_3} \qquad \frac{3 \text{ mol C}}{\text{mol Al}_2(CO_3)_3} \qquad \frac{9 \text{ mol O}}{\text{mol Al}_2(CO_3)_3}$$

SOLUTION

Al $0.345 \text{ mol Al}_2(CO_3)_3 \times \dfrac{2 \text{ mol Al}}{\text{mol Al}_2(CO_3)_3} = \underline{0.690 \text{ mol Al}}$

C $0.345 \text{ mol Al}_2(CO_3)_3 \times \dfrac{3 \text{ mol C}}{\text{mol Al}_2(CO_3)_3} = \underline{1.04 \text{ mol C}}$

O $0.345 \text{ mol Al}_2(CO_3)_3 \times \dfrac{9 \text{ mol O}}{\text{mol Al}_2(CO_3)_3} = \underline{3.10 \text{ mol O}}$

Total moles of atoms present:

0.690 mol Al + 1.04 mol C + 3.10 mol O = $\underline{4.83 \text{ mol atoms}}$

EXAMPLE 8-9

What is the mass of 0.345 mol of $Al_2(CO_3)_3$, and how many individual ionic formula units does this amount represent?

PROCEDURE

This problem is worked much like the examples in which we were dealing with moles of atoms rather than compounds. First, find the formula weight of the compound as illustrated in Examples 8-6 and 8-7. The formula weight of the compound is 234 amu, so the molar mass is 234 g/mol.

SOLUTION

$$0.345 \text{ mol Al}_2(\text{CO}_3)_3 \times \frac{234 \text{ g}}{\text{mol Al}_2(\text{CO}_3)_3} = \underline{80.7 \text{ g}}$$

$$0.345 \text{ mol Al}_2(\text{CO}_3)_3 \times \frac{6.02 \times 10^{23} \text{ formula units}}{\text{mol Al}_2(\text{CO}_3)_3}$$

$$= \underline{2.08 \times 10^{23} \text{ formula units}}$$

It is important to be specific when discussing moles. For example, with the element chlorine be careful to distinguish whether you are discussing a mole of atoms (Cl) or a mole of molecules (Cl_2). *Note that there are two Cl atoms per Cl_2 molecule and 2 mol of Cl atoms per mole of Cl_2 molecules.*

$$1 \text{ mol of Cl atoms} \begin{cases} 35.5 \text{ g} \\ 6.02 \times 10^{23} \text{ atoms} \end{cases}$$

$$1 \text{ mol Cl}_2 \text{ molecules} \cdot \begin{cases} 71.0 \text{ g } (35.5 \text{ g} + 35.5 \text{ g}) \\ 6.02 \times 10^{23} \text{ molecules } (1.20 \times 10^{24} \text{ atoms}) \end{cases}$$

SEE PROBLEMS 8-24 THROUGH 8-32.

This situation is analogous to the difference between a dozen sneakers and a dozen *pairs* of sneakers. Both the mass and the actual number of the dozen pair of sneakers are double the dozen sneakers.

8-4 PERCENT COMPOSITION

Farmers must supplement the natural fertility of the soil with nitrogen fertilizers. In the use of such fertilizers, the actual mass of nitrogen added to the soil is more important than the total mass of the fertilizer, which contains other elements as well. Ammonia gas is the most effective as it is 82% by mass nitrogen but it must be injected into the soil. Solid ammonium nitrate is more convenient to apply but is only 35% by mass nitrogen. *The percent by mass, which is known as the* **percent composition** *of a compound, represents the mass of each element per 100 g (or any other mass unit) of compound.* It can be calculated from the formula of a compound. For example, the molar mass of CO_2 is 44.0 g/mol, which is composed of 12.0 g of carbon (one mole) and 32.0 g of oxygen (two moles). The percent composition is calculated by dividing the total mass of each component element by the total mass (molar mass) of the compound times 100%:

$$\frac{\text{total mass of component atom}}{\text{total mass (molar mass)}} \times 100\% = \text{percent composition}$$

In CO_2, the percent composition of C is

$$\frac{12.0 \text{ g C}}{44.0 \text{ g } CO_2} \times 100\% = \underline{27.3\% \text{ C}}$$

and the percent composition of O is

$$\frac{32.0 \text{ g O}}{44.0 \text{ g } CO_2} \times 100\% = \underline{72.7\% \text{ O}}$$

(Or, since there are only two components, if one is 27.3% the other must be $100 - 27.3 = 72.7\%$. Right?)

EXAMPLE 8-10

What is the percent composition of all of the elements in limestone, $CaCO_3$?

PROCEDURE

Find the molar mass and convert the *total* mass of each element to percent of molar mass.

SOLUTION

For $CaCO_3$

$$\text{Ca} \quad 1 \text{ mol Ca} \times \frac{40.1 \text{ g}}{\text{mol Ca}} = 40.1 \text{ g}$$

$$\text{C} \quad 1 \text{ mol C} \times \frac{12.0 \text{ g}}{\text{mol C}} = 12.0 \text{ g}$$

$$\text{O} \quad 3 \text{ mol O} \times \frac{16.0 \text{ g}}{\text{mol O}} = \underline{48.0 \text{ g}}$$

$$\text{molar mass} = 100.1 \text{ g/mol } CaCO_3$$

$$\% \text{ Ca} \quad \frac{40.1 \text{ g Ca}}{100.1 \text{ g } CaCO_3} \times 100\% = \underline{40.0\%}$$

$$\% \text{ C} \quad \frac{12.0 \text{ g C}}{100.1 \text{ g } CaCO_3} \times 100\% = \underline{12.0\%}$$

$$\% \text{ O} \quad \frac{48.0 \text{ g O}}{100.1 \text{ g } CaCO_3} \times 100\% = \underline{48.0\%}$$

EXAMPLE 8-11

What is the percent composition of all the elements in borazine, $B_3N_3H_6$?

PROCEDURE

Find the molar mass and convert the *total* mass of each element to percent of molar mass.

SOLUTION

For $B_3N_3H_6$

$$B \quad 3 \text{ mol B} \times \frac{10.8 \text{ g}}{\text{mol B}} = 32.4 \text{ g}$$

$$N \quad 3 \text{ mol N} \times \frac{14.0 \text{ g}}{\text{mol N}} = 42.0 \text{ g}$$

$$H \quad 6 \text{ mol H} \times \frac{1.01 \text{ g}}{\text{mol H}} = \underline{6.1 \text{ g}}$$

$$\text{molar mass} = 80.5 \text{ g/mol } B_3N_3H_6$$

$$\% B \quad \frac{32.4 \text{ g B}}{80.5 \text{ g } B_3N_3H_6} \times 100\% = \underline{40.2\%}$$

$$\% N \quad \frac{42.0 \text{ g N}}{80.5 \text{ g } B_3N_3H_6} \times 100\% = \underline{52.2\%}$$

$$\% H \quad \frac{6.1 \text{ g H}}{80.5 \text{ g } B_3N_3H_6} \times 100\% = \underline{7.6\%}$$

The percent composition can now be used to calculate the mass of a certain element in a quantity of compound.

EXAMPLE 8-12

What is the percent composition of nitrogen in ammonium nitrate and ammonia? What is the mass of nitrogen in 150 lb of each of these compounds?

PROCEDURE

Express the percent composition in appropriate fraction form (i.e., in this case, lb of N per 100 lb of compound). The mass is calculated as follows:

$$\text{mass of N} = \text{total mass of compound} \times \text{fraction of N}$$

SOLUTION

For ammonium nitrate (NH_4NO_3)

$$molar\ mass = (2 \times N) + (4 \times H) + (3 \times 0)$$
$$= 80.0\ g/mol$$

$$percent\ nitrogen = \frac{28.0\ g\ N}{80.0\ g\ NH_4NO_3} \times 100\% = \underline{\underline{35.0\%}}$$

$$fraction\ of\ nitrogen = \frac{35.0\ lb\ N}{100\ lb\ NH_4NO_3}$$

$$mass\ of\ nitrogen = 150\ lb\ \cancel{NH_4NO_3} \times \frac{35.0\ lb\ N}{100\ lb\ \cancel{NH_4NO_3}} = \underline{\underline{52.5\ lb\ N}}$$

For ammonia (NH_3)

$$molar\ mass = (N) + (3 \times H) = 17.0\ g/mol$$

$$percent\ nitrogen = \frac{14.0\ g\ N}{17.0\ g\ NH_3} \times 100\% = \underline{\underline{82.4\%}}$$

$$fraction\ of\ nitrogen = \frac{82.4\ lb\ N}{100\ lb\ NH_3}$$

$$mass\ of\ nitrogen = 150\ \cancel{lb\ NH_3} \times \frac{82.4\ lb\ N}{100\ \cancel{lb\ NH_3}}$$

$$= \underline{\underline{124\ lb\ N}}$$

SEE PROBLEMS
8-33 THROUGH
8-45.

8-5
EMPIRICAL
AND
MOLECULAR
FORMULAS

We have shown that we can calculate the percent composition given the formula. Is the reverse possible? Not quite. Note that in Example 8-11 the formula of borazine is $B_3N_3H_6$. This is called the **molecular formula** *since it represents the actual number of atoms in each molecular unit.* It is not the simplest whole-number ratio of atoms, however. *The simplest whole-number ratio of atoms in a compound is called the* **empirical formula**. In the case of borazine, the numerical subscripts can be divided through by 3 to leave the empirical formula, BNH_2. Since the percent composition is the *same* for an empirical formula as for a molecular formula, the best we can do is calculate the empirical formula from the percent composition or the mass composition.

To calculate the empirical formula from the percent composition, three steps are necessary:

1   Convert percent composition to an actual mass.
2   Convert mass to moles.
3   Find the whole-number ratio of the moles of different atoms. This is best illustrated by the following examples.

EXAMPLE 8-13

What is the empirical formula of laughing gas, which is 63.6% nitrogen and 36.4% oxygen?

PROCEDURE

To convert percent to a specific mass, remember that percent means parts per 100. Therefore, if we simply assume that we have a 100-g quantity of compound, the percent converts to a specific mass as follows:

$$63.6\% \text{ N} = \frac{63.6 \text{ g N}}{100 \text{ g compound}} \qquad 36.4\% \text{ O} = \frac{36.4 \text{ g O}}{100 \text{ g compound}}$$

We now convert the masses to moles of each element.

$$63.6 \text{ g N} \times \frac{1 \text{ mol N}}{14.0 \text{ g N}} = 4.54 \text{ mol N in 100 g}$$

$$36.4 \text{ g O} \times \frac{1 \text{ mol O}}{16.0 \text{ g O}} = 2.28 \text{ mol O in 100 g}$$

The ratio of N to O atoms will be the same as the ratio of N to O moles (or the ratio of number of dozens or any other unit). The formula cannot remain fractional, since only whole numbers of atoms are present in a compound. To find the whole-number ratio of moles, divide through by the smallest number of moles, which in this case is 2.28 mol of O.

$$\text{N} \quad \frac{4.54}{2.28} = 2.0 \qquad \text{O} \quad \frac{2.28}{2.28} = 1.0$$

The empirical formula of the compound is

$$\underline{\underline{N_2O}}$$

EXAMPLE 8-14

A sample of ordinary rust is composed of 2.78 g of iron and 1.19 g of oxygen. What is the empirical formula of rust?

PROCEDURE

Convert the mass of each element to moles.

$$2.78 \text{ g Fe} \times \frac{1 \text{ mol Fe}}{55.8 \text{ g Fe}} = 0.0498 \text{ mol Fe}$$

$$1.19 \text{ g O} \times \frac{1 \text{ mol O}}{16.0 \text{ g O}} = 0.0744 \text{ mol O}$$

Divide through by the smallest number of moles.

$$\text{Fe} \quad \frac{0.0498}{0.0498} = 1.0 \qquad \text{O} \quad \frac{0.0744}{0.0498} = 1.5$$

This time we're not quite finished, since $FeO_{1.5}$ still has a fractional number that must be cleared. (You should keep at least two significant figures in these numbers, so as not to round off a number like 1.5 to 2.) This fractional number can be cleared by multiplying both subscripts by a number that produces whole numbers only. In this case, multiply both subscripts by 2 and you have the empirical formula.

$$Fe_{(1\times2)}O_{(1.5\times2)} = \underline{\underline{Fe_2O_3}}$$

■
SEE PROBLEMS
8-46 THROUGH
8-59.

To determine the molecular formula of a compound, the molar mass (g/mol) and the mass of one empirical unit (g/emp. unit) are needed. The ratio of these two quantities must be a whole number. This whole number (a) is the number of empirical units in 1 mol.

$$a = \frac{\text{molar mass}}{\text{emp. mass}} = \frac{X \text{ g/mol}}{Y \text{ g/emp. unit}} = 1, 2, 3, \text{etc., emp. unit/mol}$$

Multiply the subscripts of the atoms in the empirical formula by a and you have the molecular formula. For example, the empirical formula of benzene is CH and its molar mass is 78 g/mol. The empirical mass (formula mass of CH) is 13.0 g/emp. unit.

$$a = \frac{78 \text{ g/mol}}{13 \text{ g/emp. unit}} = 6 \text{ emp. unit/mol}$$

The molecular formula is

$$C_{(1\times6)}H_{(1\times6)} = C_6H_6$$

EXAMPLE 8-15

A pure phosphorus–oxygen compound is 43.7% phosphorus and the remainder oxygen. The molar mass is 284 g/mol. What are the empirical and molecular formulas of this compound?

PROCEDURE

First, find the empirical formula. Convert percent to mass.

$$43.7\% \text{ P} = \frac{43.7 \text{ g P}}{100 \text{ g compound}}$$

$$\% \text{ O} = 100\% - 43.7\% = 56.3\%$$

$$56.3\% \text{ O} = \frac{56.3 \text{ g O}}{100 \text{ g compound}}$$

Convert mass to moles.

$$43.7 \, \cancel{g \, P} \times \frac{1 \, \text{mol P}}{31.0 \, \cancel{g \, P}} = 1.41 \, \text{mol P}$$

$$56.3 \, \cancel{g \, O} \times \frac{1 \, \text{mol O}}{16.0 \, \cancel{g \, O}} = 3.52 \, \text{mol O}$$

Divide through by the smallest number of moles.

$$P \quad \frac{1.41}{1.41} = 1.0 \qquad O \quad \frac{3.52}{1.41} = 2.5$$

Multiply both numbers by "2" to remove the fraction.

$$PO_{2.5} = P_{(1 \times 2)} O_{(2.5 \times 2)} = \underline{\underline{P_2O_5}}$$

To find the molecular formula, first compute the empirical mass.

$$\begin{array}{lll} P & 2 \times 31.0 \, \text{g} = & 62.0 \, \text{g} \\ O & 5 \times 16.0 \, \text{g} = & \underline{80.0 \, \text{g}} \\ & & 142 \, \text{g/emp. unit} \end{array}$$

$$a = \frac{284 \, \cancel{g}/\text{mol}}{142 \, \cancel{g}/\text{emp. unit}} = 2 \, \text{emp. unit/mol}$$

The molecular formula is

$$P_{(2 \times 2)} O_{(2 \times 5)} = \underline{\underline{P_4 O_{10}}}$$

■
SEE PROBLEMS
8-60 THROUGH
8-66.

8-6 THE USE OF EMPIRICAL AND MOLECULAR FORMULAS

One of the biggest thrills for the researcher in chemistry, from undergraduate to professor, is to discover and identify a new chemical compound. Such discoveries are actually frequent occurrences. (In March 1983 the American Chemical Society recorded the six millionth unique chemical compound.) To gain acceptance, a new compound must be characterized and reported in the chemical literature. (See Figure 8-3.) Perhaps the most fundamental part of any discovery is establishment of the molecular formula of the compound. To do so requires the type of calculations performed in this chapter. However, let us back up and go through the steps that are followed before a compound is accepted as a unique new substance. The following must be established in the order presented: (1) purity, (2) empirical formula, (3) molecular formula, (4) the structural formula (types and arrangement of bonds between atoms in the molecule). A list of physical and chemical properties of the new compound is needed to serve as a means of identification.

Before we move on, let's discuss a little more about how (1) through (4) are established.

1   Is It a Pure Substance?

This can be determined by examination of physical properties. A mixture of salt and sugar looks pure (as any joker who has put salt in the sugar jar knows), but heating the mixture soon reveals the truth. The sugar crystals soon melt and decompose, but the salt crystals do not. Pure substances have definite and sharp melting and boiling points; mixtures do not (see Chapter 1).

If the substance is pure and the chemists have established the melting and/or boiling point, they should then take a look in a handbook listing physical properties of known compounds to make sure they haven't just rediscovered something old. If, as often happens, they have, then back to the lab. If the compound still appears to be original, then on to the next step.

2   What Is the Empirical Formula of the New Compound?

Most chemists take the easy but smart way out and send a sample of their new pure substance to a commercial analytical laboratory, which reports back on percent composition of the elements requested. To actually obtain percent composition, many methods are used depending on the particular element, but the most commonly requested elements, carbon and hydrogen, are determined by analysis of combustion products. Such a determination is illustrated in Problems 8-59 and 8-66.

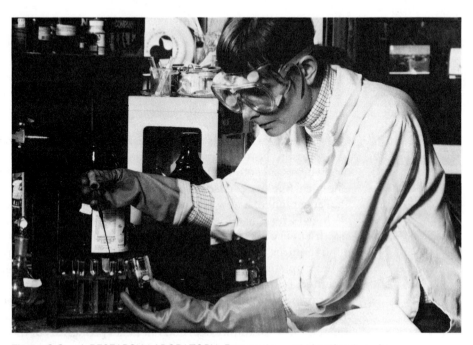

Figure 8-3   A RESEARCH LABORATORY. Preparation and identification of new compounds occur frequently in chemical laboratories.

3   What Is the Molecular Formula of the New Compound?

With the empirical formula now available, one next needs the molar mass to obtain the molecular formula. The molar mass can be obtained commercially like percent composition, or it may be obtained by several straightforward laboratory experiments. Determination of the molar mass of an unknown substance is usually performed in the general chemistry laboratory.

4   What Is the Structure of the Molecule?

Molecules can have the same molecular formula but different structures. For example, C_2H_6O is the formula for both dimethyl ether and ethyl alcohol (see Problem 6-27e). Determination of structure can take a few minutes or a few years depending on the nature of the molecule, its complexity, and the instruments necessary. Nobel prizes have been awarded for determination of structure of molecules (e.g., Watson and Crick for the structure of DNA).

■
*SEE PROBLEMS
8-52 THROUGH
8-57.*

With all of the above information in hand, chemists should feel confident enough to face their peers with their discovery.

8-7 CHAPTER REVIEW

The atom is so small that it takes literally an uncountable number of atoms to work with in a laboratory. One person could not count the atoms in a single tiny crystal of sugar even if several billion years were available. Regardless of the problem, we still need to be able to measure equivalent numbers of atoms of elements. Since an actual count is out of the question, we use a mass that contains a known quantity of atoms. The standard quantity used is the SI unit known as the mole (mol) and it represents 6.02×10^{23} atoms, molecules, or formula units of the element or compound. By use of atomic weights from the periodic table the mass of one mole can be determined. The information on the mole is summarized as follows.

Unit	Number	Mass
1.00 mol ⎰	6.02×10^{23} atoms	Atomic weight of element in grams
	6.02×10^{23} molecules	Formula weight of molecular compound in grams
	6.02×10^{23} formula units	Formula weight of ionic compound in grams

A major emphasis of this chapter is to become comfortable with conversions between moles and number and, more importantly, between moles and mass. The conversions are summarized as follows. Note that moles can be converted to either a number (using Avogadro's number as a conversion factor) or a mass (using the molar mass as a conversion factor).

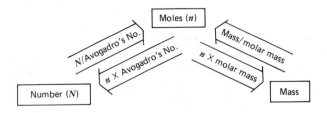

Percent composition can be obtained from the formula of the compound and the atomic weights of the elements. If the percent composition or mass composition is given, the empirical formula can be calculated. The empirical formula is the simplest whole number ratio of atoms in a compound. The empirical formula is determined according to the following steps.

percent composition → mass in grams of elements →
            moles of atoms → mole ratio of atoms →
                whole-number mole ratio of atoms → empirical formula

The molecular formula may be the same or a multiple of the empirical formula. It is obtained by comparing the molar mass of the compound with the empirical mass.

■ **EXERCISES**

THE MAGNITUDE OF THE MOLE

8-1 If you could count two numbers per second, how many years would it take you to count Avogadro's number? If you were helped by the whole human race of five billion people, how long would it take?

8-2 A small can of soda contains 350 mL. How many cans of soda would it take to equal all of the water on earth? The earth contains 326 million cubic miles (3.5×10^{20} gal) of water. How many moles of cans is this?

RELATIVE NUMBERS OF ATOMS

8-3 A piece of pure gold has a mass of 145 g. What would be the mass of the same number of silver atoms?

8-4 A chunk of pure aluminum has a mass of 212 lb. What is the mass of the same number of carbon atoms?

8-5 Some copper pennies have a mass of 16.0 g; the same number of atoms of a precious metal has a mass of 49.1 g. What is the metal?

8-6 In the compound CuO, what is the mass of copper present for each 18.0 g of oxygen?

8-7 In the compound NaCl, what mass of sodium is present for each 425 g of chlorine?

*8-8 In a compound containing one atom of carbon and one atom of another element, it is found that 25.0 g of carbon is combined with 33.3 g of the other element. What is the element and the formula of the compound?

8-9 In the compound $MgBr_2$, what mass of bromine is present for each 46.0 g of magnesium? (Remember, there are two bromines per magnesium.)

*8-10 In the compound SO_3, what mass of sulfur is present for each 60.0 lb of oxygen?

MOLES OF ATOMS

8-11   Fill in the blanks. Use the factor-label method to determine the answers.

Element	Mass in Grams	Number of Moles	Number of Atoms
S	8.00	0.250	1.50×10^{23}
(a) P	14.5		
(b) Rb		1.75	
(c) Al			6.02×10^{23}
(d)	363		3.01×10^{24}
(e) Ti			1

8-12   Fill in the blanks.

Element	Mass in Grams	Number of Moles	Number of Atoms
(a) Na	0.390		
(b) Cr		3.01×10^{4}	
(c)	43.2		2.41×10^{24}
(d) K	4.25×10^{-6}		
(e) Ne			3.66×10^{21}

8-13   What is the mass in grams of each of the following?
(a) 1.00 mol of Cu
(b) 0.50 mol of S
(c) 6.02×10^{23} atoms of Ca

8-14   What is the mass in grams of each of the following?
(a) 4.55 mol of Be
(b) 6.02×10^{24} atoms of Ca
(c) 3.40×10^{5} mol of B

8-15   How many individual atoms are in each of the following?
(a) 32.1 mol of sulfur
(b) 32.1 g of sulfur
(c) 32.0 g of oxygen

8-16   How many moles are in each of the following?
(a) 281 g of silicon
(b) 7.34×10^{25} atoms of phosphorus
(c) 19.0 atoms of fluorine

8-17   Which has more atoms: 50.0 g of Al or 50.0 g of Fe?

8-18   Which contains more Ni: 20.0 g, 2.85×10^{23} atoms, or 0.450 mol?

8-19   Which contains more Cr: 0.025 mol, 6.0×10^{21} atoms, or 1.5 g of Cr?

FORMULA WEIGHT

8-20   What is the formula weight of each of the following? Express your answer to three significant figures.
(a) $KClO_2$ (e) Na_2CO_3
(b) SO_3 (f) CH_3COOH
(c) N_2O_5 (g) $Fe_2(CrO_4)_3$
(d) H_2SO_4

8-21   What is the formula weight of each of the following?
(a) $CuSO_4 \cdot 6H_2O$ (include the H_2O's)
(b) Cl_2O_3
(c) $Al_2(C_2O_4)_3$
(d) $Na_2BH_3CO_2$
(e) P_4O_6

8-22   A compound is composed of three sulfate ions and two chromium ions. What is the formula weight of the compound?

8-23   What is the formula weight of strontium perchlorate?

MOLES OF COMPOUNDS

8-24   Fill in the blanks. Use the factor-label method to determine answers.

Molecules	Mass in Grams	Number of Moles	Number of Molecules or Formula Units
N_2O	23.8	0.542	3.26×10^{23}
(a) H_2O		10.5	
(b) BF_3			3.01×10^{21}
(c) SO_2	14.0		
(d) K_2SO_4		1.20×10^{-4}	
(e) SO_3			4.50×10^{24}
(f) $N(CH_3)_3$	0.450		

8-25   Fill in the blanks.

Molecules	Mass in Grams	Number of Moles	Number of Molecules or Formula Units
(a) O_3	176		
(b) NO_2		3.75×10^{-3}	
(c) Cl_2O_3			150
(d) UF_6			8.50×10^{22}

8-26   How many moles of each type of atom are present in 2.55 mol of grain alcohol, C_2H_6O? What is the total number of moles of atoms present? What is the mass of each element present? What is the total mass?

8-27   How many moles are in 28.0 g of $Ca(ClO_3)_2$? How many moles of each element are present? How many total moles of atoms are present?

8-28   How many moles are in 84.0 g of $K_2Cr_2O_7$? How many moles of each element are present? How many total moles of atoms are present?

8-29   What mass of each element is present in 1.50 mol of H_2SO_3?

8-30   What mass of each element is present in 2.45 mol of boric acid (H_3BO_3)?

8-31   How many moles of O_2 are in 1.20×10^{22} O_2 molecules? How many moles of oxygen atoms? What is the mass of oxygen molecules? What is the mass of oxygen atoms?

8-32   How many moles of Cl_2 molecules are in 985 g of Cl_2? How many moles of Cl atoms are present? What is the mass of Cl atoms present?

PERCENT COMPOSITION

8-33   What is the percent composition of a compound composed of 1.375 g of N and 3.935 g of O?

8-34   A sample of a compound has a mass of 4.86 g and is composed of silicon and oxygen. What is the percent composition if 2.27 g of the mass is silicon?

8-35   The mass of a sample of a compound is 7.44 g. Of that mass 2.88 g is potassium, 1.03 g is nitrogen, and the remainder is oxygen. What is the percent composition of the elements?

8-36   What is the percent composition of all of the elements in the following compounds?
(a) C_2H_6O (d) Na_2SO_4
(b) C_3H_6 (e) $(NH_4)_2CO_3$
(c) C_9H_{18}

8-37   What is the percent composition of all of the elements in the following compounds?
(a) H_2CO_3 (c) $Al(NO_3)_3$
(b) Cl_2O_7 (d) $NH_4H_2PO_4$

8-38   What is the percent composition of all of the elements in borax $(Na_2B_4O_7 \cdot 10H_2O)$?

8-39   What is the percent composition of all of the elements in acetaminophen $(C_8H_9O_2N)$? (Acetaminophen is an aspirin substitute.)

8-40   What is the percent composition of all of the elements in saccharin $(C_7H_5SNO_3)$?

8-41   What is the percent composition of all of the elements in amphetamine (C_9NH_{13})? (Amphetamine is a stimulant.)

8-42   What mass of carbon is in a 125-g quantity of sodium oxalate?

8-43   What is the mass of phosphorus in a 25.0-lb quantity of sodium phosphate?

8-44   What mass of chromium is in a 275-kg quantity of chromium(III) carbonate?

8-45   Iron is recovered from Fe_2O_3. How many pounds of iron can be recovered from each ton of the iron ore, Fe_2O_3 (1 ton = 2000 lb)?

EMPIRICAL FORMULAS

8-46   Which of the following are not empirical formulas?
(a) N_2O_4 (d) $H_2C_2O_4$
(b) Cr_2O_3 (e) Mn_2O_7
(c) $H_2S_2O_3$

8-47   Convert the following mole ratios of elements to empirical formulas.
(a) 0.25 mol of Fe and 0.25 mol of S
(b) 1.88 mol of Sr and 3.76 mol of I
(c) 0.32 mol of K, 0.32 mol of Cl, and 0.96 mol of O
(d) 1.0 mol of I and 2.5 mol of O
(e) 2.0 mol of Fe and 2.66 mol of O

(f) 4.22 mol of C, 7.03 mol of H, and 4.22 mol of Cl

8-48 Convert the following mole ratios of elements to empirical formulas.
(a) 1.20 mol of Si and 2.40 mol of O
(b) 0.045 mol of Cs and 0.022 mol of S
(c) 1.0 mol of X and 1.20 mol of Y
(d) 3.11 mol of Fe, 4.66 mol of C, and 14.0 mol of O

8-49 What is the empirical formula of a compound that has the composition 63.1% oxygen and 36.8% nitrogen?

8-50 What is the empirical formula of a compound that has the composition 41.0% K, 33.7% S, and 25.3% O?

8-51 In an experiment it was found that 8.25 g of potassium combines with 6.75 g of O_2. What is the empirical formula of the compound?

8-52 Orlon is composed of very long molecules with a composition of 26.4% N, 5.66% H, and 67.9% C. What is the empirical formula for orlon?

8-53 A compound is 21.6% Mg, 21.4% C, and 57.0% O. What is the empirical formula of the compound?

8-54 A compound is composed of 9.90 g of carbon, 1.65 g of hydrogen, and 29.3 g of chlorine. What is the empirical formula of the compound?

8-55 A compound is composed of 0.46 g of Na, 0.52 g of Cr, and 0.64 g of O. What is the empirical formula of the compound?

8-56 A compound is composed of 24.1% nitrogen, 6.90% hydrogen, 27.6% sulfur, and the remainder oxygen. What is the empirical formula of the compound?

*8-57 Methyl salicylate is also known as "oil of wintergreen." It is composed of 63.2% carbon, 31.6% oxygen, and 5.26% hydrogen. What is its empirical formula?

*8-58 Nitroglycerin is used as an explosive and as a heart medicine. It is composed of 15.9% carbon, 18.5% nitrogen, 63.4% oxygen, and 2.20% hydrogen. What is its empirical formula?

*8-59 A hydrocarbon (a compound that contains only carbon and hydrogen) was burned and the products of the combustion were collected and weighed. All of the carbon present in the original compound is now present in 1.20 g of CO_2. All of the hydrogen is present in 0.489 g of H_2O. What is the empirical formula of the compound? (Hint: Remember that all of the moles of C atoms in CO_2 and of H atoms in H_2O came from the original compound.)

MOLECULAR FORMULAS

8-60 A compound has the following composition: 20.0% C, 2.2% H, and 77.8% Cl. The molar mass of the compound is 545 g/mol. What is the molecular formula of the compound?

8-61 A compound is composed of 1.65 g of nitrogen and 3.78 g of sulfur. If its molar mass is 184 g/mol, what is its molecular formula?

8-62 A compound was reported in 1967 to have a composition of 18.7% B, 20.7% C, 5.15% H, and 55.4% O. Its molar mass was found to be about 115 g/mol. What is the molecular formula of the compound?

8-63 A new compound reported in 1970 has a composition of 34.9% K, 21.4% C, 12.5% N, 2.68% H, and 28.6% O. It has a molar mass of about 224 g/mol. What is its molecular formula?

8-64 Fructose is also known as fruit sugar. It has a molar mass of 180 g/mol and is composed of 40.0% carbon, 53.3% oxygen, and 6.7% hydrogen. What is its molecular formula?

8-65 A new compound reported in 1982 has a molar mass of 834 g/mol. A 20.0-g sample of the compound contains 18.3 g of iodine and the remainder carbon. What is the molecular formula of the compound?

*8-66 A 0.500-g sample of a compound containing C, H, and O was burned and the products collected. The combustion produced 0.733 g of CO_2 and 0.302 g of H_2O. The molar mass of the compound is 60.0 g/mol. What is the molecular formula of this compound? (Hint: Find the mass of C and H in the original compound; the remainder of the 0.500 g will be O.)

IRON

Hot, molten iron is being processed in a steel mill. Iron is found in the earth combined with other elements. Chemical reactions free the iron from the other elements.

CHEMICAL
REACTIONS:
EQUATIONS
AND
QUANTITIES

PURPOSE OF CHAPTER

In Chapter 9 we use balanced chemical equations to illustrate types of reactions and quantitative calculations.

OBJECTIVES FOR CHAPTER

After completion of this chapter, you should be able to:

1 Describe the information represented by a chemical equation. (9-1)

2 Balance simple equations. (9-1)

3 Classify chemical reactions into one of the five types listed. (9-2)

4 Use the balanced equation to obtain mole relationships among reactants and products (equation factors). (9-3)

5 Make the following stoichiometric conversions: (9-3)
   (a) Mole to mole   (c) Mass to mass
   (b) Mole to mass   (d) Mass to number

6 Calculate the percent yield from the actual yield and the theoretical yield. (9-4)

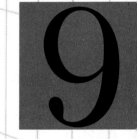

7 Calculate the percent purity of a sample from the mass of a product. (9-4)

8 Identify the limiting reactant, the reactant in excess, and the yield from given amounts of reactants. (9-5)

9 Use the balanced equation to calculate the amount of heat energy involved in a specified reaction. (9-6)

As you sit here calmly trying to read this page keep in mind that all of the activity of a small chemical plant is going on in your body. Many of the compounds in the food you had for lunch are undergoing combustion to produce other compounds and releasing life-giving energy. No wonder it seems so hard to concentrate with all that going on inside. Fortunately, these chemical reactions (collectively known as metabolism) are proceeding at a controlled rate so that you don't spontaneously ignite and go up in smoke. Combustion is a type of chemical reaction that describes a certain chemical change. As mentioned in Chapter 1, chemical changes produce other compounds and in so doing produce or absorb energy. We are now ready to examine these changes in more detail.

Two aspects of chemical changes concern us in this chapter. The first is the representation of these changes with what is known as the chemical equation. If the chemical symbol of an element is the chemist's analog to the alphabet, and the formulas of compounds are equivalent to words, then the chemical equation is like a complete sentence. After discussion of how chemical reactions are properly represented by equations we will see how these representations illustrate various types of chemical reactions.

The second aspect concerns the significant amount of quantitative information implied by the chemical equation. For example, a question of utmost importance to the chemist is how much of a product compound may be produced by a specific amount of a reactant compound. As in any other endeavor, the amount of a product is limited by the input of ingredients. We will see how such calculations are made in chemistry.

9-1 THE CHEMICAL EQUATION

In an earlier chapter we had occasion to describe certain chemical reactions. For example, we have mentioned that gasoline burns to produce carbon dioxide and water. Iron rusts to produce iron(III) oxide, and sugar (sucrose) ferments to produce alcohol and carbon dioxide gas. We are now ready to represent such information with chemical symbols. *A **chemical equation** is the*

TABLE 9-1	Symbols in the Chemical Equation
Symbol	Use
+	Between the formulas of reactants or products
$\longrightarrow$	Means "yields" or "produces"; separates reactants from products
=	Same as arrow
$\rightleftharpoons$	Used in place of a single arrow for reversible reactions (see Chapter 14)
(g)	Indicates a gaseous reactant or product
↑	Sometimes used to indicate a gaseous product
(s)	Indicates a solid reactant or product
↓	Sometimes used to indicate a solid product
(l)	Indicates a liquid reactant or product
(aq)	Indicates that the reactant or product is in aqueous solution (dissolved in water)
$\xrightarrow{\Delta}$	Indicates that heat must be supplied to reactants before a reaction occurs
$\xrightarrow{MnO_2}$	An element or compound written above the arrow is a *catalyst; a catalyst speeds up a reaction but is not consumed in the reaction*

representation of a chemical reaction using elemental symbols and formulas.
Let us take, as an example, a simple chemical reaction. In words, the reaction is
described as follows: "hydrogen combines with oxygen to produce water."
This information may appear as

$$H + O \longrightarrow H_2O$$

In this format note that "combines with" (or "reacts with") is represented by a
"+". When there is more than one reactant or product, the formulas on each
side of the equation are separated by a "+". The word "produce," also referred to as "yields," is represented by a →. Note in Table 9-1 that there are
other representations for the yield sign, depending on the situation.

The above chemical equation is in need of considerable refinement. First, if
an element exists as molecules under normal conditions then the formula of
the element is shown. Recall from Chapter 5 that hydrogen and oxygen [along
with nitrogen and all group 17 (VIIA) elements] are diatomic molecules under
normal conditions. Including this information the equation is

$$H_2 + O_2 \longrightarrow H_2O$$

We now must consider the law of conservation of mass, which states that
matter can be neither created nor destroyed. In Dalton's atomic theory, this
law was explained in terms of chemical reactions. He suggested that reactions
were simply rearrangements of the same number of atoms. (See Figure 9-1.) A
close look at the equation above shows that there is one more oxygen on the
left side of the equation than on the right. Equations must be balanced, which
means that all atoms on the left side of the equation (the reactants) must be

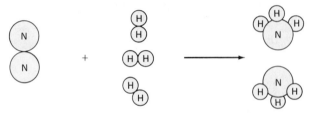

Figure 9-1   NITROGEN PLUS HYDROGEN YIELDS AMMONIA. In the chemical reaction, the atoms are simply rearranged into different molecules.

present on the right side of the equation (the products). The equation cannot be balanced by changing the subscripts of a compound, because that would change the compound's identity. For example, the equation would be balanced if the H_2O were changed to H_2O_2. However, H_2O_2 is hydrogen peroxide, which is not the same thing as water. *An equation is balanced by introducing* **coefficients.** In this case a "2" in front of the H_2O solves the original oxygen problem but unbalances the hydrogens:

$$H_2 + O_2 \longrightarrow 2H_2O$$

This problem can be solved easily. Simply return to the left and put a coefficient of "2" in front of the H_2 and the equation is balanced:

$$2H_2 + O_2 \longrightarrow 2H_2O$$

Finally, the physical states of the reactants and products under the reaction conditions are sometimes added in parentheses after the formula for each compound. Hydrogen and oxygen are gases and water is a liquid under the conditions of this reaction. Using the proper letters shown in Table 9-1, we have the balanced chemical equation in proper form:

$$2H_2(g) + O_2(g) \longrightarrow 2H_2O(l)$$

Note that if we describe this reaction in words, we have quite a mouthful. "Two molecules of gaseous hydrogen react with one molecule of gaseous oxygen to produce two molecules of liquid water."

We will now consider the balancing of chemical equations in more detail. This endeavor is important to us because it is a prerequisite to an understanding of the quantitative aspects of chemical reactions presented later in this chapter. Before we consider some guidelines there are three points that one should keep in mind concerning balanced equations.

1   The subscripts of a compound are fixed; they cannot be changed to balance an equation.

2   The coefficients used should be the smallest whole numbers possible.

3   The coefficient multiplies every number in the formula. For example, $2KClO_2$ indicates the presence of two atoms of K, two atoms of Cl, and four atoms of O.

In this chapter, equations will be balanced by inspection. Certainly, there are many complex equations that are extremely tedious to balance by this method but such equations will be held until later chapters in which more systematic methods can be employed. The following rules are helpful in balancing simple equations by inspection.

1 Consider first the compound with the most atoms and then the element in that compound (other than hydrogen or oxygen) with the largest number of atoms. Consider polyatomic ions that are the same on both sides of the equation as single units. Balance the atom in question on the other side of the equation.

2 Balance all other atoms except hydrogen and oxygen.

3 Generally, it is easiest to balance hydrogen next, followed by oxygen.

4 Check to see that the atoms of all elements are balanced and the coefficients are expressed as the smallest whole-number ratio.

EXAMPLE 9-1

Write a balanced chemical equation from the following word equation: "Nitrogen gas reacts with hydrogen gas to produce ammonia gas.

PROCEDURE

The unbalanced chemical equation is

$$N_2(g) + H_2(g) \longrightarrow NH_3(g)$$

First consider the NH_3 molecule since it has the largest number of atoms. Because the rules suggest that we hold hydrogen until later, we first balance the nitrogens. There is one on the right and two on the left. If "2" is used as the coefficient for NH_3, the nitrogens are balanced:

$$N_2(g) + H_2(g) \longrightarrow 2NH_3(g)$$

Now consider the hydrogens. There are six on the right but only two on the left. Place the coefficient "3" before the H_2 and the equation is balanced. (See Figure 9-1.)

SOLUTION

$$N_2(g) + 3H_2(g) \longrightarrow 2NH_3(g)$$

EXAMPLE 9-2

Balance the following equation:

$$B_2H_6(g) + H_2O(l) \longrightarrow H_3BO_3(aq) + H_2(g)$$

PROCEDURE

First consider the B_2H_6 molecule because it has the largest number of atoms. The B should be balanced first, so add the coefficient "2" before the H_3BO_3.

$$B_2H_6(g) + H_2O(l) \longrightarrow 2H_3BO_3(aq) + H_2(g)$$

Now consider the hydrogens. Note that they are balanced so next consider oxygen which is not balanced. The coefficient "6" is needed before the H_2O but this unbalances the hydrogens. There are now 18 on the left but only 8 on the right. If the coefficient "6" is placed before the H_2, however, the equation is balanced.

SOLUTION

$$\underline{B_2H_6(g) + 6H_2O(l) \longrightarrow 2H_3BO_3(aq) + 6H_2(g)}$$

EXAMPLE 9-3

Balance the following equation:

$$C_3H_8O(l) + O_2(g) \longrightarrow CO_2(g) + H_2O(l)$$

PROCEDURE

First, consider the C in the C_3H_8O molecule. Add a "3" before the CO_2 on the right to balance the three carbons on the left:

$$C_3H_8O(l) + O_2(g) \longrightarrow 3CO_2(g) + H_2O(l)$$

Balance hydrogen next by adding a "4" before the H_2O to balance the eight hydrogens in C_3H_8O:

$$C_3H_8O(l) + O_2(g) \longrightarrow 3CO_2(g) + 4H_2O(l)$$

Now note that there are 10 oxygens on the right. On the left, one oxygen is in C_3H_8O so nine are needed from O_2. To get an odd number of oxygens from O_2 we would need to use a fractional coefficient, in this case "$\frac{9}{2}$":

$$C_3H_8O(l) + \tfrac{9}{2}O_2(g) \longrightarrow 3CO_2(g) + 4H_2O(l)$$

Since we are expressing only whole numbers in a balanced equation, all coefficients should be multiplied by "2" to clear the fraction. The final balanced equation is:

$$\underline{2C_3H_8O(l) + 9O_2(g) \longrightarrow 6CO_2(g) + 8H_2O(l)}$$

SEE PROBLEMS 9-1 THROUGH 9-5.

9-2 TYPES OF CHEMICAL REACTIONS

Equations represent all sorts of chemical reactions that occur. Some involve putting substances together, others involve breaking them apart, and still others involve more complex processes. Whatever happens, it has long been realized that it is possible to group most chemical reactions into some simple classifications. In this chapter, we describe the common characteristics of reactions that allow us to assign most reactions to five basic types. In later chapters other common characteristics of reactions that we will discuss at the time allow other ways of classification. The five types of reactions are (1) combination reactions, (2) decomposition reactions, (3) combustion reactions, (4) single-replacement reactions, and (5) double-replacement reactions. A brief discussion of each type follows.

Combination Reactions

The first classification of chemical reactions concerns the synthesis of one compound from elements or other simpler compounds. Whatever the reactants, the common characteristic of this class is the production of one compound. An example, is the synthesis of MgO from its elements as shown below with a balanced equation and an illustration of the atoms concerned. (See also Figure 9-2.)

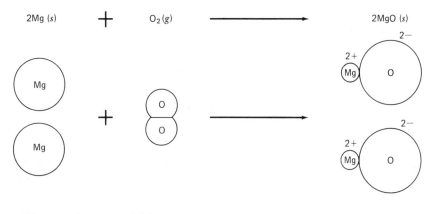

$$2Mg\,(s) \quad + \quad O_2\,(g) \quad \longrightarrow \quad 2MgO\,(s)$$

Figure 9-2 COMBINATION OR COMBUSTION REACTION. When magnesium burns in air, the reaction can be classified as either a combination or a combustion reaction.

The following equations represent other combination reactions.

$$2Na(s) + Cl_2(g) \longrightarrow 2NaCl(s)$$
$$C(s) + O_2(g) \longrightarrow CO_2(g)$$
$$CaO(s) + CO_2(g) \longrightarrow CaCO_3(s)$$
$$SO_2(g) + H_2O(l) \longrightarrow H_2SO_3(aq)$$

Decomposition Reactions

The next type of reaction is simply the reverse of the first. In this type of reaction, there is only one reactant, which decomposes to form elements or other simpler compounds. Many of these reactions take place only at a high temperature (indicated by the Δ). The decomposition of carbonic acid (H_2CO_3) is illustrated below and is followed by other examples. (See Figure 9-3.)

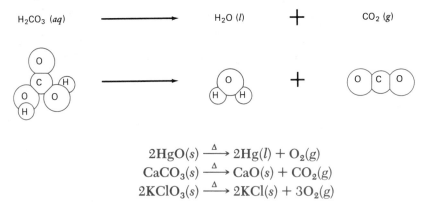

$$2HgO(s) \xrightarrow{\Delta} 2Hg(l) + O_2(g)$$
$$CaCO_3(s) \xrightarrow{\Delta} CaO(s) + CO_2(g)$$
$$2KClO_3(s) \xrightarrow{\Delta} 2KCl(s) + 3O_2(g)$$

Combustion Reactions

When a substance combines with oxygen it is said to undergo combustion. We know these reactions simply as "burning." There is some overlap of this classification with another type of reaction. Note that when an element com-

Figure 9-3 DECOMPOSITION REACTION. The fizz of carbonated water is a result of a decomposition reaction.

bines with oxygen, only one product is formed, so the reaction can also be classified as a combination reaction. (See Figure 9-2.)

$$C(s) + O_2(g) \longrightarrow CO_2(g)$$
$$2Mg(s) + O_2(g) \longrightarrow 2MgO(s)$$

When compounds containing carbon and hydrogen burn in sufficient oxygen, carbon dioxide and water are formed.

$$CH_4(g) + 2O_2(g) \longrightarrow CO_2(g) + 2H_2O(l)$$
$$2C_3H_8O(l) + 9O_2(g) \longrightarrow 6CO_2(g) + 8H_2O(l)$$

When insufficient oxygen is present (as in the combustion of gasoline in an automobile engine), some carbon monoxide (CO) also forms.

Single-Replacement Reactions

These reactions are also referred to as **substitution reactions** because the reaction involves the substitution of a free element for another element in a compound. These reactions usually occur in aqueous solution. The substitution of zinc metal for the Cu^{2+} ion in an aqueous solution of $CuCl_2$ is illustrated below; other examples follow. (See also Figure 9-4.)

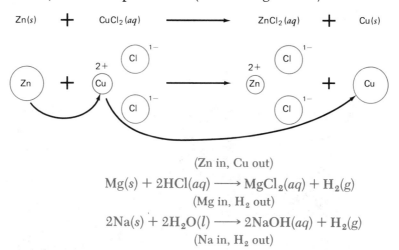

(Zn in, Cu out)
$$Mg(s) + 2HCl(aq) \longrightarrow MgCl_2(aq) + H_2(g)$$
(Mg in, H_2 out)
$$2Na(s) + 2H_2O(l) \longrightarrow 2NaOH(aq) + H_2(g)$$
(Na in, H_2 out)

Figure 9-4 SINGLE-REPLACEMENT REACTION. The formation of a layer of copper on a piece of zinc is a single-replacement reaction.

For single-replacement reactions, tables are available that allow one to predict which element is able to replace another. Predictions of what chemical reactions may or may not occur is an important endeavor in chemistry but will not be discussed at this time.

Double-Replacement Reactions

These reactions are also known as **metathesis** reactions. Like single-replacement reactions, these reactions occur mainly in aqueous solution between ionic compounds or compounds such as acids that can produce ions in solution. In these reactions the cations and anions exchange partners. What type of product forms from this exchange leads to a classification into two subgroups. In the first, the products are also ionic compounds but one is insoluble in water so it forms a solid product known as a **precipitate.** These reactions are thus known as **precipitation reactions.** The example below illustrates the mixing of aqueous solutions of $AgNO_3$ and $NaCl$ to form a precipitate of $AgCl$. (See also Figure 9-5.)

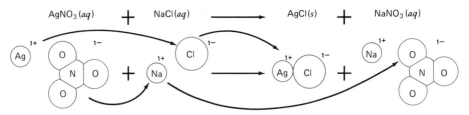

Figure 9-5   DOUBLE-REPLACEMENT REACTION. The formation of solid silver chloride when solutions of silver nitrate and sodium chloride are mixed is a double-replacement reaction.

A second example of a precipitation reaction is

$$(NH_4)_2S(aq) + Pb(NO_3)_2(aq) \longrightarrow PbS(s) + 2NH_4NO_3(aq)$$

A second type of double-replacement reaction involves the formation of a molecular compound from the reactants. The most important reactions of this classification are known as *neutralization reactions.* They involve the reaction of an **acid** *(a compound that produces H⁺ ions in water)* with a **base** *(a compound that produces OH⁻ ions in water).* The result of the union of H⁺ from the acid and OH⁻ from the base is the molecular compound water. The other product is an ionic compound known as a **salt.** Two examples of neutralization reactions are

$$HCl(aq) + NaOH(aq) \longrightarrow NaCl(aq) + H_2O(l)$$
$$2HNO_3(aq) + Ca(OH)_2(aq) \longrightarrow Ca(NO_3)_2(aq) + 2H_2O(l)$$

■
*SEE PROBLEMS
9-6 THROUGH
9-9.*
Double-replacement reactions deserve a more thorough discussion and, therefore, are discussed later in the text. Precipitation reactions are discussed in Chapter 12 and neutralization reactions are discussed in Chapter 13.

9-3 STOICHI-OMETRY

We are now ready to return to the balanced chemical equation and consider its quantitative implications. Earlier in this chapter we described the meaning of the balanced equation on the molecular level. This is shown in line 1 below the following balanced equation. But note in lines 2, 3, and 4 that the equation implies any multiple of the basic molecular ratio. In lines 5 and 6, numbers have been changed to mole units and to mass.

$N_2(g)$	$+ 3H_2(g)$	$\longrightarrow 2NH_3(g)$
1 1 molecule	+ 3 molecules	⟶ 2 molecules
2 12 molecules	+ 36 molecules	⟶ 24 molecules
3 1 dozen molecules	+ 3 dozen molecules	⟶ 2 dozen molecules
4 6.02×10^{23} molecules	+ 18.1×10^{23} molecules	⟶ 12.0×10^{23} molecules
5 1 mol	+ 3 mol	⟶ 2 mol
6 28 g	+ 6 g	⟶ 34 g

We will be concerned primarily with the last three relationships, as they concern laboratory situations.

Balanced equations give the necessary relationships to convert moles, grams, or number of molecules of one reactant or product into the equivalent number of moles, grams, or number of molecules of another reactant or product. *The quantitative relationship among reactants and products is known as* **stoichiometry.**

First we examine the relationship of moles in the above equation.

1 mol of N_2 produces 2 mol of NH_3
3 mol of H_2 produces 2 mol of NH_3
1 mol of N_2 reacts with 3 mol of H_2

From these three relationships six conversion factors can be written. *Conversion factors that originate from balanced equations are referred to as* **equation factors** *in this text.* They are, of course, exact numbers.

SEE PROBLEMS
9-10 THROUGH
9-12.

$$(1) \; \frac{1 \text{ mol } N_2}{2 \text{ mol } NH_3} \quad (3) \; \frac{1 \text{ mol } N_2}{3 \text{ mol } H_2} \quad (5) \; \frac{3 \text{ mol } H_2}{2 \text{ mol } NH_3}$$

$$(2) \; \frac{2 \text{ mol } NH_3}{1 \text{ mol } N_2} \quad (4) \; \frac{3 \text{ mol } H_2}{1 \text{ mol } N_2} \quad (6) \; \frac{2 \text{ mol } NH_3}{3 \text{ mol } H_2}$$

We can now use these factors to make the following stoichiometry conversions:

1   Mole ⟷ mole
2   Mole ⟷ mass
3   Mass ⟷ mass
4   Mass ⟷ number

This is illustrated by the following examples.

EXAMPLE 9-4: MOLE–MOLE

How many moles of NH_3 can be produced from 5.00 mol of H_2?

PROCEDURE

Convert moles of what's given (H_2) to moles of what's requested (NH_3). Use equation factor 6, which relates moles of NH_3 to moles of H_2.

SCHEME

Given		*Requested*
moles (H_2)	$\xrightarrow{\text{equation factor 6}}$	moles (NH_3)

SOLUTION

$$5.00 \; \cancel{\text{mol } H_2} \times \frac{2 \text{ mol } NH_3}{3 \; \cancel{\text{mol } H_2}} = \underline{3.33 \text{ mol } NH_3}$$

EXAMPLE 9-5: MOLE–MASS

How many moles of NH_3 can be produced from 33.6 g of N_2?

PROCEDURE

First convert mass of what's given to moles (N_2) and then use equation factor 2 to convert moles of N_2 to moles of NH_3.

SCHEME

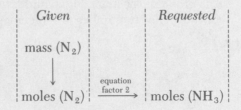

SOLUTION

$$33.6 \text{ g } N_2 \times \frac{1 \text{ mol } N_2}{28.0 \text{ g } N_2} \times \frac{2 \text{ mol } NH_3}{1 \text{ mol } N_2} = \underline{\underline{2.40 \text{ mol } NH_3}}$$

EXAMPLE 9-6: MASS–MASS

What mass of H_2 is needed to produce 119 g of NH_3?

PROCEDURE

This is a three-step conversion. In step 1, convert mass to moles of NH_3. In step 2, use equation factor 5 to convert moles of NH_3 to moles of H_2. Finally, in step 3, convert moles of H_2 to mass of H_2.

SCHEME

	Given		Requested
	mass (NH_3)		mass (H_2)
	↓		↑
	moles (NH_3)	$\xrightarrow{\text{equation factor 5}}$	moles (H_2)

SOLUTION

$$119 \text{ g } NH_3 \times \frac{1 \text{ mol } NH_3}{17.0 \text{ g } NH_3} \times \frac{3 \text{ mol } H_2}{2 \text{ mol } NH_3} \times \frac{2.02 \text{ g } H_2}{\text{mol } H_2} = \underline{\underline{21.2 \text{ g } H_2}}$$

EXAMPLE 9-7: MASS–NUMBER

How many molecules of N_2 are needed to react with 17.0 g of H_2?

PROCEDURE

This problem is similar to Example 9-6 except that in the final step we convert moles to number of molecules instead of mass.

SCHEME

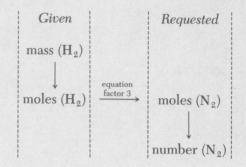

SOLUTION

$$17.0 \text{ g } H_2 \times \frac{1 \text{ mol } H_2}{2.02 \text{ g } H_2} \times \frac{1 \text{ mol } N_2}{3 \text{ mol } H_2}$$
$$\times \frac{6.02 \times 10^{23} \text{ molecules } N_2}{\text{mol } N_2} = \underline{1.69 \times 10^{24} \text{ molecules } N_2}$$

EXAMPLE 9-8

What mass of NH_3 is produced by 4.65×10^{22} molecules of H_2?

PROCEDURE

This example is similar to Example 9-7 except that in the first step we convert number of molecules to moles of what's given.

SCHEME

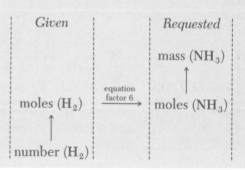

SOLUTION

$$4.65 \times 10^{22} \text{ molecules } H_2 \times \frac{1 \text{ mol } H_2}{6.02 \times 10^{23} \text{ molecules } H_2}$$

$$\times \frac{2 \text{ mol } NH_3}{3 \text{ mol } H_2} \times \frac{17.0 \text{ g } NH_3}{\text{mol } NH_3} = \underline{\underline{0.875 \text{ g } NH_3}}$$

The general procedure for working stoichiometry problems covering all of the possibilities discussed so far is shown in Figure 9-6. In words, the general procedure is as follows:

1 Write down (a) what is given and (b) what is requested.
2 If mass or a number of molecules is given, convert to moles in the first step.
3 Using the correct equation factor from the *balanced* equation, convert moles of what is given to moles of what is requested in the second step.
4 If mass or number of molecules is requested, convert from moles of what is requested in the third step.

As you work the problems, first map out a procedure as shown in the examples. Set up the problems, making sure that all of the factors are correct and the appropriate units cancel, and finally *carefully* do the math. If you have worked stoichiometry problems previously using proportions, you may find the procedures outlined here slightly more time consuming; but you will also find these procedures more dependable and organized in obtaining the correct answer. In chemistry that's important.

Before proceeding, we will follow the general procedure with two more examples of stoichiometry problems.

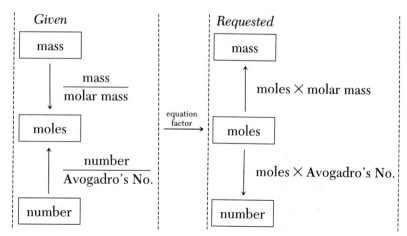

Figure 9-6   GENERAL PROCEDURE FOR STOICHIOMETRY PROBLEMS.

EXAMPLE 9-9

Some sulfur is present in coal in the form of pyrite, FeS_2 (also known as "fool's gold"). When it burns, it pollutes the air with the combustion product SO_2, as shown by the following chemical equation:

$$4FeS_2(s) + 11O_2(g) \longrightarrow 2Fe_2O_3(s) + 8SO_2(g)$$

How many moles of Fe_2O_3 are produced from 145 g of O_2?

PROCEDURE

Given: 145 g of O_2.   Requested: ___?___ mol of Fe_2O_3.
Convert mass of O_2 to moles of Fe_2O_3.

SCHEME

```
┌──────────────┐
│  mass of O₂  │
└──────────────┘
       │ 32.0 g/mol
       ↓
moles of O₂ ──→ ┌──────────────────┐
                │ moles of Fe₂O₃   │
                └──────────────────┘
```

From the equation, an equation factor relating O_2 to Fe_2O_3 is needed. The factor will have what's requested in the numerator.

$$\frac{2 \text{ mol Fe}_2\text{O}_3}{11 \text{ mol O}_2}$$

SOLUTION

$$145 \text{ g } O_2 \times \frac{1 \text{ mol } O_2}{32.0 \text{ g } O_2} \times \frac{2 \text{ mol Fe}_2\text{O}_3}{11 \text{ mol } O_2} = \underline{\underline{0.824 \text{ mol Fe}_2\text{O}_3}}$$

EXAMPLE 9-10

From the equation used in the preceding example, determine the mass of SO_2 that is produced from the combustion of 38.8 g of FeS_2.

PROCEDURE

Given: 38.8 g of FeS_2.   Requested: ___?___ g of SO_2.
Convert grams of FeS_2 to grams of SO_2.

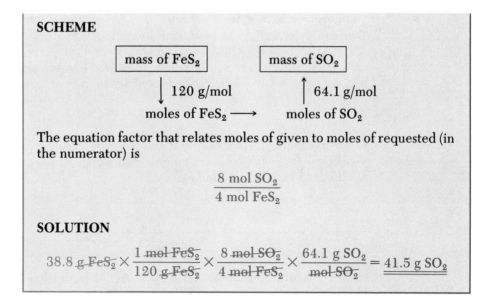

SCHEME

$$\boxed{\text{mass of FeS}_2} \qquad \boxed{\text{mass of SO}_2}$$

$$\downarrow 120 \text{ g/mol} \qquad \uparrow 64.1 \text{ g/mol}$$

$$\text{moles of FeS}_2 \longrightarrow \text{moles of SO}_2$$

The equation factor that relates moles of given to moles of requested (in the numerator) is

$$\frac{8 \text{ mol SO}_2}{4 \text{ mol FeS}_2}$$

SOLUTION

$$38.8 \text{ g FeS}_2 \times \frac{1 \text{ mol FeS}_2}{120 \text{ g FeS}_2} \times \frac{8 \text{ mol SO}_2}{4 \text{ mol FeS}_2} \times \frac{64.1 \text{ g SO}_2}{\text{mol SO}_2} = \underline{\underline{41.5 \text{ g SO}_2}}$$

■
SEE PROBLEMS
9-13 THROUGH
9-26.

9-4 PERCENT YIELD

An efficient automobile engine burns gasoline (mainly a hydrocarbon, C_8H_{18}) to form carbon dioxide and water. Untuned engines, however, do not burn gasoline efficiently and produce carbon monoxide and may even exhaust unburned fuel. The two combustion reactions are shown below.

Complete combustion $2C_8H_{18}(g) + 25O_2(g) \longrightarrow 16CO_2(g) + 18H_2O(l)$
Incomplete combustion $2C_8H_{18}(g) + 17O_2(g) \longrightarrow 16CO(g) + 18H_2O(l)$

Note that if we were asked to calculate the mass of CO_2 produced from a certain amount of C_8H_{18} our answer would not be correct if all the hydrocarbon were not converted to CO_2. *The measured amount of product obtained in any reaction is known as the* **actual yield.** *The* **theoretical yield** *is the calculated amount of product that would be obtained if all of the reactant were converted to a certain product. The* **percent yield** *is the actual yield in grams or moles divided by the theoretical yield in grams or moles times 100%:*

$$\frac{\text{actual yield}}{\text{theoretical yield}} \times 100\% = \text{percent yield}$$

In other reactions, there is another reason for the incomplete conversion of reactants to products. An example is the reaction illustrated previously in which nitrogen and hydrogen are converted to ammonia. In this reaction, all of the hydrogen and nitrogen do not react to form ammonia. Actually, what is happening in such a reaction is that while reactants are forming products, products are reacting to reform reactants. The two reactions (forward and reverse) offset each other at some point, leaving measurable quantities of both reactants and products present. Such a reaction is *reversible* and is known as an *equilibrium reaction;* the reversibility is indicated by a double arrow rather than the yield sign. As in the case of the combustion of octane, the actual yield and the theoretical yield are not the same so that the percent yield is less than 100%. (Equilibrium reactions will be discussed in more detail in Chapters 13 and 14.) The reversible reaction involving the formation of ammonia is

$$N_2(g) + 3H_2(g) \rightleftharpoons 2NH_3(g)$$

To determine the percent yield, it is necessary to determine the theoretical yield (which is what we've been doing all along) and compare this with the actual yield. An example follows.

EXAMPLE 9-11

In a certain experiment a 4.70-g quantity of H_2 is allowed to react with N_2; a 12.5-g quantity of NH_3 is formed. What is the percent yield based on the H_2?

PROCEDURE

Find the mass of NH_3 that would form if all the 4.70 g of H_2 is converted to NH_3 (the theoretical yield). Using the actual yield (12.5 g), find the percent yield.

SCHEME

$$\boxed{\text{mass of } H_2} \qquad \boxed{\text{mass of } NH_3}$$

$$\downarrow\ 2.02 \text{ g/mol} \qquad \uparrow\ 17.0 \text{ g/mol}$$

$$\text{moles of } H_2 \longrightarrow \text{moles of } NH_3$$

The equation factor needed is

$$\frac{2 \text{ mol } NH_3}{3 \text{ mol } H_2}$$

SOLUTION

$$4.70 \text{ g } H_2 \times \frac{1 \text{ mol } H_2}{2.02 \text{ g } H_2} \times \frac{2 \text{ mol } NH_3}{3 \text{ mol } H_2} \times \frac{17.0 \text{ g } NH_3}{\text{mol } NH_3}$$

$$= \underline{26.4 \text{ g } NH_3} \quad \text{(theoretical yield)}$$

$$\frac{12.5 \text{ g}}{26.4 \text{ g}} \times 100\% = \underline{47.3\% \text{ yield}}$$

There is a third reason that a sample of a particular reactant compound may not be completely converted into products—the original sample is impure. Percent purity of compounds is often determined by chemical analysis. Chemical analysis involves finding the amount of compound present by measuring the mass of a product formed by a specific reaction. Using a balanced equation, the mass of the product can then be converted to the mass of the original compound present. The percent purity can then be calculated from the mass of compound present and the total mass of the impure sample. In such an analysis, we assume that the compound in question undergoes a complete reaction. An example of chemical analysis follows.

EXAMPLE 9-12

A 125-g sample of impure calcium carbonate is heated to drive off all of the CO_2 according to the equation

$$CaCO_3(s) \longrightarrow CaO(s) + CO_2(g)$$

If 50.6 g of CO_2 is collected, what is the purity of the original sample?

PROCEDURE

Assume that all $CaCO_3$ in the impure sample decomposes to form CaO and CO_2.

1   Find the mass of $CaCO_3$ needed to produce 50.6 g of CO_2 using the following scheme.

mass of CO_2	mass of $CaCO_3$

$$\downarrow 44.0 \text{ g/mol} \qquad \uparrow 100 \text{ g/mol}$$

$$\text{moles of } CO_2 \longrightarrow \text{moles of } CaCO_3$$

The equation factor needed is

$$\frac{1 \text{ mol } CaCO_3}{1 \text{ mol } CO_2}$$

$$50.6 \text{ g } CO_2 \times \frac{1 \text{ mol } CO_2}{44.0 \text{ g } CO_2} \times \frac{1 \text{ mol } CaCO_3}{1 \text{ mol } CO_2}$$

$$\times \frac{100 \text{ g } CaCO_3}{\text{mol } CaCO_3} = \underline{115 \text{ g } CaCO_3}$$

2   Find the percent of the original sample that the mass of $CaCO_3$ found in step 1 represents.

$$\frac{115 \text{ g } CaCO_3}{125 \text{ g sample}} \times 100\% = \underline{92.0\% \text{ pure}}$$

■
SEE PROBLEMS
9-27 THROUGH
9-37.

9-5 LIMITING REACTANT

Assume that your instructor is trying to put together a three-page exam for the class. He or she may have 85 copies of page 1, 85 copies of page 2, but only 80 copies of page 3. Obviously, only 80 copies of the test can be put together even though there is an excess of pages 1 and 2. The number of complete tests is limited by the page with the least copies. The others are in excess. Likewise, if certain amounts of reactants are mixed the amount of product formed is limited by the reactant that is completely consumed. We can illustrate this with the simple example of the production of water from its elements, hydrogen and oxygen.

$$2H_2 + O_2 \longrightarrow 2H_2O$$

A 4.0-g quantity of H_2 reacts completely with 32.0 g of O_2. In fact, whenever H_2 and O_2 are mixed in a $4.0 : 32(1 : 8)$ mass ratio and a reaction is initiated, all reactants are consumed and only products appear. (This reaction goes to completion, which is a 100% yield.) *When reactants are mixed in exactly the mass ratio determined from the balanced equation, the mixture is said to be* **stoichiometric.**

$$\text{4.0 g of } H_2 + 32.0 \text{ g of } O_2 \longrightarrow 36.0 \text{ g of } H_2O \quad \text{(stoichiometric)}$$

If a 6.0-g quantity of H_2 and a 32.0-g quantity of O_2 are mixed, there is still only 4.0 g of H_2 used so that 2.0 g of H_2 remains after the reaction is complete. Thus H_2 is in excess, and the amount of H_2O formed is limited by the amount of O_2 originally present. The O_2 is the limiting reactant. *The* **limiting reactant** *is the reactant that would produce the least amount of products if completely consumed.*

$$\text{6.0 g of } H_2 + 32.0 \text{ g of } O_2 \longrightarrow 36.0 \text{ g of } H_2O + \boxed{2.0 \text{ g of } H_2 \text{ unreacted}}$$
$$\text{(}H_2 \text{ in excess, } O_2 \text{ limiting reactant)}$$

If a 4.0-g quantity of H_2 and a 38.0-g quantity of O_2 are mixed, 36.0 g of H_2O is produced. Now O_2 is in excess and H_2 is the limiting reactant:

$$\text{4.0 g of } H_2 + 38.0 \text{ g of } O_2 \longrightarrow 36.0 \text{ g of } H_2O + \boxed{6.0 \text{ g of } O_2 \text{ unreacted}}$$
$$\text{(}O_2 \text{ in excess, } H_2 \text{ limiting reactant)}$$

When quantities of two or more reactants are given, it is necessary to determine which is the limiting reactant (unless they are mixed in exactly stoichiometric amounts). This can be simplified by the following procedures.

1   Determine the number of moles of a product produced by each reactant using the general procedure discussed earlier.

2   The reactant producing the *smallest* yield is the limiting reactant.

EXAMPLE 9-13

Silver tarnishes (turns black) in homes because of the presence of small amounts of H_2S (a rotten-smelling gas that originates from the decay of food). The reaction is

$$4Ag(s) + 2H_2S(g) + O_2(g) \longrightarrow 2Ag_2S(s) + 2H_2O(l)$$
(black)

If 0.145 mol of Ag is present with 0.0872 mol of H_2S and excess O_2:

(a) What is the limiting reactant?

(b) What mass of Ag_2S is produced?

(c) What mass of the other reactant in excess remains?

PROCEDURE

(a) Convert moles of Ag and H_2S to moles of Ag_2S produced by each.

$$\text{moles of reactant} \longrightarrow \text{moles of } Ag_2S$$

$$Ag \quad 0.145 \text{ mol Ag} \times \frac{2 \text{ mol } Ag_2S}{4 \text{ mol Ag}} = 0.0725 \text{ mol } Ag_2S$$

$$H_2S \quad 0.0872 \text{ mol } H_2S \times \frac{2 \text{ mol } Ag_2S}{2 \text{ mol } H_2S} = 0.0872 \text{ mol } Ag_2S$$

Since Ag produces the smaller yield of Ag_2S, Ag is the limiting reactant.

(b) To find the mass of Ag_2S produced convert moles of Ag_2S to mass.

$$0.0725 \text{ mol } Ag_2S \times \frac{248 \text{ g } Ag_2S}{\text{mol } Ag_2S} = 18.0 \text{ g } Ag_2S$$

(c) To find the mass of H_2S in excess, first convert moles of Ag to moles of H_2S. This will be moles of H_2S reacted.

$$0.145 \text{ mol Ag} \times \frac{2 \text{ mol } H_2S}{4 \text{ mol Ag}} = 0.0725 \text{ mol } H_2S \text{ reacted}$$

Find the moles of H_2S remaining or unreacted by subtracting moles reacted from original moles present.

$$0.0872 \text{ mol} - 0.0725 \text{ mol} = 0.0147 \text{ mol } H_2S \text{ unreacted}$$

Finally, convert moles of unreacted H_2S to mass.

$$0.0147 \text{ mol } H_2S \times \frac{34.1 \text{ g } H_2S}{\text{mol } H_2S} = 0.501 \text{ g } H_2S \text{ in excess}$$

EXAMPLE 9-14

Methanol (CH_3OH) is used as a fuel for racing cars. It burns in the engine according to the equation

$$2CH_3OH(l) + 3O_2(g) \longrightarrow 2CO_2(g) + 4H_2O(g)$$

If 40.0 g of methanol is mixed with 46.0 g of O_2:

(a) What is the limiting reactant?

(b) What is the theoretical yield of CO_2?

(c) If 38.0 g of CO_2 is actually produced, what is the percent yield?

PROCEDURE

(a) Determine the moles of CO_2 formed from each reactant.

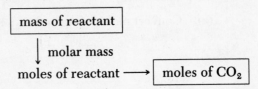

$$CH_3OH \quad 40.0 \text{ g CH}_3\text{OH} \times \frac{1 \text{ mol CH}_3\text{OH}}{32.0 \text{ g CH}_3\text{OH}} \times \frac{2 \text{ mol CO}_2}{2 \text{ mol CH}_3\text{OH}}$$
$$= 1.25 \text{ mol CO}_2$$

$$O_2 \quad 46.0 \text{ g O}_2 \times \frac{1 \text{ mol O}_2}{32.0 \text{ g O}_2} \times \frac{2 \text{ mol CO}_2}{3 \text{ mol O}_2} = 0.958 \text{ mol CO}_2$$

Therefore, O_2 is the limiting reactant.

(b) The theoretical yield is determined from the amount of product formed *from the limiting reactant.* Thus we simply convert the 0.958 mol of CO_2 produced by the O_2 to grams.

$$0.958 \text{ mol CO}_2 \times \frac{44.0 \text{ g CO}_2}{\text{mol CO}_2} = 42.2 \text{ g CO}_2$$

(c) From the theoretical yield and the actual yield, determine the percent yield.

SEE PROBLEMS
9-38 THROUGH
9-46.

$$\frac{38.0 \text{ g}}{42.2 \text{ g}} \times 100\% = 90.0\% \text{ yield of CO}_2$$

9-6 HEAT ENERGY IN CHEMICAL REACTIONS

Matter is not all that is involved in chemical reactions. As stated previously heat energy is also an important component of chemical changes. *The study of heat and its relationship to chemical changes is known as chemical thermodynamics.*

Most chemical reactions that occur spontaneously do so for the same reason that a pencil drops to the floor when it falls off a table or that a car rolls down a hill when the brake is released. The pencil on the floor, the car at the bottom of the hill, and the compounds formed as products in a reaction represent states of lower potential energy than the initial states. For a chemical reaction the potential energy is in the form of chemical energy. The difference in potential energy (chemical energy) between reactants and products is usually transformed into heat energy by a reaction. When a reaction leads to products of lower potential energy it evolves heat and is said to be *exothermic*. Not all chemical reactions lead to products of lower potential energy than reactants, however. Surprisingly, a few reactions occur spontaneously that lead to products of higher potential energy and thus absorb heat from the surroundings.* Still other chemical reactions do not occur spontaneously but a reaction occurs if heat energy is continually supplied to the reactants. The latter two cases, in which heat is absorbed by the reactants, are known as *endothermic* reactions. (Some exothermic reactions can be initiated only at high temperature, however, but once started the reactions usually continue unaided. An example is the combustion of gasoline.) The amount of heat (in kilocalories or kilojoules) involved in a reaction is a constant amount that depends on the amount of reactants present. For example, if one mole of hydrogen is burned to produce liquid water, 286 kJ of heat is evolved.

A balanced equation that includes heat energy is referred to as a **thermochemical equation.** A thermochemical equation can be represented in either of two ways. In the first, the heat is shown separately from the reaction using the symbol ΔH. This term is defined as the **change in enthalpy** or simply "**the heat of the reaction.**" (In this case, the Δ refers to a difference and is not the same as the Δ in $\overset{\Delta}{\longrightarrow}$). Enthalpy is the heat content in the form of chemical energy for a particular compound. The change in enthalpy is simply the change in heat content in going from reactants to products. By convention, a negative sign for ΔH corresponds to an exothermic reaction and a positive sign, to an endothermic reaction. (Note that since there are two moles of hydrogen undergoing combustion, the heat evolved is twice that for one mole of hydrogen.)

$$2H_2(g) + O_2(g) \longrightarrow 2H_2O(l) \quad \Delta H = -572 \text{ kJ}$$

The second way a thermochemical equation is represented shows the amount of heat energy as if it were a reactant or product.

$$2H_2(g) + O_2(g) \longrightarrow 2H_2O(l) + 572 \text{ kJ} \quad \text{(exothermic reaction)}$$
$$N_2(g) + O_2(g) + 181 \text{ kJ} \longrightarrow 2NO(g) \quad \text{(endothermic reaction)}$$

---

* This is like releasing the brake on a car and finding that it spontaneously rolls uphill. In chemistry, however, there is a good reason why a reaction occurs that is in a way "uphill," but this is beyond the scope of this discussion.

The amount of heat energy involved in a chemical reaction may be treated quantitatively in a manner similar to the amount of a compound. The following examples illustrate the calculations implied by a thermochemical equation.

EXAMPLE 9-15

Oxygen is often prepared in the laboratory by heating potassium chlorate. In addition to oxygen gas, potassium chloride is produced. The process requires the addition of 45.0 kJ per mole of potassium chlorate.

(a) Write the balanced thermochemical equation in both ways.

(b) How much heat is required to decompose 1.00 g of potassium chlorate?

SOLUTION

(a) $2KClO_3(s) \longrightarrow 2KCl(s) + 3O_2(g)$ $\Delta H = 90.0$ kJ

Note that 2 mole of $KClO_3$ is in the balanced equation. Thus,

$$\Delta H = 2 \text{ mol } KClO_3 \times \frac{45.0 \text{ kJ}}{\text{mol } KClO_3} = 90.0 \text{ kJ}$$

$$2KClO_3(s) + 90.0 \text{ kJ} \longrightarrow 2KCl(s) + 3O_2(g)$$

(b)   g $KClO_3 \longrightarrow$ mol $KClO_3 \longrightarrow$ kJ

$$1.00 \text{ g } KClO_3 \times \frac{1 \text{ mol } KClO_3}{123 \text{ g } KClO_3} \times \frac{45.0 \text{ kJ}}{\text{mol } KClO_3} = 0.366 \text{ kJ}$$

EXAMPLE 9-16

Acetylene, which is used in welding torches, burns according to the following thermochemical equation:

$$2C_2H_2(g) + 5O_2(g) \longrightarrow 4CO_2(g) + 2H_2O(l) \quad \Delta H = -2602 \text{ kJ}$$

If 550 kJ of heat is evolved in the combustion of a sample of C_2H_2, what is the mass of CO_2 formed?

PROCEDURE

$$kJ \longrightarrow \text{mol } CO_2 \longrightarrow \text{mass } CO_2$$

SOLUTION

$$550 \text{ kJ} \times \frac{4 \text{ mol } CO_2}{2602 \text{ kJ}} \times \frac{44.0 \text{ g } CO_2}{\text{mol } CO_2} = 37.2 \text{ g } CO_2$$

SEE PROBLEMS
9-47 THROUGH
9-54.

9-7 CHAPTER REVIEW

This chapter is concerned mainly with the chemist's equivalent to a complete sentence, that is, the chemical equation. Balanced equations tell us several important things. They tell us the formulas of the elements and compounds involved in the chemical changes, their physical states under the reaction conditions, and the ratio of molecules in the reactants and products.

Chemical equations represent many types of reactions. The five types discussed in this chapter are illustrated below.

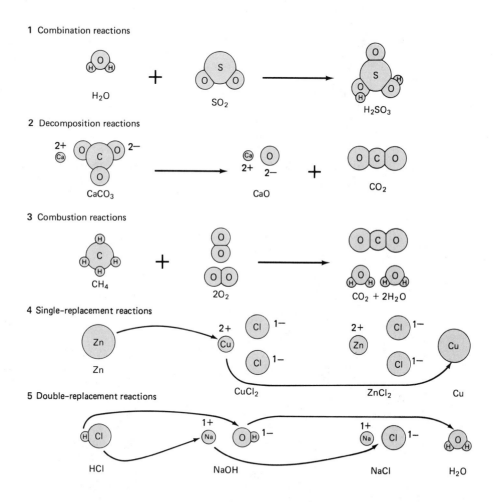

1 Combination reactions

H_2O SO_2 H_2SO_3

2 Decomposition reactions

$CaCO_3$ CaO CO_2

3 Combustion reactions

CH_4 $2O_2$ $CO_2 + 2H_2O$

4 Single-replacement reactions

Zn $CuCl_2$ $ZnCl_2$ Cu

5 Double-replacement reactions

HCl   NaOH   NaCl H_2O

Besides describing the various types of reactions, chemical equations also tell us about the mole ratios between reactants and products. These ratios open the door to important calculations that come under the heading of stoichiometry. In this text, mole ratios are referred to as equation factors and serve us by allowing conversions between amounts of reactants and products. The conversions performed are summarized as follows.

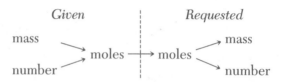

We also discussed three cases in which a certain sample is not completely converted to one product:

1 Reactions that form more than one set of products

2 Reactions that are incomplete because they reach a point of equilibrium

3 Reactions in which the original sample contains impurities

We also emphasized that reactants combine in definite, stoichiometric amounts. If the proportions of reactants are not stoichiometric, one or more reactants will be in excess and a portion will remain unreacted. The reactant that is either completely consumed or produces the smallest theoretical yield of products is the limiting reactant.

Finally, we briefly discussed how heat energy can be included in chemical equations and how calculations involving heat are similar to those involving matter.

■ EXERCISES

CHEMICAL EQUATIONS

9-1 Each of the following compounds is either gas, solid, or liquid under normal conditions. Indicate the proper physical state by adding either (g), (s), or (l) after the formula.

(a) Cl_2 (e) P_4 (i) S_8
(b) C (f) H_2 (j) Na
(c) K_2SO_4 (g) Br_2 (k) Hg
(d) H_2O (h) $NaBr$ (l) CO_2

9-2 Balance the following equations.

(a) $CaCO_3 \xrightarrow{\Delta} CaO + CO_2$
(b) $Na + O_2 \rightarrow Na_2O$
(c) $H_2SO_4 + NaOH \rightarrow Na_2SO_4 + H_2O$
(d) $H_2O_2 \rightarrow H_2O + O_2$
(e) $Si_2H_6 + H_2O \rightarrow Si(OH)_4 + H_2$
(f) $Al + H_3PO_4 \rightarrow AlPO_4 + H_2$
(g) $Ca(OH)_2 + HCl \rightarrow CaCl_2 + H_2O$
(h) $Na_2NH + H_2O \rightarrow NH_3 + NaOH$
(i) $Mg + N_2 \rightarrow Mg_3N_2$
(j) $CaC_2 + H_2O \rightarrow C_2H_2 + Ca(OH)_2$
(k) $C_2H_6 + O_2 \rightarrow CO_2 + H_2O$

9-3 Balance the following equations

(a) $NaBr + Cl_2 \rightarrow NaBr + Br_2$
(b) $KOH + H_3AsO_4 \rightarrow K_2HAsO_4 + H_2O$
(c) $Ti + Cl_2 \rightarrow TiCl_4$
(d) $PCl_5 + H_2O \rightarrow H_3PO_4 + HCl$
(e) $Al + H_2SO_4 \rightarrow Al_2(SO_4)_3 + H_2$
(f) $Ca(CN)_2 + HBr \rightarrow CaBr_2 + HCN$
(g) $C_3H_6 + O_2 \rightarrow CO + H_2O$
(h) $P_4 + S_8 \rightarrow P_4S_3$

9-4 Write balanced chemical equations from the following word equations. Include the physical state of each compound.

(a) Sodium metal plus water yields hydrogen gas and an aqueous sodium hydroxide solution.

(b) Potassium chlorate when heated yields potassium chloride plus oxygen gas. (Ionic compounds are solids.)

(c) An aqueous sodium chloride solution plus an aqueous silver nitrate solution yields a silver chloride precipitate (solid) and a sodium nitrate solution.

(d) An aqueous phosphoric acid solution plus an aqueous calcium hydroxide solution yields water and solid calcium phosphate.

9-5 Write balanced chemical equations from the following word equations.

(a) Solid phenol (C_6H_6O) reacts with oxygen to form carbon dioxide gas and liquid water.

(b) An aqueous calcium hydroxide solution reacts with gaseous sulfur trioxide to form a precipitate of calcium sulfate and water.

(c) Lithium combines with nitrogen to form lithium nitride.

(d) Magnesium dissolves in an aqueous chromium(III) nitrate solution to form chromium and a magnesium nitrate solution

TYPES OF CHEMICAL REACTIONS

9-6 Which of the five types of reactions is each reaction in Problems 9-2 and 9-4?

9-7 Which of the five types of reactions is each reaction in Problems 9-3 and 9-5?

*9-8 Write balanced equations by predicting the products of the following reactions.

(a) combination of potassium and chlorine

(b) single-replacement reaction of calcium metal with water to produce calcium hydroxide solution and a gas

(c) combustion of liquid benzene (C_6H_6)

(d) double-replacement reaction between an aqueous sodium sulfide solution and an aqueous copper(II) nitrate solution (The sulfide product is a solid.)

(e) decomposition of gold(III) oxide into its elements by heating

*9-9 Write balanced equations by predicting the products of the following reactions.

(a) combustion of liquid butane (C_4H_{10})

(b) single-replacement reaction involving chlorine gas and a potassium bromide solution

(c) decomposition of a sulfurous acid solution to water and a gas

(d) double-replacement reaction involving aqueous perchloric acid and aqueous cesium hydroxide

(e) combination of sodium oxide and gaseous sulfur trioxide

EQUATION FACTORS

9-10 Given the balanced equation

$$Mg + 2HCl \longrightarrow MgCl_2 + H_2$$

provide the proper equation factor that makes the following mole conversions.

(a) Mg to H_2 (c) HCl to H_2
(b) Mg to HCl   (d) $MgCl_2$ to HCl

9-11 Given the balanced equation

$$2C_4H_{10} + 13O_2 \longrightarrow 8CO_2 + 10H_2O$$

provide the proper equation factor that makes the following mole conversions.

(a) CO_2 to C_4H_{10} (c) CO_2 to O_2
(b) O_2 to C_4H_{10} (d) O_2 to H_2O

9-12 Given the balanced equation

$$Cu + 4HNO_3 \longrightarrow$$
$$Cu(NO_3)_2 + 2NO_2 + 2H_2O$$

provide the proper equation factor that makes the following mole conversions.

(a) Cu to NO_2
(b) HNO_3 to Cu
(c) H_2O to NO_2
(d) $Cu(NO_3)_2$ to HNO_3
(e) NO_2 to Cu
(f) Cu to H_2O

STOICHIOMETRY

9-13 Consider the reaction

$$2H_2 + O_2 \longrightarrow 2H_2O$$

(a) How many moles of H_2 are needed to produce 0.400 mol of H_2O?

(b) How many moles of H_2O will be produced from 0.640 g of O_2?

(c) How many moles of H_2 are needed to react with 0.032 g of O_2?

(d) What mass of H_2O would be produced from 0.400 g of H_2?

9-14 Propane burns according to the equation

$$C_3H_8 + 5O_2 \longrightarrow 3CO_2 + 4H_2O$$

(a) How many moles of CO_2 are produced from the combustion of

0.450 mol of C_3H_6? How many moles of H_2O? How many moles of O_2 are needed?

(b) What mass of H_2O is produced if 0.200 mol of CO_2 is also produced?

(c) What mass of C_3H_8 is required to produce 1.80 g of H_2O?

(d) What mass of C_3H_8 is required to react with 160 g of O_2?

(e) What mass of CO_2 is produced by the reaction of 1.20×10^{23} molecules of O_2?

(f) How many moles of H_2O are produced if 4.50×10^{22} molecules of CO_2 are produced?

9-15   The alcohol component of "gasohol" burns according to the equation

$$C_2H_6O(l) + 3O_2(g) \longrightarrow 2CO_2(g) + 3H_2O(g)$$

(a) What mass of alcohol is needed to produce 5.45 mol of H_2O?

(b) How many moles of CO_2 are produced along with 155 g of H_2O?

(c) What mass of CO_2 is produced from 146 g of C_2H_6O?

(d) What mass of C_2H_6O reacts with 0.898 g of O_2?

(e) What mass of H_2O is produced from 5.85×10^{24} molecules of O_2?

9-16   In the atmosphere N_2 and O_2 do not react. In the high temperatures of an automobile engine, however, the following reaction occurs:

$$N_2(g) + O_2(g) \longrightarrow 2NO(g)$$

When the NO reaches the atmosphere through the engine exhaust, a second reaction takes place:

$$2NO(g) + O_2(g) \longrightarrow 2NO_2(g)$$

The NO_2 is a somewhat brownish gas that contributes to the haze of smog and is irritating to the nasal passages and lungs. What mass of N_2 is required to produce 155 g of NO_2?

9-17   Elemental iron is produced in what is called the "thermite reaction" because it produces enough heat that the iron is

initially in the molten state. For this reason it is used in welding.

$$2Al(s) + Fe_2O_3(s) \longrightarrow Al_2O_3(s) + 2Fe(l)$$

What mass of Al is needed to produce 750 g of Fe? How many formula units of Fe_2O_3 are used in the process?

*9-18   Liquid iron is made from iron ore (Fe_2O_3) in a three-step process in a blast furnace:

1   $3Fe_2O_3(s) + CO(g) \longrightarrow 2Fe_3O_4(s) + CO_2(g)$

2   $Fe_3O_4(s) + CO(g) \longrightarrow 3FeO(s) + CO_2(g)$

3   $FeO(s) + CO(g) \longrightarrow Fe(l) + CO_2(g)$

What mass of iron would eventually be produced from 125 g of Fe_2O_3?

9-19   Antacids ($CaCO_3$) react with "stomach acid" according to the equation

$$CaCO_3(s) + 2HCl(aq) \longrightarrow CaCl_2(aq) + CO_2(g)$$

What mass of stomach acid reacts with 1.00 g of $CaCO_3$?

9-20   Elemental copper can be recovered from the mineral chalcocite (Cu_2S). From the following equation, determine what mass of Cu is formed from 7.82×10^{22} molecules of O_2.

$$Cu_2S + O_2 \longrightarrow 2Cu + SO_2$$

9-21   Fool's gold (pyrite) is so named because it looks much like gold. When it is placed in aqueous HCl, however, it dissolves and gives off a rotten-smelling gas (H_2S). Gold itself does not react with aqueous HCl. From the following equation determine how many individual molecules of H_2S are formed from 0.520 mol of FeS_2.

$$FeS_2(s) + 2HCl(aq) \longrightarrow FeCl_2(aq) + H_2S(g) + S(s)$$

9-22   Nitrogen dioxide may form "acid rain" by reaction with water in the air according to the equation

$$3NO_2(g) + H_2O(l) \longrightarrow$$
$$2HNO_3(aq) + NO(g)$$

What mass of nitric acid is produced from 18.5 kg of NO_2?

9-23 Elemental chlorine can be generated in the laboratory according to the equation

$$MnO_2(s) + 4HCl(aq) \longrightarrow$$
$$MnCl_2(aq) + 2H_2O + Cl_2(g)$$

What mass of Cl_2 is produced from the reaction of 665 g of HCl?

9-24 The fermentation of sugar to produce ethyl alcohol is represented by the equation

$$C_6H_{12}O_6 \longrightarrow 2C_2H_5OH + 2CO_2$$

What mass of alcohol is produced from 25.0 mol of sugar?

9-25 Methane gas can be made from carbon monoxide gas according to the equation

$$2CO + 2H_2 \longrightarrow CH_4 + CO_2$$

What mass of CO is required to produce 8.75×10^{25} molecules of CH_4?

9-26 Calcium cyanamide ($CaCN_2$) is used as a fertilizer. When it reacts with water it produces $CaCO_3$ (which counteracts acidity) and ammonia (which fertilizes the soil). Write the balanced equation illustrating the reaction and calculate the mass of NH_3 produced from a 1.00-kg quantity of $CaCN_2$.

THEORETICAL AND PERCENT YIELD

9-27 Sulfur trioxide (SO_3) is prepared from SO_2 according to the equation

$$2SO_2(g) + O_2(g) \rightleftharpoons 2SO_3(g)$$

In this reaction not all SO_2 is converted to SO_3 even with excess O_2 present. In a certain experiment, 21.2 g of SO_3 was produced from 24.0 g of SO_2. What is the theoretical yield of SO_3 and the percent yield?

9-28 The following is a reversible decomposition reaction:

$$2N_2O_5 \rightleftharpoons 4NO_2 + O_2$$

When 25.0 g of N_2O_5 is allowed to decompose it is found that 10.0 g of NO_2 forms. What is the percent yield?

9-29 The following equation represents a reversible combination reaction:

$$P_4O_{10} + 6PCl_5 \rightleftharpoons 10POCl_3$$

If 25.0 g of PCl_5 is allowed to react, it is found that there is a 45.0% yield of $POCl_3$. What is the actual yield in grams?

*9-30 Given the reaction

$$2NO_2 + 4H_2 \rightleftharpoons N_2 + 4H_2O$$

what mass of hydrogen is required to produce 250 g of N_2 if the yield is 70.0%?

9-31 Octane in gasoline burns in the automobile engine according to the equation

$$2C_8H_{18}(l) + 25O_2(g) \longrightarrow$$
$$16CO_2(g) + 18H_2O(g)$$

If a 57.0-g sample of octane is burned, 152 g of CO_2 is formed. What is the percent yield of CO_2?

*9-32 In Problem 9-31, the C_8H_{18} that is *not* converted to CO_2 forms CO. What is the mass of CO formed? (CO is a poisonous pollutant that is converted to CO_2 in the car's catalytic converter.)

9-33 When benzene reacts with bromine, the principal reaction is

$$C_6H_6 + Br_2 \longrightarrow C_6H_5Br + HBr$$

If the yield of bromobenzene (C_6H_5Br) is 65.2%, what mass of bromobenzene is produced from 12.5 g of C_6H_6?

*9-34 A second reaction between C_6H_6 and Br_2 (see Problem 9-33) produces dibromobenzene ($C_6H_4Br_2$).
(a) Write the balanced equation illustrating this reaction.
(b) If the remainder of the benzene from Problem 9-33 reacts to form dibromobenzene, what is the mass of $C_6H_4Br_2$ produced?

9-35 A 50.0-g sample of impure $KClO_3$ is decomposed to KCl and O_2. If a 12.0-g quantity of O_2 is produced, what percent of the sample is $KClO_3$? (Assume that all of the $KClO_3$ present decomposes.)

9-36 In Example 9-9, SO_2 was formed from the burning of pyrite (FeS_2) in coal. If a 312-g quantity of SO_2 was collected from the burning of 6.50 kg of coal, what percent of the original sample was pyrite?

*9-37 Copper metal can be recovered from an ore ($CuCO_3$) by the decomposition reaction

$$2CuCO_3 \longrightarrow 2Cu + 2CO_2 + O_2$$

What is the mass of a sample of ore if it is 47.5% $CuCO_3$ and produces 350 g of Cu? (Assume complete decomposition of $CuCO_3$.)

LIMITING REACTANT

9-38 Consider the equation

$$2Al + 3H_2SO_4 \longrightarrow Al_2(SO_4)_3 + 3H_2$$

If 0.800 mol of Al is mixed with 1.00 mol of H_2SO_4, how many moles of H_2 are produced? How many moles of one of the reactants remain?

9-39 Consider the equation

$$2C_5H_6 + 13O_2 \longrightarrow 10CO_2 + 6H_2O$$

If 3.44 mol of C_5H_6 is mixed with 20.6 mol of O_2, what mass of CO_2 is formed?

9-40 Elemental fluorine is usually prepared by electrolysis reactions. In 1986, however, a chemical reaction was reported that produces fluorine:

$$2K_2MnF_6 + 4SbF_5 \longrightarrow$$
$$4KSbF_6 + 2MnF_3 + F_2$$

If a 525-g quantity of K_2MnF_6 is mixed with 900 g of SbF_5 what is the yield (in grams) of F_2?

9-41 Consider the equation

$$4NH_3(g) + 3O_2(g) \longrightarrow$$
$$2N_2(g) + 6H_2O(l)$$

If a 40.0-g sample of O_2 is mixed with 1.50 mol of NH_3, which is the limiting reactant? What is the theoretical yield (in moles) of N_2?

9-42 Consider the equation

$$2AgNO_3(aq) + CaCl_2(aq) \longrightarrow$$
$$2AgCl(s) + Ca(NO_3)_2(aq)$$

If a solution containing 20.0 g of $AgNO_3$ is mixed with a solution containing 10.0 g of $CaCl_2$, which compound is the limiting reactant? What is the theoretical yield (in grams) of $AgCl$? What mass of one of the reactants remains?

9-43 Limestone ($CaCO_3$) dissolves in hydrochloric acid as shown by the equation

$$CaCO_3(s) + 2HCl(aq) \longrightarrow$$
$$CaCl_2(aq) + CO_2(g) + H_2O(l)$$

If 20.0 g of $CaCO_3$ and 25.0 g of HCl are mixed, what mass of CO_2 is produced? What mass of one of the reactants remains?

9-44 Consider the balanced equation

$$2HNO_3(aq) + 3H_2S(aq) \longrightarrow$$
$$2NO(g) + 4H_2O(l) + 3S(s)$$

If a 10.0-g quantity of HNO_3 is mixed with 5.00 g of H_2S, what is the mass of each product and of the reactant present in excess after reaction occurs? Assume a complete reaction.

9-45 Consider the equation

$$4NH_3(g) + 5O_2(g) \longrightarrow$$
$$4NO(g) + 6H_2O(l)$$

When an 80.0-g sample of NH_3 is mixed with 200 g of O_2, a 40.0-g quantity of NO is formed. What is the percent yield?

*9-46 Consider the equation

$$3K_2MnO_4 + 4CO_2 + 2H_2O \longrightarrow$$
$$2KMnO_4 + 4KHCO_3 + MnO_2$$

How many moles of MnO_2 are produced if 9.50 mol of K_2MnO_4, 6.02×10^{24} molecules of CO_2, and 90.0 g of H_2O are mixed?

HEAT IN CHEMICAL REACTIONS

9-47 When one mole of magnesium undergoes combustion to form magnesium oxide, 602 kJ of heat energy is evolved. Write the thermochemical equation in both forms.

9-48 In Problem 9-16 the nonspontaneous reaction between N_2 and O_2 was discussed. In fact, 90.5 kJ of heat energy must be supplied per mole of NO formed. Write the balanced thermo-chemical equation in both forms. Is the reaction exothermic or endothermic?

9-49 To decompose one mole of $CaCO_3$ to CaO and CO_2, 176 kJ must be supplied. Write the balanced thermochemical equation in both forms.

9-50 The complete combustion of one mole of octane in gasoline (C_8H_{18}) evolves 5480 kJ of heat. The complete combustion of one mole of methane is natural gas (CH_4) evolves 890 kJ. How much heat is evolved per 1.00 g for each of these fuels. Which is the better fuel on a mass basis?

9-51 Methyl alcohol (CH_4O) is used as a fuel in racing cars. It burns according to the equation

$$2CH_4O(l) + 3O_2(g) \longrightarrow$$
$$2CO_2(g) + 4H_2O(l) + 1750 \text{ kJ}$$

What amount of heat is evolved per 1.00 g of alcohol? How does this compare with the amount of heat per gram of gasoline? (See Problem 9-50.)

9-52 The thermite reaction was discussed in Problem 9-17. For the balanced equation, $\Delta H = -850$ kJ. What mass of aluminum is needed to produce 35.8 kJ of heat energy?

9-53 Photosynthesis is an endothermic reaction that forms glucose ($C_6H_{12}O_6$) from carbon dioxide, water, and energy from the sun. The balanced equation is

$$6CO_2(g) + 6H_2O(l) \longrightarrow$$
$$C_6H_{12}O_6(aq) + 6O_2(g)$$
$$\Delta H = +2519 \text{ kJ}$$

What mass of glucose is formed from 975 kJ of energy?

9-54 When butane (C_4H_{10}) in a cigarette lighter burns it evolves 2880 kJ per mole of butane. What is the mass of water formed if 1250 kJ of heat evolves?

AN OCEAN OF GAS

The hot-air balloon seems to defy gravity by rising effortlessly in the ocean of gases that we know as the atmosphere. An understanding of the gaseous state gives us insight into this phenomenon.

THE GASEOUS STATE
OF MATTER

6 Calculate the effect of a change of temperature on the pressure of a gas by application of Gay-Lussac's law. (10-4)

7 Apply the kinetic molecular theory to describe the basis for Boyle's, Charles', and Gay-Lussac's laws. (10-1, 10-3, 10-4)

8 Use the combined gas law for a quantity of gas under two sets of conditions. (10-5)

9 Apply Graham's law to problems involving the relationship between average velocities and molar masses of two gases. (10-6)

10 Perform calculations involving application of Dalton's law and Avogadro's law. (10-7, 10-8)

11 Use the ideal gas law to calculate an unknown property of a sample of gas (P, V, T, or n) when the other properties are known. (10-9)

12 Convert between moles of a gas and the volume at STP using the molar volume as a conversion factor. (10-10)

13 Calculate the density of a gas from its molar mass and the molar volume. (10-10)

14 Apply the ideal gas law or the molar volume relationship to stoichiometric calculations involving gases. (10-11)

Humans and all other land animals share one characteristic with the fish of the ocean. We too live in a sea, but our sea is composed of a mixture of gases. Outside of our sea we are as helpless as a fish out of water (literally). Indeed, travelers into space must take some of this sea of gases with them in a space suit to survive. We now have a good understanding of the homogeneous gaseous mixture we know as air but it has not always been so. In fact, early scientists were profoundly ignorant of the nature of gases although they did consider air to be one of the four elements (along with earth, fire, and water). The lack of understanding of the nature of gases actually held up the development of modern chemistry. It wasn't until the experiments of Antoine Lavoisier in France and Joseph Priestly in England in the late 1700s that much of the mystery was solved. These scientists proved that air is not just one substance but is composed of a mixture of mainly two elements, oxygen and nitrogen. Their research not only enlightened science on the nature of air but also laid the foundation for the development of the law of conservation of mass. This was the beginning of modern quantitative chemistry.

There is good news and bad news about the study of the gaseous state. First, the bad news. Most gases are invisible so we must infer their presence indirectly from measurements. In contrast, we can see and easily distinguish the presence of solid and liquid states. There is comfort in seeing what you are working with. The good news is that many properties are the same for all gases regardless of their composition. Thus, their properties are subject to many convenient generalizations including the quantitative relationships called "the gas laws." On the other hand, few generalizations apply to liquids and solids. As a result, they must be studied individually.

In this chapter, we discuss some qualitative generalizations that lead us to the modern understanding of the states of matter known as the kinetic molecular theory. We also see that the gas laws that follow are logical consequences of this theory. Finally, we take advantage of the fact that the volume of any gas relates to the moles of that gas present so that we can include gas volumes in stoichiometry calculations.

As background for this chapter you should be familiar with the concept of the mole, discussed in Chapter 8, and the procedure for solving stoichiometry problems, discussed in Chapter 9.

10-1 THE NATURE OF THE GASEOUS STATE

Sometimes it is difficult to realize that the gaseous state is a form of matter, equal to the liquid and solid states. For example, a person can move with little resistance through a gas such as air but not through a liquid such as water and certainly not through a solid. Also, air and most other gases are colorless and invisible; exceptions include chlorine and nitrogen dioxide. All in all, except in a strong wind, evidence that a gas is matter is not always obvious and we just have to accept that it is indeed there. We can certainly appreciate why gases so confused early scientists. Gases do have unique properties, however. Many of these properties we take for granted, so you may find them familiar. Five properties are described here.

1   Gases Are Compressible.

Unlike the other two states of matter, additional amounts of gas can be added to the same volume. For example, the volume of an automobile tire is essentially constant but we can keep cramming more and more air into this same volume with a high-pressure pump (at least until the tire bursts). In another example, if we had some air in a bicycle pump we could force the same quantity of air into a smaller volume by pressing on the handle. Liquids and solids, by comparison, are essentially incompressible.

2   Gases Have Low Densities.

In Chapter 2 we discussed the densities of liquids and solids. As you will recall, the density of a typical solid was about 2 g/mL. The density of a typical gas, however, is about 2 g/L. That is, a gas is roughly 1000 times less dense than a solid or liquid. That is a big difference.

3   Gases Fill a Container Uniformly.

When we blow air into a round balloon, the balloon becomes spherical and uniform. The air that we blow into the balloon is evenly distributed and is not affected by gravity. A liquid, however, fills only the bottom part of a container and a solid has a fixed shape.

4   Gases Mix Thoroughly.

When an experiment involving the noxious gas hydrogen sulfide is being performed in the chemistry building, it doesn't take long for the surrounding campus to know about it. It can quickly saturate an environment with the odor of rotten eggs. All gases mix thoroughly and rapidly. In contrast, some liquids do not mix at all (e.g., oil and water). If they do mix (e.g., water and alcohol), the mixing process occurs quite slowly. Two solids, of course, essentially do not mix at all.

5   A Gas Exerts Pressure Uniformly on All Sides of a Container.

Let's return to the balloon. If the pressure exerted by the gas inside the balloon were not uniform it would not take a perfectly spherical shape. If pressure were exerted only downward, the balloon would reflect this and look more like a balloon filled with water. A container of water is different. Because water responds to gravity it stays in the bottom part of a container and the pressure on the sides increases as the depth of the water increases.

Obviously, gases do have many common properties that distinguish them from the other two physical states. Fortunately, there is a very simple explanation of the nature of gases that covers not only all of these unique properties but others as well. This explanation is a set of assumptions collectively known as the **kinetic molecular theory**, or simply **kinetic theory**. The major points of this theory as applied to gases are as follows.

1 A gas is composed of very small particles called molecules (atoms in the case of noble gases), which are widely spaced and occupy negligible volume. A gas is thus mostly empty space.

2 The molecules of the gas are in rapid, random motion, colliding with each other and the sides of the container. Pressure is a result of these collisions with the sides of the container. (See Figure 10-1.)

3 All collisions involving gas molecules are elastic. (The total energy of two colliding molecules is conserved. A ball bouncing off the pavement undergoes inelastic collisions since it does not bounce as high each time.)

4 The gas molecules have negligible attractive (or repulsive) forces between them.

5 The temperature of a gas is related to the average kinetic energy of the gas molecules. Also, at the same temperature, different gases have the same average kinetic energy (**K.E.**):

$$\text{K.E.} = \tfrac{1}{2}mv^2$$
$$m = \text{mass}$$
$$v = \text{velocity}$$

On the basis of the kinetic theory, the characteristics of gases discussed previously now seem reasonable and predictable. For example, since a volume of gas is mostly empty space it is easily compressible and has a low density. Because gas molecules are not attracted to each other, they do not "stick together" as do molecules in liquids and solids. Thus they fill a container uniformly. The rapid motion explains why gases mix rapidly and thoroughly. Finally, since gas molecules collide with the sides of a container randomly, pressure is exerted evenly in all directions.

SEE PROBLEMS 10-1 THROUGH 10-6.

Besides the qualitative properties of gases discussed above, there are the common quantitative properties of gases known as the "gas laws" which will be discussed shortly. We will find that these relationships are also reasonable and predictable from the kinetic theory. We note, however, that most gas laws were firmly established long before any explanation was advanced. In this text we will note how the gas laws are obvious extensions of the assumptions of the kinetic theory.

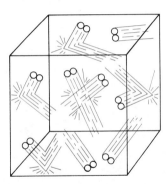

Figure 10-1 GASES. Gas molecules are in rapid, random motion.

Our first order of business is to look deeper into what is meant by "pressure." We have mentioned pressure previously and most of us have a notion of what it means. However, it is important that its actual definition be understood in the context of the ocean of gas in which we live.

10-2 THE PRESSURE OF A GAS

In 1643, an Italian scientist named Evangelista Torricelli experimented with an apparatus that is now known as a **barometer.** Torricelli filled a long glass tube with the dense, liquid metal mercury, and inverted the tube into a bowl of mercury so that no air would enter the tube. Torricelli found that the mercury in the tube would stay suspended at a height of about 76 cm no matter how long or wide the tube. (See Figure 10-2.)

At the time scientists thought that since "nature abhors a vacuum," there was a vacuum in the top of the tube that held up the column of mercury. Torricelli had a different idea. He suggested that it was actually the force of the air on the outside that pushed up the level of mercury. Otherwise, since tubes 2 and 3 in Figure 10-2 would have "more vacuum" than tube 1, the level would rise higher in those tubes. To prove his point, Torricelli took his barometer up a mountain, where, he reasoned, the atmosphere would be less dense and would thus support less mercury in the tube. He was correct; the level of the barometer fell as he ascended the mountain.

The weight of a quantity of matter pressing on a surface is an example of **force. Pressure** *is defined as the force applied per unit area.* This can be expressed mathematically as

$$P \text{ (pressure)} = \frac{F \text{ (force)}}{A \text{ (area)}}$$

In a barometer, the pressure exerted by the atmosphere on the outside is equal to the pressure exerted by the column of mercury on the inside.

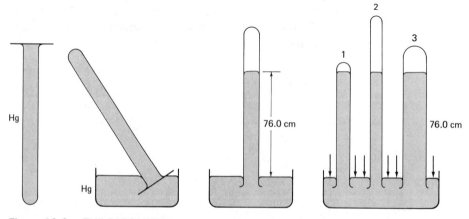

Figure 10-2 THE BAROMETER. When a long tube is filled with mercury and inverted in a bowl of mercury, the atmosphere supports the column to a height of 76.0 cm.

TABLE 10-1 One Atmosphere	
Unit of Pressure	Special Use
760 mm Hg or 760 torr	Most chemistry laboratory measurements for pressures in the neighborhood of one atmosphere
14.7 lb/in.²	U.S. pressure gauges
29.9 in. Hg	U.S. weather reports
101.325 kPa (kilopascals)	The SI unit of pressure [1 N (newton)/m²]
1.013 bars	Used in physics and astronomy mainly for very low pressures (millibars) or very high pressures (kilobars)

The human body has roughly 10,000 cm² of surface area (with, obviously, much individual variation). This means that we feel about 10^7 g (about 20,000 lb) of force from the air under normal atmospheric conditions. On days when it feels like that whole force is only on our heads, it seems like we are "under a lot of pressure." Despite how it may feel at times, the pressure is reasonable because the force is spread out over the total body area. Thus, we should feel quite comfortable at the bottom of the ocean of gas where we live.

The average pressure of the atmosphere at sea level is defined as exactly one atmosphere (1 atm) and is used as the standard. As we have seen, this is equivalent to the pressure exerted by a column of mercury 76.0 cm (760 mm) high.

$$1.00 \text{ atm} = 76.0 \text{ cm Hg} = 760 \text{ mm Hg} = 760 \text{ torr}$$

(The unit mm Hg is defined as *torr* in honor of Evangelista Torricelli.)

In addition to torr, there are several other units of pressure that have special uses. The relations of the units to 1 atm and their special uses are listed in Table 10-1.

EXAMPLE 10-1

What is 485 torr expressed in atmospheres?

PROCEDURE

The conversion factors are

$$\frac{760 \text{ torr}}{\text{atm}} \quad \text{and} \quad \frac{1 \text{ atm}}{760 \text{ torr}}$$

SOLUTION

$$485 \text{ torr} \times \frac{1 \text{ atm}}{760 \text{ torr}} = \underline{\underline{0.638 \text{ atm}}}$$

■
SEE PROBLEMS
10-7 THROUGH
10-14.

10-3 BOYLE'S LAW

Every time we take a breath, we demonstrate how pressure affects the volume of a gas. When the diaphragm under our rib cage relaxes it moves up squeezing our lungs and decreasing their volume. The decreased volume increases the air pressure inside the lungs relative to the outside atmosphere and we expel air. When the diaphragm contracts it moves down, increasing the volume of the lungs and, as a result, decreasing the pressure. Air rushes in from the atmosphere until the pressures are equal.

■

REVIEW APPENDIX B ON PROPORTION-ALITIES AND APPENDIX F ON GRAPHS.

The relationship between the volume of a gas and the pressure on it was the first quantitative relationship concerning gases that was accepted. In 1660, the British scientist Sir Robert Boyle studied this relationship with an apparatus similar to that shown in Figure 10-3. **Boyle's law** *states that there is an inverse relationship between the pressure exerted on a quantity of gas and its volume if the temperature is held constant.*

Boyle's law can be illustrated quite simply with the apparatus shown in Figure 10-3. In experiment 1, a certain quantity of gas ($V_1 = 10.0$ mL) is trapped in a U-shaped tube by some mercury. Since the level of mercury is the same in both sides of the tube and the right side is open to the atmosphere, the pressure on the trapped gas is the same as the atmospheric pressure ($P_1 = 760$ torr).

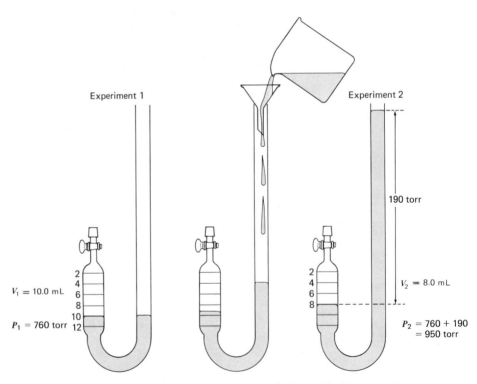

Figure 10-3   BOYLE'S LAW APPARATUS. Addition of mercury in the apparatus causes an increase in pressure on the trapped gas. This leads to a reduction in the volume.

When mercury is added to the tube, the pressure on the trapped gas is increased to 950 torr (760 torr originally plus 190 torr from the added mercury). Note in experiment 2 that the *increase in pressure* has caused a *decrease in volume* to 8.0 mL.

The inverse relation of Boyle is represented as

$$V \propto \frac{1}{P}$$

or as an equality with k the constant of proportionality:

$$V = \frac{k}{P} \quad \text{or} \quad PV = k$$

Note in experiment 1 in Figure 10-3 that

$$\underbrace{P_1V_1 = 760 \text{ torr} \times 10.0 \text{ mL}}_{\text{indicates one set of conditions}} = 7600 \text{ torr} \cdot \text{mL} = k$$

In experiment 2,

$$\underbrace{P_2V_2 = 950 \text{ torr} \cdot 8.0 \text{ mL}}_{\text{indicates a second set of conditions}} = 7600 \text{ torr} \cdot \text{mL} = k$$

As predicted by Boyle's law, note that in both experiments PV equals the same value. Therefore, for a quantity of gas under two sets of conditions at the same temperature,

$$P_1V_1 = P_2V_2 = k$$

We can use this equation to calculate how a volume of a gas changes when the pressure changes. For example, if V_2 is a new volume that is to be found at a certain new pressure, P_2, the equation becomes

$$V_2 = V_1 \times \frac{P_1}{P_2}$$

final volume = initial volume × pressure correction factor

Boyle's and many other gas law problems can be worked by substituting values in the proper formula (e.g., $P_1V_1 = P_2V_2$) or by logic. In the long run, logic is a more dependable method. For example, let us consider a Boyle's law problem where the pressure on a certain volume of gas has been *increased*. We should reason that this would lead to a *decrease* in the volume of the gas. *We therefore choose a pressure correction factor that decreases the volume and is a fraction less than one.* In this case, the lower pressure would be in the numerator of the fraction. The following two examples illustrate this reasoning.

EXAMPLE 10-2

Inside a certain automobile engine, the volume of a cylinder is 457 mL when the pressure is 1.05 atm. When the gas is compressed, the pressure increases to 5.65 atm at the same temperature. What is the volume of the compressed gas?

PROCEDURE

The final volume equals the initial volume times the pressure factor. Since the pressure increases, the volume decreases and the pressure factor must be less than one.

SOLUTION

$$V_2 = V_1 \times \frac{P_1}{P_2}$$

$$V_2 = 475 \text{ mL} \times \frac{1.05 \text{ atm}}{5.65 \text{ atm}} = 88.3 \text{ mL}$$

(Note that the units of pressure are the same. *If the initial and final pressures are given in different units, one must be converted to the other.*)

EXAMPLE 10-3

If the volume of a gas is 3420 mL at a pressure of 2.17 atm, what is the pressure if the gas expands to 8.75 L?

PROCEDURE

The Boyle's law relationship can be solved for P_2:

$$P_2 = P_1 \times \frac{V_1}{V_2}$$

final pressure = initial pressure × volume correction factor

In this case, the volume increases from 3420 mL (3.42 L) to 8.75 L. An increase in volume means a decrease in pressure, so the volume factor is less than one.

SOLUTION

$$P_2 = P_1 \times \frac{V_1}{V_2}$$

$$P_2 = 2.17 \text{ atm} \times \frac{3.42 \text{ L}}{8.75 \text{ L}} = 0.848 \text{ atm}$$

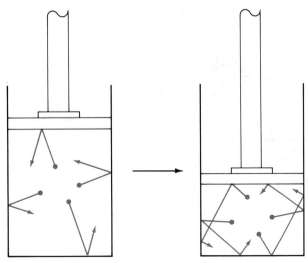

Figure 10-4   BOYLE'S LAW AND KINETIC MOLECULAR THEORY. When the volume decreases, the pressure increases because of the more frequent collisions with the walls of the container.

Boyle's law can now be seen as a direct result of the kinetic theory of gases. In Figure 10-4, the behavior of four average molecules of a certain gas has been illustrated. The distance traveled in a unit of time is represented by the length of the arrow. Since these are average molecules,* they all travel the same distance per unit time at the same temperature. Note that if the volume is decreased there are more frequent collisions with the walls, which in turn means a higher pressure.

■
SEE PROBLEMS
10-15 THROUGH
10-22.

**10-4
CHARLES'
LAW AND
GAY-LUSSAC'S
LAW**

Pressure is not all that affects the volume of gas. Our practical experience tells us that temperature can also change the volume of a gas. For example, if you have ever heated a plastic container of a frozen vegetable in boiling water you may have noticed how the bag expands somewhat. Yet it was more than a century after Boyle's law was advanced before the exact relationship between temperature and volume was understood. In 1787, a French scientist, Jacques Charles, showed that there was a mathematical relationship between volume of a gas and temperature if the pressure is held constant. Charles showed that any gas expands by a definite fraction as the temperature rises. He found that the volume increases by a fraction of $\frac{1}{273}$ for each 1 °C rise in the temperature. (See Figure 10-5.)

---

* *In fact, individual molecules have a range of velocities at a certain temperature. Temperature relates to the average velocity of a large number of molecules of a certain compound or element.*

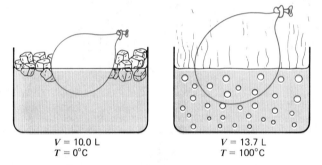

Figure 10-5   EFFECT OF TEMPERATURE ON VOLUME. When the temperature increases, the volume of the balloon increases.

In Figure 10-6, the data from four experiments listed on the left are plotted in the graph on the right. The result is a linear relationship. Note in the graph that the straight line connecting the experimental points (the solid line) has been extended to lower temperatures (the dashed line). (The procedure of extending data beyond experimental results is known as *extrapolation*.) If extended all of the way to where a gas would *theoretically* have zero volume, the line would intersect the temperature scale at about −273 °C. Certainly matter could never have zero volume and it is impossible to cool gases indefinitely. At some point, all gases condense to become liquids or solids. However, this temperature does have significance. This temperature (−273.15 °C to be more precise) turns out to be the lowest possible temperature. As noted in the kinetic molecular theory, the average velocity or kinetic energy of molecules is related to their temperature. *The lowest possible temperature, which is referred to as* **absolute zero,** *is the temperature at which translational motion (motion from point to point) of molecules ceases. The* **Kelvin scale** *assigns zero*

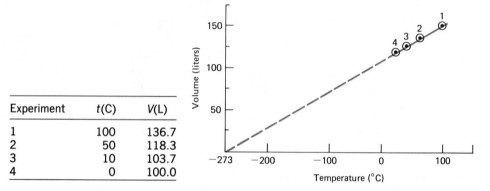

Experiment	t(C)	V(L)
1	100	136.7
2	50	118.3
3	10	103.7
4	0	100.0

Figure 10-6   VOLUME AND TEMPERATURE.

as absolute zero. Thus, there are no negative values possible on the Kelvin scale just as there are no negative values on any pressure scale. Since the magnitudes of the Celsius and Kelvin degrees are the same, we have the following simple relationship between scales, where *T* is the number of kelvins and *t*(C) represents the Celsius temperature.

$$T = [t(C) + 273]K$$

We can now restate more simply the observation of Charles, which is known as **Charles' law:** *The volume of a gas is directly proportional to the temperature in kelvins (T) at constant pressure.* This can be expressed mathematically as

$$V \propto T$$

As an equality with *k* as the constant of proportionality.

$$V = kT \quad \text{or} \quad \frac{V}{T} = k$$

For a quantity of gas under two sets of conditions at the same pressure,

$$\frac{V_1}{T_1} = \frac{V_2}{T_2}$$

This equation can be used to calculate how a volume of gas changes when the temperature changes. For example, if V_2 is a new volume that we are to find at a certain temperature T_2, the equation becomes

$$V_2 = V_1 \times \frac{T_2}{T_1}$$

final volume = initial volume × temperature correction factor

In this case, remember that if the temperature decreases, the volume must also decrease, which means that the temperature factor is less than one. On the other hand, if the temperature increases, the volume must increase, which means that the temperature factor is more than one.

EXAMPLE 10-4

A certain quantity of gas in a balloon has a volume of 185 mL at a temperature of 52 °C. What is the volume of the balloon if the temperature is lowered to −17 °C? Assume that the pressure remains constant.

PROCEDURE

Since the temperature decreases, the volume decreases. The temperature factor must therefore be less than one.

SOLUTION

$$V_2 = V_1 \times \frac{T_2}{T_1} \quad \begin{aligned} T_2 &= (-17 + 273)\ \text{K} = 256\ \text{K}^* \\ T_1 &= (52 + 273)\ \text{K} = 325\ \text{K} \end{aligned}$$

$$V_2 = 185\ \text{mL} \times \frac{256\ \cancel{\text{K}}}{325\ \cancel{\text{K}}} = \underline{\underline{146\ \text{mL}}}$$

---

* Note that temperature must be expressed in kelvins. Also, a Celsius reading with two significant figures (e.g., −17 °C) becomes a Kelvin reading with three significant figures (e.g., 256 K).]

SEE PROBLEMS 10-23 THROUGH 10-29.

Again, Charles' law is now easily predicted from kinetic theory. In Figure 10-7, the action of four molecules has been illustrated. If the temperature is increased, the molecules travel farther per unit time and their collisions are also more powerful. If the pressure is to remain constant, the volume must increase correspondingly. On the other hand, if the volume is held constant when the temperature is increased, it is obvious that the more frequent and more powerful collisions will lead to an increase in pressure. This relationship between pressure and temperature at constant volume is known as **Gay-Lussac's law**. It was proposed in 1802 by Gay-Lussac as a result of his experiments. Gay-Lussac's law can be stated as follows:

$$P \propto T \quad P = kT$$

We may be familiar with an ominous example of Gay-Lussac's law. A pressurized can containing anything from paint to deodorant has a caution: **DO NOT INCINERATE OR STORE NEAR HEAT**. Since the container has a constant volume, heating causes an increase in pressure. At some high temperature, the seals on the can fail and it explodes.

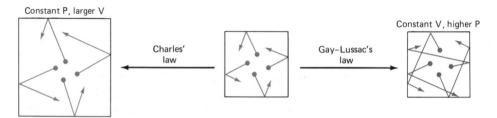

Constant P, larger V   Charles' law   Gay-Lussac's law   Constant V, higher P

Figure 10-7   GAY-LUSSAC'S AND CHARLES' LAWS AND KINETIC MOLECULAR THEORY. When the temperature increases, the molecules travel faster. If the volume is constant, the pressure increases as a result of the more frequent collisions. If the pressure is constant, the volume increases in order that the frequency of collisions remains the same.

For a sample of gas under two sets of conditions at the same volume:

$$\frac{P_1}{T_1} = \frac{P_2}{T_2} \quad P_2 = P_1 \times \frac{T_2}{T_1}$$

EXAMPLE 10-5

A quantity of gas in a steel container has a pressure of 760 torr at 25 °C. What is the pressure in the container if the temperature is increased to 50 °C?

PROCEDURE

Since the temperature increases, the pressure increases. The temperature correction factor must therefore be greater than one.

SOLUTION

$$P_2 = P_1 \times \frac{T_2}{T_1} \quad \begin{array}{l} T_2 = (50 + 273) \text{ K} = 323 \text{ K} \\ T_1 = (25 + 273) \text{ K} = 298 \text{ K} \end{array}$$

$$P_2 = 760 \text{ torr} \times \frac{323 \text{ K}}{298 \text{ K}} = \underline{824 \text{ torr}}$$

SEE PROBLEMS
10-30 THROUGH
10-36.

10-5 THE COMBINED GAS LAW

Boyle's, Charles', and Gay-Lussac's laws are each useful when one condition changes and one is constant. In many cases two conditions change at the same time. *If all three laws are combined into one (understandably called* **the combined gas law***), we have a relationship appropriate to any and all changes of conditions of a sample of gas:*

$$\frac{PV}{T} = k$$

It is apparent that the volume of a gas is very much dependent on its conditions. This is unlike the volumes of liquids and solids, which are almost incompressible by comparison. When discussing gas volumes, then, it is useful to define certain conditions that are universally accepted as standard. These conditions are called **standard temperature and pressure (STP)**:

Standard temperature   0 °C or 273 K
Standard pressure   760 torr or 1 atm

The following examples illustrate the use of the combined gas law and STP.

EXAMPLE 10-6

A 25.8-L quantity of gas has a pressure of 690 torr and a temperature of 17 °C. What is the volume if the pressure is changed to 1.85 atm and the temperature to 345 K?

SOLUTION

Initial Conditions	Final Conditions
$V_1 = 25.8$ L	$V_2 = ?$

$$P_1 = 690 \text{ torr} \times \frac{1 \text{ atm}}{760 \text{ torr}} \qquad P_2 = 1.85 \text{ atm}$$

$$= 0.908 \text{ atm}$$

$$T_1 = (17 + 273) \text{ K} = 290 \text{ K} \qquad T_2 = 345 \text{ K}$$

$$\frac{P_1 V_1}{T_1} = \frac{P_2 V_2}{T_2} \qquad V_2 = V_1 \times \frac{P_1}{P_2} \times \frac{T_2}{T_1}$$

$$\text{final volume} = \text{initial volume} \times \underset{\substack{\text{pressure} \\ \text{correction} \\ \text{factor}}}{} \times \underset{\substack{\text{temperature} \\ \text{correction} \\ \text{factor}}}{}$$

In this problem note that the pressure increases, which decreases the volume. The pressure factor should therefore be less than one. On the other hand, the temperature increases, which increases the volume. The temperature factor is greater than one.

$$V_2 = 25.8 \text{ L} \times \frac{0.908 \text{ atm}}{1.85 \text{ atm}} \times \frac{345 \text{ K}}{290 \text{ K}} = \underline{\underline{15.1 \text{ L}}}$$

EXAMPLE 10-7

A 5850-ft^3 quantity of natural gas measured at STP was purchased from the Gas Company. Only 5625 ft^3 was received at the house. Assuming that all of the gas was delivered, what was the temperature at the house if the delivery pressure was 1.10 atm?

SOLUTION

Initial Conditions	Final Conditions
$V_1 = 5850$ ft^3	$V_2 = 5625$ ft^3
$P_1 = 1.00$ atm	$P_2 = 1.10$ atm
$T_1 = 273$ K	$T_2 = ?$

$$\frac{P_1 V_1}{T_1} = \frac{P_2 V_2}{T_2} \qquad T_2 = T_1 \times \frac{P_2}{P_1} \times \frac{V_2}{V_1}$$

In this case, the final temperature is corrected by a pressure and a volume correction factor. Since the final pressure is higher, the pressure factor must be greater than one to increase the temperature. The final volume is lower, so the volume correction factor must be less than one to decrease the temperature.

SEE PROBLEMS
10-37 THROUGH
10-48.

$$T_2 = 273 \text{ K} \times \frac{1.10 \text{ atm}}{1.00 \text{ atm}} \times \frac{5625 \text{ ft}^3}{5850 \text{ ft}^3} = 289 \text{ K}$$

$$t(\text{C}) = 289 \text{ K} - 273 = \underline{\underline{16 \text{ °C}}}$$

10-6 GRAHAM'S LAW

Experiments have shown that if helium gas is substituted for nitrogen in the air in a closed room, a normally comfortable temperature feels quite cool. (Air is roughly 80% nitrogen and 20% oxygen.) Let us explain what makes us feel warm, comfortable, or cool. By metabolism the human body generates excess heat, which is carried away to the surrounding air through collisions of the air molecules with our skin. We are most comfortable when the excess heat is removed at the same rate that it is generated by air that is slightly cooler than our bodies. In a normal atmosphere, this occurs at a temperature of about 76 °F. So why does an exchange of helium atoms for nitrogen molecules make such a difference? The answer lies in the relative velocities of the two species at the same temperature. Also related to the same phenomenon are the rates at which gases **diffuse** *(mix)* and **effuse** *(move through an opening or hole from a region of high pressure to a region of lower pressure)*. According to kinetic molecular theory, the molecules of two different gases at the same temperature have the same average kinetic energy.

Gas 1   K.E. $= \frac{1}{2}m_1v_1^2$ (m_1 and v_1 = mass and velocity of gas 1)
Gas 2   K.E. $= \frac{1}{2}m_2v_2^2$ (m_2 and v_2 = mass and velocity of gas 2)

Since $(\text{K.E.})_1 = (\text{K.E.})_2$, we have the following relationship:

$$\frac{1}{2}m_1v_1^2 = \frac{1}{2}m_2v_2^2$$

$$\frac{v_1}{v_2} = \sqrt{\frac{m_2}{m_1}}$$

This relationship is known as **Graham's law of effusion.** *Thus, lighter gases diffuse, effuse, or just move faster on the average than heavier gases at the same temperature.* (See Figure 10-8.) So how does this relate to the fact that a helium–oxygen mixture feels cooler at the same temperature than a nitrogen–oxygen mixture? Graham's law provides the answer. Helium atoms are much lighter than nitrogen molecules so they travel faster and thus collide more frequently with the body (similar to what happens in the right part of the illustration in Figure 10-7). The more frequent collisions means that heat is carried away faster and, as a result, we feel cooler. The relative velocities of the two molecules are calculated in Example 10-8.

Figure 10-8   KINETIC ENERGY. To have the same kinetic energy, a light molecule travels at a higher velocity than a heavy molecule.

EXAMPLE 10-8

How much faster does a He atom travel than an N_2 molecule at the same temperature?

SOLUTION

At the same temperature, $(K.E.)_{He} = (K.E.)_{N_2}$. Therefore,

$$m_{He} = 4.00 \text{ amu} \quad m_{N_2} = 28.0 \text{ amu}$$

$$\frac{v_{He}}{v_{N_2}} = \sqrt{\frac{m_{N_2}}{m_{He}}} = \sqrt{\frac{28.0 \text{ amu}}{4.00 \text{ amu}}} = \sqrt{7.00} = 2.65$$

$$v_{He} = 2.65 \, v_{N_2}$$

On the average, He atoms travel <u>2.65 times faster</u> than N_2 molecules.

EXAMPLE 10-9

Nitrogen dioxide (NO_2) effuses at a rate 1.73 times that of an unknown gas. What is the molar mass of the unknown gas?

PROCEDURE

The rate of effusion is a function of the average velocity of the molecules. Also, molar mass (in g/mol) can be used in place of formula weight (in amu). Thus, Graham's law is often expressed as

$$\frac{r_1}{r_2} = \sqrt{\frac{MM_2}{MM_1}} \quad (r = \text{rate of effusion}, \, MM = \text{molar mass})$$

SOLUTION

Let

$$r_1 = \text{rate of effusion of } NO_2 \, (r_1 = r_{NO_2})$$
$$r_2 = \text{rate of effusion of the unknown gas}$$

From the problem

$$r_{NO_2} = 1.73r_2 \quad MM_{NO_2} = 46.0 \text{ g/mol}$$

$$\frac{r_{NO_2}}{r_2} = \frac{1.73\,\cancel{t_2}}{\cancel{t_2}} = \sqrt{\frac{MM_2}{MM_{NO_2}}}$$

$$1.73 = \sqrt{\frac{MM_2}{46.0 \text{ g/mol}}}$$

Square both sides of the equation to remove the square root sign on the right.

$$(1.73)^2 = \left(\sqrt{\frac{MM_2}{46.0 \text{ g/mol}}}\right)^2 \quad 2.99 = \frac{MM_2}{46.0 \text{ g/mol}}$$

$$MM_2 = \underline{138 \text{ g/mol}}$$

■
SEE PROBLEMS
10-49 THROUGH
10-56.

10-7 DALTON'S LAW OF PARTIAL PRESSURES

One of the fascinating things about gases is that so many of their properties do not depend on the identity of the particular gas. For example, consider the pressure exerted by a sample of a gas in a closed container. We would predict from kinetic theory that if the pressure depends only on the kinetic energy of the molecules then it doesn't matter whether they are neon atoms or carbon dioxide molecules. This is illustrated in Figure 10-9. The container on the left contains 0.10 mol of a pure gas and the container on the right contains 0.10 mol of a mixture of gases. As predicted from kinetic theory, the pressure is the same in both containers.

Let's focus on container 2 in Figure 10-9. The total pressure is the result of the pressure exerted by the component gases. Since the pressure is independent of the identity of the gases, the pressure exerted by each gas must relate

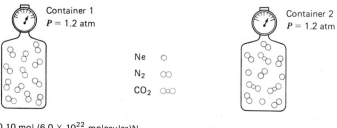

0.10 mol (6.0 × 10²² molecules)N₂

0.050 mol (3.0 × 10²² molecules) N₂
0.025 mol (1.5 × 10²² molecules) CO₂
0.025 mol (1.5 × 10²² atoms) Ne
0.100 mol (6.0 × 10²² particles) total

Figure 10-9   PRESSURES OF A PURE GAS AND A MIXTURE OF GASES. Pressure depends only on the number of molecules at a certain temperature and not on their identity.

to the relative abundance of that particular gas. For example, if 50% of the molecules are nitrogen, then 50% of the pressure is due to that gas. *The pressure exerted by a particular component gas in a mixture is known as the* **partial pressure** *of that gas.* In container 2, since 50% of the molecules are nitrogen, the partial pressure of nitrogen is 50% of the total pressure or 0.60 atm.

The phenomenon that we have just reasoned from kinetic theory was actually proposed by the author of the modern atomic theory, John Dalton, in the early 1800s. *What is now known as* **Dalton's law** *states that the total pressure in a system is the sum of the partial pressures of each component gas.*

$$P_T = P_1 + P_2 + P_3, \text{ etc.}$$

EXAMPLE 10-10

Three gases, Ar, N_2, and H_2, are mixed in a 5.00-L container. Ar has a pressure of 255 torr, N_2 has a pressure of 228 torr, and H_2 has a pressure of 752 torr. What is the total pressure in the container?

SOLUTION

$$P_{Ar} = 255 \text{ torr} \quad P_{N_2} = 228 \text{ torr} \quad P_{H_2} = 752 \text{ torr}$$
$$P_T = P_{Ar} + P_{N_2} + P_{H_2} = 255 \text{ torr} + 228 \text{ torr} + 752 \text{ torr}$$
$$= \underline{1235 \text{ torr}}$$

The pressure of our atmosphere is exerted by a mixture of many gases, some necessary (N_2, O_2, CO_2, H_2O) and some unhealthy (CO, NO_2, SO_2). The major components, however, are N_2 (about 78% of the molecules) and O_2 (about 21%). According to Dalton's law, 21% of the pressure of the atmosphere is due to O_2. Therefore, the partial pressure of O_2 at sea level is

$$P_{O_2} = 0.21 \times 760 \text{ torr} = 160 \text{ torr}$$

On top of the highest mountain, Mt. Everest, the total pressure is 270 torr, so the partial pressure of O_2 is only 57 torr or about one-third of normal. A person could not get enough oxygen to survive for long at such a low pressure. At that altitude, even the most conditioned climber must use an oxygen mask, which gives an increased partial pressure of oxygen to the lungs.

In chemistry laboratory experiments, gases are often collected by displacement of water. (See Figure 10-10.) In this method, the gas is not pure but contains some gaseous water molecules (water vapor). To obtain the pressure of the gas that was collected, the pressure exerted by the H_2O (the vapor pressure of water) must be subtracted from the total pressure. The vapor pressure of water at a certain temperature is always the same and can be obtained from a table, as discussed in Chapter 11.

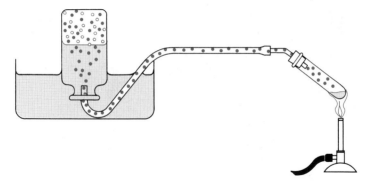

Figure 10-10 COLLECTION OF A GAS OVER WATER. When a certain gas is collected over water, it is mixed with water vapor.

EXAMPLE 10-11

A sample of hydrogen gas is collected over water. The pressure of the sample is 752 torr. If the vapor pressure of water at the temperature of the experiment is 26 torr, what is the pressure exerted by the pure hydrogen gas?

PROCEDURE

Subtract the vapor pressure of H_2O from the total pressure.

SOLUTION

$$P_T = P_{H_2} + P_{H_2O}$$
$$P_{H_2} = P_T - P_{H_2O} = 752 \text{ torr} - 26 \text{ torr} = \underline{726 \text{ torr}}$$

EXAMPLE 10-12

On a humid day the normal atmosphere ($P = 760$ torr) is composed of 2.10% water molecules. Of the remaining gas, 78.0% is N_2 and 21.0% is O_2 (the other 1.0% is a mixture of many gases). What is the partial pressure of H_2O, N_2, and O_2?

PROCEDURE

1 Find the partial pressure of H_2O.
2 Find the pressure of the remaining gas.
3 Find P_{N_2} and P_{O_2}.

SOLUTION

1   2.10% of P_T is $0.0210 \times 760 = \underline{16.0 \text{ torr } (H_2O)}$

2                $P_T = P_{H_2O} + P_{N_2} + P_{O_2} + \overline{P_{\text{other}}}$

     $(P_{N_2} + P_{O_2} + P_{\text{other}}) = P_T - P_{H_2O} = 760 - 16 = 744$ torr

3       $P_{N_2} = 0.780 \times 744$ torr $= \underline{580 \text{ torr } (N_2)}$

       $P_{O_2} = 0.210 \times 744$ torr $= \underline{156 \text{ torr } (O_2)}$

■
SEE PROBLEMS
10-57 THROUGH
10-66.

10-8
AVOGADRO'S LAW

It seems obvious that the more air we blow into a balloon the bigger it gets. Actually, that simple observation leads us to still another gas law. **Avogadro's law** *states that equal volumes of gases at the same pressure and temperature contain equal numbers of molecules.* This was originally advanced as a hypothesis to explain how volumes of gases react. Since the original hypothesis, however, the kinetic molecular theory has become accepted, and this "hypothesis" is generally considered to be a "law." According to kinetic theory, gas molecules do not interact; thus the total number of molecules is all that is important, not their identity. Note that both containers in Figure 10-9 have the same total number of molecules, although container 1 has a pure gas present and container 2 has a mixture. Otherwise, all measurable properties are identical (e.g., P, V, T, and n_{total}).

A corollary of Avogadro's law is that *the volume of a gas is proportional to the total number of molecules (moles) of gas present at constant P and T.* This is expressed as follows, where n equals the number of moles:

$$V \propto n \quad \text{(as a proportion)}$$
$$V = kn \quad \text{(as an equality)}$$
$$\frac{V_1}{n_1} = \frac{V_2}{n_2} \quad \text{(for two conditions)}$$

EXAMPLE 10-13

A balloon that is not inflated but is full of air has a volume of 275 mL and contains 0.0120 mol of air. As shown in Figure 10-11, a piece of Dry Ice (solid CO_2) weighing 1.00 g is placed in the balloon and the neck tied. What is the volume of the balloon after the Dry Ice has vaporized? (Assume constant T and P.)

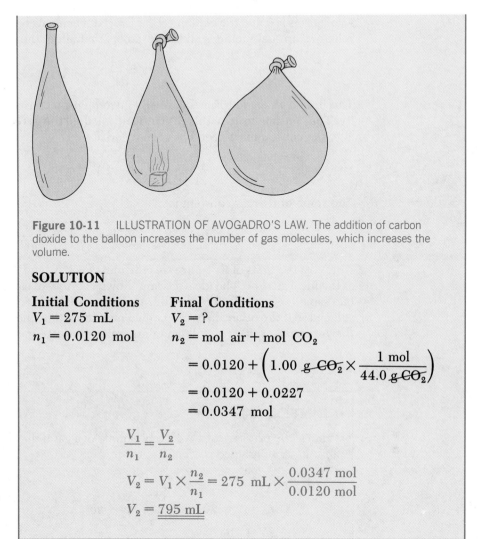

Figure 10-11   ILLUSTRATION OF AVOGADRO'S LAW. The addition of carbon dioxide to the balloon increases the number of gas molecules, which increases the volume.

SOLUTION

Initial Conditions   **Final Conditions**
$V_1 = 275$ mL $V_2 = ?$
$n_1 = 0.0120$ mol $n_2 = $ mol air + mol CO_2

$$= 0.0120 + \left(1.00 \text{ g } CO_2 \times \frac{1 \text{ mol}}{44.0 \text{ g } CO_2} \right)$$

$$= 0.0120 + 0.0227$$

$$= 0.0347 \text{ mol}$$

$$\frac{V_1}{n_1} = \frac{V_2}{n_2}$$

$$V_2 = V_1 \times \frac{n_2}{n_1} = 275 \text{ mL} \times \frac{0.0347 \text{ mol}}{0.0120 \text{ mol}}$$

$$V_2 = \underline{795 \text{ mL}}$$

■
*SEE PROBLEMS
10-67 THROUGH
10-71.*

**10-9 THE
IDEAL GAS LAW**

In Section 10-5 we found that Boyle's, Charles', and Gay-Lussac's laws could be combined into one relationship called the combined gas law. Rearrangement of the combined gas law to emphasize the effect of the other two variables on volume results in the relationship shown on the left. The effect of amount on volume (Avogadro's law) is shown on the right.

$$V \propto \frac{T}{P} \qquad\qquad V \propto n$$

Combined gas law   Avogadro's law

Avogadro's law and the combined gas law itself can now be combined into one relationship that includes all the variables that affect the volume of a gas:

$$V \propto \frac{nT}{P}$$

This can be changed to an equality by introducing a constant of proportionality. The constant used in this case is R and is called the **gas constant.** Traditionally, the constant is placed between n and T.

$$V = \frac{nRT}{P} \quad \text{or} \quad PV = nRT$$

The value of the gas constant is

$$R = 0.0821 \, \frac{L \cdot atm}{K \cdot mol} = 62.4 \, \frac{L \cdot torr}{K \cdot mol}$$

Note that the units of R require specific units for P, V, and T. This law is known as the **ideal gas law.** The ideal gas law allows a variety of calculations involving one sample of gas. Each of the four possible variables (P, V, n, and T) can be calculated if the other three are known (assuming that one knows the value of the gas constant, R). The following examples illustrate the use of the ideal gas law.

EXAMPLE 10-14

What is the pressure of a 1.45-mol sample of a gas if the volume is 20.0 L and the temperature is 25 °C?

SOLUTION

$$P = ? \quad V = 20.0 \, L \quad n = 1.45 \, mol \quad \begin{aligned} T &= (25 + 273) \, K \\ &= 298 \, K \end{aligned}$$

$$PV = nRT$$

$$P = \frac{nRT}{V} = \frac{1.45 \, \cancel{mol} \times 0.0821 \, \frac{\cancel{L} \cdot atm}{\cancel{K} \cdot \cancel{mol}} \times 298 \, \cancel{K}}{20.0 \, \cancel{L}}$$

$$= \underline{1.77 \, atm}$$

EXAMPLE 10-15

What is the temperature (in °C) of a 9.65-g quantity of O_2 in a 4560-mL container if the pressure is 895 torr?

SOLUTION

$$P = 895 \text{ torr} \quad T = ? \quad V = 4560 \text{ mL} = 4.56 \text{ L}$$

$$n = 9.65 \text{ g } O_2 \times \frac{1 \text{ mol}}{32.0 \text{ g } O_2} = 0.302 \text{ mol}$$

$$PV = nRT \quad T = \frac{PV}{nR}$$

$$T = \frac{895 \text{ torr} \times 4.56 \text{ L}}{0.302 \text{ mol} \times 62.4 \dfrac{\text{L} \cdot \text{torr}}{\text{K} \cdot \text{mol}}} = 217 \text{ K}$$

$$t(C) = 217 - 273 = \underline{-56 \,^\circ C}$$

EXAMPLE 10-16

What is the volume of 1.00 mol of a gas at STP?

SOLUTION

$$P = 1.00 \text{ atm} \quad V = ? \quad T = 273 \text{ K} \quad n = 1.00 \text{ mol}$$

$$PV = nRT \qquad V = \frac{nRT}{P}$$

$$V = \frac{1.00 \text{ mol} \times 0.0821 \dfrac{\text{L} \cdot \text{atm}}{\text{K} \cdot \text{mol}} \times 273 \text{ K}}{1.00 \text{ atm}} = \underline{22.4 \text{ L}}$$

Note that the ideal gas law is used mostly to calculate a missing variable under one set of conditions. The combined gas law is most useful for calculations that apply to a sample of gas under two sets of conditions with one missing variable.

This law is called ideal because it follows from the assumptions of the kinetic theory, which describes an ideal gas. The molecules of an ideal gas have no volume and have no attraction for each other. The molecules of a "real" gas obviously have a volume and there is some interaction between molecules, especially at high pressures and low temperatures where the molecules are pressed close together. Fortunately, at normal temperatures and pressures found on the surface of the earth, gases have close to "ideal" behavior. Therefore, the use of the ideal gas law is justified. If we lived on the planet Jupiter, however, where pressure is measured in thousands of earth atmospheres, the ideal gas law would not provide accurate answers. Other relationships would have to be used that take into account the volume of the molecules and the interaction between molecules.

SEE PROBLEMS 10-72 THROUGH 10-83.

10-10 THE MOLAR VOLUME AND DENSITY OF A GAS

We are now ready to take advantage of the fact that volumes of gases contain equal numbers of particles under specified conditions. If this is true, one mole of a gas under the same conditions has a specific volume much like one mole has a specific number of particles (Avogadro's number). The volume of one mole of gas at STP was calculated in Example 10-16. *It is known as the* **molar volume** *and has a value of 22.4 L* (STP). It is now apparent that for a *gas*, one mole describes three quantities. Two are independent of the gas (molar number and molar volume) and one depends on the identity of the gas (molar mass).

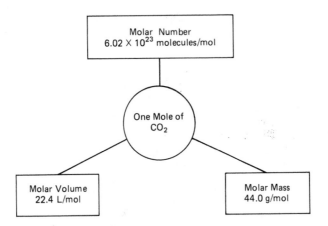

Before working sample problems illustrating the use of molar volume, let's summarize the relationships discussed between moles and volume of a gas. In Figure 10-12, note that moles can be converted to volume at STP using the molar volume relationship (path 1). Moles of a gas can also be converted to

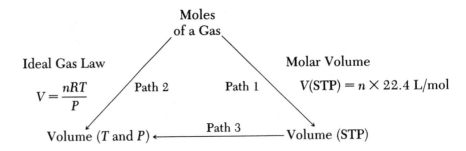

$$V(T,P) = V(\text{STP}) \times \frac{1 \text{ atm}}{P} \times \frac{T}{273 \text{ K}}$$

Figure 10-12   MOLES AND THE VOLUME OF A GAS. The number of moles relates to the volume by two paths.

volume at some other temperature and pressure using the ideal gas law (path 2). Finally, the volume at STP of a quantity of gas and the volume at some other temperature and pressure relate by the combined gas law (path 3).

EXAMPLE 10-17

What is the mass of 4.55 L of O_2 measured at STP?

PROCEDURE

Utilizing path 1 in Figure 10-12, convert volume to moles and then moles to mass. The two possible conversion factors for converting volume to moles are

$$\frac{1 \text{ mol}}{22.4 \text{ L}} \quad \text{and} \quad \frac{22.4 \text{ L}}{\text{mol}}$$

$$\text{volume (STP)} \longrightarrow \text{moles} \longrightarrow \text{mass}$$

SOLUTION

$$4.55 \text{ L} \times \frac{1 \text{ mol}}{22.4 \text{ L}} \times \frac{32.0 \text{ g}}{\text{mol}} = \underline{\underline{6.50 \text{ g}}}$$

EXAMPLE 10-18

A sample of gas has a mass of 3.20 g and occupies 2.00 L at 17 °C and 380 torr. What is the molar mass of the gas?

PROCEDURE

There are two ways that one can work this problem. The first uses paths 1 and 3 in Figure 10-12.

1 Convert given volume to volume at STP.

2 Convert volume at STP to moles.

3 Convert moles and mass to molar mass as follows: From Chapter 8,

$$n \text{ (number of moles)} = \text{mass (in g)} \times \frac{1 \text{ mol}}{\text{molar mass (MM)}}$$

or

$$n = \frac{\text{mass}}{MM} \qquad MM = \frac{\text{mass}}{n}$$

SOLUTION

Initial Conditions	Final Conditions (STP)
$V_1 = 2.00$ L	$V_2 = ?$
$P_1 = 380$ torr	$P_2 = 760$ torr
$T_1 = (17 + 273)$ K $= 290$ K	$T_2 = 273$ K

1
$$\frac{P_1 V_1}{T_1} = \frac{P_2 V_2}{T_2} \qquad V_2 = V_1 \times \frac{P_1}{P_2} \times \frac{T_2}{T_1}$$

$$V_2 = 2.00 \text{ L} \times \frac{380 \text{ torr}}{760 \text{ torr}} \times \frac{273 \text{ K}}{290 \text{ K}} = \underline{\underline{0.941 \text{ L (STP)}}}$$

2
$$0.941 \text{ L} \times \frac{1 \text{ mol}}{22.4 \text{ L}} = 0.0420 \text{ mol of gas}$$

3
$$\text{molar mass} = \frac{\text{mass}}{n} = \frac{3.20 \text{ g}}{0.0420 \text{ mol}} = \underline{\underline{76.2 \text{ g/mol}}}$$

The alternative procedure is more direct, since it uses the ideal gas law as shown in path 2 in Figure 10-12. To find the molar mass using the ideal gas law, substitute the relationship for n ($n = $ mass/MM) into the gas law.

$$PV = nRT$$

$$n = \frac{\text{mass}}{MM}$$

$$PV = \frac{\text{mass}}{MM} RT$$

Solving for *MM*, we have

$$MM = \frac{(\text{mass})(RT)}{PV}$$

$$= \frac{3.20 \text{ g} \times 62.4 \dfrac{\text{L} \cdot \text{torr}}{\text{K} \cdot \text{mol}} \times 290 \text{ K}}{380 \text{ torr} \times 2.00 \text{ L}}$$

$$= \underline{\underline{76.2 \text{ g/mol}}}$$

TABLE 10-2   Densities of Some Gases

Gas	Density [g/L (STP)]	Gas	Density [g/L (STP)]
H_2	0.090	O_2	1.43
He	0.179	CO_2	1.96
N_2	1.25	SF_6	6.52
Air (average)	1.29	UF_6	15.7

In Chapter 2 the densities of solids and liquids were given in units of g/mL. Since gases are much less dense, units in this case are usually given in g/L (STP). The density of a gas at STP can be calculated by dividing the molar mass by the molar volume.

$$CO_2 \quad 44.0 \text{ g/mol} \approx 22.4 \text{ L/mol}$$

$$\frac{44.0 \text{ g/mol}}{22.4 \text{ L/mol}} = \underline{1.96 \text{ g/L (STP)}}$$

The densities of several gases are listed in Table 10-2.

10-11 STOICHIO-METRY INVOLVING GASES

Some of the earliest experimentation that led to the understanding of the formation of certain compounds and their formulas involved the volumes of gases. Gases are formed or consumed in many chemical reactions. Since the volume of a gas relates directly to the number of moles (through the ideal gas law), we can now include gas volumes in the general scheme for stoichiometry problems as first shown in Figure 9-6. In Figure 10-13, the number of individual molecules has been deleted from the earlier scheme, since problems involving actual numbers of molecules are not often encountered. We have also included in this figure the relationships used to convert from one unit to another. Note that the molar volume relationship ($n = V$ (STP)/22.4 L) can be used to convert from volume to moles when the gas is at STP.

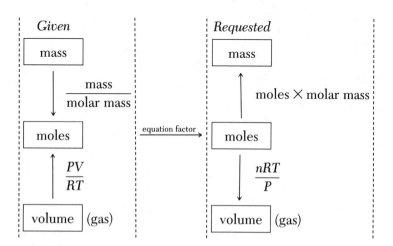

Figure 10-13   GENERAL PROCEDURE FOR STOICHIOMETRY PROBLEMS.

The following examples illustrate the relationship of gas laws to stoichiometry.

EXAMPLE 10-19

Given the balanced equation

$$2Al(s) + 6HCl(aq) \longrightarrow 2AlCl_3(aq) + 3H_2(g)$$

what mass of Al is needed to produce 50.0 L of H_2 measured at STP?

PROCEDURE

The general procedure is shown below. In this case, it is easier to use the molar volume to convert volume directly to moles.

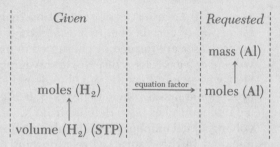

SOLUTION

$$50.0 \;\cancel{L\text{ (STP)}} \times \frac{1 \;\cancel{\text{mol } H_2}}{22.4 \;\cancel{L\text{ (STP)}}} \times \frac{2 \;\cancel{\text{mol } Al}}{3 \;\cancel{\text{mol } H_2}} \times \frac{27.0 \text{ g}}{\cancel{\text{mol } Al}} = \underline{\underline{40.2 \text{ g}}}$$

EXAMPLE 10-20

Given the balanced equation

$$4NH_3(g) + 5O_2(g) \longrightarrow 4NO(g) + 6H_2O(l)$$

what volume of NO gas measured at 550 torr and 25 °C will be produced from 19.5 g of O_2?

PROCEDURE

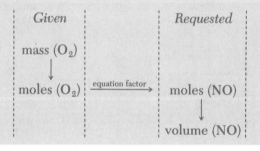

SOLUTION

$$19.5 \, \cancel{g \, O_2} \times \frac{1 \, \cancel{mol \, O_2}}{32.0 \, \cancel{g \, O_2}} \times \frac{4 \, mol \, NO}{5 \, \cancel{mol \, O_2}} = \underline{0.488 \, mol \, NO}$$

1 Using the ideal gas law

$$V = \frac{nRT}{P} = \frac{0.488 \, \cancel{mol} \times 62.4 \, \dfrac{L \cdot \cancel{torr}}{\cancel{K} \cdot \cancel{mol}} \times 298 \, \cancel{K}}{550 \, torr} = \underline{16.5 \, L}$$

2 Convert moles of NO to volume at STP and then volume at STP to volume at 550 torr and 25 °C.

$$0.488 \, mol \times 22.4 \, L/mol = 10.9 \, L \, (STP)$$

$$V_2 = V_1 \times \frac{P_1}{P_2} \times \frac{T_2}{T_1}$$

Initial Conditions	Final Conditions
$V_1 = 10.9 \, L$	$V_2 = ?$
$P_1 = 760 \, torr$	$P_2 = 550 \, torr$
$T_1 = 273 \, K$	$T_2 = (25 + 273) \, K = 298 \, K$

$$V_2 = 10.9 \, L \times \frac{760 \, \cancel{torr}}{550 \, \cancel{torr}} \times \frac{298 \, \cancel{K}}{273 \, \cancel{K}} = \underline{16.4 \, L}$$

■
*SEE PROBLEMS
10-96 THROUGH
10-104.*

**10-12
CHAPTER
REVIEW**

Beginning in the 1660s prominent scientists reported their observations concerning the behavior of air and other gases. The first observation by Torricelli using his barometer was that gases exert pressure and are therefore legitimate forms of matter. The other observations or "laws" that followed provided the foundation of our modern understanding of gases in particular but all matter in general. We now realize from application of the kinetic molecular theory to gases that these laws are all reasonable and predictable. Since the kinetic theory tells us, among other things, that gases are composed of moving molecules (or atoms of noble gases), we predict that the identity of gas molecules is not important with regard to most of the properties studied. Exceptions are the average velocity of a molecule and the density, which are dependent on the identity of the gas.

The gas laws presented in this chapter and their applications are summarized in the table on the next page.

In addition to these relationships another relationship presented was the molar volume. This relationship [22.4 L/mol (STP)] can be used as a conversion factor between moles and volume at STP. To convert between moles and

Gas Law	Relationship	Meaning	Constant Conditions	Application
Boyle's	$V \propto \dfrac{1}{P}$	$V\uparrow P\downarrow$	T, n	Relates V and P of a gas under two different sets of conditions
Charles'	$V \propto T$	$V\uparrow T\uparrow$	P, n	Relates V and T of a gas under two different sets of conditions
Gay-Lussac's	$P \propto T$	$P\uparrow T\uparrow$	V, n	Relates P and T of a gas under two different sets of conditions
Combined	$PV \propto T$	$PV\uparrow T\uparrow$	n	Relates P, V, and T of a gas under two different sets of conditions
Graham's	$v \propto \dfrac{1}{\sqrt{MM}}$	$v\uparrow$ MM$\downarrow$	T	Relates MM and v of two different gases at a certain T
Dalton's	$P_T = P_1 + P_2$, etc.	$P_T\uparrow n_T\uparrow$	T, V	Relates P_T to partial pressures of component gases
Avogadro's	$V \propto n_T$	$V\uparrow n_T\uparrow$	P, T	Relates V and n of a gas under two different sets of conditions
Ideal	$PV \propto nT$	$PV\uparrow nT\uparrow$	—	Relates P, V, T, or n to the other three variables

V = volume n = moles v = average velocity
T = kelvin temperature n_T = total moles $\uparrow$ = quantity increases
P = pressure   MM = molar mass $\downarrow$ = quantity decreases
P_T = total pressure

volume under conditions other than STP the use of the ideal gas law is convenient.

$$\text{volume } (T, P) \xleftarrow{\text{ideal gas law}} \text{moles of gas} \xrightarrow{\text{molar volume}} \text{volume (STP)}$$

The molar volume can also be used to calculate the density of a gas.

Finally, since the volume of a gas relates to moles, its application in stoichiometric calculations can be illustrated as follows:

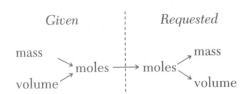

■ EXERCISES

THE KINETIC THEORY OF GASES

10-1 It is harder to move your arms in water than in air. Explain on the basis of the kinetic molecular theory.

10-2 A balloon filled with water is pear-shaped but a balloon filled with air is spherical. Explain.

10-3 When a gasoline tank is filled, no more gasoline can be added. When a tire is "filled," however, more air can be added. Explain.

10-4 The pressure inside an auto tire is the same regardless of the location of the nozzle (i.e., up, down, or to the side). Explain.

10-5 A sunbeam forms when light is reflected from dust suspended in the air. Even if the air is still, the dust particles can be seen to bounce around randomly. Explain.

***10-6** The kinetic theory assumes that the volume of the molecules and their interactions are negligible for gases. Explain why these assumptions may not be true when the pressure is very high and the temperature is very low.

UNITS OF PRESSURE

10-7 Make the following conversions.
(a) 1650 torr to atm
(b) 3.50×10^{-5} atm to torr
(c) 185 lb/in.^{2} to torr
(d) 5.65 kPa to atm
(e) 190 torr to lb/in.^{2}
(f) 85 torr to kPa

10-8 Make the following conversions.
(a) 30.2 in. of Hg to torr
(b) 25.7 kilobars to atm
(c) 57.9 kPa to lb/in.^{2}
(d) 0.025 atm to torr

10-9 Complete the following table.

torr	lb/in.^{2}	in. Hg	kPa	atm
455				
	2.45			
		117		

torr	lb/in.^{2}	in. Hg	kPa	atm
			783	
				0.0768

***10-10** A column of mercury (density 13.6 g/mL) is 15.0 cm high. A cross section of the column has an area of 12.0 cm^2. What is the force (weight) of the mercury at the bottom of the tube? What is the pressure in grams per square centimeter and in atmospheres?

10-11 The atmospheric pressure on the planet Mars is 10.3 millibars. What is this pressure in earth atmospheres?

10-12 The atmospheric pressure on the planet Venus is 0.0920 kilobar. What is this pressure in earth atmospheres?

10-13 The density of water is 1.00 g/mL. If water is substituted for mercury in the barometer, how high (in feet) would be a column of water supported by 1 atm? A water well is 40 ft deep. Can suction be used to raise the water to ground level?

10-14 A tube containing an alcohol (density 0.890 g/mL) is 1.00 m high and has a cross section of 15.0 cm^2. What is the total force at the bottom of the tube? What is the pressure? How high would be an equivalent amount of mercury assuming the same cross section?

BOYLE'S LAW

10-15 A gas has a volume of 6.85 L at a pressure of 0.650 atm. What is the volume of the gas if the pressure is decreased to 0.435 atm?

10-16 If a gas has a volume of 1560 mL at a pressure of 81.2 kPa, what is its volume if the pressure is increased to 2.50 atm?

10-17 At sea level, a balloon has a volume of 785 mL. What is its volume if it is taken to Colorado where the atmospheric pressure is 610 torr?

10-18 A gas has a volume of 125 mL at a pressure of 62.5 torr. What is the pressure if the volume is decreased to 115 mL?

*10-19 A gas in a piston engine is compressed by a ratio of 15 : 1. If the pressure before compression is 0.950 atm, what pressure is required to compress the gas? (Assume constant temperature.)

10-20 A few miles above the surface of the earth the pressure drops to 1.00×10^{-5} atm. What would be the volume of a 1.00-L sample of gas at sea level pressure (1.00 atm) if it were taken to that altitude? (Assume constant temperature.)

*10-21 The volume of a gas is measured as the pressure is varied. The four measurements are reported as follows:

Experiment	Volume (mL)	Pressure (torr)
1	125	450
2	145	385
3	175	323
4	220	253

Make a graph of the volume on the x axis and the pressure on the y axis. What is the average value of the constant of proportionality, k?

10-22 How does the kinetic molecular theory explain Boyle's law?

CHARLES' LAW

10-23 A balloon has a volume of 1.55 L at 25 °C. What would be the volume if the balloon is heated to 100 °C? (Assume constant P.)

10-24 A sample of gas has a volume of 677 mL at 63 °C. What is the volume of the gas if the temperature is decreased to 46 °C?

10-25 A balloon has a volume of 325 mL at 17 °C. What is the temperature if the volume increases to 392 mL?

*10-26 The temperature of a sample of gas is 0 °C. When the temperature is increased, the volume increases by a factor of 1.25 (i.e., $V_2 = 1.25\ V_1$).

What is the final temperature in degrees Celsius?

10-27 A quantity of gas has a volume of 3.66×10^4 L. What will be the volume if the temperature is changed from 455 K to 50 °C?

10-28 The volume of a gas is measured as the temperature is varied. The four measurements are reported as follows:

Experiment	Volume (L)	Temperature (°C)
1	1.54	20
2	1.65	40
3	1.95	100
4	2.07	120

Make a graph of the volume on the x axis and the Kelvin temperature on the y axis. What is the average value of the constant of proportionality, k?

10-29 How does the kinetic molecular theory explain Charles' law?

GAY-LUSSAC'S LAW

10-30 A confined quantity of gas is at a pressure of 2.50 atm and a temperature of -22 °C. What is the pressure if the temperature increases to 22 °C?

10-31 A quantity of gas has a volume of 3560 mL at a temperature of 55 °C and a pressure of 850 torr. What is the temperature if the volume remains unchanged but the pressure is decreased to 0.652 atm?

10-32 A metal cylinder contains a quantity of gas at a pressure of 558 torr at 25 °C. At what temperature does the pressure inside the cylinder equal 1 atm pressure?

10-33 An aerosol spray can has gas under a pressure of 1.25 atm at 25 °C. The can explodes when the pressure reaches 2.50 atm. At what temperature will this happen? (Do not throw these cans into a fire!)

10-34 The pressure in an automobile tire is 28.0 lb/in.² on a chilly morning of 17 °C. After it is driven awhile, the temperature of the tire rises to 40 °C. What is the pressure in the tire if the volume remains constant?

10-35 The pressure of a confined volume of gas is measured as the temperature is raised. The four measurements are reported as follows:

Experiment	Pressure (torr)	Temperature (K)
1	550	295
2	685	372
3	745	400
4	822	445

Make a graph of the pressure on the x axis and the temperature on the y axis. What is the average value of the constant of proportionality, k?

10-36 How does the kinetic molecular theory explain Gay-Lussac's law?

THE COMBINED GAS LAW

10-37 Which of the following are legitimate expressions of the combined gas law?

(a) $PV = kT$ (d) $\dfrac{P}{T} \propto \dfrac{1}{V}$

(b) $PT \propto V$ (e) $VT \propto P$

(c) $\dfrac{P_1T_1}{V_1} = \dfrac{P_2T_2}{V_2}$

10-38 Which of the following are not STP conditions?
(a) $T = 273$ K
(b) $P = 760$ atm
(c) $T = 0$ K
(d) $P = 1$ atm
(e) $t(C) = 273$ °C
(f) $P = 760$ torr
(g) $t(C) = 0$ °C

10-39 In the following table indicate whether the pressure, volume, or temperature increases or decreases.

Experiment	P	V	T
1	increases	constant	
2	constant		decreases
3		decreases	constant
4	increases	increases	

10-40 In the following table, indicate whether the pressure, volume, or temperature increases or decreases.

Experiment	P	V	T
1	decreases		constant
2	constant		T(initial) = 350 K T(final) = 40 °C
3	P(initial) = 1.75 atm P(final) = 2200 torr	constant	
4		increases	decreases

10-41 A 5.50-L volume of gas has a pressure of 0.950 atm at 0 °C. What is the pressure if the volume decreases to 4.75 L and the temperature increases to 35 °C?

10-42 A quantity of gas has a volume of 17.5 L at a pressure of 6.00 atm and temperature of 100 °C. What is its volume at STP?

10-43 A quantity of gas has a volume of 88.7 mL at STP. What is its volume at 0.845 atm and 35 °C?

10-44 A quantity of gas has a volume of 4.78×10^{-4} mL at a temperature of -50 °C and a pressure of 78.0 torr. If the volume changes to 9.55×10^{-5} mL and the pressure to 155 torr, what is the temperature?

10-45 A gas has a volume of 64.2 L at STP. What is the temperature if the volume decreases to 58.5 L and the pressure increases to 834 torr?

10-46 A quantity of gas has a volume of 6.55×10^{-5} L at 7 °C and 0.882 atm. What is the pressure if the volume changes to 4.90×10^{-3} L and the temperature to 273 K?

10-47 A balloon has a volume of 1.55 L at 25 °C and 1.05 atm pressure. If it is cooled in the freezer, the volume shrinks to 1.38 L and the pressure drops to 1.02 atm. What is the temperature in the freezer?

10-48 A bubble from a deep-sea diver in the ocean starts with a volume of 35.0 mL at a temperature of 17 °C and a pressure of 11.5 atm. What is the volume of the bubble when it reaches the surface? Assume that the pressure at the surface is 1 atm and the temperature is 22 °C?

GRAHAM'S LAW

10-49 A bowling ball weighs 6.00 kg and a bullet weighs 1.50 g. If the bowling ball is rolled down an alley at 20.0 miles/hr, what is the velocity of a bullet having the same kinetic energy?

10-50 Arrange the following gases in order of increasing average speed (rate of effusion) at the same temperature.
(a) CO_2 (c) N_2 (e) N_2O
(b) SO_2 (d) SF_6 (f) H_2

10-51 What is the rate of effusion of N_2 molecules compared with Ar atoms?

10-52 Compare the rates of effusion of H_2 molecules and krypton atoms at the same temperature.

10-53 A certain gas effuses twice as fast as SF_6 molecules. What is the molar mass of the unknown gas?

10-54 Carbon monoxide effuses 2.13 times faster than an unknown gas. What is the molar mass of the unknown gas?

10-55 Gas A has a molar mass 3.28 times greater than that of gas B. How fast does gas A effuse compared with gas B?

*10-56 To make enriched uranium for use in nuclear reactors or for weapons, ^{235}U must be separated from ^{238}U. Although ^{235}U is the isotope needed for fission, only 0.7% of U atoms are this isotope. Separation is a difficult and expensive process. Since UF_6 is a gas, Graham's law can be applied to separate the isotopes. How much faster does a $^{235}UF_6$ molecule travel on the average compared with a $^{238}UF_6$ molecule?

DALTON'S LAW

10-57 Three gases are mixed in a 1.00-L container. The partial pressure of CO_2 is 250 torr, that of N_2 375 torr, and that of He 137 torr. What is the pressure of the mixture of gases?

10-58 The total pressure in a cylinder containing a mixture of two gases is 1.46 atm. If the partial pressure of one gas is 750 torr, what is the partial pressure of the other gas?

10-59 Air is about 0.90% Ar. If the barometric pressure is 756 torr, what is the partial pressure of Ar?

10-60 A sample of oxygen is collected over water in a bottle. If the pressure inside the bottle is made equal to the barometric pressure, which is 752 torr, and the vapor pressure of water at the temperature of the experiment is 24 torr, what is the pressure of pure oxygen?

*10-61 A mixture of two gases is composed of CO_2 and O_2. The partial pressure of O_2 is 256 torr, and it represents 35% of the molecules of the mixture. What is the total pressure of the mixture?

10-62 A container holds two gases, A and B. Gas A has a partial pressure of 325 torr and gas B has a partial pressure of 488 torr. What percent of the molecules in the mixture is gas A?

10-63 A volume of gas is composed of N_2, O_2, and SO_2. If the total pressure is 1050 torr, what is the partial pressure of each gas if the gas is 72.0% N_2 and 8.00% O_2?

*10-64 A volume of gas has a total pressure of 2.75 atm. If the gas is composed of 0.250 mol of N_2 and 0.427 mol of CO_2, what is the partial pressure of each gas?

*10-65 The following gases are all combined into a 2.00-L container: a 2.00-L volume of N_2 at 300 torr, a 4.00-L volume of O_2 at 85 torr, and a 1.00-L volume of CO_2 at 450 torr. What is the total pressure.

*10-66 The total pressure of a mixture of two gases is 0.850 atm in a 4.00-L container. Before mixing, gas A was in a 2.50-L container and had a pressure of 0.880 atm. What is the partial pressure of gas B in the 4.00-L container?

AVOGADRO'S LAW

10-67   A 0.112-mol quantity of gas has a volume of 2.54 L at a certain temperature and pressure. What is the volume of 0.0750 mol of gas under the same conditions?

10-68   A balloon has a volume of 188 L and contains 8.40 mol of gas. How many moles of gas would be needed to expand the balloon to 275 L? Assume the same temperature and pressure in the balloon.

10-69   A balloon has a volume of 275 mL and contains 0.0212 mol of CO_2. What mass of N_2 must be added to expand the balloon to 400 mL?

10-70   A balloon has a volume of 75.0 mL and contains 2.50×10^{-3} mol of gas. What mass of of N_2 must be added to the balloon for the volume to increase to 164 mL at the same temperature and pressure?

10-71   A 48.0-g quantity of O_2 in a balloon has a volume of 30.0 L. What is the volume if 48.0 g of SO_2 is substituted for O_2 in the same balloon?

THE IDEAL GAS LAW

10-72   What is the temperature (in degrees Celsius) of 4.50 L of a 0.332-mol quantity of gas under a pressure of 2.25 atm?

10-73   A quantity of gas has a volume of 16.5 L at 32 °C and a pressure of 850 torr. How many moles of gas are present?

10-74   What mass of NH_3 gas has a volume of 16,400 mL, a pressure of 0.955 atm, and a temperature of -23 °C?

10-75   What is the pressure (in torr) exerted by 0.250 g of O_2 in a 250-mL contaner at 29 °C?

10-76   What is the pressure (in atm) in a 825-mL container at 33 °C if it contains 6.25 g of N_2 and 12.6 g of CO_2?

10-77   A container of Cl_2 gas has a volume of 750 mL and is at a temperature of 19 °C. If there is 7.88 g of Cl_2 in the container what is the pressure in atmospheres?

10-78   What mass of Ne is contained in a large neon light if the volume is 3.50 L, the pressure 1.15 atm, and the temperature 23 °C?

10-79   A sample of H_2 is collected in a bottle over water. The volume of the sample is 185 mL at a temperature of 25 °C and a pressure of 760 torr. The vapor pressure of water at this temperature is 24 torr. What is the mass of H_2 in the bottle?

10-80   One liter of a gas weighs 8.37 g. The gas volume is measured at 1.45 atm pressure and 35 °C. What is the molar mass of the gas?

*10-81   The Goodyear blimp has a volume of about 2.5×10^7 L. What is the mass of He (in lb) in the blimp at 27 °C and 780 torr? The average molar mass of air is 29.0 g/mol. What mass of air (in lb) would the blimp contain? The difference between these two values is the lifting power of the blimp. What mass could the blimp lift? If H_2 is substituted for He, what is the lifting power? Why isn't H_2 used?

10-82   A gaseous compound is 85.7% C and 14.3% H. A 6.58-g quantity of this gas occupies 4500 mL at 77.0 °C and a pressure of 1.00 atm. What is the molar mass of the compound, and what is its molecular formula?

*10-83   A good vacuum pump on earth can produce a vacuum with a pressure as low as 1.00×10^{-8} torr. How many molecules are present in each milliliter at a temperature of 27.0 °C?

MOLAR VOLUME AND DENSITY

10-84   What is the volume of 15.0 g of CO_2 measured at STP?

10-85   What is the mass (in kg) of 850 L of CO measured at STP?

10-86   What is the volume of 3.01×10^{24} molecules of N_2 at STP?

10-87   A 6.50-L quantity of a gas measured

at STP has a mass of 39.8 g. What is the molar mass of the compound?

10-88 What is the mass of 6.78×10^{-4} L of NO_2 measured at STP?

10-89 What is the density in g/L (STP) of B_2H_6?

10-90 What is the density in g/L (STP) of BF_3?

10-91 A gas has a density of 1.52 g/L (STP). What is the molar mass of the gas?

10-92 A gas has a density of 6.14 g/L (STP). What is the molar mass of the gas?

*10-93 A gas has a density of 3.60 g/L at a temperature of 25 °C and a pressure of 1.20 atm. What is its density at STP?

*10-94 What is the density (in g/L) of N_2 measured at 500 torr and 22 °C?

*10-95 What is the density (in g/L) of SF_6 measured at 0.370 atm and 37 °C?

STOICHIOMETRY INVOLVING GASES

10-96 Limestone is dissolved by CO_2 according to the equation

$$CaCO_3(s) + H_2O(l) + CO_2(g) \longrightarrow Ca(HCO_3)_2(aq)$$

What volume of CO_2 measured at STP would dissolve 115 g of $CaCO_3$?

10-97 Magnesium in flashbulbs burns according to the equation

$$2Mg(s) + O_2(g) \longrightarrow 2MgO(s)$$

What mass of Mg combines with 5.80 L of O_2 measured at STP?

10-98 Oxygen gas can be prepared in the laboratory by decomposition of potassium nitrate according to the equation

$$2KNO_3(s) \xrightarrow{\Delta} 2KNO_2(s) + O_2(g)$$

What mass of KNO_2 forms along with 14.5 L of O_2 measured at 1 atm and 25 °C?

10-99 Acetylene is produced from calcium carbide as shown by the reaction

$$CaC_2(s) + 2H_2O(l) \longrightarrow Ca(OH)_2(s) + C_2H_2(g)$$

What volume of acetylene (C_2H_2) measured at 25.0 °C and 745 torr would be produced from 5.00 g of H_2O?

10-100 Nitrogen dioxide is an air pollutant. It is produced from NO (from car exhaust) as follows:

$$2NO(g) + O_2(g) \longrightarrow 2NO_2(g)$$

What volume of NO measured at STP is required to react with 5.00 L of O_2 measured at 1.25 atm and 17 °C?

10-101 Butane (C_4H_{10}) burns according to the equation

$$2C_4H_{10}(g) + 13O_2(g) \longrightarrow 8CO_2(g) + 10H_2O(l)$$

(a) What volume of CO_2 measured at STP would be produced by 85.0 g of C_4H_{10}?

(b) What volume of O_2 measured at 3.25 atm and 127 °C would be required to react with 85.0 g of C_4H_{10}?

(c) What volume of CO_2 measured at STP would be produced from 45.0 L of C_4H_{10} measured at 25 °C and 0.750 atm?

10-102 In March 1979 a nuclear reactor overheated, producing a dangerous hydrogen gas bubble at the top of the reactor core. The following reaction occurring at the high temperatures (about 1500 °C) accounted for the hydrogen (Zr alloys hold the uranium pellets in long rods).

$$Zr(s) + 2H_2O(g) \longrightarrow ZrO_2(s) + 2H_2(g)$$

If the bubble had a volume of about 28,000 L at 250 °C and 70.0 atm, what mass (in kg and tons) of Zr had reacted?

10-103 Nitric acid is produced according to the equation

$$3NO_2(g) + H_2O(l) \longrightarrow$$
$$2HNO_3(aq) + NO(g)$$

What volume of NO_2 measured at −73 °C and 1.56×10^{-2} atm would be needed to produce 4.55×10^{-3} mol of HNO_3?

*10-104 Natural gas (CH_4) burns according to the equation

$$CH_4(g) + 2O_2(g) \longrightarrow$$
$$CO_2(g) + 2H_2O(l)$$

What volume of CO_2 measured at 27 °C and 1.50 atm is produced from 27.5 L of O_2 measured at −23 °C and 825 torr?

REVIEW TEST ON CHAPTERS 8–10

The following multiple choice questions have one correct answer.

1. A quantity of aluminum has a mass of 54.0 g. What is the mass of the same number of magnesium atoms?
   - (a) 12.1 g
   - (c) 48.6 g
   - (e) 6.0 g
   - (b) 24.3 g
   - (d) 97.2 g

2. To what is the number 6.02×10^{22} equivalent?
   - (a) 0.100 mol
   - (d) 0.500 mol
   - (b) 1.00 mol
   - (e) no such number exists
   - (c) 10.0 mol

3. What is the mass in grams of 0.250 mol of oxygen atoms?
   - (a) 16.0 g
   - (d) 4.00 g
   - (b) 1.50×10^{23} g
   - (e) 32.0 g
   - (c) 8.00 g

4. What is the mass of 3.01×10^{24} He atoms?
   - (a) 20.0 g
   - (d) 12.0×10^{24} g
   - (b) 4.00 g
   - (e) 2.00 g
   - (c) 200 g

5. What is the approximate mass of one carbon atom?
   - (a) 12 g
   - (d) 0.50×10^{-23} g
   - (b) 2.0×10^{-23} g
   - (e) 6.0 g
   - (c) 2.0×10^{23} g

6. What is the mass of one mole of $H_2C_2O_4$?
   - (a) 90.0 amu
   - (d) 58.0 g
   - (b) 46.0 g
   - (e) 46.0 amu
   - (c) 90.0 g

7. How many moles of oxygen atoms are in 0.50 mol of $Ca(ClO_3)_2$?
   - (a) 3.0
   - (c) 0.50
   - (e) 1.50
   - (b) 1.0
   - (d) 6.0

8. How many moles of oxygen atoms are in 3.01×10^{23} molecules of O_2?
   - (a) 0.50
   - (d) 0.25
   - (b) 6.02×10^{23}
   - (e) 1.50×10^{23}
   - (c) 1.0

9. A compound is composed of 0.24 mol of Fe, 0.36 mol of S, and 1.44 mol of O. What is its empirical formula?
   - (a) FeS_3O_6
   - (d) $Fe_2S_3O_{12}$
   - (b) $Fe_4S_6O_{24}$
   - (e) $Fe_2S_3O_8$
   - (c) $Fe_2S_3O_6$

10. A compound has a molar mass of 84.0 g/mol and an empirical formula of CH_2N. What is its molecular formula?
    - (a) $C_3H_4N_2O$
    - (d) $C_3H_6N_2$
    - (b) $C_3H_6N_3$
    - (e) $C_4H_8N_4$
    - (c) $C_2H_4N_2$

11. Which of the following elements should be represented in an equation as a diatomic molecule?
    - (a) C   (b) He   (c) P   (d) B   (e) Br

12. What type of reaction does the following equation represent?

    $$Ba(s) + 2H_2O(l) \longrightarrow Ba(OH)_2(aq) + H_2(g)$$

    - (a) combination
    - (b) decomposition
    - (c) single-replacement
    - (d) double-replacement
    - (e) combustion

13. What types of reaction does the following equation represent?

    $$SiH_4(g) + 2O_2(g) \longrightarrow SiO_2(s) + 2H_2O(l)$$

    - (a) combination
    - (b) decomposition
    - (c) combustion
    - (d) single-replacement
    - (e) double-replacement

14. Determine which of the answers represents the proper balanced equation for the following statement: Calcium reacts with aqueous hydrocholoric acid to produce aqueous calcium chloride and hydrogen.
    - (a) $Ca(g) + 2HCl(aq) \rightarrow CaCl_2(l) + H_2(g)$
    - (b) $Ca(s) + 2HCl(aq) \rightarrow$
      $\qquad CaCl_2(aq) + 2H(g)$
    - (c) $2Ca(s) + 2HCl(aq) \rightarrow$
      $\qquad 2CaCl(aq) + H_2(g)$
    - (d) $Ca(s) + 2HCl(aq) \rightarrow$
      $\qquad CaCl_2(aq) + H_2(g)$
    - (e) $2Ca(s) + H_2Cl(aq) \rightarrow$
      $\qquad CaCl(aq) + H_2(g)$

15. Given the equation

    $$2C_2H_2 + 5O_2 \longrightarrow 4CO_2 + 2H_2O$$

    how many moles of O_2 are needed to produce 4 mol of CO_2?

(a) 4 mol   (c) 3 mol   (e) 5 mol
(b) 2 mol   (d) 1 mol

16. Sulfur trioxide is prepared by the following two reactions:

$$S_8(s) + 8O_2(g) \longrightarrow 8SO_2(g)$$
$$2SO_2(g) + O_2(g) \longrightarrow 2SO_3(g)$$

How many moles of SO_3 are produced from 1 mol of S_8?
(a) 1   (b) 2   (c) 4   (d) 8   (e) 16

17. In the combustion of a certain hydrocarbon 16.0 g of CO_2 is produced, which represents a 75% yield. What is the theoretical yield?
(a) 12.0 g   (d) 32.0 g
(b) 8.0 g   (e) 44.0 g
(c) 21.3 g

18. Given the combustion reaction

$$CH_4 + 2O_2 \longrightarrow CO_2 + 2H_2O$$

when 16.0 g of CH_4 is burned with 32.0 g of O_2, which statement is true?
(a) CH_4 is the limiting reactant.
(b) CO_2 is present in excess.
(c) CH_4 is present in excess.
(d) O_2 is present in excess.
(e) H_2O is the limiting reactant.

19. Which of the following is the SI unit of pressure?
(a) atm   (d) in. of Hg
(b) torr   (e) lb/in.²
(c) Pa

20. Which of the following is *not* an assumption of the kinetic molecular theory applied to gases?
(a) Molecules have negligible volume.
(b) Molecules of all gases have the same average velocity at a certain temperature.
(c) Gas molecules have negligible interactions.
(d) Temperature is related to the average kinetic energy of the system.
(e) Molecules are in rapid, random motion.

21. Which of the following is a representation of Boyle's law?

(a) $P \propto \dfrac{1}{V}$ (c) $V \propto T$ (e) $V \propto \dfrac{1}{T}$

(b) $V \propto P$ (d) $P \propto \dfrac{1}{T}$

22. The temperature of a volume of gas is increased from 20 to 40 °C at constant pressure. Its volume:
(a) doubles
(b) decreases by half
(c) increases by a factor of $\frac{313}{293}$
(d) decreases by a factor of $\frac{20}{273}$
(e) decreases by a factor of $\frac{293}{313}$

23. Which of the following is a representation of Gay-Lussac's law?
(a) $P_1T_1 = P_2T_2$
(b) $P_1V_1 = P_2V_2$
(c) $V_1T_2 = V_2T_1$
(d) $P_1V_2 = P_2V_1$
(e) $P_1T_2 = P_2T_1$

24. Which of the following is the set of conditions known as standard temperature and pressure (STP)?
(a) 0 K and 1 atm
(b) 0 °F and 760 torr
(c) 0 °C and 760 atm
(d) 273 °C and 1 atm
(e) 273 K and 760 torr

25. Which of the following gases has the highest average velocity at a certain temperature?
(a) oxygen
(b) carbon monoxide
(c) neon
(d) sulfur dioxide
(e) hydrogen chloride

26. A mixture of gases has a total pressure of 2.00 atm. If one gas has a partial pressure of 0.50 atm, what part of the mixture is this gas?
(a) 50%   (d) 1.00 atm
(b) 75%   (e) 1.50 atm
(c) 25%

27. A 22.4-L quantity of O_2 at STP:
(a) contains 1 mol of oxygen atoms
(b) has a mass of 16.0 g
(c) contains 1.20×10^{24} oxygen atoms
(d) contains 2 mol of O_2 molecules
(e) has a mass of 48.0 g

28. A gas has a density of 2.68 g/L (STP). What is the gas?
(a) CO_2 (d) COS
(b) SO_2 (e) He
(c) NO_2

29. Which of the following is a value for the gas constant R?

(a) $62.4 \dfrac{L \cdot atm}{mol \cdot K}$

(b) $62.4 \dfrac{L \cdot torr}{mol \cdot K}$

(c) $82.1 \dfrac{L \cdot atm}{mol \cdot K}$

(d) $0.0821 \dfrac{L \cdot atm}{mol \cdot °C}$

(e) $0.0821 \dfrac{L \cdot torr}{mol \cdot K}$

30. Given the equation

$$C(s) + H_2O(l) \longrightarrow CO(g) + H_2(g)$$

what volume of gas measured at STP would be produced from 24.0 g of carbon?

(a) 22.4 L   (c) 44.8 L   (e) 4.0 L
(b) 89.6 L   (d) 11.2 L

PROBLEMS

1. Fill in the blanks.

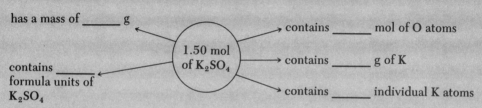

has a mass of _____ g

contains _____ mol of O atoms

1.50 mol of K_2SO_4

contains _____ g of K

contains _____ formula units of K_2SO_4

contains _____ individual K atoms

2. Fill in the blanks.

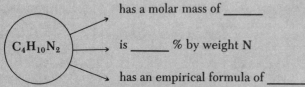

$C_4H_{10}N_2$

has a molar mass of _____

is _____ % by weight N

has an empirical formula of _____

3. A compound is composed of 1.75 g of Fe and 0.667 g of O. What is its empirical formula?

4. A compound is 9.60% H, 16.4% C, and 74.0% B by weight. Its molar mass is 146 g/mol. What is its molecular formula?

5. Complete the following equations. Add subscripts on formulas of elements, coefficients to balance the equations, and physical states where needed.

(a) ___K(___) + $Al_2Cl_6(s) \xrightarrow{\Delta}$
___KCl(___) + ___Al(s)

(b) ___$C_6H_6(l)$ + ___O(___) →
___CO_2(___) + ___$H_2O(l)$

(c) __$Cl_2O_3(g)$ + __$H_2O(l)$ → __$HClO_2(aq)$

(d) __$K_2S(aq)$ + __$H_3PO_4(aq)$ →
__$H_2S($__$)$ + __$K_3PO_4(aq)$

(e) __$B_4H_{10}(g)$ + __$H_2O(l)$ →
__$B(OH)_3(aq)$ + __$H($__$)$

6. Freon 12 (CCl_2F_2) is a gas used as a refrigerant. It is prepared according to the equation

$$3CCl_4(l) + 2SbF_3(s) →$$
$$3CCl_2F_2(g) + 2SbCl_3(s)$$

(a) How many moles of SbF_3 are needed to
produce 0.0350 mol of freon?
(b) How many moles of $SbCl_3$ are produced if
150 g of freon is also produced?
(c) What mass of CCl_4 is needed to react with
850 g of SbF_3?
(d) How many individual molecules of CCl_4
are needed to produce 12.0 kg of $SbCl_3$?
(e) What volume of freon measured at 25 °C
and 0.760 atm is produced from 14.9 g of
CCl_4?

7. Fill in the blanks.

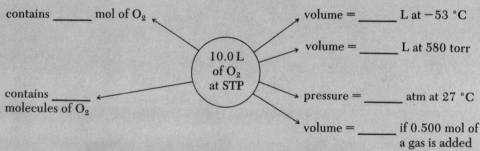

contains _____ mol of O_2

contains _____
molecules of O_2

10.0 L
of O_2
at STP

volume = _____ L at −53 °C

volume = _____ L at 580 torr

pressure = _____ atm at 27 °C

volume = _____ if 0.500 mol of
a gas is added

8. A compound can be vaporized at 100 °C. At
that temperature it is found that 1.19 g of the
compound occupies 250 mL at 756 torr.
Analysis of the compound shows that it is
49.0% C, 48.3% Cl, and 2.72% H. What is
the molecular formula of the compound?

WATER—LIQUID, SOLID, AND GAS
The earth is a unique planet among those that circle the sun. Conditions are such that water can exist in all three physical states. The clouds form from the condensation of the vapor state.

WATER: THE LIQUID AND THE SOLID STATES

PURPOSE OF CHAPTER

In Chapter 11 we examine the properties of the liquid and solid states of matter in general and the properties of water and ice in particular.

OBJECTIVES FOR CHAPTER

After completion of this chapter, you should be able to:

1 List the general properties of the liquid and solid states and describe how the kinetic molecular theory explains these properties. (11-1)

2 Describe the difference in the motion of water molecules in its three physical states. (11-1)

3 Compare the three types of forces by which molecules are attracted to each other. (11-2)

4 Describe the geometry and polarity of water and how it interacts by hydrogen bonding. (11-2)

11

5   Explain the relationship of temperature to the physical state of a compound. (11-3)

6   Describe the changes that occur as water, initially in the form of ice, is heated from below 0 °C to above 100 °C. (11-3)

7   Describe how melting points and heats of fusion relate to intermolecular forces. (11-4)

8   Describe how boiling points of hydrogen compounds of groups 15 (VA), 16 (VIA), and 17 (VIIA) reflect hydrogen bonding and London forces. (11-4)

9   Describe how boiling points and heats of vaporization reflect the degree of intermolecular attraction. (11-4)

10   Solve problems in which heat energy is involved in changes of state. (11-4)

11   Describe the process and effects of the evaporation of a liquid below its boiling point. (11-5)

12   Describe the relationships among the equilibrium vapor pressure of a liquid, its boiling point, and its normal boiling point. (11-5)

13   Write equations illustrating the formation of water and the reactions of water with certain metals and nonmetals. (11-6)

14   Write the formulas of hydrates and determine their compositions. (11-6)

Water — the cool and refreshing liquid. This simple compound is so abundant and familiar that we may lose sight of the fact that it is truly an amazing and unique compound. In fact, it so unique that many scientists feel it is doubtful that any form of life could exist without it. As an example, when we searched for life on the planet Mars some years ago, we looked for traces of free water in the soil. It is certainly one of the main ingredients of life as we know it. On earth, water, together with carbon dioxide in the air, minerals from the soil, and energy from the sun, is changed into chemical energy in the carbon compounds of vegetation. When this chemical energy is used for fuel or food, the combustion or metabolism process cycles the water back to the environment. Water supports life on this planet in even more ways. Large bodies of water in the form of oceans and lakes moderate the climate by storage and distribution of heat. In the next chapter, we will examine the unique ability of water to act as a solvent and thus serve as a medium for a host of chemical reactions.

There are two main goals in this chapter: (1) to examine important properties of the liquid and solid states in general and (2) to become more familiar with the properties of water in particular. Thus, we combine our goals by emphasizing ordinary water as an example of a liquid and ice as an example of a solid.

To prepare for the discussion of water and the properties of the solid and liquid states, the following topics should be reviewed before beginning the chapter:

1   Lewis structure of water (Section 6-5)
2   Electronegativity, bond polarity, and molecular polarity (Sections 6-10 and 6-11)
3   Kinetic molecular theory (Section 10-1)
4   Specific heat (Section 2-8)

11-1 THE NATURE OF THE LIQUID AND THE SOLID STATES

The gaseous state of matter received ample coverage in the previous chapter. In the interest of fairness, it is now appropriate to give due regard to the liquid and solid states. As mentioned, these states are easier to deal with in a way because they are visible so that portions can be conveniently isolated and measured. On the other hand, not all solids or liquids respond to conditions such as heat and pressure in a common or predictable manner. Thus, unlike gases, there are no "liquid laws" or "solid laws." Still, the two states do have some common characteristics:

1   They Have a High Density.

Solids and liquids are about 1000 times denser than a typical gas.

2   They Are Essentially Incompressible.

Tall buildings can be supported by bricks and other solids because they don't compress as would a gas. Likewise, a hydraulic jack uses a liquid to support weights such as that of a huge truck. Unlike the behavior of a gas, an increase in pressure on a solid or liquid does not result in a significant decrease in volume.

3   They Undergo Little Thermal Expansion.

When a bridge is constructed, a small space must be left between sections for expansion on a hot day. Still, this space amounts to only a few inches for a bridge span many yards long. In addition, the degree of expansion varies for different solids and liquids. Gases, on the other hand, expand significantly as the temperature rises and all gases expand by the same factor.

4   They Have a Fixed Volume.

The volume of a gas is the volume of the container. Also, the volume is the same for the same number of particles under the same conditions. There is no such convenient relationship for solids and liquids. The same volumes of different liquids have no relationship to the number of molecules present.

In addition to these common characteristics, liquids do not have a fixed shape and thus flow. Solids have a fixed shape and are thus rigid.

The characteristics of gases were adequately explained by the kinetic molecular theory. Two of the basic assumptions of the kinetic theory are also applicable to the other states. That is, solids and liquids are composed of basic particles that have kinetic energy. The average kinetic energy of the particles is related to the temperature. However, to explain the characteristics of the other two states there are obviously some assumptions related to gases that no longer apply and must be modified. In the solid and liquid states:

1   The basic particles have significant attractions for each other and so are held close together.
2   The basic particles are not in random motion; their motion is restricted by interactions with the other particles.

Again, the properties of the solid and liquid states are understandable on the basis of these assumptions. Since the basic particles are already "stuck together" they cannot be pressed together to any extent, so they are incompressible and have a high density. In fact, both liquids and solids are referred to as *condensed states*. The attraction of the particles for each other holds them together, which essentially counteracts the tendency of additional heat to move them apart. Thus, liquids and solids undergo little thermal expansion.

In Figure 11-1, we have attempted to illustrate the fundamental difference in behavior of molecules in the three states of matter. We emphasize as an example the water molecule. This simple but amazing compound exists in all three physical states on earth: gas (vapor), liquid, and solid (ice). In fact, in a thermos of ice water all three states exist at once although the presence of some H_2O molecules in the gaseous state above the ice water may not be apparent.

First, let's consider the solid state of water which we know as ice. In this case, the forces of attraction between molecules hold them in fixed positions relatively close together. In the solid state, the molecules have kinetic energy, meaning that they do have motion. But the motion is restricted to various types of vibrations within a confined space. This is much like a dancer whose motions are confined to shaking and vibrating rather than movement from point to point (translational motion). In the liquid state, the water molecules are also held close together by forces of attraction but the molecules are not held in

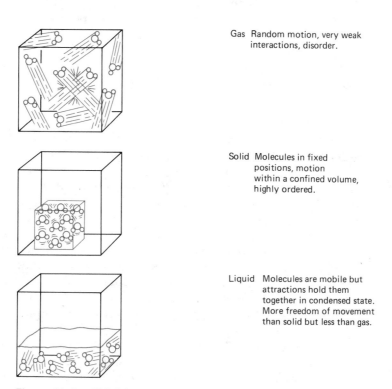

Gas Random motion, very weak interactions, disorder.

Solid Molecules in fixed positions, motion within a confined volume, highly ordered.

Liquid Molecules are mobile but attractions hold them together in condensed state. More freedom of movement than solid but less than gas.

Figure 11-1 PHYSICAL STATES OF WATER. Interactions between water molecules are different in the three physical states.

fixed positions and thus have more freedom of motion. That is, individual molecules or groups of molecules have translational motion as well as vibrational motion. Since the molecules can move past one another, liquids can flow and take the shape of the bottom of the container. Finally, we are already familiar with the behavior of water molecules in the gaseous state. In this case, the molecules have so much translational motion that they move freely throughout the whole container unaffected by the attractions to other molecules. Their motion scatters them as far apart as possible.

We are now ready to consider the characteristics of molecules of various compounds that give rise to forces of attraction between them.

■
SEE PROBLEMS
11-1 THROUGH
11-5.

11-2 INTER-MOLECULAR FORCES AND PHYSICAL STATES

It should be obvious by now that molecular compounds can exist as either solids, liquids, or gases at room temperature. The physical state of the compound depends to a large degree on how strongly the molecules "stick together." Those with the stronger attractions for each other tend to lock each other into fixed positions and thus exist as solids at room temperature. Those with very weak attractions for each other have little tendency to stick together so they remain free in the gaseous state. Those with intermediate interactions often exist as liquids. In this case, forces of attraction keep the molecules close but not strongly enough to hold them in fixed positions. Our next goal is to understand why certain molecules have a strong tendency to coalesce while others do not.

Molecules of covalent compounds display three types of attraction toward each other: London forces, dipole–dipole forces, and hydrogen bonding.

London Forces

Gravity is a force of attraction between portions of matter. As we know, the larger the portion, the greater the gravity (attraction). The molecules of a compound also have inherent forces of attraction that are like gravity but arise from different principles. *These forces of attraction are known as* **London forces.** The origin of these forces is not discussed in this text but it is important to note that compared with other forces of attraction, such as those between two oppositely charged ions, London forces are generally considered weak. London forces are dependent on the size or molar mass of the compounds. Since the forces increase as the molar mass increases, these interactions become significant for heavier molecules. An example of such a trend is found with the following hydrocarbons (carbon–hydrogen compounds) where only London forces exist between molecules. At room temperature, CH_4 (molar mass = 16 g/mol) is a gas, C_8H_{18} (molar mass = 114 g/mol) is a liquid, and $C_{18}H_{38}$ (molar mass = 244 g/mol) is a solid. *In nonpolar compounds, only London forces act between molecules.* Such compounds that also have low molar mass are found in the gaseous state at room temperature. Some examples are N_2, O_2, F_2, and CO_2.

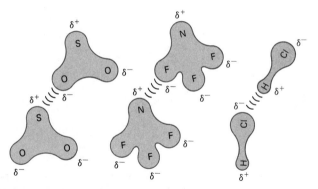

Figure 11-2 DIPOLE–DIPOLE INTERACTIONS. SO_2, NF_3, and HCl are all polar molecules and have dipole–dipole interactions.

Dipole–Dipole Interactions

In Chapter 6 (Section 6-11) we discussed how geometry affects the polarity of a molecule. For example, in a molecule of carbon dioxide the two oxygens lie at an angle of 180°. The two C—O bond dipoles cancel and the molecule is nonpolar. On the other hand, The OCS molecule is polar because the two unequal dipoles do not cancel. The SO_2 molecule is also polar but for a different reason. The two equal S—O bond dipoles of SO_2 do not cancel because they lie at an angle less than 180°. The OCS and the SO_2 molecules are thus polar in that they have a negative pole and a positive pole (a dipole). *Polar molecules can align themselves so that the negative end of one molecule is attracted to the positive end of the other.* (See Figure 11-2.) *These attractions are known as* **dipole–dipole attractions.** Polar molecules thus have two forces of attraction: London forces as described previously and dipole–dipole attractions. A polar compound has a greater tendency to exist in a condensed state than a nonpolar compound of similar molar mass because of the added dipole–dipole interactions. For example, CO_2 (44 g/mol) is nonpolar and a gas at room temperature while CH_3CN (41 g/mol) is polar and a liquid at room temperature. For most polar compounds, however, we find that the molar mass has a greater effect on their physical states than the presence of dipole–dipole interactions.

Hydrogen Bonding

The atoms of oxygen, nitrogen, and fluorine are not only the smallest (other than hydrogen) but also the three most electronegative atoms. When these atoms are bound to atoms of other elements, especially hydrogen, they tend to produce significantly polar bonds. As an example, let us consider molecules of water. First, hydrogen and oxygen have a large difference in electronegativity ($3.5 - 2.1 = 1.4$). It is not enough to indicate an ionic bond but certainly enough to indicate that it is a polar covalent bond. Also, H_2O is not a linear molecule. In fact, there is an angle of 105° between atoms in the molecule.

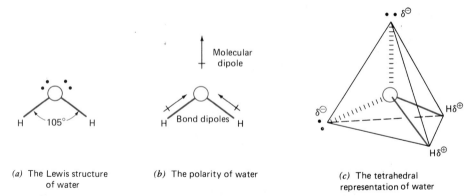

(a) The Lewis structure
    of water

(b) The polarity of water

(c) The tetrahedral
    representation of water

Figure 11-3   STRUCTURE OF WATER.

Thus, the large bond dipoles do not cancel and the molecule itself is polar. Experimental evidence indicates that the water molecule can be thought of as having a tetrahedral geometry if we assume that unshared electron pairs occupy two of the vertices and hydrogen atoms occupy the other two. (See Figure 11-3.) The partial positive charge is centered on the hydrogen atoms and the partial negative charge is centered at the location of the electron pairs at the other two vertices. As shown in Figure 11-4, in solid ice, hydrogens on two different water molecules interact with the two electron pairs. In liquid water, the structure is less orderly in that the interactions between molecules are more random.

The interaction of a hydrogen on one water molecule with the electron pair of the oxygen on another molecule is an example of what is known as a hydrogen bond. **A hydrogen bond** *is an electrostatic attraction between a hydrogen bonded to a N, O, or F atom on one molecule and an unshared electron pair on a N, O, or F atom on another.* Only these three atoms are involved in hydrogen

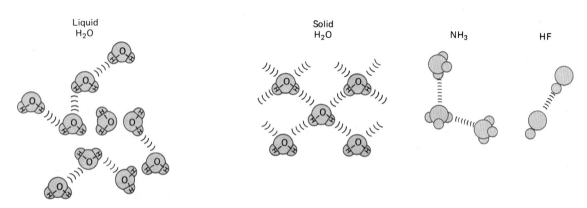

Figure 11-4   HYDROGEN BONDING. H_2O molecules in both liquid and solid states, as well as NH_3 and HF molecules, are attracted by hydrogen bonds.

bond interactions because of their high electronegativity and small size. This interaction at first may appear as a case of an extreme dipole–dipole interaction but it is actually more complex than that. Whereas regular dipole–dipole attractions are often not as important as London forces, hydrogen bonding has a significant effect on the properties of molecules. Molecules that have hydrogen bonding have a much greater tendency to exist in a condensed state than molecules in which only London forces or regular dipole–dipole interactions are present. Certainly, for water the hydrogen bonding accounts for why it exists as a liquid at room temperature when otherwise it should be a gas under even the coldest conditions found on the surface of the earth. Besides water, ammonia and hydrogen fluoride molecules are attracted by hydrogen bonding. (See Figure 11-4.)

The phenomenon of hydrogen bonding is not restricted to molecules of the same compound. Interactions between molecules of different biologically important compounds are especially important. For example, DNA, "the messenger of life," is composed of two strands of atoms offset from each other in a helical arrangement. The two strands are held together strictly by hydrogen bonds. The hydrogen bonds are not as strong as regular covalent bonds, so the two strands can separate and replicate themselves. Such a hydrogen bond between different molecules is shown below.

$$\begin{array}{ccc} H & & H \\ | & & | \\ Y-N-H & \cdots & O-X \end{array}$$

So far, we have restricted our discussion to the interactions that occur in compounds composed of molecules. There is another large group of compounds, however, the basic particles of which are made up of ions. *The force of attraction between oppositely charged ions is a very strong force compared with the forces between molecules described above.* Thus ionic compounds are always found in the *solid state* at room temperature.

In this discussion, we have used room temperature as our point of reference to a particular physical state. However, what is a gas at room temperature becomes a liquid at a lower temperature and what is a solid at room temperature changes to a liquid at a higher temperature. In the next section, we consider how temperature affects the physical state.

■
SEE PROBLEMS 11-6 THROUGH 11-17.

**11-3
TEMPERATURE
AND
PHYSICAL
STATE**

The physical state of a compound at a certain temperature depends on the subtle competition between the intermolecular or interionic forces in the compound and the kinetic energy of the molecules or ions. Kinetic energy is the energy of motion, tending to randomize particles, moving them past or away from each other. It is like their "nervous energy." The forces of attraction between the particles of a compound counteract the kinetic energy by giving them a reason to "stay put." The higher the temperature, the more "nervous energy" the particles have and the harder it is for molecules or ions to stay put in the solid or even the liquid state. At room temperature, we can assume that a substance exists as a gas because the kinetic energy of the molecules is too great for them to stick together. However, at a lower temperature, where the kinetic energy is lower, condensation to the liquid or solid state does occur. The extreme example of a gas with very weak forces between particles (atoms in this case) is helium. The temperature must be lowered all the way to −269 °C (4 K) for the gas to condense to a liquid. On the other hand, substances with strong interactions (e.g., ionic compounds) that are solids at room temperature do change to a less ordered state (liquid or gas) at a higher temperature. An extreme example is the case of zirconium nitride (ZrN). Because of the strong interactions between ions, it does not change to the liquid state until heated to about 3000 °C.

Let's now take a closer look at how kinetic energy (temperature) affects the physical states of matter. For our example, we will discuss how a change of temperature affects the three physical states of water. We start with an ice cube cooled to −10 °C in a freezer and discuss what happens as the ice is heated uniformly at a constant rate. (See Figure 11-5.) As heat is added to the ice, the temperature of the solid slowly rises, meaning that the molecules are vibrating faster and faster about their fixed positions (Figure 11-5a). At 0 °C and 1 atm pressure something begins to happen. The motion of an average molecule becomes great enough so that it can overcome some of the forces (hydrogen bonds) holding it in a fixed position. Thus, individual molecules and groups of molecules begin to move past one another. The solid ice is melting (Figure 11-5b). Despite the breakdown of the rigid structure of the solid,

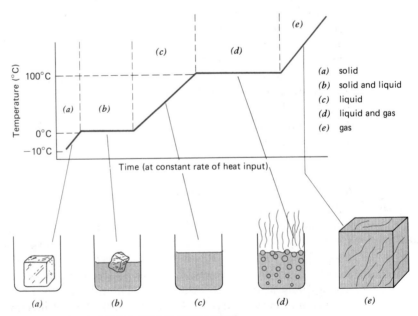

Figure 11-5   HEATING CURVE FOR WATER.

groups of molecules still interact with each other, so they remain in a condensed state. The melting process is endothermic (absorbs heat), and the added heat is consumed by the breaking of some of the hydrogen bonds between molecules. The breaking of hydrogen bonds increases the potential energy of the system by moving the molecules out of their fixed positions. Since potential energy is *not* related to temperature, the temperature remains constant as long as the melting process is still occurring.

An unusual property of water compared with most substances is that the solid state is less dense than the liquid. As a result, ice floats. Without this property, lakes and oceans would freeze from the bottom up and remain frozen all summer except for the surface. In the case of water, the H_2O molecules have a relatively open arrangement in the solid state. When ice melts, some of the hydrogen bonds holding the molecules in the open structure break and the structure collapses somewhat to form the liquid state.

After all of the ice has melted, the added heat again increases the motion or velocity of the molecules, thus increasing the temperature of the liquid water (Figure 11-5c). The velocity of the molecules continues to increase until the temperature reaches 100 °C at 1 atm pressure. At this temperature bubbles of steam form in the liquid, indicating that boiling has begun (Figure 11-5d). In fact, the average velocity of the water molecules has become so great that all remaining attractions to other molecules holding them in the liquid state are overcome and the water turns to gas. Again, the added heat is being consumed to break attractive forces between molecules and move them apart, which increases the potential energy. The temperature remains constant until all of the water has changed to steam. After that, added heat increases the average velocity of the now independent gas molecules and the temperature of the gas rises (Figure 11-5e). In the gaseous state the water molecules are still polar and there are potential interactions between them. However, their kinetic energy is too great and they are moving too fast to "stick together."

The changes in state that a substance undergoes while being heated are somewhat analogous to a dance floor crowded with gyrating, vibrating couples. At first the beat of the music is slow. Everyone just gyrates (moves about a fixed position) and vibrates in one spot, held there, presumably, by the attraction of a partner. This is equivalent to the motion of molecules in the solid state. As the beat of the music increases, however, motion increases (heats up) in response. Eventually, there comes a point where it is impossible for everyone to keep up the beat and still dance in one spot. The attractions are overcome, and the people move around the floor. The dance floor is still crowded, but the motion of the people is translational (from position to position). This is similar to the movement of molecules in the liquid state. If we now wished to extend the analogy to the ridiculous, we could foresee a point where the beat and the corresponding motion become so intense that the dancers can no longer be confined to the floor and thus move freely and independently throughout the whole room from floor to ceiling. Just as molecules of a gas are spread far apart, the dancers would no longer be crowded since the whole volume of the room would be utilized. (See Figure 11-6.)

SEE PROBLEMS 11-18 THROUGH 11-22

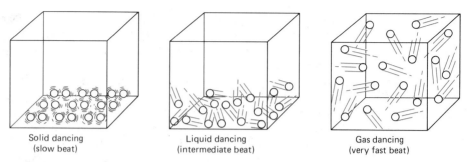

Solid dancing
(slow beat)

Liquid dancing
(intermediate beat)

Gas dancing
(very fast beat)

Figure 11-6   EFFECT OF THE BEAT ON DANCING. Temperature is like the beat of music. The faster the beat, the more the dancers move.

11-4 ENERGY AND CHANGES IN STATE

We have seen how the temperatures at which solids melt and liquids boil depend to a large degree on the intermolecular interactions. The energy required to effect these changes is also dependent on the attractions between particles. We will first discuss the transitions between the solid and liquid states (freezing and melting) and then the transitions between the liquid and gaseous states (boiling and condensation).

Melting and Freezing

As we discussed in the previous section the process of melting (fusion) is endothermic. On the other hand, consider the opposite process. When a liquid freezes to form the solid state, the process is exothermic, giving off the same amount of energy as the melting process required. Each compound requires a definite amount of heat energy to melt a specific mass of sample. *The* **heat of fusion** *of a substance is the amount of heat in calories or joules required to melt one gram of the substance.* Table 11-1 lists the heats of fusion and the melting points of several substances. Note that sodium chloride, which is ionic, has the strongest attractions between particles, the highest melting point, and the highest heat of fusion.

TABLE 11-1   Heats of Fusion and Melting Points

| Compound | Type of Compound | Heat of Fusion | | Melting Point (°C) |
		cal/g	J/g	
NaCl	Ionic	124	519	801
H_2O	Polar covalent (hydrogen-bonding)	79.8	334	0
Ethyl alcohol	Polar covalent (hydrogen-bonding)	24.9	104	−114
Ethyl ether	Polar covalent	22.2	92.5	−116
Benzene	Nonpolar covalent	30.4	127	5.5
Carbon tetrachloride	Nonpolar covalent	4.2	17.6	− 24

Considering that water is a molecular compound of low molar mass, it has a comparatively high melting point and heat of fusion. The high heat of fusion has profound consequences for the climate of this planet, especially in the vicinity of the American Great Lakes. When a lake freezes in winter, it releases heat to the environment. In the winter, this heat released by fusion has the same effect as a giant heater and helps keep the thermometer from falling as much as it otherwise would. Since the Great Lakes do not usually completely freeze over, heat is released all through the winter from the lake. On the other hand, spring is delayed because the melting ice absorbs the heat and keeps the thermometer from rising as much as it would without the lake. Siberia in Russia lies about as far north as Minnesota but has few large lakes. As a result, it is much colder there in the winter and also much hotter in the summer.

EXAMPLE 11-1

How many kilojoules of heat are released when 185 g of water freezes?

PROCEDURE

The heat of fusion can be used as a conversion factor relating mass in grams to joules.

SOLUTION

$$185 \text{ g} \times \frac{334 \text{ J}}{\text{g}} \times \frac{1 \text{ kJ}}{10^3 \text{ J}} = \underline{61.8 \text{ kJ}}$$

Boiling and Condensation

The temperature at which a liquid begins to boil also depends on the intermolecular forces holding the molecules or ions in the liquid state. Perhaps one of the most dramatic effects on boiling points is demonstrated by compounds that have hydrogen bonding between molecules. For example, consider the boiling points of the group 16 (VIA) hydrogen compounds: H_2O, H_2S, H_2Se, and H_2Te. The molar masses of these compounds increase in the order listed. Thus, London forces should increase in the same order and, as a result, so should the intermolecular attractions. From this result alone, we may predict steadily higher boiling points for these compounds. Note in Table 11-2 that this is

TABLE 11-2 Boiling Points (°C) of Some Binary Hydrides

Group	2nd Period		3rd Period		4th Period		5th Period	
17 (VIIA)	HF	17	HCl	−84	HBr	−70	HI	−37
16 (VIA)	H_2O	100	H_2S	−61	H_2Se	−42	H_2Te	−2.0
15 (VA)	NH_3	−33	PH_3	−88	AsH_3	−62	SbH_3	−18

indeed the case for H_2S, H_2Se, and H_2Te. However, the boiling point of H_2O is out of line as it is by far the highest. The explanation is that there is hydrogen bonding in H_2O that provides significant intermolecular interactions. These interactions are obviously much more important than the weak London forces for H_2O. The other three compounds are polar but the dipole – dipole interactions do not seem to have a significant effect on the boiling points compared with the London forces. In Table 11-2, the boiling points of the group 15 (VA) hydrogen compounds and the group 17 (VIIA) hydrogen compounds are also listed. Hydrogen bonding explains the unusually high boiling points of NH_3 and HF compared with other hydrogen compounds in the group.

The process of boiling a liquid also requires energy to break the forces holding the molecules in a condensed state and separate them to the freedom of the gaseous state. The amount of energy required is known as the heat of vaporization and once again depends on the strength of the intermolecular forces holding the molecules or other particles together in the liquid state. The opposite of boiling is **condensation.** An equal amount of energy is released when a given amount of substance condenses to form a liquid. *The* **heat of vaporization** *is the amount of heat in calories or joules required to vaporize one gram of the substance.* The heats of vaporization and the boiling points of several substances are given in Table 11-3.

Note again that water has an unusually high heat of vaporization compared with other covalent compounds. The number and strength of the hydrogen bonds between H_2O molecules account for its high heat of vaporization.

We can now add the information concerning heat of fusion and heat of vaporization of water to the information discussed in Chapter 2 on specific heats of ice and water (Table 2-6). Specific heat refers to the amount of heat required to raise one gram of a substance one degree Celsius. At that time we observed that water also has an unusually high specific heat compared with other compounds or elements. The heats of fusion and vaporization refer to the heat required to melt one gram of ice and vaporize one gram of water, respectively. For example, consider the energy required to change one gram of ice at 0 °C to one gram of steam at 100 °C. It takes 334 J to melt the ice,

TABLE 11-3 Heats of Vaporization and Boiling Points

| Compound | Type of Compound | Heat of Vaporization | | Normal Boiling Point (°C) |
		cal/g	J/g	
NaCl	Ionic	3130	13,100	1465
H_2O	Polar covalent (hydrogen-bonding)	540	2,260	100
Ethyl alcohol	Polar covalent (hydrogen-bonding)	204	854	78.5
Ethyl ether	Polar covalent	89.6	375	34.6
Benzene	Nonpolar	94	393	80
Carbon tetrachloride	Nonpolar	46	192	76

REVIEW SECTION 2-8 ON SPECIFIC HEAT

418 J to heat the water from 0 to 100 °C, and 2260 J to vaporize the water. The total heat requirement is 3012 joules of heat. The following examples illustrate the use of heat of fusion, heat of vaporization, and specific heat.

EXAMPLE 11-2

Steam causes more severe burns than an equal mass of water. Compare the heat released when 3.00 g of steam at 100 °C condenses and then cools to 60 °C with the heat released when the same mass of water at 100 °C cools to 60 °C.

PROCEDURE

Use the heat of vaporization as a conversion factor between mass in grams and calories. Also, use the specific heat of water as a conversion factor between mass, calories, and degrees Celsius.

SOLUTION

Steam

$$\text{heat released in condensation} = 3.00 \cancel{g} \times \frac{540 \text{ cal}}{\cancel{g}} = 1620 \text{ cal}$$

$$\text{heat released in cooling} = 3.00 \cancel{g} \times 40 \text{ }\cancel{°C} \times \frac{1.00 \text{ cal}}{\cancel{g} \text{ }\cancel{°C}} = 120 \text{ cal}$$

$$\text{total heat released} = 1620 \text{ cal} + 120 \text{ cal} = \underline{1740 \text{ cal released}}.$$

Water

$$\text{heat released in cooling} = \underline{120 \text{ cal}}$$

Note that almost 15 times more heat is released by the steam at 100 °C than the water at the same temperature.

EXAMPLE 11-3

How many kilojoules of heat are required to convert 250 g of ice at −15 °C to steam at 100 °C?

PROCEDURE

The heats of fusion and vaporization as well as the specific heats of ice and water (Table 2-6) are needed.

SOLUTION

To heat ice from -15 °C to its melting point (0 °C), or a total of 15 °C:

$$250 \text{ g} \times 15 \text{ °C} \times 2.06 \frac{J}{g \cdot \text{°C}} \times \frac{1 \text{ kJ}}{10^3 \text{ J}} = 7.7 \text{ kJ}$$

To melt the ice at 0 °C:

$$250 \text{ g} \times 334 \frac{J}{g} \times \frac{1 \text{ kJ}}{10^3 J} = 83.5 \text{ kJ}$$

To heat the water from 0 °C to its boiling point (100 °C), or a total of 100 °C:

$$250 \text{ g} \times 100 \text{ °C} \times 4.18 \frac{J}{g \cdot \text{°C}} \times \frac{1 \text{ kJ}}{10^3 J} = 104 \text{ kJ}$$

To vaporize the water at 100 °C:

$$250 \text{ g} \times 2260 \frac{J}{g} \times \frac{1 \text{ kJ}}{10^3 J} = 565 \text{ kJ}$$

$$\text{total} = 7.7 + 83.5 + 104 + 565 = \underline{\underline{760 \text{ kJ}}}$$

■
SEE PROBLEMS 11-23 THROUGH 11-48.

**11-5
EVAPORATION
AND VAPOR
PRESSURE**

If we leave water in a glass, we know what happens. It disappears. We say that it has evaporated. **Evaporation** *is the vaporization of a liquid below its boiling point.* Water in the form of ice can also go directly to the gaseous state even though the temperature remains below freezing. *The vaporization of a solid is known as* **sublimation.** In the previous section, we mentioned that molecules must attain a minimum kinetic energy to go from the liquid to the gaseous state. How can this happen if the temperature of the liquid water is still below the boiling point? To answer this question, we must consider the distribution of energies of molecules at a certain temperature. The temperature of a substance relates to the *average* kinetic energy of its molecules or other particles. In fact, at any temperature molecules have a wide range of energies — some below average and others above average. In Figure 11-7 the distribution of energies of molecules at two different temperatures is represented in graphical form. The unbroken vertical lines indicate the average kinetic energy of the molecules at the two temperatures. The average of the curve representing the molecules at T_2 is higher than the average of the curve representing the molecules at T_1. This, of course, means that T_2 is a higher temperature than T_1. The broken vertical line represents the minimum kinetic energy necessary for a molecule to be able to escape to the vapor state. Note that even at T_1 a small fraction of molecules have the energy needed to vaporize. If the molecule with this energy lies at or near the surface it is likely to escape as a gas. As expected, a higher fraction of molecules have the energy needed to escape at the higher temperature, T_2.

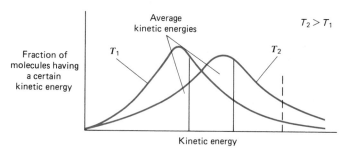

Figure 11-7   DISTRIBUTION OF KINETIC ENERGIES AT TWO TEMPERATURES. The average kinetic energy is higher at T_2 than at T_1.

Now we examine an important effect of this information. In Figure 11-8 a beaker of water has been placed inside a glass container to which a mercury manometer is attached. The manometer allows us to measure the pressure inside the container. Initially, the air is dry inside the container and the pressure is the same inside and out. As the more energetic molecules in the liquid find their way to the surface and escape to the vapor, the pressure above the liquid increases (Dalton's law). As more molecules escape, some other molecules return to the liquid state. Eventually, a point is reached where molecules escaping from the surface of the liquid are all replaced by others condensing from the gas. At this point, when the rate of vaporization equals the rate of condensation, the pressure stabilizes and does not change. The system is said to have reached a point of equilibrium.

Equilibrium

rate of evaporation = rate of condensation

The pressure exerted by the vapor above a liquid at a certain temperature is called its **equilibrium vapor pressure.** Note that at a higher temperature more molecules have the minimum energy needed to escape and, as a result, the vapor pressure is higher. Solids can also have an equilibrium vapor pressure. Dry Ice (solid CO_2) has a high vapor pressure and sublimes rapidly. Regular ice (solid H_2O) has a very low vapor pressure but enough that snow slowly evaporates or sublimes even though the temperature does not rise above freezing.

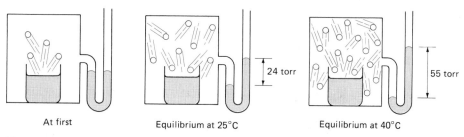

At first   Equilibrium at 25°C   Equilibrium at 40°C

Figure 11-8   EQUILIBRIUM VAPOR PRESSURE. A certain fraction of molecules escape to the vapor state above a liquid.

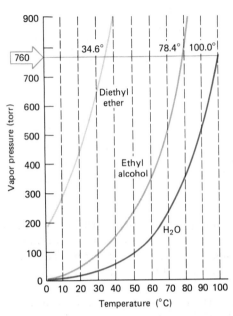

Figure 11-9   VAPOR PRESSURE AND TEMPERATURE. Reprinted from Brady and Holum, *Fundamentals of Chemistry*, 3rd ed. © 1987 John Wiley & Sons, Inc., p. 365.

The vapor pressures of water and two other liquids are shown as a function of temperature in Figure 11-9. Note that at 100 °C the vapor pressure of water is 760 torr, or 1 atm. *When the vapor pressure of a liquid equals the restraining or the atmospheric pressure, the liquid boils. The* **normal boiling point** *of a liquid is the temperature at which the vapor pressure is equal to 760 torr.* Note that the vapor pressures of both alcohol and ether reach 1 atm at temperatures below that of water. Thus they boil at around 78 and 34 °C, respectively. The actual boiling point of water where you live depends on the elevation of your community. At high elevations, where the air pressure is significantly lower than that at sea level, the boiling point of water is noticeably lower. On top of Pike's Peak in Colorado, the atmospheric pressure is about 450 torr and the boiling point of water is only 86 °C. On the highest mountain, Mt. Everest, the boiling point is 76 °C. Cooking would obviously take much longer (if it can be done at all) at these elevations. On the other hand, a pressure cooker confines the vapor so that the pressure inside the cooker is greater than atmospheric pressure. The result is a higher boiling point for water and a shorter cooking time.

Now consider the water left behind after some of the molecules escape to the vapor state. To understand what happens, picture the wonderful possibility that all those who average above 90% on the next chemistry test are not included in the average. For most of us that would certainly make our scores look better compared with the average because the average of those remaining would be lower. A similar phenomenon occurs with the water left behind after molecules with the highest kinetic energies escape. That is, the average kinetic energy of the remaining molecules is lower. This in turn means that the temperature of the water is lower. *Evaporation causes a liquid to cool.* If it is assumed that the air over the water was originally at the same temperature as the water, the addition of gaseous water molecules with above-average energies means that the air becomes warmer.

Cooling by evaporation is well known. The cool feeling after even a warm shower is not just a feeling; it is actually cooler. When we get out of the water, evaporation quickly cools the water on our bodies and us along with it. In fact, this is how our bodies are cooled on hot days. Perspiration evaporates and carries off excess heat from the body. On humid days, this cooling mechanism is less efficient. The atmosphere on humid*days already holds a large portion of the water vapor that it can hold (the vapor pressure at that temperature), so evaporation is much slower and the cooling effect is greatly diminished. When they say "It's not the heat, it's the humidity," they're right.

■
SEE PROBLEMS
11-49 THROUGH
11-62.

11-6 THE FORMATION AND REACTIONS OF WATER

We have now become quite familiar with many of the physical properties of the unique compound known as water. Finally, we consider some of its chemical properties. First, however, we summarize three types of reactions that produce water as a product.

Combination $2H_2(g) + O_2(g) \longrightarrow 2H_2O(l)$

Combustion $CH_4(g) + 2O_2(g) \longrightarrow CO_2(g) + 2H_2O(l)$
                  Methane

                  $C_6H_{12}O_6(aq) + 6O_2(g) \longrightarrow 6CO_2(g) + 6H_2O(l)$
                  Glucose

■
REVIEW SECTION
9-2 ON TYPES
OF REACTIONS.

Double replacement $HCl(aq) + NaOH(aq) \longrightarrow NaCl(aq) + H_2O(l)$
(acid–base)

All of these reactions are exothermic, which means that energy is released along with the products. Hydrogen burns smoothly unless the hydrogen and oxygen are allowed to mix and then ignited. This, of course, causes a potentially dangerous explosion. This reaction finds use, however, in certain space rockets because of the low mass of hydrogen and the large amount of energy released by the reaction. Methane and glucose are both burned as fuels. Methane (natural gas) is used to heat some of our homes and for cooking, and glucose combustion is the principal source of energy for our bodies. With the

---

* Relative humidity is the percent of saturation of water vapor in the atmosphere. For example, at 25 °C the vapor pressure of water is 24 torr (100% saturation). If the actual pressure of water in the air is 20 torr, the relative humidity is

$$\frac{20 \ torr}{24 \ torr} \times 100\% = 83\% \ of \ saturation$$

Figure 11-10   REACTIONS OF METALS WITH WATER. Sodium and calcium react with water in a similar manner but at different rates.

help of certain enzymes the combination of glucose and oxygen takes place at a controlled rate, giving us a steady source of energy. This process is called metabolism. Production of water by the reaction of aqueous acids and hydroxides (called bases) is a type of double-replacement reaction called a neutralization reaction. This type of reaction is discussed in detail in Chapter 13.

Water is also involved as a reactant in many and varied types of reactions. Three types are mentioned here: (1) reaction with certain elements (single-replacement reactions), (2) reaction with oxides (combination reactions), and (3) formation of hydrates (combination reactions).

Water reacts with both metals and nonmetals. Its reactions with several metals are illustrated by the following equations.

$$2Na(s) + 2H_2O(l) \longrightarrow 2NaOH(aq) + H_2(g)$$
$$Ca(s) + 2H_2O(l) \longrightarrow Ca(OH)_2(aq) + H_2(g)$$
$$Zn(s) + H_2O(steam) \xrightarrow{\Delta} ZnO(s) + H_2(g)$$

Sodium reacts with water so rapidly that the sudden evolution of heat often ignites the hydrogen as it is evolved. Other alkali metals react in a similar fashion. Calcium reacts very slowly with water with a steady evolution of hydrogen gas. (See Figure 11-10.) Other alkaline earth metals react slowly with water except for beryllium. Beryllium and zinc react with steam at a high temperature. The chemical reactivity of a metal is usually judged by its activity toward water. Thus, sodium and other alkali metals are considered very reactive metals. Alkaline earth metals are somewhat less reactive, and other metals such as copper, silver, and gold, are considered comparatively unreactive since they do not react with water under any conditions.

Certain nonmetals also react with water:

$$2F_2(g) + 2H_2O(l) \longrightarrow 4HF(aq) + O_2(g)$$
$$Cl_2(g) + H_2O(l) \rightleftharpoons HCl(aq) + HOCl(aq)$$
$$C(s) + H_2O(steam) \longrightarrow CO(g) + H_2(g)$$

The reaction of fluorine is violent, as are many reactions involving fluorine, the most reactive nonmetal. Chlorine, a yellow-green gas, reacts with water, but the reaction is incomplete (reaches a state of equilibrium). The unreacted Cl_2 accounts for the green color seen when chlorine is used as a disinfectant in swimming pools. The reaction of carbon with steam produces H_2 and CO, which together are known as "water gas." It is a method used to convert coal (mostly elemental carbon) into a relatively cheap combustible fuel gas.

Water also reacts with certain nonmetal oxides to form acids

$$SO_3(g) \quad + H_2O(l) \longrightarrow H_2SO_4(aq)$$

Sulfur trioxide   Sulfuric acid

This reaction is one of those that account for what is known as "acid rain." Water reacts with certain metal oxides to form metal hydroxides, which are known as bases:

$$CaO(s) + H_2O(l) \longrightarrow Ca(OH)_2(s)$$

These latter two reactions will be studied in more detail in Chapter 13.

Water reacts with certain compounds to form hydrates:

$$CuSO_4(s) + 5H_2O(l) \longrightarrow CuSO_4 \cdot 5H_2O$$

Pale green   Dark blue
solid   solid

Water in a hydrate is called water of hydration. Hydrates are solid compounds that are distinctly different from the anhydrous compound (the form without water of hydration). In the hydrate the water molecules are attached to either the cation, by attractive forces between the positive ion and the negative dipole of water, or the anion, by hydrogen bonds. Since these attractions are not nearly as strong as covalent bonds, the water molecules can usually be removed without decomposition of the rest of the compound. As a result, many hydrates can be "dehydrated" by heating.

$$CaCl_2 \cdot 2H_2O(s) \xrightarrow{\Delta} CaCl_2(s) + 2H_2O(g)$$

Hydrates are named by adding the word "hydrate" to the regular compound name as discussed in Chapter 7. Greek prefixes are added to designate the number of H_2O molecules. (See Table 7-4.)

SEE PROBLEMS
11-63 THROUGH
11-70.

$CuSO_4 \cdot 5H_2O$ copper(II) sulfate pentahydrate
$CaCl_2 \cdot 2H_2O$ calcium chloride dihydrate

In the calculation of molar mass of a hydrate, waters of hydration are included.

11-7 CHAPTER REVIEW

This chapter is dedicated to the liquid and solid states of matter. These two states share many properties, but both differ significantly from the gaseous state. The basis of the differences involves the fact that the basic particles of liquids and solids "stick together." For this reason, liquids and solids (referred to as condensed states) are essentially incompressible, have high densities, and undergo little thermal expansion. Since the basic particles of the condensed states do have kinetic energy, however, they are in some sort of motion. Motion of molecules in solids is restricted to vibrations about fixed positions. Motion in liquids includes translational motion but is restricted by other molecules. As discussed in the previous chapter, motion in gases is also translational but totally unrestricted and random.

The physical state of a particular compound at room temperature is determined by intermolecular forces. There are three types of forces by which molecules of the same compound may be mutually attracted. London forces exist between molecules of all compounds. These forces are dependent on the molar mass of the compound, becoming more significant for heavier molecules. Molecules whose bonds are polar and whose geometry is such that the bond dipoles do not cancel are polar molecules. Such compounds have dipole–dipole forces of attraction between molecules in addition to London forces. Generally, dipole–dipole forces do not have a large effect on the physical state of the compound unless the compound has particularly polar molecules. The properties of compounds that have hydrogen bonding between molecules, on the other hand, are affected significantly. Ionic compounds have very strong interactions between the basic particles, which are ions. For this reason, all ionic compounds exist as solids at room temperature.

Kinetic energy (temperature) and intermolecular (or interionic) forces work in opposite directions in determining the physical state of a compound. Kinetic energy causes the basic particles to move about while intermolecular forces cause them to stay put. As a result, the higher the temperature, the more likely that the intermolecular forces are overcome and we find the compound in the liquid or even gaseous state. We examined what happens when we heat a sample of ice through two phase changes all the way to the gas phase. In the melting and boiling processes, added heat is converted to potential energy so that the temperature remains constant during these intervals. Adding heat to just one phase increases the kinetic energy, thus increasing the temperature of the sample. The changes in phase that a substance undergoes are summarized on the next page.

Intermolecular or interionic attractions between basic particles relate not only to the melting and boiling points of a compound but also to its heat of fusion and heat of vaporization. The heat of fusion refers to the amount of heat required to melt one gram of solid and the heat of vaporization refers to the amount of heat required to vaporize one gram of liquid. Ionic compounds, with their strong ion–ion interactions, have high melting and boiling points and correspondingly high heats of fusion and vaporization. If water molecules were attracted to each other only by London forces, water would have proper-

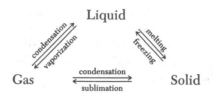

Gas	Liquid	Solid
There is negligible association between individual H_2O molecules and random motion.	Molecules are mobile, but significant association holds molecules in condensed state.	Water molecules held in fixed positions by hydrogen bonding; motion restricted to vibrations within a fixed volume.

ties similar to those of methane because of their similar molar masses (boiling point of $CH_4 = -164\ °C$). However, water has properties that indicate significant intermolecular attractions. Water is known to form hydrogen bonds which accounts for its unique properties. The four types of interactions that occur in compounds are summarized as follows:

Type of Compound	Predominant Type of Interaction	Melting and Boiling Points, Heats of Fusion and Vaporization	Examples
Ionic	Ion–ion	High	$NaCl$, $CaSO_4$
Polar covalent	Hydrogen bonding	Intermediate	H_2O, NH_3
Polar covalent	Dipole–dipole and London forces	Intermediate to low[a]	SO_2, CO, HBr
Nonpolar	London forces only	Intermediate to low[b]	CO_2, N_2, CCl_4

[a] Depends on the extent of molecular polarity and molar mass.
[b] Depends on molar mass.

A molecule must have a certain minimum kinetic energy to escape from the interactions in the liquid to the vapor. Since molecules have a range of energies at a certain temperature, some of the molecules with above-average energy may escape from the surface, creating an equilibrium vapor pressure. The higher the temperature, the larger the fraction of molecules having this minimum energy, and thus the higher the vapor pressure. The liquid boils when the vapor pressure equals the atmospheric pressure. The normal boiling point is the temperature at which the vapor pressure equals 1 atm pressure. When evaporation is allowed to occur so that the vapor escapes and no heat enters the liquid from the outside, the liquid cools. This is due to the loss of molecules with above-average kinetic energy, lowering the average kinetic energy (and the temperature) of the remaining molecules.

Water is both a product and a reactant in many important chemical reactions. It is formed mainly from the combustion of hydrogen or hydrogen-containing compounds. It reacts with many elements, both metals and nonmetals, and with many oxides. Water is also associated with many ionic compounds in the form of "hydrates."

■ EXERCISES

THE NATURE OF THE SOLID AND LIQUID STATES

11-1 Liquids mix more slowly than gases. Why?

11-2 Why is a gas compressible and a liquid not compressible?

11-3 Describe why a drop of food coloring in a glass of water slowly becomes evenly distributed without stirring.

11-4 What properties do liquids have in common with solids? With gases?

11-5 Review the densities of the solids listed in Table 2-6. In general, which have higher densities, solids or liquids? Why is this so?

INTERMOLECULAR FORCES AND PHYSICAL STATE

11-6 If H_2O were a linear molecule, why wouldn't it be polar?

11-7 The H_2S molecule is also bent, similar to H_2O, but it has a very small molecular dipole. Why?

11-8 Write the Lewis structure of NH_3, with the electron pair and the hydrogens in a tetrahedral arrangement. (See Figure 11-2c.) Is NH_3 a polar molecule?

11-9 The PH_3 molecule can be represented in a tetrahedral arrangement similar to NH_3. Is PH_3 a polar molecule?

11-10 Which of the following molecules have dipole–dipole interactions in the liquid state?
(a) HBr
(b) SO_2 (nonlinear)
(c) CO_2 (linear)
(d) BF_3 (trigonal planar)
(e) K_2S
(f) CO

11-11 Which of the following molecules have dipole–dipole interactions in the liquid state?
(a) SCl_2 (nonlinear)
(b) PH_3
(c) CCl_4 (tetrahedral)
(d) $CaCl_2$
(e) CS_2 (linear)
(f) FCl

11-12 Which of the following molecules may have hydrogen bonding in the liquid state?
(a) HF
(b) NCl_3
(c) H_2NCl
(d) H_2O
(e) CH_4 (tetrahedral)
(f) $H-C{\Large<}^{\displaystyle O}_{\displaystyle O-H}$
(g) CH_3Cl (tetrahedral)

11-13 Which should have stronger hydrogen bonding, NH_3 or H_2O?

11-14 At room temperature Cl_2 is a gas, Br_2 is a liquid, and I_2 is a solid. Explain this trend.

11-15 At room temperature, CO_2 is a gas and CS_2 is a liquid. Why is this reasonable?

11-16 At room temperature SF_6 is a gas and SnO is a solid. Both have similar molar masses. What accounts for the difference in physical states of the two compounds?

11-17 At room temperature, CH_3OH is a liquid and H_2CO is a gas. Both are polar and have similar molar masses. What accounts for the difference in physical state of the two compounds?

TEMPERATURE AND PHYSICAL STATE

11-18 Which has the higher kinetic energy, H_2O molecules in the form of ice at 0 °C or in the form of water at 0 °C?

11-19 Which has the higher potential energy, H_2O molecules in the form of ice at 0 °C or in the form of water at 0 °C?

11-20 Which has the higher potential energy, H_2O molecules in the form of steam at 100 °C or in the form of water at 100 °C?

11-21 Which of the following processes are endothermic?
(a) freezing   (c) boiling
(b) melting   (d) condensation

11-22 If water is boiling and the flame supplying the heat is turned up, does the water become hotter? What happens?

ENERGY AND CHANGES OF STATE

11-23 CH_3F and CH_3OH have almost the same molar mass and both are polar compounds. Yet CH_3OH boils at 65 °C and CH_3F at −78 °C. What accounts for the large difference?

11-24 The boiling point of F_2 is −188 °C and that of Cl_2 is −34 °C, yet the boiling point of HF is much higher than that of HCl. Explain.

11-25 Nitrogen gas and carbon monoxide have the same molar masses. Carbon monoxide boils at a slightly higher temperature, however (−191 °C versus −196 °C for N_2). Account for the difference.

11-26 Graph the data in Table 11-2 for the group 16 (VIA) hydrogen compounds. Plot the boiling points on the y axis and the molar masses of the compounds on the x axis. What would be the expected boiling point of H_2O if only London forces were important, as in the case of the other compounds in this series? (Determine from the graph.)

11-27 Graph the data in Table 11-2 for the group 15 (VA) hydrogen compounds. Plot the boiling points on the y axis and the molar masses of the compounds on the x axis. What would be the expected boiling point of NH_3 if only London forces were important, as in the case of the other compounds in this group?

11-28 Lead forms two compounds with chlorine, $PbCl_2$ and $PbCl_4$. The melting point of $PbCl_2$ is 501 °C and that of $PbCl_4$ is −15 °C. Explain in terms of bonding.

11-29 The following three compounds are similar in molar mass: $C_2H_5NH_2$, CH_3OCH_3, and CO_2. The temperatures at which these compounds boil are −78 °C, −25 °C, and 17 °C. Match the boiling point with the compound and give the intermolecular forces present in each case that account for the order.

11-30 If 850 J of heat is added to solid H_2O, NaCl, and benzene at their respective melting points, what mass of each is changed to a liquid?

11-31 When 25.0 g of ethyl ether freezes, how many calories are liberated? When 25.0 g of water freezes, how many calories are liberated? In a large lake which liquid would be more effective in modifying climate? How many joules of heat are required to melt 125 g of ethyl alcohol at its melting point?

11-32 How many kilojoules are required to vaporize 3.50 kg of H_2O at its boiling point?

11-33 How many joules of heat are released when an 18.0-g sample of benzene condenses at its boiling point?

11-34 What mass of carbon tetrachloride can be vaporized by addition of 1.00 kJ of heat energy to the liquid at its boiling point? What mass of water would be vaporized by the same amount of heat?

11-35 Refrigerators cool because a liquid extracts heat when it is vaporized. Before the synthesis of Freon (CF_2Cl_2), ammonia was used. Freon is nontoxic, and NH_3 is a pungent and toxic gas. The heat of vaporization of NH_3 is 1.36 kJ/g and that of Freon is 161 J/g. How many joules can be extracted by the vaporization of 450 g of each of these compounds. Which is the better refrigerant on a mass basis?

11-36 Molten ionic compounds are used as a

method to store heat. How many kilojoules of heat are released when 8.37 kg of NaCl crystallizes at its melting point?

11-37 How many joules of heat are released when 275 g of steam at 100.0 °C is condensed and cooled to room temperature (25 °C)?

11-38 How many kilocalories of heat are required to melt 0.135 kg of ice and then heat the water to 75.0 °C?

11-39 How many calories are required to change 132 g of ice at -20.0 °C to steam at 100.0 °C? (The specific heat of ice is 0.492 cal/g · °C.)

11-40 How many kilojoules of heat are released when 2.66 kg of steam at 100.0 °C is condensed, cooled, frozen, and then cooled to -25.0 °C? (The specific heat of ice is 2.06 J/g · °C.)

*11-41 A sample of steam is condensed at 100.0 °C and then cooled to 75.0 °C. If 28.4 kJ of heat is released, what is the mass of the sample?

*11-42 What mass of ice at 0 °C can be changed into steam at 100 °C by 2 kJ of heat?

11-43 How many calories of heat are needed to heat 120 g of ethyl alcohol from 25.5 °C to its boiling point and then vaporize the alcohol? (The specific heat of alcohol is 0.590 cal/g · °C.)

*11-44 A 10.0-g sample of benzene is condensed from the vapor at its boiling point and the liquid is allowed to cool. If 5000 J is released, what is the final temperature of the liquid benzene? (The specific heat of benzene is 1.72 J/g · °C.)

11-45 Ethyl ether ($C_2H_5OC_2H_5$) and ethyl alcohol (C_2H_5OH) are both polar covalent molecules. What accounts for the considerably higher heat of vaporization for alcohol?

11-46 A certain compound has a heat of fusion of about 600 cal/g. Is it likely to have a comparatively high or low melting point?

11-47 A certain compound has a boiling point of -75 °C. Is it likely to have a comparatively high or low heat of vaporization?

11-48 A certain compound has a boiling point of 845 °C. Is it likely to have a comparatively high or low melting point?

VAPOR PRESSURE

11-49 Explain how molecules of a liquid may go into the vapor state if the temperature is below the boiling point.

11-50 What is implied by the word "equilibrium" in equilibrium vapor pressure?

11-51 What is the difference between boiling point and *normal* boiling point?

11-52 A liquid has a vapor pressure of 850 torr at 75 °C. Is the substance a gas or a liquid at 75 °C. and 1 atm pressure?

11-53 On top of Mt. Everest the atmospheric pressure is about 260 torr. What is the boiling point of ethyl alcohol at that pressure? If the temperature is 10 °C, is ethyl ether a gas or a liquid under conditions on Mt. Everest? (Refer to Figure 11-9.)

11-54 At a certain temperature one liquid has a vapor pressure of 240 torr and another, 420 torr. Which liquid probably has the lower boiling point? Which probably has the lower heat of vaporization?

11-55 On the planet Mars the temperature can reach as high as a comfortable 50 °F (10 °C) at the equator. The atmospheric pressure is about 8 torr on Mars, however. Can liquid water exist on Mars under these conditions? What would the atmospheric pressure have to be before liquid water could exist at this temperature? What would happen to a glass of water set out on the surface of Mars? (Refer to Figure 11-9.)

11-56 If the atmospheric pressure is 500 torr, what are the approximate boiling points of water, ethyl alcohol, and ethyl ether? (Refer to Figure 11-9.)

*11-57 On a hot, humid day the relative humidity is 70% of saturation. If the temperature is 34 °C, the vapor pressure of water is 39.0 torr (this would be 100% of saturation). What mass of water is in each 100 L of air under these conditions?

11-58 How can the boiling point of a pure liquid be raised?

11-59 Rubbing alcohol feels cool when applied to the skin even if the alcohol is initially at room temperature when first applied. Why does the alcohol feel cool?

11-60 Ethyl chloride boils at 12 °C. When it is sprayed on the skin, it freezes a small part of the skin and thus serves as a local anesthetic. Explain how it cools the skin.

11-61 A beaker of a liquid with a vapor pressure of 350 torr at 25 °C is set alongside a beaker of water, and both are allowed to evaporate. Which liquid cools faster? Why?

11-62 Which cools faster, water at 90 °C or water at 30 °C, when both are allowed to evaporate?

REACTIONS OF WATER AND HYDRATES

11-63 Complete and balance the following equations.
(a) $K(s) + H_2O(l) \rightarrow$
(b) $Br_2(l) + H_2O(l) \rightarrow$
(c) $K_2O(s) + H_2O(l) \rightarrow$
(d) $CO_2(g) + H_2O(l) \rightarrow$
(e) $C_2H_4(g) + O_2(g) \rightarrow$

11-64 Complete and balance the following equations.
(a) $Sr(s) + H_2O(l) \rightarrow$
(b) $Si(s) + H_2O(l) \rightarrow SiO_2(s) +$
(c) $ZnO(s) + H_2O(l) \rightarrow$
(d) $SO_2(g) + H_2O(l) \rightarrow$
(e) $B_4H_{10}(g) + O_2(g) \rightarrow B_2O_3(s) +$

11-65 Name the following hydrates.
(a) $MgCO_3 \cdot 5H_2O$
(b) $Na_2SO_4 \cdot 10H_2O$
(c) $FeBr_3 \cdot 6H_2O$

11-66 Name the following hydrates.
(a) $Na_2CO_3 \cdot 7H_2O$
(b) $Sr(NO_3)_2 \cdot 4H_2O$
(c) $CrC_2O_4 \cdot H_2O$

11-67 Calculate the percent composition of water in strontium hydroxide octahydrate.

11-68 Calculate the percent composition of water in sodium sulfide nonahydrate.

11-69 Epsom salts is a magnesium sulfate hydrate. If the compound is 51.1% by mass water, what is its formula?

11-70 A potassium carbonate compound is 20.7% water. What is its formula?

HOT SPRINGS
The water from these hot springs in Yellowstone Park is actually a concentrated solution that comes from deep in the earth. The beautiful terraces are formed from deposits of the dissolved compounds.

AQUEOUS

SOLUTIONS

12

PURPOSES OF CHAPTER

In Chapter 12 we study the familiar compound water in its role as a solvent. We examine how water dissolves a substance, the amount of a substance that dissolves, some chemical reactions in water, units of concentration, and how solutes affect the properties of water.

OBJECTIVES FOR CHAPTER

After completion of this chapter, you should be able to:

1   Describe the process of solution of an ionic compound in water. (12-1)

2   Write equations illustrating the solution of various ionic compounds in water. (12-1)

3   Define the terms "saturated," "unsaturated," and "supersaturated." (12-2)

4   Determine whether a specified ionic compound is soluble in water given a set of solubility rules. (12-2)

5   Determine the solubility of a specified compound at two different temperatures given the appropriate graph. (12-2)

6   Predict precipitation reactions using solubility rules. (12-3)

7   Write balanced molecular, total ionic, and net ionic equations for specified precipitation reactions. (12-3)

8   Solve problems involving the percent composition of a solution. (12-4)

9   Apply the definition of molarity to calculate concentration, quantity, or volume given appropriate data. (12-5)

10 Calculate the volume or concentration of a specific solution when a solution is diluted. (12-6)

11 Apply the definition of molarity to problems involving stoichiometry of solutions. (12-7)

12 Describe the conductivity properties and compositions of strong electrolytes, nonelectrolytes, and weak electrolytes in water. (12-8)

13 Describe how the vapor pressure, melting point, boiling point, and osmotic pressure of water are affected by the presence of a solute. (12-8)

14 Calculate the actual melting point or boiling point of a solution using molality. (12-8)

In the last chapter we discussed how water is such a unique and important compound. In this chapter, we discuss *why* this is so. Essentially, water is so important because of its extensive ability to disperse other substances, forming a homogeneous mixture known as a solution. In fact, almost all of the familiar water in our lives is a solution to some extent. The oceans, for example, contain so many dissolved substances, especially sodium chloride, that the water is unsuitable for drinking or irrigation of crops. The closest thing to pure water comes in the form of rain. But even rain contains dissolved gases from the air.

We need to look no farther than the blood flowing in our veins for a prime example of the importance of water as a solvent. This system serves us as a vital waterway transporting nutrients and oxygen to the muscles and organs. In turn, the system carries wastes to the kidneys for removal. All of these nutrients and wastes are carried in the blood in the form of an aqueous solution. We begin our discussion with how water acts so effectively as a solvent. We also see how a knowledge of the ionic substances that do not dissolve in water leads us to a particular type of chemical reaction that occurs in water. After discussion of some of the units used to express the amount of solute present we turn to stoichiometry involving solutions. Finally, we see how the presence of a solute alters the properties of a solvent.

As background to this chapter, you should be familiar with the following topics.

1   Ionic compounds and how they are identified (Section 6-4)
2   Charges, formulas, and names of ions (Table 7-3)
3   Polarity of molecules (Sections 6-10 and 6-11)
4   Stoichiometry (Sections 9-3 and 10-11)
5   Water and hydrogen bonding (Section 11-2)

12-1 THE NATURE OF AQUEOUS SOLUTIONS

Our first job is to review some of the terms relevant to solutions first mentioned in Chapter 1. A **solvent**, *such as water, is a medium that dissolves or disperses another substance called a* **solute** *to form a homogeneous mixture of solute and solvent called a* **solution.** (See Figure 12-1.) The solute is completely dispersed in the solvent down to the molecular or ion level. The solution is in the same physical state as the solvent.

Water is truly a unique and convenient solvent. It is able to dissolve a wide range of both polar covalent and ionic compounds because of the nature of its molecules. As mentioned in the previous chapter, water molecules have comparatively strong intermolecular attractions as a result of hydrogen bonding. (See Figures 11-3 and 11-4.) These interactions account for the relatively high melting and boiling points of this compound compared with compounds of similar molar mass. As we will see in this section, the ability of the polar water molecules to interact with other polar compounds or ions accounts for water's strength as a solvent.

Let us now consider what happens when we place a small crystal of an ionic

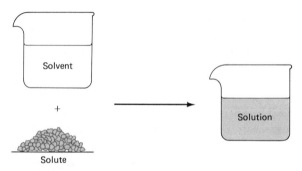

Figure 12-1   A SOLUTION. A solution is a homogeneous mixture of solute and solvent that is in the same physical state as the solvent.

compound such as NaCl in an aqueous medium. Recall that a water molecule contains a dipole. That is, the oxygen carries a partial negative charge and the hydrogen carries a partial positive charge. Immediately, water molecules are attracted to the ions on the surface of the crystal. The hydrogens on the water molecules interact with the Cl⁻ ions and the oxygens interact with the Na⁺

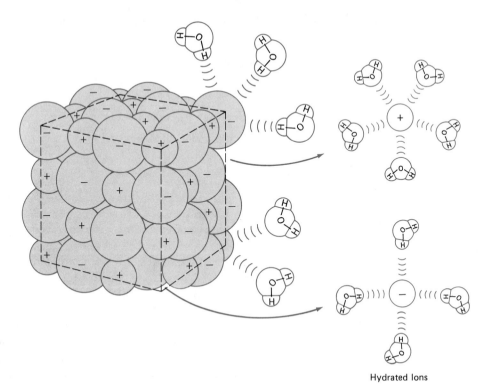

Hydrated Ions

Figure 12-2   INTERACTION OF WATER AND IONIC COMPOUNDS. There is an electrostatic interaction between the polar water molecules and the ions. This is a dipole–ion force.

ions. (See Figure 12-2.) *The electrostatic interactions between an ion and the dipoles of polar covalent molecules are known as* ion–dipole forces. For water, the forces are measurable and have known values (in units of kJ or kcal). *The quantitative value of the ion–dipole forces between water and one mole of a specific ion is known as the ion's* hydration energy.

Ion–dipole forces pull on these surface ions to lift them out of the crystal. However, these ion–dipole forces are counteracted by the *forces of attraction between oppositely charged ions in the solid which are known as* ion–ion forces. These latter forces are also measurable and have known values for each ionic substance. *The quantitative value of the ion–ion forces for one mole of a specific compound is known as that compound's* lattice energy. In effect, we have a tug-of-war between the two opposing forces: the lattice energy of the crystal and the hydration energy of the ions. If the two forces are comparable, the ions can be lifted from the crystal by the water molecules and held in the aqueous medium surrounded by an "escort" of water molecules. The ion in solution is said to be *hydrated*. (See Figure 12-2.) In solution the positive and negative ions are no longer associated with each other. The water molecules that surround the ions diminish the effects of the charges by separating them. The solution of NaCl in water is represented by the equation

$$NaCl(s) \xrightarrow{xH_2O} Na^+(aq) + Cl^-(aq)$$

The x before H_2O (above the arrow) indicates a large, undetermined number of water molecules needed to hydrate the ions. The "(aq)" after the symbol of the ion tells us that the ion is hydrated in the solution.

EXAMPLE 12-1

Write equations illustrating the solution of the following ionic compounds in water: (a) Na_2CO_3, (b) $CaCl_2$, (c) 2 mol of K_3PO_4, and (d) 3 mol of $(NH_4)_2SO_4$.

SOLUTION

Review the charges and the formulas of the ions discussed in Chapters 6 and 7. The ions present in water are the same as those present in the solid ionic compound.

(a) $Na_2CO_3(s) \xrightarrow{xH_2O} 2Na^+(aq) + CO_3^{2-}(aq)$

(b) $CaCl_2(s) \xrightarrow{xH_2O} Ca^{2+}(aq) + 2Cl^-(aq)$

(c) $2K_3PO_4(s) \xrightarrow{xH_2O} 6K^+(aq) + 2PO_4^{3-}(aq)$

(d) $3(NH_4)_2SO_4(s) \xrightarrow{xH_2O} 6NH_4^+(aq) + 3SO_4^{2-}(aq)$

Because of the highly polar nature of water molecules, the heats of hydration of ions are quite high compared with similar forces for other less polar solvents such as alcohol. Thus, water can overcome the lattice energy and dissolve a large number of ionic compounds. In some compounds, however, the crystal has a particularly strong lattice energy and little solute dissolves in water (e.g., $AgCl$, $CaCO_3$, and $PbSO_4$).

As an analogy to the interactions that lead to the solution of an ionic compound in water, there is the well-known phenomenon of the interaction of a group of young men and a group of young women. The men interact together (man–man forces) quite well and are happy, as do the women (woman–woman forces). When the two groups come together, however, the man–woman forces are usually strong enough to overcome the respective man–man and woman–woman forces. As a result, a thorough mixing occurs. It is simply a competition of forces: the strongest wins.

In addition to ionic compounds many polar covalent compounds also dissolve in water. Some do so with the formation of ions in solution and some do not. For example, when the polar covalent compound HCl is dissolved in water the dipoles of water are attracted to the opposite dipoles of HCl. Another tug-of-war develops between the covalent bond holding the atoms together and the dipole–dipole forces between molecules. In this case, the bond ruptures, forming hydrated H^+ ions and Cl^- ions in solution. This type of reaction will be discussed in more detail in the next chapter. The solution of a covalent compound to produce ions is represented as

$$HCl(g) \xrightarrow{x H_2O} H^+(aq) + Cl^-(aq)$$

Other polar covalent compounds dissolve in water but without ion formation. The dipole–dipole interactions between water and solute molecules are not strong enough to cause a bond in the solute molecule to break. Methyl alcohol (CH_3OH) is an example of a compound that dissolves without ion formation. (See Figure 12-3.)

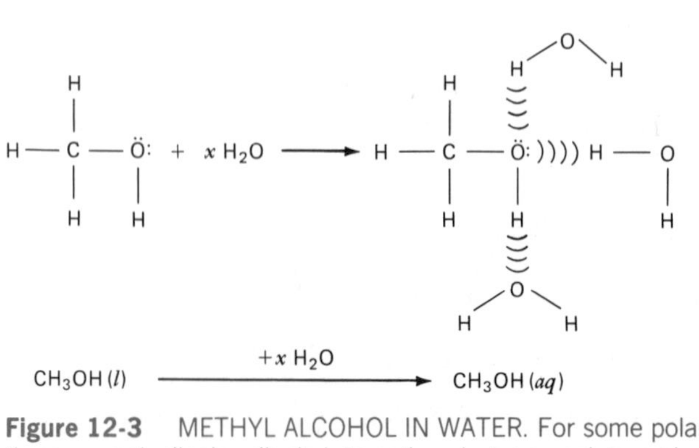

Figure 12-3   METHYL ALCOHOL IN WATER. For some polar covalent molecules in water, there are only dipole–dipole interactions between solute and water molecules.

Many nonpolar molecular compounds are not soluble in water, however. If a molecule is nonpolar, there are no centers of positive or negative charge on the molecule so there is nothing to attract the dipoles of water. On the other hand, the water molecules do interact strongly with each other by hydrogen bonding. This means that the molecules of water prefer to stick together rather than interact with a potential solute molecule. As a result, nonpolar compounds in the liquid or solid states do not generally dissolve in water. However, nonpolar compounds do dissolve in nonpolar solvents. This happens because neither solute nor solvent molecules have particularly strong forces of attraction. They mix because there isn't any reason not to.

■
SEE PROBLEMS 12-1 THROUGH 12-5.

12-2 THE SOLUBILITY OF COMPOUNDS IN WATER

As mentioned, certain ionic compounds and most nonpolar compounds do not dissolve in water to any appreciable extent. Before proceeding we need to consider some additional terms related to the amount of substance that dissolves in a solvent. *When an appreciable amount of a substance dissolves in a solvent the substance is considered* **soluble**. *If very little or none of a substance dissolves, the substance is termed* **insoluble**. *The maximum amount of a solute that dissolves in a specified amount of solvent is known as its* **solubility**.

When a portion of solvent contains the maximum amount of dissolved solute, the solution is said to be **saturated**. *If less than the maximum amount is present, the solution is* **unsaturated**. *In certain situations a solution may contain more than the maximum amount indicated by its known solubility.* Such a solution is known as a **supersaturated** solution and is unstable. For example, shaking or stirring such a solution or adding a tiny "seed" crystal of solute may cause the excess solute to suddenly drop out of solution as a solid. *A solid forming from previously dissolved substances is known as a* **precipitate**.

Several units are used to express the amount of solute present in a solution and are discussed later in this chapter. A unit frequently used to express the solubility of a saturated solution, however, simply relates the mass of solute that dissolves in 100 g of solvent (g solute/100 g H_2O). The solubilities of several compounds in water at 20 °C are listed in Table 12-1. Although there

TABLE 12-1	Solubilities of Compounds (at 20 °C)
Compound	Solubility (g solute/100 g H_2O)
Sucrose (table sugar)	205
HCl	63
NaCl	38
KNO_3	34
$PbSO_4$	0.04
$Mg(OH)_2$	0.01
AgCl	1.9×10^{-4}

TABLE 12-2   Solubility Rules for Some Ionic Compounds

Anion	Solubility Rule
Cl^-, Br^-, I^-	All cations form *soluble* compounds except Ag^+, Hg_2^{2+}, and Pb^{2+}. ($PbCl_2$ and $PbBr_2$ are slightly soluble.)
NO_3^-, ClO_3^-, ClO_4^-, $C_2H_3O_2^-$	All cations form *soluble* compounds. ($KClO_4$ and $AgC_2H_3O_2$ are slightly soluble.)
SO_4^{2-}	All cations form *soluble* compounds except Pb^{2+}, Ba^{2+}, and Sr^{2+}. Ca^{2+} and Ag^+ form slightly soluble compounds.
CO_3^{2-}, SO_3^{2-}, PO_4^{3-}	All cations form *insoluble* compounds except group 1 (IA) and 2 (IIA) metals and NH_4^+.
S^{2-}	All cations form *insoluble* compounds except group 1 (IA) and 2 (IIA) metals and NH_4^+.
OH^-	All cations form *insoluble* compounds except group 1 (IA) metals, Ba^{2+}, Sr^{2+}, and NH_4^+. [$Ca(OH)_2$ is slightly soluble.]
O^{2-}	All cations form *insoluble* compounds except group 1 (IA) metals, Ca^{2+}, Sr^{2+}, and Ba^{2+}.

is no definite dividing point, note that $PbSO_4$, $AgCl$, and $Mg(OH)_2$ have very low solubilities and are thus considered insoluble. There is some order, however, in determining which ionic compounds are comparatively insoluble and which are soluble. In Table 12-2 information on many of the compounds of the more common anions has been summarized. Note that most compounds of certain cations and anions are soluble but others are just the opposite.

The following examples illustrate the use of Table 12-2 to determine whether an ionic compound is soluble or insoluble in water.

EXAMPLE 12-2

Use Table 12-2 to predict whether the following compounds are soluble or insoluble: (a) NaI, (b) CdS, (c) $Ba(NO_3)_2$, (d) $SrSO_4$.

(a) According to Table 12-2 all alkali metal [group 1 (IA)] compounds of the ions listed are *soluble*. Therefore, <u>NaI is soluble</u>.

(b) All S^{2-} compounds are insoluble except those formed with group 1 (IA) and 2 (IIA) metals and NH_4^+. Since Cd is in group 12 (IIB), <u>CdS is insoluble</u>.

(c) All NO_3^- compounds are soluble. Therefore, <u>$Ba(NO_3)_2$ is soluble</u>.

(d) The Sr^{2+} ion forms an insoluble compound with SO_4^{2-}. Therefore, <u>$SrSO_4$ is insoluble</u>.

■
SEE PROBLEMS 12-6 AND 12-7.

Certainly the nature of the solute affects its solubility in water. In addition, the temperature is an important factor. Indeed, most of us are probably aware

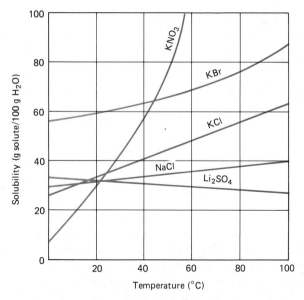

Figure 12-4 SOLUBILITY AND TEMPERATURE. The solubility of most solids increases as the temperature increases.

that hot water dissolves sugar more readily than cold water. In fact, most solid and liquid compounds are more soluble in water at higher temperatures. On the other hand, gases (e.g., CO_2, O_2, and N_2) are less soluble at higher temperatures. In Figure 12-4 the solubilities of several compounds in water are shown in graphical form as a function of temperature. Note that Li_2SO_4 is an exception in that it becomes less soluble as the temperature increases. The information shown in Figure 12-4 has important laboratory applications. For example, impure solid compounds can be purified by a process called **recrystallization.** In this procedure a saturated solution is made at a high temperature. Insoluble impurities can then be filtered from the hot solution. As the liquid cools, the solute becomes less soluble. The excess solute forms a precipitate (unless the solution becomes supersaturated) as the solution cools. The solid can then be removed by filtration. Most of the soluble impurities remain in the solution.

SEE PROBLEMS 12-8 THROUGH 12-12.

12-3 PRECIPITATION REACTIONS AND IONIC EQUATIONS

In a previous section it was mentioned that when an ionic compound dissolves in water, the cation and the anion behave independently. If the ions of two different compounds such as those of $CaCl_2$ and KNO_3 are mixed, we simply have a solution of four ions as illustrated in Figure 12-5. No cation is associated with any particular anion.

Now let us consider another example involving solutions of the soluble compounds NaCl and $AgNO_3$. In the previous section we learned that AgCl is insoluble. This is so because its lattice energy is too great to be overcome by

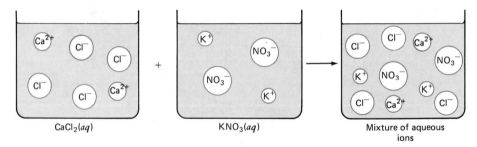

Figure 12-5   MIXTURE OF $CaCl_2$ AND KNO_3 SOLUTIONS. No reaction occurs when these solutions are mixed.

the heats of hydration of the ions and it remains in solid form. In fact, when Ag^+ and Cl^- ions find themselves in the same solution they tend to come together to form a solid precipitate. Thus when solutions of $AgNO_3$ and $NaCl$ are mixed, a chemical reaction occurs leading to the formation of a white precipitate of $AgCl$. This leaves the soluble compound $NaNO_3$ in solution. (See Figure 12-6.)

  The formation of a precipitate by mixing solutions of two soluble compounds is known as a **precipitation reaction.** It is a type of double-replacement reaction (discussed in Chapter 9). Another type leads to the formation of water and is discussed in the next chapter. Precipitation reactions find use as a method to prepare certain ionic compounds from others. Note that the reaction involves simply an exchange of partners among the four ions involved. In the example illustrated in Figure 12-6, the solid $AgCl$ can be conveniently removed by filtration. If it were desirable to recover the $NaNO_3$ still in solution, the water can be removed by boiling. The solid $NaNO_3$ remains after all the water is removed. This reaction has a very practical application. Silver is, of course, an expensive precious metal. However, it can be recovered from an aqueous solution in the form of solid $AgCl$ by adding any source of soluble choride ion.

  The following equation represents the chemical reaction illustrated in Figure 12-6.

$$NaCl(aq) + AgNO_3(aq) \longrightarrow AgCl(s) + NaNO_3(aq)$$

This is the molecular* form of the equation. *In the* **molecular equation,** *all reactants and products are shown as compounds.*

  When all cations and anions in solution are written separately, the resulting equation is known as the **total ionic equation:**

$$Na^+(aq) + Cl^-(aq) + Ag^+(aq) + NO_3^-(aq) \longrightarrow$$
$$AgCl(s) + Na^+(aq) + NO_3^-(aq)$$

---

* *"Molecular" is a misnomer in this case, since no real molecules exist in ionic compounds.*

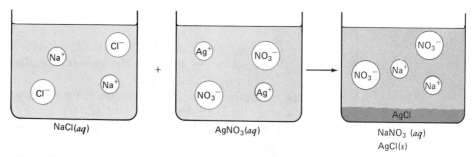

Figure 12-6 MIXTURE OF NaCl AND AgNO₃ SOLUTIONS. When these solutions are mixed, a precipitate forms.

In the total ionic equation, note that the $Na^+(aq)$ and the $NO_3^-(aq)$ ions are in the same state on both sides of the equation. From algebra you are aware that when identical numbers or variables appear on both sides of an equation, they can be subtracted from both sides of the equation as follows:

$$x + y = \quad x + 17 + z$$
$$\underline{-x = -x}$$
$$y = 17 + z$$

In the same manner, identical substances that appear on both sides of a chemical equation can be subtracted to produce the **net ionic equation.** In our example, the net ionic equation is

$$Ag^+(aq) + Cl^-(aq) \longrightarrow AgCl(s)$$

Ions that are identical on both sides of an equation are called **spectator ions,** *since they are not involved directly with the chemical reaction.* The net ionic equation is important, since it focuses on the "main action" in the solution.

Precipitation reactions can be predicted by the use of solubility rules such as those listed in Table 12-2. One simply writes the two products that would result from an exchange of ion partners and then refers to the table to see if either is insoluble in water. If the answer is "yes," a precipitation reaction occurs. The following examples illustrate the use of Table 12-2 to predict and write precipitation reactions.

Write the balanced molecular equation, the total ionic equation, and the net ionic equation for any reaction that occurs when solutions of the following are mixed.

EXAMPLE 12-3

A solution of Na_2CO_3 is mixed with a solution of $CaCl_2$.

SOLUTION

To begin, write out the formulas of the possible products resulting from a precipitation reaction. In this case the possible products from the exchange of ions are NaCl and $CaCO_3$.

If both of these compounds are soluble, no reaction occurs. In this case, however, Table 12-2 tells us that $CaCO_3$ is insoluble. Thus a reaction occurs, as illustrated by the following balanced equation written in molecular form:

$$Na_2CO_3(aq) + CaCl_2(aq) \longrightarrow CaCO_3(s) + 2NaCl(aq)$$

The equation written in total ionic form is

$$2Na^+(aq) + CO_3{}^{2-}(aq) + Ca^{2+}(aq) + 2Cl^-(aq) \longrightarrow$$
$$CaCO_3(s) + 2Na^+(aq) + 2Cl^-(aq)$$

Note that the Na^+ and the Cl^- ions are spectator ions. Elimination of the spectator ions on both sides of the equation leaves the net ionic equation

$$Ca^{2+}(aq) + CO_3{}^{2-}(aq) \longrightarrow CaCO_3(s)$$

EXAMPLE 12-4

A solution of KOH is mixed with a solution of MgI_2.

SOLUTION

The possible precipitation reaction products are KI and $Mg(OH)_2$. The information in Table 12-2 indicates that $Mg(OH)_2$ is insoluble. The balanced molecular equation for this reaction is

$$2KOH(aq) + MgI_2(aq) \longrightarrow Mg(OH)_2(s) + 2KI(aq)$$

The total ionic equation is

$$2K^+(aq) + 2OH^-(aq) + Mg^{2+}(aq) + 2I^-(aq) \longrightarrow$$
$$Mg(OH)_2(s) + 2K^+(aq) + 2I^-(aq)$$

Elimination of spectator ions gives the net ionic equation

$$Mg^{2+}(aq) + 2OH^-(aq) \longrightarrow Mg(OH^-)_2(s)$$

EXAMPLE 12-5

A solution of KNO_3 is mixed with a solution of $CaBr_2$.

■
SEE PROBLEMS
12-13 THROUGH
12-20.

SOLUTION

The possible precipitation reaction products are KBr and $Ca(NO_3)_2$. Since these compounds are both soluble, no precipitate forms and a reaction does not occur.

**12-4
CONCENTRA-
TION:
PERCENT BY
WEIGHT**

In a previous section (Section 12-3) we expressed the solubility of a solute in units of grams of solute per 100 g of solvent. There are several other ways in which amount of solute present in solution (known as concentration) is expressed. A convenient but strictly qualitative way of expressing concentration is by use of the terms "dilute" and "concentrated." Generally, these refer to the same solute and simply compare the concentrations of different solutions. **Concentrated** *means that a large amount of solute is present in solution; in a* **dilute** *solution a relatively small amount of solute is present.* For laboratory situations quantitative methods are obviously necessary.

Some quantitative units used to express concentration are percent by weight, molarity, molality, and normality. All of these have their special uses, but percent by weight and molarity are used in many laboratory situations and are discussed here. Molality is used in connection with certain properties of -solutions and will be discussed at that time.

Percent by weight, * *a rather straightforward method of expressing concentration, simply relates the mass of solute as a percent of the total mass of the solution.* Therefore, in 100 g of a solution that is 25% by weight HCl there are 25 g of HCl and 75 g of H_2O. The formula for percent by weight is

$$\% \text{ by weight (solute)} = \frac{\text{mass of solute}}{\text{mass of solution}} \times 100\%$$

---

* *Percent by mass is actually the correct term. However, we use percent by weight since it is the traditional and more commonly used term.*

EXAMPLE 12-6

What is the percent by weight of NaCl if 1.75 g of NaCl is dissolved in 5.85 g of H_2O?

PROCEDURE

Find the total mass of the solution and then the percent of NaCl.

SOLUTION

The total mass is

$$
\begin{array}{l}
1.75 \text{ g NaCl (solute)} \\
\underline{5.85 \text{ g H}_2\text{O (solvent)}} \\
7.60 \text{ g solution}
\end{array}
$$

$$\frac{1.75 \text{ g NaCl}}{7.60 \text{ g solution}} \times 100\% = \underline{23.0\% \text{ by weight NaCl}}$$

EXAMPLE 12-7

A solution is 14.0% by weight H_2SO_4. How many moles of H_2SO_4 is in 155 g of solution?

PROCEDURE

1   Find the mass of H_2SO_4 in the solution.
2   Convert mass to moles.

SOLUTION

1   Multiply the mass of compound by the percent in fraction form to find the mass of H_2SO_4:

$$155 \text{ g compound} \times \frac{14.0 \text{ g H}_2\text{SO}_4}{100 \text{ g compound}} = 21.7 \text{ g H}_2\text{SO}_4$$

2   The molar mass of H_2SO_4 is

$$2.0 \text{ g (H)} + 32.1 \text{ g (S)} + 64.0 \text{ g (O)} = 98.1 \text{ g}$$

$$21.7 \text{ g} \times \frac{1 \text{ mol}}{98.1 \text{ g}} = \underline{0.221 \text{ mol H}_2\text{SO}_4}$$

Another way of expressing percent by weight is simply parts per hundred. When concentrations are extremely low, however, two closely related units become more convenient. These units are **parts per million (ppm)** and **parts per billion (ppb)** and are particularly useful for expressing concentrations of trace amounts of a substance relative to the total amount present. For example, one hears of dangerous dioxin levels in the soil in ranges of ppm and even ppb. *Parts per million is obtained by multiplying the ratio of the mass of solute to mass of solution by 10^6 rather than 100%. Parts per billion is obtained by multiplying the same ratio by 10^9. For example, if a solution has a mass of*

1.00 kg and contains only 3.0 mg of a solute, it has the following concentrations in percent by weight, ppm, and ppb.

$$\frac{3.0 \times 10^{-3} \text{ g (solute)}}{1.0 \times 10^3 \text{ g (solution)}} \times 100\% = \underline{3.0 \times 10^{-4} \%}$$

$$\frac{3.0 \times 10^{-3} \text{ g}}{1.0 \times 10^3 \text{ g}} \times 10^6 = \underline{3.0 \text{ ppm}}$$

$$\frac{3.0 \times 10^{-3} \text{ g}}{1.0 \times 10^3 \text{ g}} \times 10^9 = \underline{3.0 \times 10^3 \text{ ppb}}$$

■
*SEE PROBLEMS
12-21 THROUGH
12-28.*

12-5 CONCENTRATION: MOLARITY

The most convenient and commonly used unit of concentration is known as molarity. **Molarity (*M*)** *relates the number of moles of solute (n) to the total volume of the solution in liters (V):*

$$M = \frac{n \text{ (moles of solute)}}{V \text{ (liters of solution)}}$$

Thus, a 1.00 *M* solution of HCl contains 1.00 mol (36.5 g) of HCl dissolved in enough H_2O to make 1.00 L of solution.

EXAMPLE 12-8

What is the molarity of H_2SO_4 in a solution made by dissolving 49.0 g of pure H_2SO_4 in enough water to make 250 mL of solution?

PROCEDURE

Write down the formula for molarity and what you have been given, and then solve for what's requested.

SOLUTION

$$M = \frac{n}{V}$$

(*n*) $49.0 \text{ g } H_2SO_4 \times \dfrac{1 \text{ mol}}{98.1 \text{ g } H_2SO_4} = 0.499 \text{ mol}$

(*V*) $250 \text{ mL} \times \dfrac{10^{-3} \text{ L}}{\text{mL}} = 0.250 \text{ L}$

$$\frac{n}{V} = \frac{0.499 \text{ mol}}{0.250 \text{ L}} = \underline{\underline{2.00 \text{ M}}}$$

EXAMPLE 12-9

What mass of HCl is present in 155 mL of a 0.540 M solution?

SOLUTION

$$M = \frac{n}{V} \quad n = M \times V$$

$$M = 0.540 \text{ mol/L} \quad V = 155 \text{ mL} = 0.155 \text{ L}$$

$$n = 0.540 \text{ mol/L} \times 0.155 \text{ L} = 0.0837 \text{ mol HCl}$$

$$0.0837 \text{ mol HCl} \times \frac{36.5 \text{ g}}{\text{mol HCl}} = \underline{\underline{3.06 \text{ g HCl}}}$$

EXAMPLE 12-10

Concentrated laboratory acid is 35.0% by weight HCl and has a density of 1.18 g/mL. What is its molarity?

PROCEDURE

Since a volume was not given, you can start with any volume you wish. The molarity will be the *same* for 1 mL as for 25 L. To make the problem as simple as possible, assume that you have exactly 1 L of solution ($V = 1.00$ L) and go from there. The number of moles of HCl (n) in 1 L can be obtained as follows:

1   Find the mass of 1 L from the density.
2   Find the mass of HCl in 1 L using the percent by weight and the mass of 1 L.
3   Convert the mass of HCl to moles of HCl.

SOLUTION

Assume that $V = 1.00$ L

1   The mass of 1.00 L (10^3 mL) is

$$10^3 \text{ mL} \times 1.18 \text{ g/mL} = 1180 \text{ g solution}$$

2   The mass of HCl in 1.00 L is

$$1180 \text{ g solution} \times \frac{35.0 \text{ g HCl}}{100 \text{ g solution}} = 413 \text{ g HCl}$$

3 The number of moles of HCl in 1.00 L is

$$413 \text{ g} \times \frac{1 \text{ mol}}{36.5 \text{ g}} = 11.3 \text{ mol HCl}$$

$$\frac{n}{V} = \frac{11.3 \text{ mol}}{1.00 \text{ L}} = \underline{11.3 \, M}$$

SEE PROBLEMS 12-29 THROUGH 12-38.

12-6 DILUTION OF CONCENTRATED SOLUTIONS

One of the common laboratory exercises for the experienced as well as the beginning chemist is to make a certain volume of a *dilute* solution from a *concentrated* solution. In this case a certain amount (number of moles) of solute is withdrawn from the concentrated solution and diluted with water to form the dilute solution. The quantity of moles needed for the dilute solution is designated n_d. It is calculated from the molarity and volume for a specified dilute solution:

$$M_d \times V_d = n_d$$

The moles of solute taken from the concentrated solution is designated n_c. The moles of solute in the concentrated solution relates to the volume and molarity of the concentrated solution:

$$M_c \times V_c = n_c$$

Since the moles of solute used in the dilute solution is the same as that taken from the concentrated solution, moles of solute remains constant:

$$\text{moles solute} = n_d = n_c$$

Because quantities equal to the same quantity are equal to each other, we have the simple relationship between the dilute solution and the volume and molarity of the concentrated solution that is used:

$$M_c \times V_c = M_d \times V_d$$

EXAMPLE 12-11

What volume of 11.3 M HCl must be mixed with water to make 1.00 L of 0.555 M HCl? (See also Figure 12-7.)

PROCEDURE

$$M_c \times V_c = M_d \times V_d$$

$$V_c = \frac{M_d \times V_d}{M_c}$$

SOLUTION

$$V_c = \frac{0.555 \ \text{mol/L} \times 1.00 \ \text{L}}{11.3 \ \text{mol/L}} = 0.0491 \ \text{L} = \underline{\underline{49.1 \ \text{mL}}}$$

EXAMPLE 12-12

What is the molarity of a solution of KCl that is prepared by dilution of 855 mL of a 0.475 M solution to a volume of 1.25 L?

PROCEDURE

$$M_c \times V_c = M_d \times V_d$$

$$M_d = \frac{M_c \times V_c}{V_d}$$

SOLUTION

$$M_c = 0.475 \ M \qquad\qquad M_d = ?$$

$$V_c = 855 \ \text{mL} = 0.855 \ \text{L} \qquad V_d = 1.25 \ \text{L}$$

$$M_d = \frac{0.475 \ M \times 0.855 \ \text{L}}{1.25 \ \text{L}} = \underline{\underline{0.325 \ M}}$$

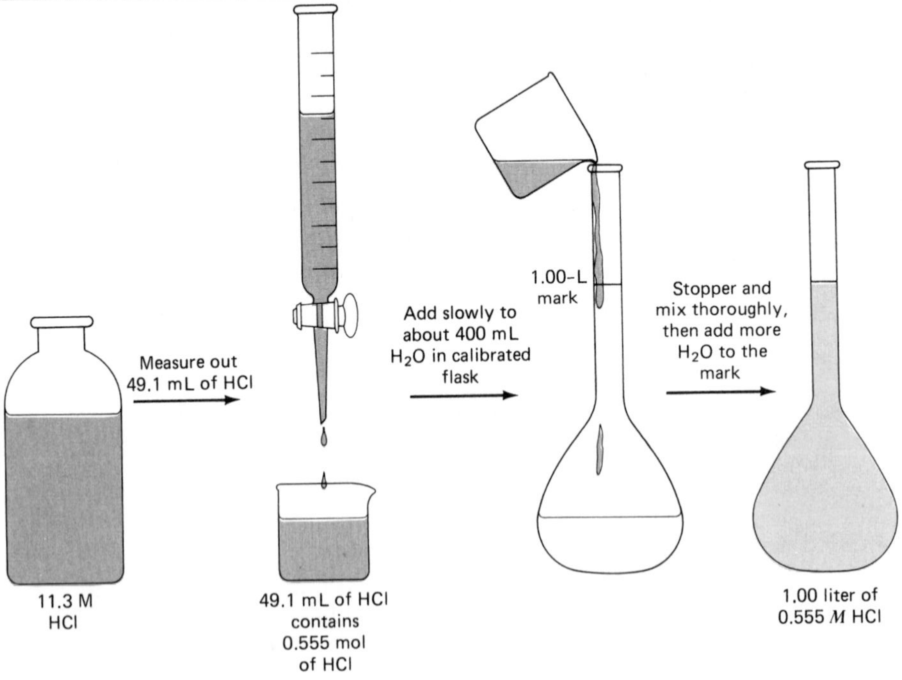

■
SEE PROBLEMS
12-39 THROUGH
12-47.

Measure out
49.1 mL of HCl

11.3 M
HCl

49.1 mL of HCl
contains
0.555 mol
of HCl

Add slowly to
about 400 mL
H₂O in calibrated
flask

1.00-L
mark

Stopper and
mix thoroughly,
then add more
H₂O to the
mark

1,00 liter of
0.555 M HCl

Figure 12-7   DILUTION OF CONCENTRATED HCl. (*Note:* Water is never added directly to concentrated acid, because it may splatter and cause severe burns.)

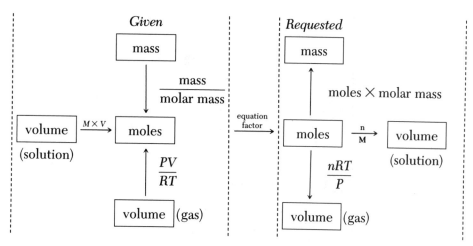

Figure 12-8 GENERAL PROCEDURE FOR STOICHIOMETRY PROBLEMS.

12-7 STOICHI-OMETRY INVOLVING SOLUTIONS

In Chapter 9 a general procedure was presented for working stoichiometry problems. This was expanded in Chapter 10 to include gases. We can now further expand the general procedure to include stoichiometry in aqueous solution, since the number of moles is related to the volume of a solution by the molarity. (See Figure 12-8.)

The following examples illustrate the role of solutions in stoichiometry.

EXAMPLE 12-13

Given the balanced equation

$$3NaOH(aq) + H_3PO_4(aq) \longrightarrow Na_3PO_4(aq) + 3H_2O(l)$$

what volume of 0.250 M NaOH is required to react completely with 4.90 g of H_3PO_4?

PROCEDURE

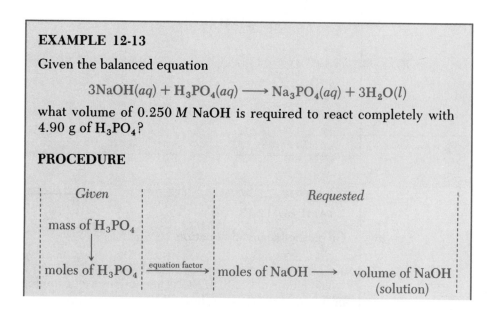

SOLUTION

$$4.90 \text{ g H}_3\text{PO}_4 \times \frac{1 \text{ mol H}_3\text{PO}_4}{98.0 \text{ g H}_3\text{PO}_4} \times \frac{3 \text{ mol NaOH}}{1 \text{ mol H}_3\text{PO}_4} = 0.150 \text{ mol NaOH}$$

$$\text{vol NaOH} = \frac{\text{mol NaOH}}{M} = \frac{0.150 \text{ mol}}{0.250 \text{ mol/L}} = \underline{\underline{0.600 \text{ L}}}$$

EXAMPLE 12-14

Given the balanced equation

$$\text{Cd(NO}_3)_2(aq) + \text{K}_2\text{S}(aq) \longrightarrow \text{CdS}(s) + 2\text{KNO}_3(aq)$$

what mass of CdS would be produced from 15.8 mL of a 0.122 M Cd(NO$_3$)$_2$ solution with excess of K$_2$S present?

PROCEDURE

	Given		*Requested*
volume of Cd(NO$_3$)$_2$ (solution)	$\longrightarrow$ moles of Cd(NO$_3$)$_2$	$\xrightarrow[\text{factor}]{\text{equation}}$	moles of CdS $\downarrow$ mass of CdS

SOLUTION

$$\text{mol Cd(NO}_3)_2 = V \times M$$
$$= 0.0158 \text{ L} \times 0.122 \text{ mol/L}$$
$$= 0.00193 \text{ mol Cd(NO}_3)_2$$

$$0.00193 \text{ mol Cd(NO}_3)_2 \times \frac{1 \text{ mol CdS}}{1 \text{ mol Cd(NO}_3)_2}$$

$$\times \frac{144 \text{ g CdS}}{\text{mol CdS}} = \underline{\underline{0.278 \text{ g CdS}}}$$

EXAMPLE 12-15

Given the balanced equation

$$2\text{KCl}(aq) + \text{Pb(NO}_3)_2(aq) \longrightarrow \text{PbCl}_2(s) + 2\text{KCl}(aq)$$

what volume of 0.200 M KCl is needed to react completely with 185 mL of 0.245 M Pb(NO$_3$)$_2$ solution?

PROCEDURE

	Given			Requested	

volume of ⟶ moles of | equation factor | moles of ⟶ volume of
$Pb(NO_3)_2$ $Pb(NO_3)_2$ KCl   KCl
(solution)   (solution)

SOLUTION

$$\text{mol } Pb(NO_3)_2 = V[Pb(NO_3)_2] \times M[Pb(NO_3)_2]$$

$$0.185 \text{ L} \times 0.245 \text{ mol/L} = 0.0453 \text{ mol } Pb(NO_3)_2$$

$$0.0453 \text{ mol } Pb(NO_3)_2 \times \frac{2 \text{ mol KCl}}{1 \text{ mol } Pb(NO_3)_2} = 0.0906 \text{ mol KCl}$$

$$V(KCl) = \frac{n(KCl)}{M(KCl)} = \frac{0.906 \text{ mol}}{0.200 \text{ mol/L}} = 0.453 \text{ L}$$

$$0.453 \text{ L} \times \frac{1 \text{ mL}}{10^{-3} \text{ L}} = \underline{453 \text{ mL}}$$

EXAMPLE 12-16

Given the balanced equation

$$2HCl(aq) + K_2S(aq) \longrightarrow H_2S(g) + 2KCl(aq)$$

what volume of H_2S measured at STP would be evolved from 1.65 L of a 0.552 *M* HCl solution with excess K_2S present?

PROCEDURE

	Given			Requested	

volume of HCl ⟶ moles of HCl | equation factor | moles of H_2S
(solution)   ↓
                                                              volume of H_2S
                                                                (gas)

SOLUTION

$$V(\text{solution}) \times M(HCl) = n(HCl)$$

$$1.65 \text{ L} \times 0.552 \text{ mol/L} = 0.911 \text{ mol HCl}$$

(Since the volume of the gas is at STP, the molar volume relationship can be used rather than the ideal gas law.)

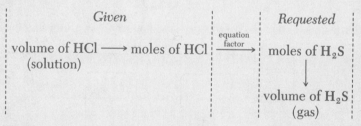

$$0.911 \text{ mol HCl} \times \frac{1 \text{ mol } H_2S}{2 \text{ mol HCl}} \times \frac{22.4 \text{ L (STP)}}{\text{mol } H_2S} = \underline{\underline{10.2 \text{ L (STP)}}}$$

■
*SEE PROBLEMS
12-48 THROUGH
12-56.*

12-8
PHYSICAL
PROPERTIES
OF SOLUTIONS

Conductivity

It has long been known that the presence of a solute in water may affect its ability to conduct electricity. *Electricity is simply a flow of electrons through a substance called a* **conductor.** Metals are the most familiar conductors and as such find use in electrical wires. Because the outer electrons of metals are loosely held they can be made to flow through a continuous length of wire. *Other substances resist the flow of electricity and are known as* **nonconductors** *or* **insulators.** Glass and wood are examples of nonconductors of electricity. Pure water is also an example of a nonconductor. (See Figure 12-9.) When wires are attached to a charged battery and then to a light bulb, the light shines brightly. If the wire is cut the light goes out of course, because the circuit is broken. If the two ends of the cut wire are now immersed in pure water the light stays out, indicating that water does not conduct electricity. Now let us dissolve certain solutes in water and examine what happens. When com-

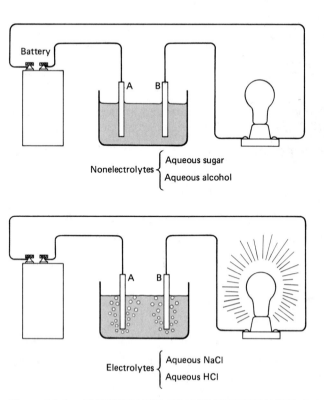

Figure 12-9   ELECTROLYTES AND NONELECTROLYTES. Solutions of electrolytes conduct electricity; solutions of nonelectrolytes do not.

pounds such as NaCl and HCl are dissolved in water the effect is obvious. The light immediately begins to shine, indicating that the solution is a good conductor of electricity. *Compounds whose aqueous solutions conduct electricity are known as* **electrolytes.** (Some ionic compounds are not soluble in water but if their molten state conducts electricity, they are also classified as electrolytes.)

We now understand that it is the presence of ions in the aqueous solution that allows the solution to conduct electricity. Almost all soluble ionic compounds form ions in solution and some polar covalent compounds also dissolve to form ions. For example, both NaCl (ionic) and HCl (polar covalent) are classified as electrolytes because they form ions in aqueous solution.

Other compounds such as sucrose (table sugar) and alcohol dissolve in water but their solutions do not conduct electricity. *Compounds whose aqueous solutions do not conduct electricity are known as* **nonelectrolytes.** Nonelectrolytes are molecular compounds that dissolve in water without formation ions. Methyl alcohol is an example of a nonelectrolyte.

There are two classes of electrolytes: strong electrolytes and weak electrolytes. Solutions of **strong electrolytes** (e.g., NaCl and HCl) *are good conductors of electricity.* Solutions of **weak electrolytes** *allow a limited amount of conduction.* When electrodes are immersed in solutions of weak electrolytes, the light glows, but very faintly. Even adding more of the solute does not help. An example of a weak electrolyte is HF. This gas is soluble in water but, unlike HCl, only a small percent of the HF molecules dissociate into ions at any one time. Because the number of ions present is small compared with the total number of HF molecules dissolved, the solution conducts only a limited amount of electricity. The ionization of HF is an example of a reversible reaction that reaches a point of equilibrium. This type of reaction will be discussed in more detail in the next chapter. The solution and dissociation of HF can be represented by the following equations.

$$HF(g) \xrightarrow{xH_2O} HF(aq)$$
$$HF(aq) \rightleftharpoons H^+(aq) + F^-(aq)$$

The presence of a solute in water may or may not affect its conductivity depending upon whether the solute is an electrolyte or a nonelectrolyte. There are other properties of water, however, that are always affected to some extent by the presence of a solute. Consider again what was mentioned in Chapter 1 as characteristic of a pure substance. Recall that a pure substance has a distinct and unvarying melting point and boiling point. Mixtures, such as aqueous solutions, freeze and boil over a range of temperatures that are lower (for freezing) and higher (for boiling) than those of the pure solvent. The more solute that is present, the more the melting and boiling points are affected. *A property that depends only on the relative amounts of solute and solvent present is known as a* **colligative** **property.** The remaining physical properties discussed in this section are all colligative properties.

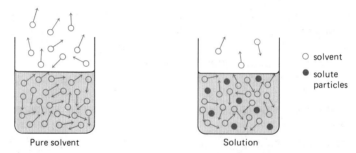

Figure 12-10   VAPOR PRESSURE LOWERING. A nonvolatile solute reduces the number of solvent molecules escaping to the vapor.

Vapor Pressure

The presence of a nonvolatile solute* in a solvent lowers the equilibrium vapor pressure from that of the pure solvent. (See Section 11-5.) A simple explanation of why this happens is illustrated in Figure 12-10. In the solution a certain proportion of the solute particles are near the surface. They do not escape to the vapor and also block access to the surface of some of the solvent molecules. Thus, fewer of the solvent molecules can escape to the vapor, which means that the solution has a lower equilibrium vapor pressure than the pure solvent. As we might predict from this model, the more solute particles present, the greater the effect of vapor pressure lowering.

An example of the effect of vapor pressure lowering is provided by the Dead Sea in Israel. This large body of water has no outlet to the ocean, so dissolved substances have accumulated, forming a very concentrated solution. Even though it exists in an arid region with high summer temperatures, it evaporates very slowly compared with a freshwater lake or even the ocean. It evaporates slowly because of its reduced vapor pressure. If this water evaporated at the same rate as fresh water, the Dead Sea would nearly dry up in a matter of a few years.

Boiling Point

A direct effect of the lowered vapor pressure of a solution is a higher boiling point compared with the solvent. Since a liquid boils at the temperature at which its vapor pressure equals the atmospheric pressure, a higher temperature is necessary to cause a solution to reach the same vapor pressure as the solvent.

The amount of boiling point elevation is given by the equation

$$\Delta T_b = K_b m$$

where ΔT_b = the number of Celsius degrees that the boiling point is raised

K_b = a constant characteristic of the solvent (for water $K_b = 0.512$ °C · kg/mol. Other values of K_b are given for particular solvents in the exercises.)

m = a concentration unit called molality.

---

* A nonvolatile solute is one that has essentially no vapor pressure at the relevant temperatures.

Molality emphasizes a relationship between moles of solute and mass of solvent rather than volume, as does molarity. The definition of molality is

$$\text{molality } (m) = \frac{\text{moles of solute}}{\text{kg of solvent}}$$

The calculation of molality and its use will be illustrated shortly.

Freezing Point

Just as the boiling point of a solution is higher than that of the pure solvent, the freezing point is lower. This effect has many applications. When water freezes, it expands, and it will likely break its container. This would occur in the automobile radiator and engine block. Antifreeze (ethylene glycol) is used to lower the freezing point of water in the radiator and prevent freezing and the damage it would cause. (It also raises the boiling point, preventing the fluid from boiling away in the summer.) The amount of freezing point lowering is given by the equation

$$\Delta T_f = K_f m$$

where ΔT_f = the number of Celsius degrees that the freezing point is lowered

K_f = a constant characteristic of the solvent (for water K_b = 1.86 °C · kg/mol.)

m = molality of the solution

Examples of these calculations follow.

EXAMPLE 12-17

What is the molality of methyl alcohol in a solution made by dissolving 18.5 g of methyl alcohol (CH_3OH) in 850 g of water?

SOLUTION

$$\text{molality} = \frac{\text{mol solute}}{\text{kg solvent}} \qquad \text{mol solute} = \frac{18.5 \text{ g}}{32.0 \text{ g/mol}} = 0.578 \text{ mol}$$

$$\text{kg solvent} = \frac{850 \text{ g}}{10^3 \text{ g/kg}} = 0.850 \text{ kg}$$

$$\text{molality} = \frac{0.578 \text{ mol}}{0.850 \text{ kg}} = \underline{\underline{0.680 \text{ m}}}$$

EXAMPLE 12-18

What is the boiling point of an aqueous solution containing 468 g of sucrose ($C_{12}H_{22}O_{11}$) in 350 g of water?

PROCEDURE

$$\Delta T_b = K_b m = 0.512 \text{ °C} \cdot \text{kg/mol} \times \frac{\text{mol solute}}{\text{kg solvent}}$$

SOLUTION

$$\text{mol solute} = \frac{468 \cancel{g}}{342 \cancel{g}/\text{mol}} = 1.37 \text{ mol}$$

$$\text{kg solvent} = \frac{350 \cancel{g}}{10^3 \cancel{g}/\text{kg}} = 0.350 \text{ kg}$$

$$\Delta T_b = 0.512 \text{ °C} \cdot \cancel{\text{kg/mol}} \times \frac{1.37 \cancel{\text{mol}}}{0.350 \cancel{\text{kg}}} = 2.0 \text{ °C}$$

Normal boiling point of water = 100.0 °C

Boiling point of the solution = 100.0 °C + ΔT_b = 100.0 + 2.0

$$= \underline{102.0 \text{ °C}}$$

EXAMPLE 12-19

What is the freezing point of the solution in Example 12-18?

PROCEDURE

$$\Delta T_f = K_f m = 1.86 \text{ °C} \cdot \text{kg/mol} \times \frac{\text{mol solute}}{\text{kg solvent}}$$

SOLUTION

$$\Delta T_f = 1.86 \text{ °C} \cdot \cancel{\text{kg/mol}} \times \frac{1.37 \cancel{\text{mol}}}{0.350 \cancel{\text{kg}}} = 7.3 \text{ °C}$$

Freezing point of water = 0.0 °C

Freezing point of the solution = 0.0 °C − ΔT_f = 0.0 °C − 7.3 °C

$$= \underline{-7.3 \text{ °C}}$$

Note that the liquid range was extended by over 9 °C for the solution.

Osmotic Pressure   An important phenomenon in the life process is osmosis. **Osmosis** *is the tendency for a solvent to move through a thin porous membrane from a dilute solution to a more concentrated solution.* The membrane is said to be **semipermeable,** which means that small solvent molecules can pass through but solute species cannot. Figure 12-11 illustrates osmosis. On the right is a pure solvent,

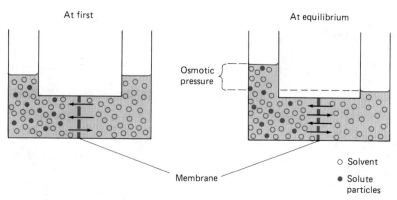

Figure 12-11 OSMOTIC PRESSURE. Osmosis causes dilution of the more concentrated solution.

and on the left, a solution. The two are separated by a semipermeable membrane. Solvent molecules can pass through the membrane in both directions, but the rate at which they diffuse to the right is lower because solute particles block some of the pores of the membrane. As a result, the water level rises on the left and drops on the right. This creates an increased pressure on the left, which eventually counteracts the osmosis, and equilibrium is established. *The extra pressure required to establish this equilibrium is known as the* **osmotic pressure.** Like other colligative properties, it is dependent on the concentration of the solute.

We see an example of the osmosis process whenever we leave our hands in a soapy water or saltwater solution. The movement of water molecules from the cells of our skin to the more concentrated solution causes them to be wrinkled. Pickles are wrinkled because the cells of the cucumber have been dehydrated by the salty brine solution. In fact, brine solutions preserve many foods because the concentrated solution of salt removes water from the cells of bacteria, thus killing the bacteria. Trees and plants obtain water from the absorption of water through the semipermeable membranes in their roots into the more concentrated solution inside the root cells. Osmosis has many important applications in addition to life processes. In Figure 12-11, if pressure greater than the osmotic pressure is applied on the left, reverse osmosis takes place and solvent molecules move from the solution to the pure solvent. This process is used in desalination plants that convert sea water (a solution) to drinkable water. This is important in areas of the world such as the Middle East, where there is a shortage of fresh water.

Electrolytes have a more pronounced effect on colligative properties than do nonelectrolytes. For example, one mole of NaCl dissolves in water to produce two moles of particles, one mole of Na^+ and one mole of Cl^-.

$$NaCl(s) \xrightarrow{H_2O} Na^+(aq) + Cl^-(aq)$$

Figure 12-12 COLLIGATIVE PROPERTIES. Antifreeze not only lowers the freezing point but raises the boiling point of the engine coolant. This salt lake does not dry up because of a low vapor pressure.

Thus, one mole of NaCl lowers the freezing point approximately twice as much as one mole of a nonelectrolyte. This effect is put to good use in the U.S. snow belt where sodium chloride is spread on snow and ice to cause melting even though the temperature is below freezing. Even more effective in melting ice is calcium chloride ($CaCl_2$). This compound produces three moles of ions ($Ca^{2+} + 2Cl^-$) per mole of solute and is therefore three times as effective per mole in lowering the freezing point as a nonelectrolyte. Calcium chloride is occasionally used on roads when the temperature is too low for sodium chloride to be effective. Electrolytes would be very effective as an antifreeze for the water in a radiator, but, unfortunately, their presence in water causes severe corrosion of most metals. (See Figure 12-12.)

■
SEE PROBLEMS 12-57 THROUGH 12-79.

12-9 CHAPTER REVIEW

Water, the most common and inexpensive chemical known, serves as the most effective medium for many chemical reactions. It does so because of its rather extensive ability to act as a solvent. Solvents and solutes mix homogeneously to form solutions.

Water is an effective solvent for both ionic compounds and polar covalent compounds. It can effectively counteract the ion–ion forces holding the crystal of an ionic compound together (lattice energy) by extensive ion–dipole interactions between water molecules and ions (hydration energy). Ionic compounds completely disperse into independent ions in solution. Many polar covalent compounds also dissolve in water. Some, such as HCl, dissociate into ions in solution. Others, such as methyl alcohol, dissolve without ion formation.

Not all ionic compounds dissolve to any extent in water and are referred to as insoluble. Table 12-2 provides information about the solubility of compounds of common cations and anions. Solubility also depends on temperature and the identity of the solute. Most solids and liquids become more soluble at higher temperatures, but gases become less soluble.

Whenever the cation and anion of an insoluble compound are mixed into the same solution, they join together to form a precipitate. Precipitation reactions, a type of double-replacement reaction, can be predicted by use of solubility rules such as those in Table 12-2. The equation representing a precipitation reaction can be shown in three ways:

$$\text{Molecular} \quad AB(aq) + CD(aq) \longrightarrow AD(s) + CB(aq)$$
$$\text{Total ionic} \quad A^+(aq) + B^-(aq) + C^+(aq) + D^-(aq) \longrightarrow$$
$$AD(s) + C^+(aq) + B^-(aq)$$
$$\text{Net ionic} \quad A^+(aq) + D^-(aq) \longrightarrow AD(s)$$

Besides mass of solute per 100 g of solvent there are additional ways of expressing concentration, each having its own use. The three presented in this text are percent by weight, molarity, and molality.

Percent by weight	Molarity	Molality
$\dfrac{\text{mass of solute}}{\text{mass of solvent}} \times 100\%$	$M = \dfrac{\text{mol solute}}{\text{L solution}}$	$m = \dfrac{\text{mol solute}}{\text{kg solvent}}$

Since molarity relates volume of a solution to moles of solute, it can be incorporated into the general scheme for stoichiometry problems along with mass of a compound (Chapter 9) and volume of a gas (Chapter 10) as follows:

	Given			*Requested*	
			equation factor		
	volume $\longrightarrow$ moles	$\xrightarrow{}$	moles $\longrightarrow$	volume	
	(of solution)			(of solution)	

In the final section, we studied the physical properties of solutions. The first property is the ability of water to conduct electricity. Water itself is a nonconductor but solutes that produce ions in solution alter this property. Aqueous solutions of ions are conductors. A dissolved substance may be classified as a nonelectrolyte, a strong electrolyte, or a weak electrolyte:

Type of Solute	Property in Water	Reason	Examples
Nonelectrolyte	Solution is a nonconductor of electricity.	Ions are not formed in solution.	$C_{12}H_{22}O_{11}$ (sugar) CH_3OH (methyl alcohol)
Strong electrolyte	Solution is a good conductor of electricity.	Ions are formed in solution.	NaCl K_2SO_4
Weak electrolyte	Solution is a weak conductor of electricity.	Limited amounts of ions are formed in solution.	HF HNO_2

The alteration of several other properties of water depends only on the amount of solute present. These are known as colligative properties and are summarized as follows:

Property	Effect	Result
Vapor pressure	Lowered	Solutions evaporate slower than pure solvents.
Boiling point	Raised	Solutions boil at higher temperatures than pure solvents.
Melting point	Lowered	Solutions freeze at lower temperatures than pure solvents.
Osmotic pressure	Raised	Solvent from dilute solutions diffuses through a semipermeable membrane into concentrated solutions.

■ EXERCISES

AQUEOUS SOLUTIONS

12-1 Write equations illustrating the solution of each of the following ionic compounds in water.
(a) Na_2S
(b) Li_2SO_4
(c) $K_2Cr_2O_7$
(d) CaS
(e) 2 mol of $(NH_4)_2S$
(f) 4 mol of $Ba(OH)_2$

12-2 Write equations illustrating the solution of each of the following ionic compounds in water.
(a) $Ca(ClO_3)_2$
(b) $CsBr$
(c) $AlCl_3$
(d) 2 mol of Cs_2SO_3

12-3 What ions and how many moles of each are present when the following dissolve in water?
(a) 1 mol of $Al_2(SO_4)_3$
(b) 2 mol of $Mg_3(PO_4)_2$

12-4 Formaldehyde (H_2CO) dissolves in water without formation of ions. Write the Lewis structure of formaldehyde and show what type of interactions between solute and solvent are involved.

12-5 Nitric acid is a covalent compound that dissolves in water to form ions as does HCl. Write the equation illustrating its solution in water.

SOLUBILITY OF IONIC COMPOUNDS

12-6 Referring to Table 12-2, determine which of the following compounds are insoluble in water.
(a) Na_2S
(b) $PbSO_4$
(c) $MgSO_3$
(d) Ag_2O
(e) $(NH_4)_2S$
(f) HgI_2

12-7 Referring to Table 12-2 determine which of the following compounds are insoluble in water.
(a) NiS
(b) Hg_2Br_2
(c) $Al(OH)_3$
(d) Rb_2SO_4
(e) CaS
(f) $BaCO_3$

TEMPERATURE AND SOLUBILITY

12-8 Referring to Figure 12-4, which of the following compounds is most soluble at 10 °C: NaCl, KCl or Li_2SO_4? Which is more soluble at 70 °C?

12-9 What mass of each of the following dissolves in 250 g of H_2O at 60 °C: KBr, KCl, and Li_2SO_4?

12-10 Tell whether each of the following solutions is saturated, unsaturated, or supersaturated (all are in 100 g of H_2O.)
(a) 40 g of KNO_3 at 40 °C
(b) 40 g of KNO_3 at 20 °C
(c) 75 g of KBr at 80 °C
(d) 20 g of NaCl at 40 °C

12-11 A 200-g sample of water is saturated with KNO_3 at 50 °C. What mass of KNO_3 forms as a precipitate if the solution is cooled to the freezing point of water?

12-12 A 500-mL portion of water is saturated with Li_2SO_4 at 0 °C. What happens if the solution is heated to 100 °C?

PRECIPITATION REACTIONS AND IONIC EQUATIONS

12-13 Write the balanced molecular equation for any reaction that occurs when the following solutions are mixed.
(a) KI and $Pb(C_2H_3O_2)_2$
(b) $AgClO_3$ and KNO_3
(c) $Sr(ClO_3)_2$ and $Ba(OH)_2$
(d) BaS and $Hg_2(NO_3)_2$
(e) $FeCl_3$ and KOH

12-14 Write the balanced molecular equation for any reaction that occurs when the following solutions are mixed.
(a) $Ba(C_2H_3O_2)_2$ and Na_2SO_4
(b) $NaClO_4$ and $Pb(NO_3)_2$
(c) $Mg(NO_3)_2$ and Na_3PO_4
(d) SrS and NiI_2

12-15 Write tht total ionic equation and the net ionic equation for each of the following reactions.
(a) $K_2S(aq) + Pb(NO_3)_2(aq) \rightarrow$
$$PbS(s) + 2KNO_3(aq)$$
(b) $(NH_4)_2CO_3(aq) + CaCl_2(aq) \rightarrow$
$$CaCO_3(s) + 2NH_4Cl(aq)$$
(c) $2AgClO_4(aq) + Na_2CrO_4(aq) \rightarrow$
$$Ag_2CrO_4(s) + 2NaClO_4(aq)$$

12-16 Write the total ionic equation for any reaction that occurred in Problem 12-13.

12-17 Write the total ionic equation for any reaction that occurred in Problem 12-14.

12-18 Write the net ionic equation for any reaction that occurred in Problem 12-13.

12-19 Write the net ionic equation for any reaction that occurred in Problem 12-14.

*12-20 Write balanced molecular equations indicating how the following ionic compounds can be prepared by a precipitation reaction using any other ionic compounds. In some cases the desired compound may be soluble and must be recovered by vaporizing the solvent water after removal of a precipitate.
(a) $CuCO_3$ (d) NH_4NO_3
(b) $PbSO_3$ (e) $KC_2H_3O_2$
(c) Hg_2I_2

PERCENT BY WEIGHT

12-21 What is the percent by weight of solute in a solution made by dissolving 9.85 g of $Ca(NO_3)_2$ in 650 g of water?

12-22 What is the percent by weight of solute if 14.15 g of NaI is mixed with 75.55 g of water?

12-23 A solution is 10.0% by weight $NaOH$. How many moles of $NaOH$ are dissolved in 150 g of solution?

12-24 A solution contains 15.0 g of NH_4Cl in water and is 8.50% NH_4Cl. What is the mass of water present?

12-25 A solution is 23.2% by weight KNO_3. What mass of KNO_3 is present in each 100 g of H_2O?

12-26 A solution contains 1 mol of $NaOH$ dissolved in 9 mol of ethyl alcohol (C_2H_5OH). What is the percent by weight $NaOH$?

12-27 Blood contains 10 mg of calcium ion in 100 g of blood serum (solution). What is this concentration in ppm?

12-28 A high concentration of mercury in fish is 0.5 ppm. What mass of mercury is present in each kilogram of fish? What is this concentration in ppb?

MOLARITY

12-29 What is the molarity of a solution made by dissolving 2.44 mol of $NaCl$ in enough water to make 4.50 L of solution?

12-30 Fill in the blanks.

Solute	M	Amount of Solute	Volume of Solution
(a) Ki	____	2.40 mol	2.75 L
(b) C_2H_5OH	____	26.5 g	410 mL
(c) $NaC_2H_3O_2$	0.255	3.15 mol	____ L
(d) $LiNO_2$	0.625	____ g	1.25 L
(e) $BaCl_2$	____	0.250 mol	850 mL
(f) Na_2SO_3	0.054	____ mol	0.45 L
(g) K_2CO_3	0.345	14.7 g	____ mL
(h) $LiOH$	1.24	____ g	1650 ml
(i) H_2SO_4	0.905	0.178 g	____ mL

12-31 What is the molarity of a solution of 345 g of Epsom salts ($MgSO_4 \cdot 7H_2O$) in 7.50 L of solution?

12-32 What is the mass of $CaCl_2$ present in 2.58 L that has a concentration of 0.0784 M?

12-33 What is the volume in liters that contains 37.5 g of KOH with a concentration of 0.250 M?

12-34 What is the molarity of a solution made by dissolving 2.50×10^{-4} g of baking soda ($NaHCO_3$) in enough water to make 2.54 mL of solution?

12-35 What is the molarity of the hydroxide ion and the barium ion if 13.5 g of $Ba(OH)_2$ is dissolved in enough water to make 475 mL of solution?

12-36 A solution is 25.0% by weight calcium nitrate and has a density of 1.21 g/mL. What is its molarity?

*12-37 A solution of concentrated NaOH is 16.4 M. If the density of the solution is 1.43 g/mL, what is the percent by weight NaOH?

*12-38 Concentrated nitric acid is 70.0% HNO_3 and 14.7 M. What is the density of the solution?

DILUTION

12-39 What volume of 4.50 M H_2SO_4 should be diluted with water to form 2.50 L of 1.50 M acid?

12-40 If 450 mL of a certain solution is diluted to 950 mL with water to form a 0.600 M solution, what was the molarity of the original solution?

12-41 One liter of a 0.250 M solution of NaOH is needed. The only available solution of NaOH is a 0.800 M solution. Describe how to make the desired solution.

12-42 What is the volume in liters of a 0.440 M solution if it was made by dilution of 250 mL of a 1.25 M solution?

12-43 What is the molarity of a solution made by diluting 3.50 L of a 0.200 M solution to a volume of 5.00 L?

12-44 What volume of water in milliliters should be added to 1.25 L of 0.860 M HCl so that its molarity will be 0.545?

*12-45 What volume in milliliters of *pure* acetic acid should be used to make 250 mL of 0.200 M $HC_2H_3O_2$? (The density of the pure acid is 1.05 g/mL.)

12-46 What volume of water in milliliters should be *added* to 400 mL of a solution containing 35.0 g of KBr to make a 0.100 M KBr solution?

*12-47 What would be the molarity of a solution made by mixing 150 mL of 0.250 M HCl with 450 mL of 0.375 M HCl?

STOICHIOMETRY INVOLVING SOLUTIONS

12-48 Given the reaction

$$3KOH(aq) + CrCl_3(aq) \longrightarrow Cr(OH)_3(s) + 3KCl(aq)$$

what mass of $Cr(OH)_3$ would be produced if 500 mL of 0.250 M KOH were added to a solution containing excess $CrCl_3$?

12-49 Given the reaction

$$2KCl(aq) + Pb(NO_3)_2(aq) \longrightarrow PbCl_2(s) + 2KNO_3(aq)$$

what mass of $Pb(NO_3)_2$ is required to react with 1.25 L of 0.550 M KCl?

12-50 Given the reaction

$$Al_2(SO_4)_3(aq) + 3BaCl_2(aq) \longrightarrow 3BaSO_4(s) + 2AlCl_3(aq)$$

what mass of $BaSO_4$ is produced from 650 mL of 0.320 M $Al_2(SO_4)_3$?

12-51 Given the reaction

$$3Ba(OH)_2(aq) + 2Al(NO_3)_3(aq) \longrightarrow 2Al(OH)_3(s) + 3Ba(NO_3)_2(aq)$$

what volume of 1.25 M $Ba(OH)_2$ is required to produce 265 g of $Al(OH)_3$?

12-52 Given the reaction

$$2AgClO_4(aq) + Na_2CrO_4(aq) \longrightarrow Ag_2CrO_4(s) + 2NaClO_4(aq)$$

what volume of a 0.600 M solution of $AgClO_4$ is needed to produce 160 g of Ag_2CrO_4?

12-53 Given the reaction

$$3Ca(ClO_3)_2(aq) + 2Na_3PO_4(aq) \longrightarrow$$
$$Ca_3(PO_4)_2(s) + 6NaClO_3(aq)$$

what volume of a 2.22 M solution of Na_3PO_4 is needed to react with 580 mL of a 3.75 M solution of $Ca(ClO_3)_2$?

12-54 Consider the reaction.

$$2HNO_3(aq) + 3H_2S(aq) \longrightarrow$$
$$2NO(g) + 3S(s) + 4H_2O$$

(a) What volume of 0.350 M HNO_3 will completely react with 275 mL of 0.100 M H_2S?

(b) What volume of NO gas measured at 27 °C and 720 torr will be produced from 650 mL of 0.100 M H_2S solution?

*12-55 Given the reaction

$$2NaOH(aq) + MgCl_2(aq) \longrightarrow$$
$$Mg(OH)_2(s) + 2NaCl(aq)$$

what mass of $Mg(OH)_2$ would be produced by mixing 250 mL of 0.240 M NaOH with 400 mL of 0.100 M $MgCl_2$?

*12-56 Given the reaction

$$CO_2(g) + Ca(OH)_2(aq) \longrightarrow$$
$$CaCO_3(s) + H_2O$$

what is the molarity of a 1.00-L solution of $Ca(OH)_2$ that would completely react with 10.0 L of CO_2 measured at 25.0 °C and 0.950 atm?

PROPERTIES OF SOLUTIONS

12-57 Three hypothetical binary compounds dissolve in water. AB is a strong electrolyte, AC is a weak electrolyte, and AD is a nonelectrolyte. Describe the extent to which each of these solutions conducts electricity and how each compound exists in solution.

12-58 Chlorous acid ($HClO_2$) is a weak electrolyte and perchloric acid ($HClO_4$) is a strong electrolyte in water. Write equations illustrating the difference in behavior of these two polar covalent molecules.

12-59 What is the molality of a solution made by dissolving 25.0 g of NaOH in (a) 250 g of water? (b) 250 g of alcohol (C_2H_5OH)?

12-60 What is the molality of a solution made by dissolving 1.50 kg of KCl in 2.85 kg of water?

*12-61 What is the molality of an aqueous solution that is 10.0% by weight $CaCl_2$?

12-62 What mass of NaOH is present in 550 g of water if the concentration is 0.720 m?

12-63 What mass of water is present in a 0.430 m solution containing 2.58 g of CH_3OH?

*12-64 A 1.00 m KBr solution has a mass of 1.00 kg. What mass of water is present?

12-65 What is the freezing point of a 0.20 m aqueous solution of a non-electrolyte?

12-66 What is the boiling point of a 0.45 m aqueous solution of a nonelectrolyte?

12-67 Ethylene glycol ($C_2H_6O_2$) is used as an antifreeze. What mass of ethylene glycol should be added to 5.00 kg of water to lower the freezing point to −5.0 °C? (Ethylene glycol is a nonelectrolyte.)

12-68 What is the boiling point of the solution in Problem 12-67?

12-69 Methyl alcohol can also be used as an antifreeze. What mass of methyl alcohol (CH_3OH) must be added to 5.00 kg of water to lower its freezing point to −5.0 °C?

12-70 What is the molality of an aqueous solution that boils at 101.5 °C?

12-71 What is the boiling point of a 0.15 m solution of a solute in liquid benzene? (For benzene, $K_b = 2.53$ and the boiling point of pure benzene is 80.1 °C.)

12-72 What is the boiling point of a solution of 75.0 g of naphthalene ($C_{10}H_8$) in 250 g of benzene? (See Problem 12-71.)

12-73 What is the freezing point of a solution of 100 g of CH_3OH in 800 g benzene? (For benzene, $K_f = 5.12$ and the freezing point of pure benzene is 5.5 °C.)

12-74   What is the freezing point of a 10.0% by weight solution of CH_3OH in benzene? (See Problem 12-73.)

*12-75   Give the freezing point of each of the following in 100 g of water.
(a) 10.0 g of CH_3OH
(b) 10.0 g of NaCl
(c) 10.0 g of $CaCl_2$

12-76   When one is immersed in the salty ocean water for an extended period one gets unusually thirsty. Explain.

12-77   Dehydrated fruit is wrinkled and shriveled up. When put in water, the fruit expands and becomes smooth again. Explain.

12-78   Explain how pure water can be obtained from a solution without boiling.

12-79   In industrial processes it is often necessary to concentrate a dilute solution (much harder than diluting a concentrated solution). Explain how the principle of reverse osmosis can be applied.

A DYING FOREST
It is suspected that the presence of acids in rain has had a
destructive impact on certain forests in America and Europe.

ACIDS,
BASES,
AND SALTS

PURPOSE OF CHAPTER

In Chapter 13 we classify substances as acids, bases, or salts, and describe chemical reactions illustrating this behavior.

OBJECTIVES FOR CHAPTER

After completion of this chapter, you should be able to:

1 Apply the Arrhenius definition to identify compounds as acids or bases and write equations illustrating this behavior. (13-1)

2 Give the names and formulas of some common acids and bases derived from specified anions or cations. (13-1)

3 Write equations for neutralization reactions and identify the salts formed. (13-2)

4 Write equations for the stepwise neutralization of a polyprotic acid by a base. (13-2)

5 Distinguish between the behavior of a strong acid and a weak acid in water. (13-3)

13

6 Describe the dynamic equilibrium involved in the partial ionization of a weak acid or a weak base in water. (13-3)

7 Calculate $[OH^-]$ from a specified $[H_3O^+]$ and vice versa by use of K_w. (13-4)

8 Distinguish among acidic, basic, and neutral solutions in terms of $[H_3O^+]$ and $[OH^-]$. (13-4)

9 Convert $[H_3O^+]$ to pH and vice versa. (13-5)

10 Distinguish among acidic, basic, and neutral solutions in terms of pH. (13-5)

11 Determine the conjugate base of a specified acid or the conjugate acid of a specified base. (13-6)

12 Write specified proton exchange reactions by application of the Brønsted–Lowry definition. (13-6)

13 Use a table of acid–base strengths to predict favorable proton exchange reactions. (13-7)

14 From the location of an acid in a table of acid–base strengths predict whether its ionization is complete, is partial, or does not occur. (13-7)

15 From the location of an ion in a table of acid–base strengths predict whether its hydrolysis is complete, is partial, or does not occur. (13-7)

16 Predict whether the solutions of specified salts are acidic, basic, or neutral. (13-7)

17 Describe how a solution acts as a buffer. (13-8)

18 Determine whether specified solutions may act as buffers. (13-8)

19 Write equations illustrating the reactions of oxides as acids or bases. (13-9)

It should be obvious by now that chemists have a hopeless habit of classifying or grouping as much information as possible. But with more than 100 elements and millions of compounds and chemical reactions, any time we can detect a common thread in any mass of information it does make sense to take advantage of it. Thus we've put elements into groups and periods, classified chemical reactions into types, and grouped compounds as ionic or covalent, electrolytes or nonelectrolytes, and soluble or insoluble. Classification is as helpful in chemistry as it is in human relations. If we classify a fellow human being as a "nerd" or a "fox," it means that that person's appearance or behavior falls into certain expected patterns (in our minds, anyway). Likewise, if we classify a substance as an "acid" or a "base," it means that the compound has a certain behavior common to other compounds in that particular classification. The classification of substances as acids, bases, and salts is among the oldest classifications in chemistry and is still one of the most important. These compounds and their behavior are the subjects of this chapter.

Our first goal is simply to identify acids and bases and understand the reasons why they are classified as such. We then do some finer classifying as to acid and base strength and how the pH scale relates to this strength. Finally, we will look at acids and bases through a slightly wider-angle lens so as to expand our understanding of acid–base behavior.

Background for this chapter includes the following:

1 Ionic and covalent compounds (Sections 6-4 and 6-5)
2 Nomenclature of acids (Section 7-5)
3 Ionic and net ionic equations (Section 12-3)
4 Molarity (Section 12-5)

Figure 13-1   ZINC AND LIMESTONE IN ACID. Both zinc (right) and limestone ($CaCO_3$) (left) react with acid to liberate a gas.

13-1 THE NATURE OF ACIDS AND BASES

More than half a millennium ago in the Middle Ages alchemists worked with a curious group of chemicals that seemed to have spectacular if not downright magical powers. Harmless-looking aqueous solutions of these chemicals could make the strongest metals disappear into solution with dramatic fizzing and hissing. Even gold, the noblest and least reactive metal, would dissolve in a combination of these compounds known as *aqua regia* (royal water). These awesome compounds were known as acids (from the Latin *acidus*, meaning "sour," or *acetum*, meaning "vinegar"). Obviously, acids were originally classified as such because of their common chemical behavior rather than some unique composition. We first describe the long-known chemical behavior of acids and bases and then move to a straightforward model that explains the chemical reactions accounting for this behavior.

Acids are compounds that:

1   Taste sour*
2   Dissolve certain metals (e.g., Zn) with the liberation of a gas (see Figure 13-1)
3   Cause certain organic dyes to change color (an example is litmus, which turns from blue to red in acids)
4   Dissolve limestone ($CaCO_3$) with the liberation of a gas (see Figure 13-1)
5   React with bases to form salts and water

---

* Tasting is certainly not a recommended laboratory procedure. It is all right for vinegar but not for battery acid or caustic lye.

Bases are compounds that:

1 Taste bitter*
2 Are slippery or soapy feeling
3 Dissolve oils and grease
4 Cause certain organic dyes to change color (red litmus turns blue in bases)
5 React with acids to form salts and water

As one can see, classification of a substance as either an acid or a base can be achieved with simple laboratory tests. It wasn't until a little more than 100 years ago, however, that a currently accepted model explaining this behavior was advanced. *In 1884, the Swedish chemist Svante Arrhenius suggested that* **acids** *are substances that produce H^+ ions and* **bases** *are substances that produce OH^- ions in aqueous solution.*

First let us consider the action of acids. Four common acids that most of us recognize are hydrochloric acid (also known as muriatic acid and available in most hardware stores), sulfuric acid (oil of vitriol or battery acid), acetic acid (the source of the sour taste of vinegar), and carbonic acid (carbonated water). The following equations illustrate how these four acids produce the common ingredient, the H^+ ion, but different anions.

Muriatic or hydrochloric acid	$HCl \longrightarrow H^+ + Cl^-$
Oil of vitriol or sulfuric acid	$H_2SO_4 \longrightarrow H^+ + HSO_4^-$
Acetic acid in vinegar	$HC_2H_3O_2 \rightleftharpoons H^+ + C_2H_3O_2^-$
Carbonated water or carbonic acid	$H_2CO_3 \rightleftharpoons H^+ + HCO_3^-$

These acids are not ionic compounds when pure. They form ions only when mixed with water. In fact, a chemical reaction occurs between the acid molecules and the H_2O molecules that results in removal of a H^+ ion from the rest of the molecule. As discussed in Chapters 6 and 12, acids (such as HCl) and H_2O are polar covalent compounds. When HCl dissolves in water, there is a dipole–dipole electrostatic attraction between the negative dipole of one molecule and the positive dipole of the other as illustrated in Figure 13-2. (Actually, there are many more H_2O molecules involved with each HCl than those shown in the figure.)

A tug-of-war develops. On one side the H_2O's pull on the HCl molecule, attempting to split the molecule into a H^+ ion and a Cl^- ion. On the other side, the HCl covalent bond tries mightily to hold the molecule together. For HCl, the H_2O molecules are clearly the winners, since the interaction is strong enough to break the bonds of all dissolved HCl molecules. Therefore, HCl is an acid since an H^+ is produced in the solution as a result of the **dissociation** (breaking apart) of HCl. (Since dissociation produces ions in this case, the process is also called **ionization.**)

To illustrate the importance of water in the ionization process, the reaction of an acid with water can be represented as

$$HCl + H_2O \longrightarrow H_3O^+(aq) + Cl^-(aq)$$

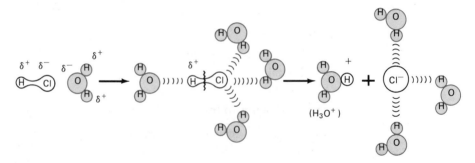

Figure 13-2   INTERACTION OF HCl AND H_2O. The dipole–dipole interaction between H_2O and HCl leads to breaking of the HCl bond.

Instead of H^+, the acid species is often represented as H_3O^+, which is known as **the hydronium ion.** The hydronium ion is simply a representation of the H^+ ion in a hydrated form. The acid species is represented as H_3O^+ rather than H^+ because it is somewhat closer to what is believed to be the actual species. In fact, the nature of H^+ in aqueous solution is even more complex than H_3O^+ (i.e., $H_5O_2^+$, $H_7O_3^+$, etc.). In any case, the acid species is represented as H^+, $H^+(aq)$, or H_3O^+ (aq) depending on the convenience of the particular situation. *Just remember that all refer to the same species in aqueous solution.* If $H^+(aq)$ is used, it should be understood that it is not just a bare proton in aqueous solution but is associated with water molecules (it is hydrated).

The common property of acids is that they produce $H^+(aq)$ ions in aqueous solution. It is this ion that is responsible for the common chemical reactions associated with acids. The following equations focus on the role of the H^+ ion in the reactions of acids with metals, limestone, and bases. The anion of the acid is not included in the first two reactions.

1   Acids react with zinc and give off a gas (H_2):

$$Zn(s) + \underline{2H^+(aq)} \longrightarrow Zn^{2+}(aq) + H_2(g)$$

2   Acids react with limestone ($CaCO_3$) and give off a gas (CO_2):

$$CaCO_3(s) + \underline{2H^+(aq)} \longrightarrow Ca^{2+}(aq) + H_2O + CO_2(g)$$

3   Acids react with bases:

$$HCl(aq)[H^+(aq) + Cl^-(aq)] + NaOH(aq) \longrightarrow NaCl(aq) + H_2O$$

The last reaction is of prime importance and is discussed in more detail in the next section.

Other molecules containing hydrogen may not be acidic in H_2O because the hydrogens are (1) too strongly attached to the rest of the molecule to be ionized or (2) not polar enough to create a strong interaction with H_2O. Sometimes both reasons are important. The hydrogens in CH_4 and NH_3 and the three hydrogens on the carbon in acetic acid are examples of hydrogens that do not ionize in aqueous solution to form H^+ ions.

Three H's on the C in acetic acid do not ionize (the C—H bond is essentially nonpolar).

$$H-\overset{\overset{\displaystyle H}{|}}{\underset{\underset{\displaystyle H}{|}}{C}}-C\overset{\displaystyle \ddot{O}\colon \ \delta^-}{\underset{\displaystyle \underset{\overset{\displaystyle |}{\underset{\displaystyle \ddot{O}-H}{}}}{\delta^+}}{\diagup\diagdown}}$$

The O—H bond is polar, so the H can be ionized.

Now let us consider the counterpart to acids known as bases. These compounds produce OH^- ions in water, forming what is known as basic or alkaline solutions. Some commonly known bases are sodium hydroxide (also known as caustic soda or lye, used to unclog drains), potassium hydroxide (caustic potash), calcium hydroxide (slaked lime), and ammonia (used as a cleaner for grease spots). Except for ammonia these compounds are all solid ionic compounds. Solution in water simply releases the OH^- ion into the aqueous medium:

$$NaOH(s) \xrightarrow{xH_2O} Na^+(aq) + OH^-(aq)$$
$$Ba(OH)_2(s) \xrightarrow{xH_2O} Ba^{2+}(aq) + 2OH^-(aq)$$

■
SEE PROBLEMS
13-1 THROUGH
13-5.

The action of ammonia (NH_3) as a base is somewhat different from that of the ionic hydroxides and is discussed in Section 13-3.

13-2 NEUTRALIZATION AND SALTS

One of the important characteristics of acids and bases is that they react with each other to produce salts and water. Such a reaction is known as a **neutralization reaction.** Perhaps the most familiar neutralization involves the reaction of hydrochloric acid with sodium hydroxide:

$$HCl(aq) + NaOH(aq) \longrightarrow NaCl(aq) + H_2O$$

In a neutralization reaction the active ingredient of acids (H^+ ions) combines with the active ingredient of bases (OH^- ions) to form the molecular compound H_2O. *The salt (in this case NaCl) is what is left over. It is composed of the cation from the base (Na^+) and the anion from the acid (Cl^-).* Although acids or bases by themselves are dangerous and corrosive compounds, mixing the two in stoichiometric amounts tames the beast in both, leaving a harmless solution of a salt. *The species that cause the characteristic behavior of acids and of bases combine to form ordinary water.

A neutralization reaction is a type of double-replacement reaction first described in Section 9-2. In the previous chapter, we discussed another type of double-replacement reaction known as a precipitation reaction, in which two ions combine to form an insoluble ionic compound.

---

* Don't try this without supervision: Much heat evolution, with boiling and splattering, can occur.

The neutralization reaction can also be written in total ionic and net ionic forms:

$$Na^+(aq) + OH^-(aq) + H^+(aq) + Cl^-(aq) \longrightarrow Na^+(aq) + Cl^-(aq) + H_2O$$

$$H^+(aq) + OH^-(aq) \longrightarrow H_2O$$

The word "salt" is often thought of as referring to just one substance, sodium chloride, as formed in the previous reaction. Actually, a salt can result from many different combinations of anions and cations from a variety of neutralizations. The following neutralization reactions illustrate the formation of some other salts.

Acid	+	Base	$\longrightarrow$	Salt	+	Water
1 $HNO_3(aq)$	+	$KOH(aq)$	$\longrightarrow$	$KNO_3(aq)$	+	H_2O
2 $H_2SO_4(aq)$	+	$2NaOH(aq)$	$\longrightarrow$	$Na_2SO_4(aq)$	+	$2H_2O$
3 $2HClO_4(aq)$	+	$Ca(OH)_2(aq)$	$\longrightarrow$	$Ca(ClO_4)_2(aq)$	+	$2H_2O$
4 $2H_3PO_4(aq)$	+	$3Ba(OH)_2(aq)$	$\longrightarrow$	$Ba_3(PO_4)_2(s)$	+	$6H_2O$

Hydrochloric acid (HCl) and nitric acid (HNO_3) are examples of acids that produce *one mole of H^+ ions in solution per mole of acid.* Such acids are known as **monoprotic acids.** On the other hand, note in example 2 above that one mole of sulfuric acid (H_2SO_4) requires two moles of sodium hydroxide for complete neutralization. Sulfuric acid is an example of a **diprotic acid,** *which means that it can produce two moles of H^+ per mole of acid:*

$$H_2SO_4 \longrightarrow 2H^+(aq) + SO_4{}^{2-}(aq)$$

There are several examples of **triprotic acids** *(three potential H^+ ions per molecule),* but the most important one is phosphoric acid (H_3PO_4).

The alkaline earth hydroxides are analogous to the diprotic acids except that they produce two OH^- ions per formula unit dissolved, for example, the solution of calcium hydroxide in water:

$$Ca(OH)_2 \longrightarrow Ca^{2+}(aq) + 2OH^-(aq)$$

Note in example 3 that two moles of a monoprotic acid such as $HClO_4$ is required to neutralize one mole of calcium hydroxide.

The neutralizations of sulfuric acid (example 2 above) and calcium hydroxide (example 3 above) are illustrated further with the total ionic equations and the net ionic equations.

$$2H^+(aq) + \cancel{SO_4{}^{2-}}(aq) + \cancel{2Na^+(aq)} + \cancel{2OH^-(aq)} \longrightarrow$$
$$\cancel{2Na^+(aq)} + \cancel{SO_4{}^{2-}} + \cancel{2}H_2O$$

$$H^+(aq) + OH^-(aq) \longrightarrow H_2O$$

$$\cancel{2}H^+(aq) + \cancel{2ClO_4{}^-}(aq) + \cancel{Ca^{2+}(aq)} + \cancel{2OH^-}(aq) \longrightarrow$$
$$\cancel{Ca^{2+}(aq)} + \cancel{2ClO_4{}^-} + \cancel{2}H_2O$$

$$H^+(aq) + OH^-(aq) \longrightarrow H_2O$$

Polyprotic acids *(acids with more than one acidic hydrogen)* can be partially neutralized. The complete ionization of H_2SO_4 represented previously actually takes place in two steps.

$$H_2SO_4 \longrightarrow H^+(aq) + HSO_4^-(aq) \quad \text{(first ionization of } H_2SO_4\text{)}$$
$$HSO_4^-(aq) \rightleftharpoons H^+(aq) + SO_4^{2-}(aq) \quad \text{(second ionization of } H_2SO_4\text{)}$$

If *one mole* of NaOH is added to *one mole* of H_2SO_4, only the H^+ from the first ionization is neutralized:

$$H_2SO_4(aq) + NaOH(aq) \longrightarrow NaHSO_4(aq) + H_2O$$
$$H^+(aq) + \cancel{HSO_4^-(aq)} + \cancel{Na^+(aq)} + OH^-(aq) \longrightarrow$$
$$\cancel{Na^+(aq)} + \cancel{HSO_4^-(aq)} + H_2O$$
$$H^+(aq) + OH^-(aq) \longrightarrow H_2O$$

Note that the salt formed in the partial neutralization ($NaHSO_4$) still contains a potentially acidic hydrogen on the anion. The acid ionization of the HSO_4^- ion (hydrogen sulfate or bisulfate ion) is shown above by the second ionization of H_2SO_4. Such a salt is known as an **acid salt**. *Acid salts are ionic compounds containing one or more acidic hydrogens on the anion.* They result from the partial neutralization of a diprotic or triprotic acid. When additional base is added to a solution of an acid salt, further neutralization takes place. For example, neutralization of a solution of $NaHSO_4$ is illustrated as follows:

$$NaHSO_4(aq) + NaOH(aq) \longrightarrow Na_2SO_4(aq) + H_2O$$

The neutralization of the triprotic acid H_3PO_4 with KOH can occur in three steps. Consider the reaction when 1 mol of KOH is added to 1 mol of H_3PO_4:

1
$$H{-}PO_4 \, (aq) + KOH(aq) \longrightarrow KH_2PO_4(aq) + H_2O$$

If 1 mol of KOH is then added to the KH_2PO_4 solution, reaction 2 occurs:

2
$$K \quad PO_4 (aq) + KOH(aq) \longrightarrow K_2HPO_4(aq) + H_2O$$

Finally, by adding 1 mol of KOH to the K_2HPO_4 solution, reaction 3 occurs:

3
$$K_2 \, \text{(H)}{-}\text{(} PO_4 \text{)}(aq) + KOH \longrightarrow K_3PO_4(aq) + H_2O$$

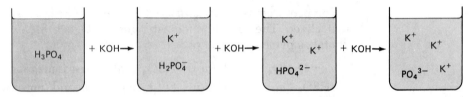

Figure 13-3   NEUTRALIZATION OF H_3PO_4. The hydrogens of H_3PO_4 can be neutralized one at a time.

Note that each additional OH^- removes one H^+ from the acid or acid salt. The total reaction, which is the algebraic sum of all three reactions, is

$$H_3PO_4(aq) + 3KOH(aq) \longrightarrow K_3PO_4(aq) + 3H_2O$$

This is the reaction that occurs regardless of whether the KOH is added one mole at a time or all three moles at once. In these reactions, the H_3PO_4 is identified as an acid, KH_2PO_4 and K_2HPO_4 as acid salts, and K_3PO_4 as a normal salt. (See Figure 13-3.)

EXAMPLE 13-1

Identify each of the following as either an acid, a base, a normal salt, or an acid salt:

(a) $KClO_4$,   **(b)** H_2S,   **(c)** $Ba(HSO_3)_2$,   **(d)** $Al(OH)_3$.

ANSWER

(a)   normal salt
(b)   acid
(c)   acid salt
(d)   base

EXAMPLE 13-2

Write the balanced equation in molecular form illustrating the complete neutralization of $Al(OH)_3$ with H_2SO_4.

PROCEDURE

Each mole of base produces *3 mol* of OH^-; each mole of acid produces 2 *mol* of H^+. Therefore, 2 mol of base (producing 6 mol of OH^-) reacts with 3 mol of acid (producing 6 mol of H^+) for complete neutralization.

ANSWER

$$2Al(OH)_3 + 3H_2SO_4 \longrightarrow Al_2(SO_4)_3 + 6H_2O$$

EXAMPLE 13-3

Write the balanced molecular equation illustrating the reaction of 1 mol of H_3PO_4 with 1 mol of $Ca(OH)_2$.

PROCEDURE

One mole of base produces 2 mol of OH^-, which reacts with 2 mol of H^+. The removal of 2 mol of H^+ from 1 mol of H_3PO_4 leaves the HPO_4^{2-} ion in solution.

ANSWER

$$H_3PO_4 + Ca(OH)_2 \longrightarrow CaHPO_4 + 2H_2O$$

■
SEE PROBLEMS 13-6 THROUGH 13-15.

13-3 THE STRENGTHS OF ACIDS AND BASES

Most of us enjoy the sour taste provided by the vinegar on a salad. The tangy taste of a carbonated beverage is also pleasurable. On the other hand, we must handle battery (sulfuric) acid with extreme caution. Battery acid would not be very appetizing on a salad as it would rapidly dissolve the lettuce as well as the fork and maybe even the bowl. The difference in behavior between acetic and sulfuric acids is not necessarily related to concentration. For example, even a concentrated solution of acetic acid (1.0 *M*) is still not nearly as corrosive as a comparatively dilute solution of sulfuric acid (e.g., 0.10 *M*).

The difference in the degree of acid behavior between acetic and sulfuric acid relates to their respective acid *strengths*. Acid strength in turn relates to the concentration of hydronium ions produced in solution.

Sulfuric and hydrochloric acids are known as strong acids. **Strong acids** *are essentially 100% ionized in aqueous solutions.* Thus, strong acids are also strong electrolytes. The ionization of HCl is illustrated in the following equation.

$$HCl(g) + H_2O \longrightarrow H_3O^+(aq) + Cl^-(aq)$$

Besides HCl and H_2SO_4 other common strong acids are HNO_3, $HClO_4$, HBr, and HI. Most other common acids behave as weak acids in water. **A weak acid** *is only partially ionized (usually less than 5% at typical molar concentrations).* Because of the small concentration of ions, weak acids behave as weak electrolytes in water. (See Section 12-8.) The ionization of a weak acid is incomplete because it is a reversible reaction. A reaction in which the reverse

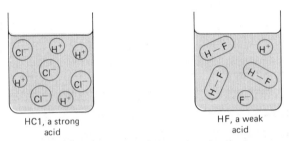

HC1, a strong
acid

HF, a weak
acid

Figure 13-4   STRONG ACIDS AND WEAK ACIDS. Strong acids are completely ionized in water; weak acids are only partially ionized.

reaction also occurs to an appreciable extent is illustrated by double arrows ($\rightleftharpoons$) rather than a single arrow, which implies a reaction that is essentially complete such as was previously shown for HCl. Ionization of the weak acid HF is illustrated as follows:

$$HF(aq) + H_2O \rightleftharpoons H_3O^+(aq) + F^-(aq)$$

The partial ionization of a weak acid is one example of a chemical reaction that reaches a state of equilibrium. *In a reaction at* **equilibrium,** *two reactions are occurring simultaneously.* In the ionization of HF, for example, a forward reaction occurs to the right, producing ions (H_3O^+ and F^-), and a reverse reaction occurs to the left, producing covalent compounds (HF and H_2O).

$$\text{Forward} \quad HF(aq) + H_2O \longrightarrow H_3O^+(aq) + F^-(aq)$$
$$\text{Reverse} \quad H_3O^+(aq) + F^-(aq) \longrightarrow HF(aq) + H_2O$$

At equilibrium, the forward and the reverse reactions occur at the same rate. For weak acids, the point of equilibrium lies far to the left side of the original ionization equation. This means that most of the fluorine is present in the form of covalent HF molecules rather than fluoride ions. (See Figure 13-4.)

When a system is at equilibrium, the concentrations of all species (reactants and products) remain the same but the identities of the individual molecules change. The reaction thus *appears* to have gone so far to the right and then stopped. In fact, at equilibrium, a **dynamic** *(constantly changing)* situation exists in which two reactions going in opposite directions at the same rate keep the concentrations of all species constant.

There is a simple analogy to a situation in which an amount remains constant as a result of equal rates of change. As shown in Figure 13-5, if the rate at which water enters a lake equals the rate at which it leaves over the spillway, the amount of water in the lake remains constant. In a situation similar to our chemical system the amount of water stays the same but its identity is changing. We will have much more to say about equilibrium in Chapter 15. The subject is introduced now so that we may appreciate the reason for such incomplete reactions as the ionization of a weak acid or base.

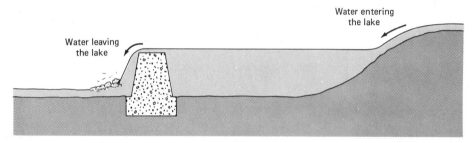

Figure 13-5 EQUILIBRIUM. The level of the lake is at equilibrium when the flow of water entering the lake equals the flow of water leaving the lake.

The following examples illustrate the difference in acidity (the difference in H_3O^+ concentration) between a strong acid and a weak acid. In these examples the appearance of a species in brackets (e.g., $[H_3O^+]$) represents the numerical value of the concentration of that species in moles per liter (M).

EXAMPLE 13-4

What is $[H_3O^+]$ in a 0.100 M HNO_3 solution?

SOLUTION

HNO_3 is a strong acid, so the following reaction goes 100% to the right.

$$HNO_3 + H_2O \longrightarrow H_3O^+(aq) + NO_3^-(aq)$$

The initial concentration of HNO_3 is 0.100 M, and all of the HNO_3 ionizes to form H_3O^+. The reaction stoichiometry tells us that 1 mol of H_3O^+ is formed for every mole of HNO_3 initially present. Therefore, 0.100 mol of HNO_3 forms 0.100 mol of H_3O^+ in 1 L. Thus,

$$[H_3O^+] = \underline{\underline{0.100\ M}}$$

EXAMPLE 13-5

What is $[H_3O^+]$ in a 0.100 M $HC_2H_3O_2$ solution that is 1.34% ionized?

SOLUTION

Since $HC_2H_3O_2$ is a weak acid, the following ionization reaches equilibrium when 1.34% of the initial $HC_2H_3O_2$ ionizes.

$$HC_2H_3O_2(aq) + H_2O \rightleftharpoons H_3O^+(aq) + C_2H_3O_2^-(aq)$$

The $[H_3O^+]$ is calculated by multiplying the original concentration of acid by the percent **expressed in fraction form.**

$$[H_3O^+] = [\text{original concn of acid}] \times \frac{\%\ \text{ionization}}{100\%}$$

In this case,

$$[H_3O^+] = [0.100] \times \frac{1.34\%}{100\%} = 1.34 \times 10^{-3}\ M$$

■
REVIEW
APPENDIX A
ON PERCENT
PROBLEMS.

Note in the above examples that the concentration of H_3O^+ is about 100 times greater in the nitric acid solution than in the acetic acid solution. Yet both acids were at the same original concentration. *The large difference in acid strength lies in the proportion of the acid molecules originally present that ionize.* We should take note that the concentration of an acid in water does not necessarily correspond to the concentration of H_3O^+ ions present in solution.

A common property of bases is that they dissolve grease and other organic material. When a drain is stopped up we use caustic soda (NaOH) to dissolve the hair and grease. On the other hand, if we have a simple grease spot on the floor, caustic soda would dissolve the grease but it may also dissolve other organic materials such as a rubber glove, the floor, and skin. Ammonia, however, is perfect for this job as its dissolving ability is more moderate and controlled. Like acids, bases can also be categorized as to strength.

Strong bases are those compounds that completely ionize in water to produce OH^- ions. All alkali hydroxides are strong bases and are quite soluble in water. The solution of the ionic compound NaOH is illustrated by the equation

$$NaOH(s) \longrightarrow Na^+(aq) + OH^-(aq)$$

All of the alkaline earth hydroxides [except $Be(OH)_2$] also completely dissociate into ions in solution. However, $Mg(OH)_2$ has a very low solubility in water and so in fact produces a very small concentration of aqueous OH^-. Because of its low solubility it can be taken internally to combat excess stomach acid (Milk of Magnesia).

As mentioned, ammonia exhibits much more moderate basic properties than caustic soda. This is because ammonia is a **weak base** in water:

$$NH_3(aq) + H_2O \rightleftharpoons NH_4^+(aq) + OH^-(aq)$$

Only a small portion of the dissolved ammonia molecules react to form ions so the concentration of OH^- remains low. *In this equilibrium, like that of weak acids, the position of equilibrium lies far to the left.* Ammonia is the most familiar weak base and this reaction is encountered often. Many other neutral nitrogen compounds such as methylamine (CH_3NH_2) and pyridine (C_5H_5N) also react with water in a manner similar to that of ammonia to produce weakly basic solutions.

■
SEE PROBLEMS
13-16 THROUGH
13-23.

Before expanding our view of acids and bases in water, let's go back and look at the composition of pure water. In Chapter 12, pure water was classified as a nonconductor of electricity, which meant that ions were not detectable by an ordinary conductivity experiment. With very sensitive instruments, however, we find that there actually is a small concentration of ions present in pure water. The ions arise from the dissociation of water molecules in a reversible reaction:

$$H_2O + H_2O \rightleftharpoons H_3O^+ + OH^-$$

This process whereby a pure covalent compound dissociates into positive and negative ions is known as **autoionization.** Although the point of this equilibrium lies very far to the left, there are small but important concentrations of H_3O^+ and OH^- that do coexist in pure water. Like other equilibria that we have discussed, this is a dynamic situation, with the forward and reverse reactions occurring simultaneously.

The concentration of each ion at 25 °C has been found by experiment to be 1.0×10^{-7} mol/L. This means that only about one out of every 500 million water molecules is actually ionized at any one time. Other experimental results tell us that the product of the ion concentrations is a constant. This phenomenon will be explained in more detail in Chapter 15, but for now we accept it as fact. Therefore, at 25 °C

$$[H_3O^+][OH^-] = K_w \quad \text{(a constant)}$$

Substituting the actual concentrations of the ions, we can now find the numerical value of the constant:

$$[1.0 \times 10^{-7}][1.0 \times 10^{-7}] = 1.0 \times 10^{-14}$$

K_w (1.0×10^{-14}) is known as the **ion product** *of water* at 25 °C. The importance of this constant is that it tells us the concentrations of H_3O^+ and OH^- not only in pure water but also in acidic and basic solutions. The following example illustrates this relationship.

EXAMPLE 13-6

In a certain solution, $[H_3O^+] = 1.5 \times 10^{-2}$ *M.* What is $[OH^-]$ in this solution?

PROCEDURE

Use the relationship for K_w, $[H_3O^+][OH^-] = 1.0 \times 10^{-14}$, and solve for $[OH^-]$.

SOLUTION

$$[H_3O^+][OH^-] = 1.0 \times 10^{-14}$$

$$[OH^-] = \frac{1.0 \times 10^{-14}}{[H_3O^+]} = \frac{1.0 \times 10^{-14}}{1.5 \times 10^{-2}}$$

$$= \underline{6.7 \times 10^{-13}\ M}$$

As you can see, there is an inverse relationship between $[H_3O^+]$ and $[OH^-]$. Although inverse relationships have been discussed and illustrated several times previously in this text, the relationship between $[H_3O^+]$ and $[OH^-]$ is illustrated in Figures 13-6 and 13-7. In these figures, concentrations of $[H_3O^+]$ and $[OH^-]$ between 10^{-4} and 10^{-10} M are listed (higher and lower concentrations exist, however). Concentrations of both ($[H_3O^+]$ on the left and $[OH^-]$ on the right) are shown to decrease from top to bottom. In between, there is a rigid pointer with an arrow at each end. Figure 13-6 illustrates the situation in pure water, which is neutral. In pure water or a **neutral solution,** the pointer is exactly balanced, indicating that the *concentrations of both ions are equal to* $10^{-7}\ M$.

We will now add an *acid* to a neutral solution. By definition, H_3O^+ is added to the solution, and the pointer should go up on the left. At the same time, however, the rigid pointer goes *down on the right,* indicating a correspondingly smaller $[OH^-]$. If, instead, a base had been added, the situation would have been just the opposite. Figure 13-7 illustrates the relationship in an acid solution. A calculation such as we made in Example 13-6 would confirm that when $[H_3O^+]$ is *increased* from 10^{-7} to 10^{-5} M, $[OH^-]$ is *decreased* from 10^{-7} to 10^{-9} M.

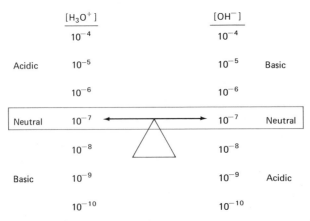

Figure 13-6   A NEUTRAL SOLUTION. In a neutral solution $[H_3O^+]$ and $[OH^-]$ are both equal to $10^{-7}\ M$.

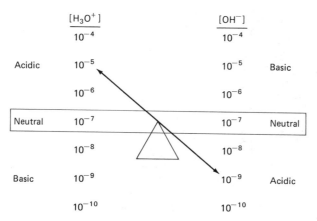

Figure 13-7   AN ACID SOLUTION. There is an inverse relationship between $[H_3O^+]$ and $[OH^-]$.

In summary, an acidic, basic, or neutral solution can now be defined in terms of concentrations of ions:

Neutral $[H_3O^+] = [OH^-] = 1.0 \times 10^{-7}$
Acidic $[H_3O^+] > 1.0 \times 10^{-7}$ and $[OH^-] < 1.0 \times 10^{-7}$
Basic $[H_3O^+] < 1.0 \times 10^{-7}$ and $[OH^-] > 1.0 \times 10^{-7}$

We should now incorporate this information into our understanding of acids and bases. A modern version of the Arrhenius definition of an acid is a substance that produces H_3O^+ ions in aqueous solution. But now we see that H_3O^+ is present in neutral and even basic solutions as well. A slight modification of the definitions solves this problem. *An **acid** is any substance that increases $[H_3O^+]$ in water, and a **base** is any substance that increases $[OH^-]$ in water.* With our new understanding of the equilibrium, note that a substance can be an acid by directly donating $H^+(aq)$ to the solution (e.g., HCl, H_2S), or a substance can be an acid by reacting with OH^- ions, thus removing them from the solution. As illustrated in Figure 13-7, lowering $[OH^-]$ on the right has the effect of raising $[H_3O^+]$ on the left.

■
SEE PROBLEMS 13-24 THROUGH 13-33.

13-5 THE pH SCALE

It is difficult to watch commercial television for long without hearing some reference to pH. Presumably, the whole population understands what is meant (or is at least impressed) when a hair rinse has a high, low, or controlled pH. (See Figure 13-8.) In fact, pH is an important and convenient method for expressing $[H_3O^+]$ in aqueous solution. In a previous section, we found that the $[H_3O^+]$ of a 0.100 M acetic acid solution is 1.34×10^{-3} M. Expressing concentrations such as this in scientific notation is certainly easier than using many zeros, but it is still awkward. The solution to this problem is the use of a logarithmic scale called pH. **pH** *is a logarithmic definition of acidity:*

$$pH = -\log[H_3O^+]$$

Figure 13-8 pH AND PRODUCTS. Most of us are quite familiar with the use of pH in commercial products.

REFER TO APPENDIX E FOR DISCUSSION OF LOGARITHMS.

A much less popular but valid way of expressing [OH⁻] is **pOH**:

$$pOH = -\log[OH^-]$$

A simple relationship between **pH** and **pOH** can be derived from the ion product of water:

$$[H_3O^+][OH^-] = 1.0 \times 10^{-14}$$

If we now take $-\log$ of both sides of the equation, we have

$$-\log[H_3O^+][OH^-] = -\log(1.0 \times 10^{-14})$$

Since $\log(A \times B) = \log A + \log B$, the equation can be written as

$$-\log[H_3O^+] - \log[OH^-] = -\log 1.0 - \log 10^{-14}$$

Since $\log 1.0 = 0$ and $\log 10^{-14} = -14$, the equation is

$$pH + pOH = 14$$

Generally, pOH is not used extensively since pH relates to the OH⁻ concentration as well as the H_3O^+ concentration. However, the relationships among $[H_3O^+]$, $[OH^-]$, pH, and pOH can be summarized:

$$pH + pOH = 14$$

pH — pOH

$-\log[H_3O^+]$ — $-\log[OH^-]$

$$[H_3O^+][OH^-] = 10^{-14}$$

$[H_3O^+]$ — $[OH^-]$

EXAMPLE 13-7

What is the pH of a neutral solution? What is the pOH?

SOLUTION

In a neutral solution, $[H_3O^+] = 1.0 \times 10^{-7}$.

$$pH = -\log[H_3O^+] = -\log(1.0 \times 10^{-7})$$
$$= -\log 1.0 - \log 10^{-7}$$
$$= 0.00 - (-7.00) = \underline{7.00}^*$$

Also in a neutral solution, $[OH^-] = 1.0 \times 10^{-7}$.

$$pOH = \underline{7.00}$$

---

*At first glance it appears that the pH is expressed to three significant figures and the $[H_3O^+]$ to only two. In fact, only the number to the right of the decimal (the mantissa) in the pH should be expressed to two significant figures. The number to the left of the decimal (the characteristic) in the pH refers to the exponent of 10.

EXAMPLE 13-8

What is the pH of a 0.015 M solution of $HClO_4$? What is the pOH of this solution?

SOLUTION

$$HClO_4 + H_2O \longrightarrow H_3O^+ + ClO_4^-$$

Since this is a strong acid, all of the $HClO_4$ is ionized to produce H_3O^+. Therefore, $[H_3O^+] = 0.015$.

$$pH = -\log[H_3O^+] = -\log(1.5 \times 10^{-2})$$
$$= -\log 1.5 - \log 10^{-2}$$
$$= -0.18 + 2.00 = \underline{1.82}$$

$$pH + pOH = 14.00$$
$$pOH = 14.00 - pH$$
$$= 14.00 - 1.82 = \underline{12.18}$$

EXAMPLE 13-9

In a certain weakly basic solution, pH = 9.62. What is the $[H_3O^+]$?

SOLUTION

$$pH = 9.62 = -\log[H_3O^+]$$
$$\log[H_3O^+] = -9.62 = 0.38 - 10*$$

Since the antilog of 0.38 is 2.4 and the antilog of -10 is 10^{-10},

$$[H_3O^+] = \underline{2.4 \times 10^{-10}\ M}$$

---

* See Appendix E-4.

Acidic, basic, and neutral solutions were previously defined in terms of concentrations. We can now do the same thing in terms of pH and pOH.

Neutral	pH = pOH = 7.00
Acidic	pH < 7.00 and pOH > 7.00
Basic	pH > 7.00 and pOH < 7.00

Remember, then, a high pH (greater than 7) means basic and a low pH (less than 7) means acidic.

You've probably heard of the Richter scale for measuring earthquake intensity, especially if you're from California. If so, you are aware that an earthquake with a reading of 8.0 on the Richter scale is 10 times more powerful than one with a reading of 7.0. This is because the Richter scale, like pH, is a logarithmic scale. In such a scale, a difference of one integer actually represents a 10-fold change. For example, a solution of pH = 4 is 10 times more acidic than a solution of pH = 5. A pH = 1 solution is actually 100,000 times more acidic than a pH = 6 solution, although both solutions are labeled acidic. The pH values of some common substances are given in Table 13-1.

■
SEE PROBLEMS
13-34 THROUGH
13-48.

**13-6
BRØNSTED–
LOWRY ACIDS
AND BASES**

At this point, our understanding of acids and bases is simply that of substances that change the pH of an aqueous solution. We have illustrated how many compounds such as HCl, $HC_2H_3O_2$, NH_3, and NaOH can act as either an acid or a base. In addition to compounds, however, many ions such as NH_4^+, CN^-, and $C_2H_3O_2^-$ also act as acids and bases in water. To better describe the action of ions in water, it is helpful to employ a more general definition of acid and base behavior. The Brønsted–Lowry definition is useful not only for this purpose but to extend the concept of acids and bases to certain other solvents. *In the Brønsted–Lowry definition, an* **acid** *is a proton (H⁺) donor and a* **base** *is a proton acceptor.* To illustrate this definition, we again look at the equilibrium involving the common weak base ammonia (NH_3):

$$NH_3 + H_2O \rightleftharpoons NH_4^+ + OH^-$$

TABLE 13-1   pH Values of Some Common
Substances

Strong acid	1	← 0.1 M HCl (stomach acid)
	2	
		← Lemon juice, vinegar
	3	
		← Orange juice
Weak acid	4	
	5	
	6	
		← Milk
Neutral	7	← Pure water
		← Blood
	8	
		← Sea water
	9	
		← Baking soda solution
Weak base	10	
		← Milk of magnesia
	11	
		← Household ammonia
	12	
	13	← 0.1 M NaOH (dilute lye)
Strong base	14	

Ammonia is a base by the Arrhenius definition because by removing a proton from an H_2O molecule *it produces OH⁻*. It is also a base by the Brønsted–Lowry definition because *it accepts an H^+ from H_2O*:

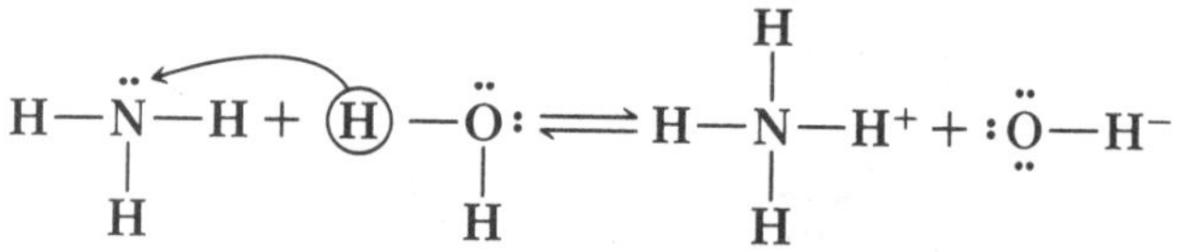

$$H-\overset{..}{N}-H + \overset{..}{H)}-\overset{..}{O}: \rightleftharpoons H-\overset{\overset{H}{|}}{\underset{|}{N}}-H^+ + :\overset{..}{\underset{..}{O}}-H^-$$

Although NH_3 is a base by both definitions, the H_2O molecule plays the role of an acid in the Brønsted–Lowry sense because it donates the H^+ to NH_3.

As mentioned previously, this reaction is a reversible reaction that reaches a point of equilibrium. This means that the reverse reaction, in which an NH_4^+ ion donates an H^+ to an OH^- to form the original reactants, is also occurring. *Thus in the reverse reaction, NH_4^+ acts as an acid and OH^- as a base.*

In the Brønsted–Lowry sense the reaction can be viewed as simply an exchange of an H^+. When the base (NH_3) reacts, it adds H^+ to form an acid (NH_4^+). When the acid (H_2O) reacts, it loses an H^+ to form a base (OH^-). The NH_3,NH_4^+ and the H_2O,OH^- pairs are known as **conjugate acid–base pairs.**

Thus a reaction of an acid with a base is said to produce a conjugate base and acid as illustrated in the following equation (A = Brønsted–Lowry acid, B = Brønsted–Lowry base).

$$NH_3 + H_2O \rightleftharpoons NH_4^+ + OH^-$$
$$B_1 \quad\ A_2 \qquad\quad A_1 \qquad B_2$$

The conjugate base of a substance can be determined by removing an H^+. The conjugate acid of a substance is determined by adding an H^+.

$$\text{conjugate acid} \underset{+H^+}{\overset{-H^+}{\rightleftharpoons}} \text{conjugate base}$$

For example

$$H_3PO_4 \underset{+H^+}{\overset{-H^+}{\rightleftharpoons}} H_2PO_4^-$$

We have seen that bases by the Arrhenius definition, such as NH_3, are also bases in the Brønsted–Lowry sense. Arrhenius acids such as HCl also maintain their status as acids when they are considered as H^+ donors. Consider the following equation, which illustrates the acid behavior of HCl.

$$HCl + H_2O \longrightarrow H_3O^+ + Cl^-$$
$$A_1 \quad\ B_2 \qquad\quad A_2 \qquad B_1$$

In this reaction H_2O is a base, since it accepts a H^+ to form H_3O^+. (Remember that H_2O is an acid when NH_3 is present.) *A compound that can act as either an acid or base, depending on what other substance is present is called* **amphoteric.** One must determine whether an amphoteric substance is an acid or a base from the particular reaction. For example, we cannot determine whether H_2O is a Brønsted–Lowry acid or base unless we know what other substance is present (HCl or NH_3).

EXAMPLE 13-10

What are the conjugate bases of **(a)** H_2SO_3 and **(b)** $H_2PO_4^-$?

PROCEDURE

$$\text{acid} - H^+ = \text{conjugate base}$$

ANSWER

(a) $H_2SO_3 - H^+ = \underline{\underline{HSO_3^-}}$ **(b)** $H_2PO_4^- - H^+ = \underline{\underline{HPO_4^{2-}}}$

EXAMPLE 13-11

What are the conjugate acids of **(a)** CN^- and **(b)** $H_2PO_4^-$?

PROCEDURE

$$base + H^+ = conjugate\ acid$$

ANSWER

(a) $CN^- + H^+ = \underline{\underline{HCN}}$ (b) $H_2PO_4^- + H^+ = \underline{\underline{H_3PO_4}}$

EXAMPLE 13-12

Write the equations illustrating the following Brønsted–Lowry acid–base reactions.

(a) H_2S as an acid with H_2O
(b) $H_2PO_4^-$ as an acid with OH^-
(c) $H_2PO_4^-$ as a base with H_3O^+
(d) CN^- as a base with H_2O

PROCEDURE

A Brønsted–Lowry acid–base reaction produces a conjugate acid and base.

ANSWER

	Acid		Base		$\longrightarrow$	Acid		Base
(a)	H_2S	+	H_2O		$\longrightarrow$	H_3O^+	+	HS^-
(b)	$H_2PO_4^-$	+	OH^-		$\longrightarrow$	H_2O	+	HPO_4^{2-}
(c)	H_3O^+	+	$H_2PO_4^-$		$\longrightarrow$	H_3PO_4	+	H_2O
(d)	H_2O	+	CN^-		$\longrightarrow$	HCN	+	OH^-

Note that equations **(b)** and **(c)** indicate that the $H_2PO_4^-$ is amphoteric.

■
*SEE PROBLEMS
13-49 THROUGH
13-56.*

**13-7
PREDICTING
ACID AND
BASE
REACTIONS IN
WATER**

The strength of an acid is determined by the degree to which it ionizes in water to produce H_3O^+ ions. Thus it is possible to construct a list of acids in order of strength. Such a ranking is illustrated in Table 13-2. The acids, listed on the left, are shown in order of decreasing strength going down the list. Note that there are three groups of acids. Those acids that are stronger H^+ donors than the H_3O^+ ion are shown as group 1; they are the very strong acids. The

TABLE 13-1　　Relative Strengths of Some Acids and Bases[a]

$$\text{Acid} \xrightarrow{\ -H^+\ } \text{Base}$$
$$\xleftarrow{\ +H^+\ }$$

		Acid	Base	
1	Very strong	$HClO_4$ HBr HCl HNO_3	ClO_4^- Br^- Cl^- NO_3^-	Very weak (negligible)
		H_3O^+	H_2O	
2	Intermediate to weak	H_2SO_3 HNO_2 HF $HC_2H_3O_2$ H_2S H_2CO_3 $HOCl$ NH_4^+ HCN	HSO_3^- NO_2^- F^- $C_2H_3O_2^-$ HS^- HCO_3^- OCl^- NH_3 CN^-	Intermediate to weak
		H_2O	OH^-	
3	Very weak (negligible)	H_2 CH_4	H^- CH_3^-	Very Strong

Increasing acid strength (left, upward arrow) — Increasing base strength (right, downward arrow)

[a] The species in the shaded region are all capable of existence in appreciable concentrations in aqueous solutions.

intermediate or weak acids in group 2 are weaker proton donors than H_3O^+ but stronger than the H_2O molecule. The acids in group 3 are the very weak acids and are all weaker proton donors than H_2O.

Now consider the conjugate bases listed to the right of the acids. They are also shown in order of increasing base strength but the trend is opposite that of the acids in that the bases become stronger down the list. In other words, there is an inverse relationship between the strength of an acid and the strength of its conjugate base. *The stronger the acid, the weaker is its conjugate base.* An inverse relationship is somewhat like a "seesaw." When one side is up, the other is down.

This table can be used to predict a whole host of acid–base chemical reactions. *In the Brønsted–Lowry concept an acid–base chemical reaction is simply competition for an* H^+, *with the stronger acid reacting with the stronger base to produce a weaker acid and base.* Therefore, a favorable reaction occurs between an acid higher in the table (on the left) and a base lower in the table (on the right). A favorable reaction implies that when reactants are mixed, most will be converted into products. (The equilibrium lies to the right.) Unless the two competing acids are very close to each other in the table, such reactions

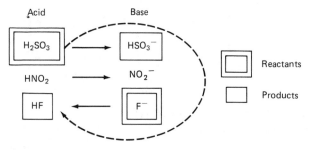

Figure 13-9   A FAVORABLE REACTION. The stronger acid (H_2SO_3) reacts with the stronger base (F^-).

are essentially 100% complete. As illustrated in Figure 13-9, a favorable reaction takes place in a clockwise direction. Consider the following question: Which reaction is favorable: the reaction of H_2SO_3 with the base F^- or the reaction of HF with the base HSO_3^-? Note that H_2SO_3 is a stronger acid than HF and that F^- is a stronger base than HSO_3^-. Therefore, the following reaction is favorable.

$$H_2SO_3(aq) + F^-(aq) \longrightarrow HF(aq) + HSO_3^-(aq)$$

EXAMPLE 13-13

Use Table 13-2 to predict which of the following reactions are favorable.

(a) $HCl + HS^- \longrightarrow H_2S + Cl^-$
(b) $NH_4^+ + HS^- \longrightarrow H_2S + NH_3$
(c) $CH_4 + OH^- \longrightarrow CH_3^- + H_2O$
(d) $HNO_2 + OH^- \longrightarrow NO_2^- + H_2O$

PROCEDURE

Consider the two acids (e.g., HCl and H_2S). If the stronger acid is on the left side of the equation, the reaction is favorable.

ANSWER

Reactions **(a)** and **(d)** are favorable.

EXAMPLE 13-14

Use Table 13-2 to write equations for favorable reactions between the following acid–conjugate base systems.

(a) HNO_2, NO_2^- and HOCl, OCl^-
(b) HBr, Br^- and H_3O^+, H_2O

SEE PROBLEMS
13-57 THROUGH
13-60.

PROCEDURE

The stronger acid (e.g., HNO_2) and the stronger conjugate base (e.g., OCl^-) are shown as reactants on the left.

ANSWER

(a) $HNO_2 + OCl^- \longrightarrow NO_2^- + HOCl$
(b) $HBr + H_2O \longrightarrow H_3O^+ + Br^-$

Now let us focus on the action of the three groups of acids specifically with water as a base. Consider the group 1 acids shown in their reaction with water *(as a base)*. As mentioned previously and predicted by this table they are all essentially 100% ionized in water. This implies that the molecular form of the acid, as shown on the left side of the following equation, does not exist in water.

$$HNO_3 + H_2O \longrightarrow H_3O^+ + NO_3^-$$

The reaction of a group 2 acid with water is different. Note that H_2O as a base lies higher in the table than these acids. Thus we predict that the ionization of these acids in water is not favorable because it leads to formation of a stronger Brønsted–Lowry acid and base. *That it is unfavorable does not mean it does not occur at all, however.* In fact, we have already indicated that these acids produce a small but measurable concentration of H_3O^+ in water. Our prediction that the reaction is unfavorable simply means that the equilibrium lies to the left in favor of the molecular form of the acid. The lower the acid in the table, however, the more unfavorable the ionization and the fewer hydronium ions formed (the higher the pH). The reaction of a typical group 2 acid is

$$HNO_2 + H_2O \rightleftharpoons H_3O^+ + NO_2^-$$

The acids in group 3 do not react with H_2O as a base. The reaction is so unfavorable that the presence of these molecules in water has no measurable effect on the pH of water. For these molecules, only the molecular form of the acid can exist in water, not its conjugate base. At least in water, these compounds are acids in name only.

Now consider the reaction of the conjugate bases with water. As most of these are ions, we assume that they are added to water in the form of some salt such as NaCl, KCN, or $Ca(NO_2)_2$. The cations of these salts do not affect the pH of water so we are free to focus on the anions. *In the reaction of a base with water, recall that water acts as an acid forming its conjugate base, the OH^- ion.* Thus, reaction of a hypothetical anion (X^-) as a base with water would proceed as follows:

$$X^- + H_2O \longrightarrow HX + OH^-$$

This type of reaction is known as a hydrolysis reaction. **A hydrolysis reaction** *is the reaction of an anion with water to produce OH^- or the reaction of a cation with water to produce H_3O^+.*

Let us consider the anions at the bottom of the table in group 3. These anions are the conjugate bases of the very weak acids. Because of the inverse relationship between acids and their conjugate bases, these anions are very strong bases. Note that H_2O *as an acid* lies higher in the table than these bases. Thus we predict that the reaction of one of these anions with water is favorable and proceeds essentially 100% to the right. In fact, these anions cannot exist in water since they react completely to form the conjugate acid. The reaction of the hydride ion (H^-) with water is

$$H^- + H_2O \longrightarrow H_2 + OH^-$$

Consider the anions in group 2. In this case, the hydrolysis reaction is unfavorable because water as an acid lies lower in the table than the anion bases. Once again, however, that does not mean that the reaction does not occur at all. *Conjugate bases of the weak or intermediate acids do undergo a limited hydrolysis, producing a small equilibrium concentration of OH^-.* In this case, the equilibrium lies to the left. A typical hydrolysis reaction involving a group 2 anion is

$$NO_2^- + H_2O \rightleftharpoons HNO_2 + OH^-$$

Since the anion bases become weaker as we go higher in the table the extent of hydrolysis decreases from the bottom to the top of group 2.

Consider the reaction of the group 1 anions with water as an acid. *In this case, the anions are such weak bases that there is essentially no hydrolysis.* The presence of these anions in water does not affect the H_3O^+ concentration (pH).

Cations can also undergo hydrolysis, for example, the conjugate acid of NH_3 which is the NH_4^+ ion. The reaction of this cation as an acid with water *as a base* is predicted to be unfavorable. However, it does occur to a limited extent:

$$NH_4^+ + H_2O \rightleftharpoons NH_3 + H_3O^+$$

Thus, if the NH_4^+ is introduced into aqueous solution in the form of a salt such as NH_4Cl, a slightly acidic solution results. In this case, the Cl^- ion does not affect the pH because it is the conjugate base of the strong acid HCl.

Cations of strong bases such as Na^+, Ca^{2+}, and K^+ do not undergo hydrolysis and thus do not affect the pH of water by themselves.

In summary, the species capable of existence in water in at least some concentration appear in the shaded area of Table 13-2. The species not in the shaded region cannot exist in water since they react completely with water as acids (group 1) or as bases (group 3). The bases in group 1 and the acids in group 3 do not react with water to any measurable extent so their presence in water does not affect the pH. When the acids in group 2 are introduced into water, the acidity is affected to a some extent. A limited concentration of H_3O^+ forms as a result of the partial ionization of the acid. When the bases in group 2 are introduced into water [e.g., as $NH_3(g)$ or KCN], the solution becomes basic to some extent because of formation of a limited concentration of OH^-. We can now use this information to write acid–base reactions and predict how the acidity of water is affected by the presence of certain solutes. Of significance in this discussion is that solutions of salts are not necessarily neutral. Because of the action of the cation or anion or both, the solution of a salt may be basic or acidic. Situations in which both the cation and the anion of a salt undergo hydrolysis or examples in which the anion is an acid salt are somewhat more complicated and have been avoided in the following examples.

EXAMPLE 13-15

Referring to Table 13-2, write an equation illustrating the reaction of each of the following as an acid or base in water. If a partial reaction occurs, indicate with double arrows. If the species does not exhibit acid or base nature in water, write "no reaction."

(a) HBr   **(b)** CH_3^- **(c)** OCl^- **(d)** ClO_4^- **(e)** H_2 **(f)** HCN

ANSWER

(a)   HBr is a very strong acid: $HBr + H_2O \rightarrow H_3O^+ + Br^-$.
(b) CH_3^- is a very strong base: $CH_3^- + H_2O \rightarrow CH_4 + OH^-$.
(c) OCl^- is a weak base (hydrolyzes); $OCl^- + H_2O \rightleftharpoons HOCl + OH^-$.
(d) ClO_4^- is a very weak base: no reaction with water.
(e) H_2 is a very weak acid: no reaction with water.
(f)   HCN is a weak acid: $HCN + H_2O \rightleftharpoons H_3O^+ + CN^-$.

In the following examples, indicate whether a solution of the salt is acidic, basic, or neutral, and write the reaction illustrating this behavior. To do this, both the cation and anion must be examined for hydrolysis behavior.

EXAMPLE 13-16: KCN

SOLUTION

K^+ is the cation of the strong base KOH and does not hydrolyze. The CN^- ion, however, is the conjugate base of the weak acid HCN and hydrolyzes as follows:

$$CN^- + H_2O \rightleftharpoons HCN + OH^-$$

Since OH^- is formed in this solution, the solution is <u>basic</u>.

EXAMPLE 13-17: Ca(NO$_3$)$_2$

SOLUTION

Ca^{2+} is the cation of the strong base Ca(OH)$_2$ and does not hydrolyze. NO$_3^-$ is the conjugate base of the strong acid HNO$_3$ and also does not hydrolyze. Since neither ion hydrolyzes, the solution is <u>neutral</u>.

EXAMPLE 13-18: H$_2$N(CH$_3$)$_2$$^+Br^-$

[HN(CH$_3$)$_2$ is a weak base like NH$_3$.]

SOLUTION

H$_2$N(CH$_3$)$_2$$^+$ is the conjugate acid of the weak base HN(CH$_3$)$_2$. It undergoes hydrolysis according to the equation

$$H_2N(CH_3)_2^+ + H_2O \rightleftharpoons HN(CH_3)_2 + H_3O^+$$

The Br$^-$ ion is the conjugate base of the strong acid HBr and does not hydrolyze. Since only the cation undergoes hydrolysis, the solution is <u>acidic</u>.

■
*SEE PROBLEMS
13-61 THROUGH
13-82.*

13-8 BUFFER SOLUTIONS

The blood coursing through our bodies is an amazing and complex system. The pH of blood (7.4) indicates that it is very slightly basic. The nourishment that we take, however, can be significantly acidic or more basic than the blood that absorbs it. Still, the pH of blood essentially does not change despite what we ingest. The pH of pure water (7.0) is about the same as that of blood, yet addition of small amounts of acid or base has drastic effects on the pH. For example, if 0.10 mol of HCl is added to 1 L of water, the pH changes from 7.0 to 1.0. If 0.10 mol of KOH is added to 1 L of pure water, the pH changes from 7.0 to 13.0. The pH of blood does not change to any great extent because it is a buffer solution. *A* **buffer solution** *resists changes in pH from the addition of limited amounts of a strong acid or a strong base.*

One type of buffer solution is formed by a solution of a weak acid which also contains a salt of its conjugate base. As an example of how a buffer is formed and works to resist change in pH, we consider a solution of the weak acid HCN and the salt NaCN. The Na$^+$ ion of the salt is a spectator ion and is not part of the reaction. Recall that the action of a weak acid in water is described by an equilibrium reaction in which the vast majority of the molecules of the acid are present in solution as un-ionized molecules. Only a small percentage of the HCN molecules dissociate into ions. The ionization of HCN is represented as

$$HCN + H_2O \rightleftharpoons H_3O^+ + CN^-$$

Now consider what happens when a small amount of OH$^-$ is added to this solution. There are several ways to look at what happens in a buffer solution, but we can easily apply what we discussed in the previous section. From Table 13-2 we can determine that the following reaction is favorable (essentially complete).

$$HCN + OH^- \longrightarrow H_2O + CN^-$$

In effect, the undissociated acid present in solution reacts with the added OH$^-$, leaving the original H$_3$O$^+$ concentration nearly unchanged. Note that a strong acid cannot behave in a similar manner because the molecular form of the acid does not exist in water.

Consider the following analogy. One person has \$20 in his pocket with no savings in the bank; another person has \$20 in her pocket and \$100 in the bank. A \$10 expense will change the first person's pocket money drastically. The second person is "buffered" from this expense and can cover it with bank savings; this person can then maintain the same amount of pocket money. The un-ionized HCN is like "money in the bank." It is available to react with added OH$^-$, leaving the concentration of H$_3$O$^+$ unchanged. The person who keeps all of his money in his pocket is analogous to a strong acid solution in which there is no un-ionized acid in reserve. In this case, since all of the acid ionizes to H$_3$O$^+$, any added OH$^-$ decreases the H$_3$O$^+$ concentration and thus increases the pH.

There is a limit to how much a buffer system can resist change. If the added amount of OH$^-$ exceeds the reserve of HCN (referred to as the *buffer capacity*) then the pH will rise. This is analogous to a \$110 expense for the person with the savings. It is more than she can cover with bank savings, so the amount in her pocket decreases.

Now let us consider what happens when a small amount of strong acid (H$_3$O$^+$) is added to a solution of HCN and its conjugate base CN$^-$. Again, if we refer to Table 13-2, the following reaction is predicted to be favorable.

$$CN^- + H_3O^+ \longrightarrow HCN + H_2O$$

This reaction is, of course, just the reverse of the ionization reaction which we understood to occur to a very limited extent. The CN$^-$ ion then acts as "money in the bank," removing added H$_3$O$^+$ and leaving the original concentration of H$_3$O$^+$ nearly unchanged.

In summary, one type of buffer solution is composed of a solution of a weak acid that contains an additional source of its conjugate base such as from a salt. The acid is mostly un-ionized and serves as a reservoir of molecules that can protect the pH from change caused by the addition of base (OH$^-$). The presence of the conjugate base in solution protects the pH from change caused by the addition of acid (H$_3$O$^+$). (See Figure 13-10.)

Solutions of weak bases and salts containing their conjugate acids (e.g., NH$_3$ and NH$_4$Cl) also serve as buffers. The pH of the buffer solution depends on the strength of the acid or base chosen. In Chapter 15 actual pH values are calculated from quantitative information related to acid strength. Certainly, the

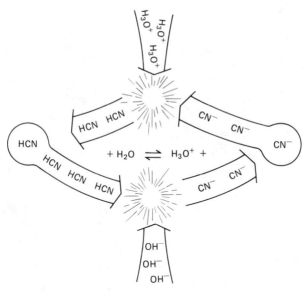

Figure 13-10   ACTION OF A BUFFER. Added OH^- is removed by the reservoir of HCN and added H_3O^+ is removed by the reservoir of CN^-.

term "buffer" is familiar to us in various commercial products. For example, it is popular to advertise that a shampoo or aspirin product is buffered. Presumably, this makes such a product easier on our head or stomach.

As mentioned, a common buffer solution is blood, which must maintain a nearly constant pH for metabolism. This is accomplished by a combination of the following two buffer systems:

$$H_2CO_3 + H_2O \rightleftharpoons H_3O^+ + HCO_3^-$$
$$H_2PO_4^- + H_2O \rightleftharpoons H_3O^+ + HPO_4^{2-}$$

If for some reason the buffer breaks down and the pH of the blood drops lower than 7.2 (acidosis) or rises higher than 7.8 (alkalosis), serious difficulties are encountered—such as dizziness, coma, and muscle spasms. Since we ingest many acidic and basic substances, a healthy buffering system is necessary to maintain a controlled pH.

■
SEE PROBLEMS 13-76 THROUGH 13-82.

13-9 OXIDES AS ACIDS AND BASES

■
REFER TO APPENDIX E FOR DISCUSSION OF LOGARITHMS.

Without doubt, one of the more important consequences of the human race's progress is acid rain. Such rain originates from exhaust gases that may be produced in one country but can be deposited in another. Probably no other issue is currently more touchy between the United States and Canada and between countries of Northern Europe than control of acid rain. Its effect on lakes and forests can be devastating and there is little doubt that the problem must be faced and solved.

Most acid rain originates from the combustion of coal or other fossil fuels that contain sulfur as an impurity. Combustion of sulfur or sulfur compounds

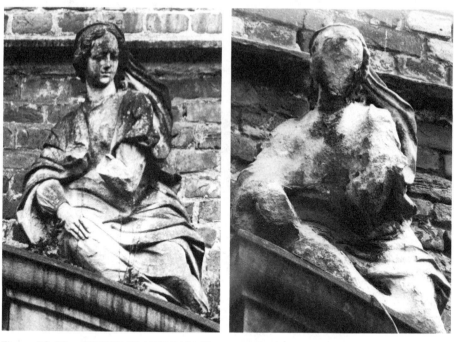

Figure 13-11   EFFECT OF ACID RAIN. The deterioration of this ancient statue is blamed on acid rain and air pollution.

produces sulfur dioxide (SO_2). In the atmosphere, sulfur dioxide somehow reacts with oxygen to form sulfur trioxide (SO_3).

$$2SO_2(g) + O_2(g) \longrightarrow 2SO_3(g)$$

The sulfur trioxide reacts with water in the atmosphere according to the equation

$$SO_3(g) + H_2O \longrightarrow H_2SO_4(aq)$$

The product is a sulfuric acid solution, which is one of the strong acids — corrosive and destructive.* As mentioned earlier, acids react with metals and limestone, which are both used externally in buildings. (See Figure 13-11.) In the above reaction, sulfur trioxide can be considered as simply the dehydrated form of sulfuric acid. It is thus known as an **acid anhydride**, *which means acid without water.* Many nonmetal oxides are acid anhydrides. When dissolved in

* On April 10, 1974, a rain fell on Pilochry, Scotland, that had a pH of 2.4, which is about the same as that of vinegar. This is the most acidic rain ever recorded. Nitrogen oxides formed in automobile engines also contribute to acid rain by forming nitric acid.

water the acid is formed. Some other reactions of nonmetal oxides to form acids follow.

$$CO_2(g) + H_2O \rightleftharpoons H_2CO_3(aq) \quad \text{(carbonic acid)}$$
$$SO_2(g) + H_2O \rightleftharpoons H_2SO_3(aq) \quad \text{(sulfurous acid)}$$
$$N_2O_5(l) + H_2O \longrightarrow 2HNO_3(aq) \quad \text{(nitric acid)}$$

Note that SO_3 is the anhydride of sulfuric acid and that SO_2 is the anhydride of sulfurous acid. To tell which acid goes with which anhydride one can refer to the oxidation state of the central nonmetal. It is the same in both the acid and the anhydride. For example, the oxidation state of sulfur in both SO_2 and H_2SO_3 is $+4$.

You are familiar with the action of carbon dioxide as an acid anhydride producing carbonic acid. The fizz associated with carbonated beverages controls the slightly tangy taste. If all of the carbon dioxide escapes, the taste is altered (it goes flat). Carbon dioxide is also present in the atmosphere and dissolves in rain water to make rain slightly acidic. Carbonic acid is such a weak acid, however, that the pH is lowered to only about 6.5 from carbon dioxide alone. The principal culprit in acid rain is sulfuric acid, although in certain areas where there is a large amount of automobile exhaust, nitrogen oxides produce appreciable amounts of nitric acid. Nitric acid is also one of the strong acids and is therefore potentially destructive.

Ionic metal oxides dissolve in water to form bases and thus are known as **base anhydrides.** Some examples of these reactions are

$$Na_2O(s) + H_2O \longrightarrow 2NaOH(aq)$$
$$CaO(s) + H_2O \longrightarrow Ca(OH)_2(aq)$$

Salt is formed by the reaction between an acid anhydride and a base anhydride. For example, the following reaction produces the same salt formed in the neutralization of H_2SO_3 with $Ca(OH)_2$ in aqueous solution.

$$SO_2(g) + CaO(s) \longrightarrow CaSO_3(s)$$
$$H_2SO_3(aq) + Ca(OH)_2(aq) \longrightarrow CaSO_3(s) + 2H_2O$$

■

SEE PROBLEMS
13-83 THROUGH
13-90.

The first reaction is typical of reactions that are being studied as a possible way of removing SO_2 from the combustion products of an industrial plant so that some of our abundant high-sulfur coal can be used without harming the environment.

13-10
CHAPTER
REVIEW

Compounds have been classified as acids or bases for hundreds of years on the basis of common sets of chemical characteristics. In this century, however, this acid character has been found to be due to formation of H^+ (aq) [also represented as the hydronium ion (H_3O^+)] in aqueous solution. The well-known base character is due to the formation of $OH^-(aq)$ in solution. When acid and base solutions are mixed, the two ions combine to form water in what is known as a neutralization reaction. Complete neutralizations also form ionic compounds called salts, as in reaction **1**. Incomplete neutralizations of polyprotic acids produce acid salts, as in reaction **2**.

$$\text{Acid} \quad + \text{Base} \quad \longrightarrow \text{Salt} \quad + \text{Water}$$

1 $HX(aq) + M^+OH^-(aq) \longrightarrow M^+X^-(aq) + H_2O$

2 $H_2Y(aq) + M^+OH^-(aq) \longrightarrow M^+HY^-(aq) + H_2O$

Acids and bases can also be classified as to strength. Strong acids and bases are 100% ionized, whereas weak acids and bases are only partially ionized. Partial ionization results when a reaction reaches a point of equilibrium in which both molecules and ions are present. For weak acids and bases the point of equilibrium favors the left or molecular side of the equation. Therefore, the H_3O^+ concentration in a weak acid solution is considerably lower than in a strong acid solution at the same initial concentration of acid.

Even in pure water there is a very small equilibrium concentration of both H_3O^+ and OH^-, which is found to be 10^{-7} M. The product of these concentrations is a constant known as the ion product of water. The ion product can be used to calculate the concentration of one ion from that of the other in any aqueous solution.

A convenient method to express the H_3O^+ or OH^- concentrations of solutions involves the use of the logarithmic definition, pH and pOH. The following are examples of solutions of various acidities in terms of $[H_3O^+]$ and pH.

Solution	$[H_3O^+]$	$[OH^-]$	pH	pOH
Strongly acidic	$>10^{-1}$	$<10^{-13}$	$<$ 1.0	>13.0
Weakly acidic	10^{-4}	10^{-10}	4.0	10.0
Neutral	10^{-7}	10^{-7}	7.0	7.0
Weakly basic	10^{-10}	10^{-4}	10.0	4.0
Strongly basic	$<10^{-13}$	$>10^{-1}$	>13.0	$<$ 1.0

Our understanding of acid–base behavior can be broadened somewhat by use of the Brønsted–Lowry definition. According to this definition, acids are H^+ donors and bases are H^+ acceptors. An acid reacts with a base to form the conjugate base and acid, respectively. In other words, an acid–base reaction is simply a H^+ exchange reaction. A favorable acid–base reaction occurs when an acid reacts with a base to form a weaker acid and a weaker base. Acids can be ranked with respect to their strength as proton donors as in Table 13-2. Because of an inverse relationship between the strengths of an acid and its conjugate base, the bases are ranked in reverse order with respect to strength as proton acceptors. Favorable reactions can be predicted from Table 13-2 by the following scheme:

Acids and their conjugate bases are divided into three groups in Table 13-2. A reaction of a group 2 acid with H_2O is not predicted to be favorable. However, such a reaction does occur to a small extent, producing a measurable

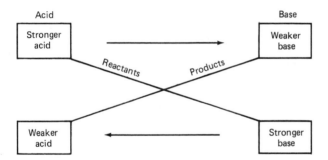

concentration of H_3O^+. Likewise, anions in this group react with water to produce a small but measurable concentration of OH^-. This latter reaction is known as hydrolysis.

The extent of ionization of the acids in each group and the extent of hydrolysis of their anions are summarized as follows.

Group	Acid in Water	Conjugate Base in Water	Comments
1	$HCl + H_2O \xrightarrow{100\%}$ $Cl^- + H_3O^+$	$Cl^- + H_2O \longrightarrow$ no reaction	100% ionization of acid No hydrolysis of anion
2	$HF + H_2O \rightleftharpoons$ $F^- + H_3O^+$	$F^- + H_2O \rightleftharpoons$ $HF + OH^-$	Partial ionization of acid Partial hydrolysis of anion
3	$H_2 + H_2O \longrightarrow$ no reaction	$H^- + H_2O \xrightarrow{100\%}$ $H_2 + OH^-$	No ionization of acid 100% hydrolysis of anion

The possibility of hydrolysis reactions means that solutions of salts may not be neutral. To predict the acidity of a solution of a salt, possible hydrolysis reactions of both cation and anion must be examined. Four possible combinations follow.

Salt	Solution	Comments
1 $NaClO_4$	Neutral	Neither cation nor anion hydrolyzes.
2 K_2S	Basic	Anion (S^{2-}) undergoes hydrolysis. $S^{2-} + H_2O \rightleftharpoons HS^- + OH^-$
3 NH_4Br	Acidic	Cation (NH_4^+) undergoes hydrolysis. $NH_4^+ + H_2O \rightleftharpoons NH_3 + H_3O^+$
4 $(NH_4)_2S$	Unable to predict	Both ions undergo hydrolysis, but more information is needed as to the extent of hydrolysis of each ion.

When a solution of a weak acid is mixed with a solution of a salt providing its conjugate base, a buffer solution is formed. Buffer solutions resist changes in pH from addition of small amounts of a strong acid or base. In a buffer the reservoir of un-ionized acid (e.g., HCN) absorbs added OH^-, while the reservoir of the conjugate base (e.g., CN^-) absorbs added H_3O^+. Weak bases and salts providing their conjugate acids also serve as buffers (e.g., NH_3 and $NH_4^+Cl^-$).

Finally, the list of acids and bases was expanded to include oxides. The acids and bases studied are summarized as follows.

Type	Example	Reaction
ACIDS		
1 Molecular hydrogen compounds (H^+ + ion)	$HClO_4$	$HClO_4 + H_2O \longrightarrow H_3O^+ + ClO_4^-$
2 Cations (conjugate acids of weak bases)	NH_4^+	$NH_4^+ + H_2O \rightleftharpoons H_3O^+ + NH_3$
3 Nonmetal oxides	SO_3	$SO_3 + H_2O \longrightarrow H_2SO_4$ $H_2SO_4 + H_2O \longrightarrow H_3O^+ + HSO_4^-$
BASES		
1 Ionic hydroxides	$Ca(OH)_2$	$Ca(OH)_2 \xrightarrow{H_2O} Ca^{2+} + 2OH^-$
2 Molecular nitrogen compounds	$NH(CH_3)_2$	$NH(CH_3)_2 + H_2O \rightleftharpoons HNH(CH_3)_2^+ + OH^-$
3 Anions (conjugate bases of weak acids)	CN^-	$CN^- + H_2O \rightleftharpoons HCN + OH^-$
4 Metal oxides	K_2O	$K_2O + H_2O \longrightarrow 2KOH$ $KOH \xrightarrow{H_2O} K^+ + OH^-$

■ EXERCISES

ACIDS AND BASES

13-1 Give the formulas and names of the acid compounds derived from the following anions.
(a) NO_3^- (c) ClO_3^-
(b) NO_2^- (d) SO_3^{2-}

13-2 Give the formulas and names of the base compounds derived from the following cations.
(a) Cs^+ (c) Al^{3+}
(b) Sr^{2+} (d) Mn^{3+}

13-3 Give the formulas and names of the acid or base compounds derived from the following ions.
(a) Ba^{2+} (d) ClO^-
(b) Se^{2-} (e) $H_2PO_4^-$
(c) Pb^{2+} (f) Fe^{3+}

13-4 Write equations illustrating the reactions with water of the acids formed in Problem 13-1.

13-5 Write equations illustrating the reactions with water of the acids and bases formed in Problem 13-3.

NEUTRALIZATION AND SALTS

13-6 Identify each of the following as an acid, base, normal salt, or acid salt.
(a) H_2S (d) $Ba(HSO_4)_2$
(b) $BaCl_2$ (e) K_2SO_4
(c) H_3AsO_4 (f) $LiOH$

13-7 Write the balanced equation showing complete neutralization of the following.
(a) KOH by $HC_2H_3O_2$
(b) $Ca(OH)_2$ by HI
(c) H_2SO_4 by $Ca(OH)_2$

13-8 Write the balanced equation showing complete neutralization of the following.

(a) HNO_2 by NaOH
(b) H_2S by CsOH
(c) $H_2C_2O_4$ by $Ba(OH)_2$

13-9 Write the total ionic equations and the net ionic equations for reactions (b) and (c) in Problem 13-7.

13-10 Write balanced acid–base neutralization reactions that would lead to formation of the following salts or acid salts.
(a) $CaBr_2$ (c) $Ba(HS)_2$
(b) $Sr(ClO_2)_2$ (d) Li_2S

13-11 Write balanced acid–base neutralization reactions that would lead to formation of the following salts or acid salts.
(a) Na_2SO_3 (c) $Mg_3(PO_4)_2$
(b) AlI_3 (d) $NaHCO_3$

13-12 Write two equations illustrating the stepwise neutralization of H_2S with LiOH.

13-13 Write three equations illustrating the stepwise neutralization of H_3AsO_4 with NaOH. Write the total reaction.

13-14 Write the equation illustrating the reaction of 1 mol of H_2S with 1 mol of NaOH.

*13-15 Write the equation illustrating the reaction between 1 mol of $Ca(OH)_2$ and 2 mol of H_3PO_4.

STRENGTHS OF ACIDS AND BASES

13-16 Describe the difference between a strong acid and a weak acid.

13-17 Dimethylamine $[NH(CH_3)_2]$ is a weak base that reacts in water like ammonia (NH_3). Write the equilibrium illustrating this reaction.

13-18 The concentration of a weak monoprotic acid (HX) in water is 0.10 M. The concentration of H_3O^+ ion in this solution is 0.010 M. Is HX a weak or a strong acid? What percent of the acid is ionized?

13-19 A 0.50-mol quantity of an acid is dissolved in 2.0 L of water. In the solution, $[H_3O^+] = 0.25$ M. Is this a strong or a weak acid? Explain.

13-20 What is $[H_3O^+]$ in a 0.55 M $HClO_4$ solution?

13-21 What is $[H_3O^+]$ in a 0.55 M solution of a weak acid, HX, that is 3.0% ionized?

13-22 What is $[OH^-]$ in a 1.45 M solution of NH_3 if the NH_3 is 0.95% ionized?

*13-23 What is $[H_3O^+]$ in a 0.354 M solution of H_2SO_4? Assume that the first ionization is complete but that the second is only 25% complete.

EQUILIBRIUM OF WATER AND K_w

13-24 If some ions are present in pure water, why isn't pure water considered to be an electrolyte?

13-25 Why doesn't $[H_3O^+] = [OH^-] = 1.00 \times 10^{-2}$ M in water? What would happen if we tried to make such a solution by mixing 10^{-2} mol/L of KOH with 10^{-2} mol/L of HCl?

13-26 (a) What is $[H_3O^+]$ when $[OH^-] = 10^{-12}$ M?
(b) What is $[H_3O^+]$ when $[OH^-] = 10$ M?
(c) What is $[OH^-]$ when $[H_3O^+] = 2.0 \times 10^{-5}$ M?

13-27 (a) What is $[OH^-]$ when $[H_3O^+] = 1.50 \times 10^{-3}$ M?
(b) What is $[H_3O^+]$ when $[OH^-] = 2.58 \times 10^{-7}$ M?
(c) What is $[H_3O^+]$ when $[OH^-] = 56.9 \times 10^{-9}$ M?

13-28 When 0.250 mol of the strong acid $HClO_4$ is dissolved in 10.0 L of water, what is $[H_3O^+]$? What is $[OH^-]$?

13-29 Lye is a very strong base. What is $[H_3O^+]$ in a 2.55 M solution of NaOH? In the weakly basic household ammonia, $[OH^-] = 4.0 \times 10^{-3}$ M. What is $[H_3O^+]$?

13-30 Identify the solutions in Problem 13-26 as acidic, basic, or neutral.

13-31 Identify the solutions in Problem 13-27 as acidic, basic, or neutral.

13-32 Identify each of the following as an acidic, basic, or neutral solution.
(a) $[H_3O^+] = 6.5 \times 10^{-3}$ M
(b) $[H_3O^+] = 5.5 \times 10^{-10}$ M
(c) $[OH^-] = 4.5 \times 10^{-8}$ M
(d) $[OH^-] = 50 \times 10^{-8}$ M

13-33 Identify each of the following as an acidic, basic, or neutral solution.

(a) $[OH^-] = 8.1 \times 10^{-8}\ M$
(b) $[H_3O^+] = 10.0 \times 10^{-8}\ M$
(c) $[H_3O^+] = 4.0 \times 10^{-3}\ M$
(d) $[OH^-] = 55 \times 10^{-8}\ M$

pH AND pOH

13-34 What is the pH of the following solutions?
(a) $[H_3O^+] = 1.0 \times 10^{-6}\ M$
(b) $[H_3O^+] = 1.0 \times 10^{-9}\ M$
(c) $[OH^-] = 1.0 \times 10^{-2}\ M$
(d) $[OH^-] = 2.5 \times 10^{-5}\ M$
(e) $[H_3O^+] = 6.5 \times 10^{-11}\ M$

13-35 What is the pH of the following solutions?
(a) $[H_3O^+] = 1.0 \times 10^{-2}\ M$
(b) $[OH^-] = 1.0 \times 10^{-4}\ M$
(c) $[H_3O^+] = 1.0\ M$
(d) $[OH^-] = 3.6 \times 10^{-9}\ M$
(e) $[OH^-] = 7.8 \times 10^{-4}\ M$
(f) $[H_3O^+] = 42.2 \times 10^{-5}\ M$

13-36 What is $[H_3O^+]$ when
(a) pH = 3.00?   (d) pOH = 6.38?
(b) pH = 3.54?   (e) pH = 12.70?
(c) pOH = 8.00?

13-37 What is $[H_3O^+]$ when
(a) pH = 9.0?   (c) pH = 2.30?
(b) pOH = 9.0?   (d) pH = 8.90?

13-38 Identify each of the solutions in Problems 13-34 and 13-36 as acidic, basic, or neutral.

13-39 Identify each of the solutions in Problems 13-35 and 13-37 as acidic, basic, or neutral.

13-40 What is the pH of a 0.075 M solution of the strong acid HNO_3?

13-41 What is the pH of a 0.0034 M solution of the strong base KOH?

13-42 What is the pH of a 0.018 M solution of the strong base $Ca(OH)_2$?

13-43 A weak monoprotic acid is 10.0% ionized in solution. What is the pH of a 0.10 M solution of this acid?

13-44 A weak base is 5.0% ionized in solution. What is the pH of a 0.25 M solution of this base? (Assume one OH^- per molecule.)

***13-45** What is the pH of a 0.0010 M solution of H_2SO_4? (Assume that the first

ionization is complete but the second is only 25% complete.)

13-46 Identify each of the following solutions as strongly basic, weakly basic, neutral, weakly acidic, or strongly acidic.
(a) pH = 1.5   (e) pOH = 7.0
(b) pOH = 13.0   (f) pH = 8.5
(c) pH = 5.8   (g) pOH = 7.5
(d) pH = 13.0   (h) pH = −1.00

13-47 Arrange the following substances in order of increasing acidity.
(a) household ammonia, pH = 11.4
(b) vinegar, $[H_3O^+] = 2.5 \times 10^{-3}\ M$
(c) grape juice, $[OH^-] = 1.0 \times 10^{-10}\ M$
(d) sulfuric acid, pOH = 13.6
(e) eggs, pH = 7.8
(f) rain water, $[H_3O^+] = 2.0 \times 10^{-6}\ M$

13-48 Arrange the following substances in order of increasing acidity.
(a) lime juice, $[H_3O^+] = 6.0 \times 10^{-2}\ M$
(b) antacid tablet in water, $[OH^-] = 2.5 \times 10^{-6}\ M$
(c) coffee, pOH = 8.50
(d) stomach acid, pH = 1.8
(e) saliva, $[H_3O^+] = 2.2 \times 10^{-7}\ M$
(f) a soap solution, pH = 8.3
(g) a solution of lye, pOH = 1.2
(h) a banana, $[OH^-] = 4.0 \times 10^{-10}\ M$

BRØNSTED–LOWRY ACIDS AND BASES

13-49 What is the conjugate base of each of the following?
(a) HNO_3 (d) CH_4
(b) H_2SO_4 (e) H_2O
(c) HPO_4^{2-} (f) NH_3

13-50 What is the conjugate acid of each of the following?
(a) NH_2CH_3 (d) O^{2-}
(b) HPO_4^{2-} (e) HCN
(c) NO_3^- (f) H_2O

13-51 Identify conjugate acid–base pairs in the following reactions.
(a) $HClO_4 + OH^- \longrightarrow H_2O + ClO_4^-$
(b) $HSO_4^- + OCl^- \longrightarrow$
$$HOCl + SO_4^{2-}$$
(c) $H_2O + NH_2^- \longrightarrow NH_3 + OH^-$
(d) $NH_4^+ + H_2O \longrightarrow NH_3 + H_3O^+$

13-52 Identify conjugate acid–base pairs in the following reactions.

(a) $HCN + H_2O \longrightarrow H_3O^+ + CN^-$
(b) $HClO_4 + NO_3^- \longrightarrow$
$$HNO_3 + ClO_4^-$$
(c) $H_2S + NH_3 \longrightarrow NH_4^+ + HS^-$
(d) $H_3O^+ + HCO_3^- \longrightarrow$
$$H_2CO_3 + H_2O$$

13-53 Write reactions indicating Brønsted–Lowry acid behavior with H_2O for the following. Indicate conjugate acid–base pairs.
(a) H_2SO_3 (c) HBr   (e) H_2S
(b) HClO   (d) HSO_3^- (f) NH_4^+

13-54 Write reactions indicating Brønsted–Lowry base behavior with H_2O for the following. Indicate conjugate acid–base pairs.
(a) NH_3 (c) HS^- (e) F^-
(b) N_2H_4 (d) H^-

13-55 Most acid salts are amphoteric, as illustrated in Example 13-12. Write equations showing how HS^- can act as a Brønsted–Lowry base with H_3O^+ and as a Brønsted–Lowry acid with OH^-.

13-56 Calcium carbonate is the principal ingredient in Tums and other products used to neutralize the H_3O^+ from excess stomach acid (HCl). Write an equation illustrating how the CO_3^{2-} ion acts as a Brønsted–Lowry base with H_3O^+. Bicarbonate of soda ($NaHCO_3$) also acts as an antacid (base) in water. Write an equation illustrating how the HCO_3^- ion reacts with H_3O^+.

PREDICTING ACID–BASE REACTIONS
13-57 Refer to Table 13-2 to predict which of the following equations represent favorable reactions.
(a) $HOCl + Cl^- \longrightarrow HCl + OCl^-$
(b) $H_2S + CN^- \longrightarrow HCN + HS^-$
(c) $CH_4 + H^- \longrightarrow CH_3^- + H_2$
(d) $HCl + H_2O \longrightarrow H_3O^+ + Cl^-$

13-58 Refer to Table 13-2 to predict which of the following equations represent favorable reactions.
(a) $H_2S + F^- \longrightarrow HS^- + HF$
(b) $HNO_3 + NO_2^- \longrightarrow$
$$HNO_2 + NO_3^-$$

(c) $NH_4^+ + OH^- \longrightarrow NH_3 + H_2O$
(d) $HC_2H_3O_2 + HCO_3^- \longrightarrow$
$$H_2CO_3 + C_2H_3O_2^-$$

13-59 Use Table 13-2 to write equations for favorable reactions between the following acid–conjugate base pairs.
(a) H_3O^+, H_2O and H_2S, HS^-
(b) HOCl, OCl^- and HCN, CN^-
(c) H_2O, OH^- and H_2SO_3, HSO_3^-
(d) HBr, Br^- and CH_4, CH_3^-

13-60 Use Table 13-2 to write equations for favorable reactions between the following acid–conjugate base pairs.
(a) HCN, CN^- and H_2, H^-
(b) NH_4^+, NH_3 and H_2CO_3, HCO_3^-
(c) $HC_2H_3O_2$, $C_2H_3O_2^-$ and HCN, CN^-
(d) H_2O, OH^- and HNO_3, NO_3^-

*13-61 Complete the following equations in which water acts as a base. If the reaction is favorable use a single arrow. If it occurs to a limited extent use double arrows. If it does not occur, write "no reaction."
(a) $H_2S + H_2O$ (c) $HNO_3 + H_2O$
(b) $CH_4 + H_2O$ (d) $NH_4^+ + H_2O$

*13-62 Complete the following equations in which water acts as an acid. If the reaction is favorable use a single arrow. If it occurs to a limited extent use double arrows. If it does not occur, write "no reaction."
(a) $NO_3^- + H_2O$
(b) $C_2H_3O_2^- + H_2O$
(c) $NH_3 + H_2O$
(d) $CH_3^- + H_2O$

13-63 Which of the following equations represents a reaction that does not occur to a measurable extent?
(a) $HOCl + OH^- \longrightarrow OCl^- + H_2O$
(b) $HNO_3 + OH^- \longrightarrow NO_3^- + H_2O$
(c) $H_2 + OH^- \longrightarrow H^- + H_2O$
(d) $NH_4^+ + OH^- \longrightarrow NH_3 + H_2O$

13-64 Which of the following equations represents a reaction that does not occur to a measurable extent?
(a) $ClO_4^- + H_3O^+ \longrightarrow HClO_4 + H_2O$
(b) $HS^- + H_3O^+ \longrightarrow H_2S + H_2O$
(c) $CH_3^- + H_3O^+ \longrightarrow CH_4 + H_2O$

13-65 Complete the following hydrolysis equilibria:

(a) $S^{2-} + H_2O \rightleftharpoons$ _____ $+ OH^-$

(b) $N_2H_5^+ + H_2O \rightleftharpoons N_2H_4 +$ _____

(c) $HPO_4^{2-} + H_2O \rightleftharpoons$

$$H_2PO_4^- +$$ _____

(d) $HNH(CH_3)_2^+ + H_2O \rightleftharpoons$

_____ $+ H_3O^+$

13-66 Complete the following hydrolysis equilibria.

(a) $CN^- + H_2O \rightleftharpoons HCN +$ _____

(b) $NH_4^+ + H_2O \rightleftharpoons$ _____ $+ H_3O^+$

(c) $B(OH)_4^- + H_2O \rightleftharpoons$

$$H_3BO_3 +$$ _____

(d) $Al(H_2O)_6^{3+} + H_2O \rightleftharpoons$

$$Al(H_2O)_5(OH)^{2+} +$$ _____

13-67 Write the hydrolysis equilibrium (if any) for the following ions.

(a) F^- (d) HPO_4^{2-}

(b) SO_3^{2-} (e) CN^-

(c) $H_2N(CH_3)_2^+$ (f) Li^+

13-68 Write hydrolysis reactions, if any occur, for the following ions.

(a) Br^- (c) ClO_4^- (e) Ca^{2+}

(b) HS^- (d) H^-

*13-69 Both C_2^{2-} and its conjugate acid HC_2^- hydrolyze 100% in water. From this information complete the following equation.

$$CaC_2(s) + 2H_2O \longrightarrow \text{\_\_\_\_\_} (g)$$
$$+ Ca^{2+}(aq) + 2 \text{\_\_\_\_\_} (aq)$$

[The gas formed (acetylene) can be burned as it is produced. This reaction was once important for this purpose as a source of light in old miners' lamps.]

13-70 Sodium hypochlorite (NaOCl) is the principal ingredient in commercial bleaches. When dissolved in water, it makes a slightly basic solution. Write the equation illustrating this basic behavior.

13-71 Aqueous NaF solutions are slightly basic whereas aqueous NaCl solutions are neutral. Write the appropriate equation that illustrates this. Why aren't NaCl solutions also basic?

13-72 Predict whether aqueous solutions of the following salts are acidic, neutral, or basic.

(a) $Ba(ClO_4)_2$

(b) $N_2H_5^+NO_3^-$ (N_2H_4 is a weak base)

(c) $LiC_2H_3O_2$

(d) KBr

(e) NH_4Cl

(f) BaF_2

13-73 Predict whether aqueous solutions of the following salts are acidic, neutral, or basic.

(a) Na_2CO_3 (c) NH_4ClO_4

(b) K_3PO_4 (d) SrI_2

*13-74 Aqueous solutions of NH_4CN are basic. Write the two hydrolysis reactions and indicate which takes place to the greater extent.

*13-75 Aqueous solutions of $NaHSO_3$ are acidic. Write the two equations (one hydrolysis and one ionization) and indicate which takes place to the greater extent.

BUFFERS

13-76 Identify which of the following form buffer solutions when 0.50 mol of each compound is dissolved in 1 L of water

(a) HNO_2 and KNO_2

(b) NH_4Cl and NH_3

(c) HNO_3 and KNO_2

(d) HNO_3 and KNO_3

(e) HOBr and $Ca(OBr)_2$

(f) HCN and KClO

(g) NH_3 and $BaBr_2$

(h) H_2S and LiHS

(i) KH_2PO_4 and K_2HPO_4

13-77 A certain solution contains dissolved HCl and NaCl. Why can't this solution act as a buffer?

*13-78 If 0.5 mol of KOH is added to a solution containing 1.0 mol of $HC_2H_3O_2$, the resulting solution is a buffer. Explain.

13-79 A solution contains 0.50 mol each of HClO and NaClO. If 0.60 mol of KOH is added, will the buffer prevent a significant change in pH? Explain.

13-80 Write the equilibrium involved in the NH_3, NH_4Cl buffer system. Write equations illustrating how this system reacts with added H_3O^+ and added OH^-.

13-81 Write the equilibrium involved in the H_2CO_3, HCO_3^- buffer system. Write equations illustrating how this system reacts with added H_3O^+ and added OH^-.

13-82 Write the equilibrium involved in the $H_2PO_4^-$, HPO_4^{2-} buffer system. Write equations illustrating how this system reacts with added H_3O^+ and added OH^-.

OXIDES AS ACIDS AND BASES

13-83 Write the formula of the acid or base formed when each of the following anhydrides is dissolved in water.
(a) SrO (c) P_4O_{10} (e) N_2O_3
(b) SeO_3 (d) Cs_2O (f) Cl_2O_5

13-84 Write the formula of the acid or base formed when each of the following anhydrides is dissolved in water.
(a) BaO (c) Cl_2O (e) K_2O
(b) SeO_2 (d) Br_2O

13-85 Carbon dioxide is removed from manned space capsules by bubbling the air through a LiOH solution. Show the reaction and the product formed.

*13-86 Complete the following reaction.

$$Li_2O(s) + N_2O_5(g) \longrightarrow \underline{\hspace{2cm}}(s)$$

GENERAL PROBLEMS

13-87 Sulfite ion (SO_3^{2-}) and sulfur trioxide (SO_3) look similar, but one forms a strongly acid solution whereas the other is weakly basic. Write equations illustrating this behavior.

13-88 Tell whether each of the following compounds forms acidic, basic, or neutral solutions when added to pure water. Write the reaction illustrating the acidic or basic behavior where appropriate.
(a) H_2S (f) $Ba(OH)_2$
(b) $KClO$ (g) $Sr(NO_3)_2$
(c) NaI (h) $LiNO_2$
(d) NH_3 (i) H_2SO_3
(e) $HN_2H_4^+Br^-$ (j) Cl_2O_3

13-89 Tell whether each of the following compounds forms acidic, basic, or neutral solutions when added to pure water. Write the reaction illustrating the acidic or basic behavior where appropriate.
(a) $HOBr$ (e) SO_2
(b) CaO (f) $Ba(C_2H_3O_2)_2$
(c) NH_4ClO_4 (g) $RbBr$
(d) N_2H_4

*13-90 In a lab there are five different solutions with pH's of 1.0, 5.2, 7.0, 10.2, and 13.0. The solutions are LiOH, $SrBr_2$, KOCl, NH_4Cl, and HI, all at the same concentration. Which pH corresponds to which compound? What must be the concentration of all of these compounds?

ELECTRICITY
Our lives have become so dependent upon this convenient source of
energy. Chemical reactions can be used as a source of electricity.

OXIDATION –
REDUCTION
REACTIONS

PURPOSES OF CHAPTER

In Chapter 14 we define oxidation–reduction reactions, balance their equations, describe some practical applications, and predict whether or not certain reactions will occur.

OBJECTIVES FOR CHAPTER

After completion of this chapter, you should be able to:

1   Apply the appropriate rules to calculate the oxidation state of the atoms of each element in a compound or ion. (14-1)

2   Identify the reactant oxidized (reducing agent) and the reactant reduced (oxidizing agent) in a specified redox reaction. (14-1)

3   Balance redox equations by either the oxidation state method or the ion-electron method in acidic or basic solutions. (14-2, 14-3).

4   Describe how the relationship of the potential energy of reactants and products

14

relates to the reactions occurring in a voltaic cell. (14-4)

5 Write the anode, cathode, and overall reactions occurring in specified voltaic cells, including the Daniell cell, the dry cell, and the lead–acid battery. (14-4)

6 Rank the strength of two competing oxidizing agents and reducing agents from observance of a single spontaneous redox reaction. (14-5)

7 Describe the inverse relationship between the oxidizing strength of a metal ion and the reducing strength of the corresponding metal. (14-5)

8 Predict the occurrence and write the equations for spontaneous redox reactions by use of an appropriate table. (14-5)

9 Give some practical applications of electrolytic cells. (14-6)

A current college student may think of an "oldtimer" as someone who remembers what it was like before television. This textbook author prefers to think of an "oldtimer" as someone who remembers what it was like before electricity became available. In fact, it wasn't until the early part of this century that this amazing tool became readily available to free us from so many ordinary but time-consuming chores. Electricity, which is simply a flow of electrons, is a source of energy that most of us now take for granted. But we have chemistry to thank for the electrical age. Many of the early experiments, such as those of Alexander Graham Bell with the telephone, used chemical reactions as a source of electrical current. Even now, when we turn on a calculator or a flashlight or start a car, the flow of electrons originates from a specific chemical reaction. In such a reaction there is an exchange of electrons between reactants and it is these electrons that can serve as a source of electricity. We will see in this chapter that the reaction occurs because of a competition for electrons between reactants and products much like acid–base reactions occur because of a similar competition for H^+ ions. Combustion, metabolism, corrosion, and decay of organic matter are also examples of electron exchange reactions. In fact, all types of reactions discussed in Chapter 9, except double-replacement reactions, usually involve an exchange of electrons.

Although this classification of chemical reactions is quite broad, it is well worth the effort to examine the common characteristics of such reactions. After defining our terms we will apply our knowledge of the electron exchange process to balance equations that are too complicated for inspection methods described in Chapter 9. We will then explore how the energy of certain favorable reactions can be harnessed in voltaic cells or batteries. Next we will examine how a knowledge of the competitive forces for electrons can be used to predict the occurrence of certain reactions in a manner similar to what was done for acid–conjugate base pairs in the previous chapter. Finally, we will see how unfavorable reactions can be made to occur by the input of electrical energy from an outside source.

As background, review the following:

1   The formation of ions and binary compounds (Sections 6-3 and 6-4)
2   The names and formulas of ions (Section 7-3)
3   Ionic equations (Section 12-3)

14-1 THE NATURE OF OXIDATION AND REDUCTION

When a small chunk of sodium metal is placed in a container of chlorine gas an obvious chemical reaction occurs. Flames leap from the sodium as a white deposit of sodium chloride forms on the walls. (See Figure 14-1.) Two toxic elements — one a solid metal, the other a gaseous nonmetal — combine to form a compound that is necessary for our good health. This reaction is represented by the equation

$$2Na(s) + Cl_2(g) \longrightarrow 2NaCl(s)$$

Figure 14-1   FORMATION OF NaCl FROM ITS ELEMENTS. An active metal reacts with a poisonous gas to form sodium chloride.

In the precipitation and acid–base reactions discussed so far, the atoms in the reactants kept their quota of electrons in changing to products. Such is not the case in the reaction shown above, however. This is best illustrated by taking the reaction apart and examining the change that each reactant undergoes as the product is formed. An electron exchange reaction can be viewed as the sum of two half-reactions. A **half-reaction** *illustrates either the loss of electrons or the gain of electrons as a separate balanced equation.* Thus, the half-reaction involving only sodium is

$$Na \longrightarrow Na^+ + e^-$$

Recall that the oxidation state of an element is zero and that of a monatomic ion is the charge of the ion. In the half-reaction note that sodium has changed from zero to $+1$ and in so doing has lost one electron. *A substance that loses electrons in a chemical reaction (as indicated by an increase in oxidation state) is said to be* **oxidized.** Originally, the term "oxidation" referred only to a substance that added oxygen atoms in a reaction. But since this meant that the substance adding oxygen underwent an increase in oxidation state, the process of oxidation now refers to any substance containing an element undergoing an increase in oxidation state. This is so regardless of whether oxygen is even involved in the reaction.

Now let's consider what happens to the chlorine in going from reactant to product.

$$2e^- + Cl_2 \longrightarrow 2Cl^-$$

In this half-reaction, note that the oxidation state of the chlorine has decreased from zero to -1 for each chloride ion formed. Two electrons are gained by the elemental chlorine in the reaction. *A substance that gains electrons in a chemical reaction (as indicated by a decrease in oxidation state) is said to be* **reduced.** Originally, the term "reduction" referred to a process in which oxygens were removed from a substance, leaving its mass "reduced." Now, any substance containing an element undergoing a decrease in oxidation state is said to be reduced or to undergo reduction.

Obviously, the two processes (oxidation and reduction) complement each other, giving us the basis of this classification of chemical reactions. *Reactions involving an exchange of electrons are known as* **oxidation–reduction** *or simply* **redox** *reactions.*

Note that the reactant that is oxidized (Na) provides the electrons for the reactant reduced. *The reactant oxidized is therefore referred to as the* **reducing agent.** Conversely, the reactant reduced (Cl_2) accepts the electrons from the reactant oxidized. *The reactant reduced is referred to as the* **oxidizing agent.**

The reaction of Na and Cl_2 is summarized as follows:

Reactant	Change	Product	Agent
Na	Oxidation	Na^+	Reducing
Cl_2	Reduction	Cl^-	Oxidizing

Let's now see how the two half-reactions add together to make a complete, balanced equation. *An important principle of a redox reaction is that the electrons gained in the reduction process must equal the electrons lost in the oxidation process.* Note in our sample reaction that the reduction process involving Cl_2 requires two electrons. Therefore, the oxidation process must involve two Na's to provide these two electrons. In adding the two half-reactions, the electrons on both sides of the equation must be equal so that they can be eliminated by subtraction.

Oxidation half-reaction	$2Na \longrightarrow 2Na^+ + 2e^-$
Reduction half-reaction	$2e^- + Cl_2 \longrightarrow 2Cl^-$
Total reaction	$2Na + Cl_2 \longrightarrow 2Na^+Cl^-$

The principle of "electrons gained equal electrons lost" provides the basis for methods to balance equations of redox reactions. We apply this principle in the next two sections which are dedicated specifically to balancing equations. For now, however, we concentrate on the identification of substances undergoing oxidation and reduction. In many reactions, the substance undergoing a change is not as obvious as in the sample reaction we discussed above. In polyatomic compounds usually only one of the atoms undergoes a change. Thus, it is often necessary to calculate the oxidation states of all of the elements in a compound so we can see which has undergone the change. Eventually, identification of oxidized and reduced species becomes somewhat automatic. The rules for calculation of oxidation states were given in Section 7-1 but are summarized again for your convenience. Sample calculations were given in Chapter 7.

1 The oxidation state of an element in its free state is zero.

2 The oxidation state of a monatomic ion is the same as the charge on that ion. (Alkali metals are $+1$, alkaline earth metals are $+2$.)

3 The halogens have a -1 oxidation state in binary compounds when bound to a less electronegative element.

4 Oxygen is usually -2. Peroxides and superoxides contain oxygen in -1 or $-\frac{1}{2}$ oxidation states, respectively. When bound to F, oxygen has a positive oxidation state.

5 Hydrogen is usually $+1$. When combined with a less electronegative element (i.e., a metal), H has a -1 oxidation state.

6 The sum of the oxidation states of all of the atoms in a neutral compound is zero. For polyatomic ions, the sum of the oxidation states equals the charge on the ion.

■
SEE PROBLEMS
14-1 AND 14.2.

EXAMPLE 14-1

In the following unbalanced equations, indicate the reactant oxidized, the reactant reduced, the oxidizing agent, and the reducing agent. Indicate the products that contain the elements that were oxidized or reduced.

(a) $Al + HCl \rightarrow AlCl_3 + H_2$
(b) $CH_4 + O_2 \rightarrow CO_2 + H_2O$
(c) $MnO_2 + HCl \rightarrow MnCl_2 + Cl_2 + H_2O$
(d) $K_2Cr_2O_7 + SnCl_2 + HCl \rightarrow CrCl_3 + SnCl_4 + KCl + H_2O$

PROCEDURE

In the equations, we wish to identify the species that contain atoms of an element undergoing a change in oxidation state. At first, it may be necessary to calculate the oxidation state of every atom in the equation until you can recognize the changes by inspection. You will notice that any substance present as a free element is involved in either the oxidation or the reduction process and that hydrogen and oxygen are generally not oxidized or reduced unless they are present as free elements.

SOLUTION

(a)   Oxidation state of element

$$\overset{0}{Al} + \overset{+1\,-1}{HCl} \longrightarrow \overset{+3\,-1}{AlCl_3} + \overset{0}{H_2}$$

Reactant	Change	Product	Agent
Al	Oxidation	$AlCl_3$	Reducing
HCl	Reduction	H_2	Oxidizing

(b)   Oxidation state of element

$$\overset{-4\,+1}{CH_4} + \overset{0}{O_2} \longrightarrow \overset{+4\,-2}{CO_2} + \overset{+1\,-2}{H_2O}$$

Reactant	Change	Product	Agent
CH_4	Oxidation	CO_2	Reducing
O_2	Reduction	CO_2, H_2O	Oxidizing

(c)   Oxidation state of element

$$\overset{+4\,-2}{MnO_2} + \overset{+1\,-1}{HCl} \longrightarrow \overset{+2\,-1}{MnCl_2} + \overset{0}{Cl_2} + \overset{+1\,-2}{H_2O}$$

Reactant	Change	Product	Agent
HCl	Oxidation	Cl_2	Reducing
MnO_2	Reduction	$MnCl_2$	Oxidizing

(d) Oxidation state of element

$$\overset{+1\ +6\ -2}{K_2Cr_2O_7} + \overset{+2\ -1}{SnCl_2} + \overset{+1\ -1}{HCl} \longrightarrow \overset{+3\ -1}{CrCl_3} + \overset{+4\ -1}{SnCl_4} + \overset{+1\ -2}{H_2O}$$

Reactant	Change	Product	Agent
$SnCl_2$	Oxidation	$SnCl_4$	Reducing
$K_2Cr_2O_7$	Reduction	$CrCl_3$	Oxidizing

■
SEE PROBLEMS
14-3 THROUGH
14-9.

**14-2
BALANCING
REDOX
EQUATIONS:
OXIDATION
STATE
METHOD**

There are two widely used procedures for balancing redox equations. The **bridge** or **oxidation state** *method discussed in this section focuses on the atoms of the elements undergoing a change in oxidation state.* This method serves as a helpful introduction to the concepts and is effective in balancing uncomplicated redox equations. *The second procedure is the* **ion-electron** *method, which focuses on the entire molecule or ion containing the atom undergoing a change in oxidation state.* This latter method is generally used in general and subsequent chemistry courses and is discussed in the next section.

The following reaction will be used to illustrate the procedures for balancing equations by the oxidation state method:

$$HNO_3(aq) + H_2S(aq) \longrightarrow NO(g) + S(s) + H_2O$$

1   Identify the atoms whose oxidation states have changed.

$$\overset{+5}{H\underline{N}O_3} + \overset{-2}{H_2\underline{S}} \longrightarrow \overset{+2}{\underline{N}O} + \overset{0}{\underline{S}} + H_2O$$

2   Draw a bridge between the same atoms whose oxidation states have changed, indicating the electrons gained or lost. This is the change in oxidation state. *Be sure that the atoms in question are balanced on both sides of the equation if they are not the same.*

$$\overset{\overset{\displaystyle +3e}{\overline{\quad\qquad\qquad}}}{\underset{\underset{\displaystyle -2e^-}{\overline{\quad\qquad\qquad}}}{\overset{+5}{HNO_3} + \overset{-2}{H_2S} \longrightarrow \overset{+2}{NO} + \overset{0}{S} + H_2O}}$$

3   Multiply the two numbers ($+3$ and -2) by whole numbers that produce a common number. For 3 and 2 the common number is 6. (For example, $+3 \times \underline{2} = +6$; $-2 \times \underline{3} = -6$.) Use these multipliers as coefficients of the respective compounds or elements.

$$\overset{+3e \times \textcircled{2} = +6e^-}{\underset{-2e^- \times \textcircled{3} = -6e^-}{2HNO_3 + 3H_2S \longrightarrow 2NO + 3S + H_2O}}$$

Note that six electrons are lost (bottom) and six are gained (top).

4   Balance the rest of the equation by inspection. Note that there are eight H's on the left, so *four* H_2O's are needed on the right. If the equation has been balanced correctly, the O's should balance. Note that they do.

$$2HNO_3 + 3H_2S \longrightarrow 2NO + 3S + 4H_2O$$

EXAMPLE 14-2

Balance the following equations by the oxidation state method.

(a) $Zn + AgNO_3 \rightarrow Zn(NO_3)_2 + Ag$

$$\overset{-2e^-}{\underset{+1e^-}{\underset{+1 \qquad\qquad\qquad 0}{\overset{0 \qquad\qquad\qquad +2}{Zn + AgNO_3 \longrightarrow Zn(NO_3)_2 + Ag}}}}$$

The oxidation process (top) should be multiplied by 1 and the reduction process (bottom) should be multiplied by 2.

$$\overset{-2e^- \times 1 = -2e^-}{\underset{+1e^- \times 2 = +2e}{Zn + 2AgNO_3 \longrightarrow Zn(NO_3)_2 + 2Ag}}$$

The final balanced equation is

$$Zn + 2AgNO_3 \longrightarrow Zn(NO_3)_2 + 2Ag$$

(b) $Cu + HNO_3 \rightarrow Cu(NO_3)_2 + H_2O + NO_2$

$$-2e^- \times 1 = -2e^-$$

$$\begin{array}{ccc} 0 & & +2 \\ Cu + HNO_3 & \longrightarrow & Cu(NO_3)_2 + H_2O + NO_2 \\ +5 & & +4 \end{array}$$

$$+1e^- \times 2 = +2e^-$$

The equation, so far, is

$$Cu + 2HNO_3 \longrightarrow Cu(NO_3)_2 + H_2O + 2NO_2$$

Note, however, that four N's are present on the right but only two on the left. The addition of two more HNO_3's balances the N's, and the equation is completely balanced with two H_2O's on the right:

$$Cu + 4HNO_3 \longrightarrow Cu(NO_3)_2 + 2H_2O + 2NO_2$$

(In this aqueous reaction, HNO_3 serves two functions. Two HNO_3's are reduced to two NO_2's and the other two HNO_3's provide anions for the Cu^{2+} ion. These latter NO_3^- ions are present in the solution as spectator ions. Spectator ions are not oxidized, reduced, or otherwise changed during the reaction.)

(c) $Al + H_2SO_4 \rightarrow Al_2(SO_4)_3 + H_2$

The atoms undergoing a change in oxidation state are Al and H. Before you calculate electrons gained or lost, both atoms in question must be balanced on both sides of the equation. In this case, two Al's lose a total of six electrons and two H's gain a total of two electrons.

$$2(-3e^-) \times 1 = -6e^-$$

$$\begin{array}{ccc} 0 & & +6 \\ 2Al + H_2SO_4 & \longrightarrow & Al_2(SO_4)_3 + H_2 \\ +2 & & 0 \end{array}$$

$$2(+1e^-) \times 3 = +6e^-$$

$$2Al + 3H_2SO_4 \longrightarrow Al_2(SO_4)_3 + 3H_2$$

■
SEE PROBLEMS
14-10 AND 14-11.

14-3 BALANCING REDOX EQUATIONS: ION-ELECTRON METHOD

In the ion-electron method (also known as the half-reaction method), the total reaction is separated into half-reactions, which are then balanced separately and added. Although this method is somewhat more involved than the oxidation state method, it is more realistic for redox reactions in aqueous solutions. The ion-electron method recognizes that the entire molecule or ion, not just one atom, undergoes a change. This method also provides the proper background for the study of electrochemistry, which involves the applications of balanced half-reactions. This will be apparent later in this chapter.

The rules for balancing equations are somewhat different in acidic solution [containing $H^+(aq)$ ion] than in basic solution [containing $OH^-(aq)$ ion]. The two solutions are considered separately, with acid solution reactions discussed first. To simplify the equations, only the net ionic equations are balanced.

The balancing of an equation in aqueous acid solution is illustrated by the following unbalanced equation.

$$Cr_2O_7{}^{2-}(aq) + Cl^-(aq) + H^+(aq) \longrightarrow Cr^{3+}(aq) + Cl_2(g) + H_2O$$

1 Separate the molecule or ion that contains an atom that has been oxidized or reduced and the product containing the atom that changed. If necessary, calculate the oxidation states of individual atoms until you are able to recognize the species that changes. *It is actually not necessary to know the oxidation state.* The reduction process is

$$Cr_2O_7{}^{2-} \longrightarrow Cr^{3+}$$

2 If necessary, balance the atom undergoing a change in oxidation state. In this case it is Cr.

$$Cr_2O_7{}^{2-} \longrightarrow 2Cr^{3+}$$

3 Balance the oxygens by adding H_2O on the side needing the oxygens (one H_2O for each O needed).

$$Cr_2O_7{}^{2-} \longrightarrow 2Cr^{3+} + 7H_2O$$

4 Balance the hydrogens by adding H^+ on the other side of the equation from the H_2O's ($2H^+$ for each H_2O added). Note that the H and O have not undergone a change in oxidation state.

$$14H^+ + Cr_2O_7{}^{2-} \longrightarrow 2Cr^{3+} + 7H_2O$$

5 The atoms in the half-reaction are now balanced. Check to make sure. The charge on both sides of the reaction must now be balanced. To do this, add the appropriate number of electrons to the *more positive* side. The total charge on the left is $(14 \times +1) + (-2) = +12$. The total charge on the right is $(2 \times +3) = +6$. By adding $6e^-$ on the left, the charges balance on both sides and the half-reaction is balanced.

$$6e^- + 14H^+ + Cr_2O_7{}^{2-} \longrightarrow 2Cr^{3+} + 7H_2O$$

6 Repeat the same procedure for the other half-reaction.

$$Cl^- \longrightarrow Cl_2$$
$$2Cl^- \longrightarrow Cl_2$$
$$2Cl^- \longrightarrow Cl_2 + 2e^-$$

7 Before the two half-reactions are added, we must make sure that electrons gained equal electrons lost. Sometimes, the half-reactions must be multiplied by factors that give the same number of electrons. In this case, if the oxidation process is multiplied by 3 (and the reduction process by 1), there will be an exchange of $6e^-$. When these two half-reactions are added, the $6e^-$ can be subtracted from both sides of the equation.

$$3[2Cl^- \longrightarrow Cl_2 + 2e^-]$$
$$6Cl^- \longrightarrow 3Cl_2 + 6e^-$$

Addition produces the balanced net ionic equation:

$$6e^- + 14H^+ + Cr_2O_7^{2-} \longrightarrow 2Cr^{3+} + 7H_2O$$
$$6Cl^- \longrightarrow 3Cl_2 + 6e^-$$
$$\overline{14H^+(aq) + 6Cl^-(aq) + Cr_2O_7^{2-}(aq) \longrightarrow 2Cr^{3+}(aq) + 3Cl_2(g) + 7H_2O(l)}$$

EXAMPLE 14-3

Balance the following equations for reactions occurring in acid solution by the ion-electron method.

(a) $\quad MnO_4^-(aq) + SO_2(g) + H_2O(l) \longrightarrow Mn^{2+}(aq) + SO_4^{2-}(aq) + H^+(aq)$

Reduction:	$MnO_4^- \longrightarrow Mn^{2+}$
H_2O:	$MnO_4^- \longrightarrow Mn^{2+} + 4H_2O$
H^+:	$8H^+ + MnO_4^- \longrightarrow Mn^{2+} + 4H_2O$
e^-:	$5e^- + 8H^+ + MnO_4^- \longrightarrow Mn^{2+} + 4H_2O$
Oxidation:	$SO_2 \longrightarrow SO_4^{2-}$
H_2O:	$2H_2O + SO_2 \longrightarrow SO_4^{2-}$
H^+:	$2H_2O + SO_2 \longrightarrow SO_4^{2-} + 4H^+$
e^-:	$2H_2O + SO_2 \longrightarrow SO_4^{2-} + 4H^+ + 2e^-$

The reduction reaction is multiplied by 2 and the oxidation by 5 to produce 10 electrons for each process as shown below.

$$2(5e^- + 8H^+ + MnO_4^- \longrightarrow Mn^{2+} + 4H_2O)$$
$$5(2H_2O + SO_2 \longrightarrow SO_4^{2-} + 4H^+ + 2e^-)$$
$$\cancel{10e^-} + 16H^+ + 2MnO_4^- \longrightarrow 2Mn^{2+} + 8H_2O$$
$$\overline{10H_2O + 5SO_2 \longrightarrow 5SO_4^{2-} + 20H^+ + \cancel{10e^-}}$$

$$10H_2O + 16H^+ + 5SO_2 + 2MnO_4^- \longrightarrow$$
$$5SO_4^{2-} + 2Mn^{2+} + 8H_2O + 20H^+$$

Note that H_2O and H^+ are present on both sides of the equation. Therefore, $8H_2O$ and $16H^+$ can be subtracted from *both sides*, leaving the final balanced net ionic equation:

$$2MnO_4^-(aq) + 5SO_2(g) + 2H_2O(l) \longrightarrow$$
$$2Mn^{2+}(aq) + 5SO_4^{2-}(aq) + 4H^+(aq)$$

(b) $Cu(s) + NO_3^-(aq) \rightarrow Cu^{2+}(aq) + H_2O + NO(g)$

Reduction:	$NO_3^- \longrightarrow NO$
H_2O:	$NO_3^- \longrightarrow NO + 2H_2O$
H^+:	$4H^+ + NO_3^- \longrightarrow NO + 2H_2O$
e^-:	$3e^- + 4H^+ + NO_3^- \longrightarrow NO + 2H_2O$
Oxidation:	$Cu \longrightarrow Cu^{2+}$
e^-:	$Cu \longrightarrow Cu^{2+} + 2e^-$

Multiply the reduction half-reaction by 2 and the oxidation half-reaction by 3, and then add the two half-reactions:

$$\cancel{6e^-} + 8H^+ + 2NO_3^- \longrightarrow 2NO + 4H_2O$$
$$3Cu \longrightarrow 3Cu^{2+} + \cancel{6e^-}$$
$$\overline{8H^+(aq) + 2NO_3^-(aq) + 3Cu(s) \longrightarrow 3Cu^{2+}(aq) + 2NO(g) + 4H_2O(l)}$$

■
SEE PROBLEMS
14-12 THROUGH
14-15.

In a basic solution, OH^- ion is predominant rather than H^+. Therefore, the procedure is adjusted to allow the half-reactions to be balanced with OH^- ions and H_2O molecules in the basic solution.

The balancing of an equation in aqueous base solution is illustrated by the following unbalanced equation.

$$MnO_4^-(aq) + C_2O_4^{2-}(aq) + OH^-(aq) \longrightarrow MnO_2(s) + CO_3^{2-}(aq) + H_2O(l)$$

1   Separate the molecule or ion that contains an atom that has been oxidized or reduced and the product containing that atom. The reduction process is

$$MnO_4^- \longrightarrow MnO_2$$

2   For every oxygen needed on the oxygen-deficient side, add *two* OH^- ions. This provides one O and one H_2O:

$$O\;\boxed{H + OH} = O + H_2O$$
$$MnO_4^- \longrightarrow MnO_2 + 4OH^-$$

3   For every *two* OH^- ions added on the one side, add *one* H_2O to the other side:

$$2H_2O + MnO_4^- \longrightarrow MnO_2 + 4OH^-$$

4   Balance the charge by adding electrons to the more positive side as before:

$$3e^- + 2H_2O + MnO_4^- \longrightarrow MnO_2 + 4OH^-$$

5 Repeat the same procedure for the other half-reaction:

$$C_2O_4^{2-} \longrightarrow CO_3^{2-}$$

(balance C) $$C_2O_4^{2-} \longrightarrow 2CO_3^{2-}$$

(balance O and H) $$4OH^- + C_2O_4^{2-} \longrightarrow 2CO_3^{2-} + 2H_2O$$

(balance charge) $$4OH^- + C_2O_4^{2-} \longrightarrow 2CO_3^{2-} + 2H_2O + 2e^-$$

6 Multiply the half-reactions so that electrons gained equal electrons lost. In this example the oxidation half-reaction is multiplied by 3 and the reduction half-reaction is multiplied by 2:

$$6e^- + 4H_2O + 2MnO_4^- \longrightarrow 2MnO_2 + 8OH^-$$
$$12OH^- + 3C_2O_4^{2-} \longrightarrow 6CO_3^{2-} + 6H_2O + 6e^-$$

$$4$$

$$4H_2O + 2MnO_4^- + 12OH^- + 3C_2O_4^{2-} \longrightarrow$$

$$2$$

$$6CO_3^{2-} + 6H_2O + 2MnO_2 + 8OH^-$$

$$2MnO_4^-(aq) + 3C_2O_4^{2-}(aq) + 4OH^-(aq) \longrightarrow$$
$$6CO_3^{2-}(aq) + 2MnO_2(s) + 2H_2O(l)$$

When hydrogen is present in a substance oxidized or reduced in basic solution, the procedure requires an additional step. Hydrogens in a compound or ion are balanced by adding an OH^- for each hydrogen *on the same side* and an H_2O on the other side. For example, consider the following process in basic solution:

$$NO_3^- \longrightarrow NH_3$$

(balance O) $$3H_2O + NO_3^- \longrightarrow NH_3 + 6OH^-$$

Now balance the H's in NH_3 by adding an additional three OH^-'s on the same side and an additional three H_2O's on the left.

$$6H_2O + NO_3^- \longrightarrow NH_3 + 9OH^-$$

(balance charge) $$8e^- + 6H_2O + NO_3^- \longrightarrow NH_3 + 9OH^-$$

EXAMPLE 14-4

Balance the following equation in basic solution by the ion-electron method.

$$Bi_2O_3(s) + OCl^-(aq) + OH^-(aq) \longrightarrow BiO_3^-(aq) + Cl^-(aq) + H_2O(l)$$

Reduction: $$OCl^- \longrightarrow Cl^-$$

OH^-: $$OCl^- \longrightarrow Cl^- + 2OH^-$$

H_2O: $$H_2O + OCl^- \longrightarrow Cl^- + 2OH^-$$

e^-: $$2e^- + H_2O + OCl^- \longrightarrow Cl^- + 2OH^-$$

Oxidation: $Bi_2O_3 \longrightarrow BiO_3^-$

Bi: $Bi_2O_3 \longrightarrow 2BiO_3^-$

OH^-: $6OH^- + Bi_2O_3 \longrightarrow 2BiO_3^-$

H_2O: $6OH^- + Bi_2O_3 \longrightarrow 2BiO_3^- + 3H_2O$

e^-: $6OH^- + Bi_2O_3 \longrightarrow 2BiO_3^- + 3H_2O + 4e^-$

Multiply the reduction half-reaction by 2 and add to the oxidation half-reaction:

$$4e^- + 2H_2O + 2OCl^- \longrightarrow 2Cl^- + 4OH^-$$
$$6OH^- + Bi_2O_3 \longrightarrow 2BiO_3^- + 3H_2O + 4e^-$$

$$2H_2O + 6OH^- + Bi_2O_3 + 2OCl^- \longrightarrow 2Cl^- + 4OH^- + 2BiO_3^- + 3H_2O$$

By eliminating H_2O and OH^- duplications, we have the balanced net ionic equation.

SEE PROBLEMS
14-16 THROUGH
14-20.

$$Bi_2O_3(s) + 2OCl^-(aq) + 2OH^-(aq) \longrightarrow$$
$$2BiO_3^-(aq) + 2Cl^-(aq) + H_2O(l)$$

14-4 VOLTAIC CELLS

Consider the case of a car parked on a steep road about halfway down a hill. If the brake is released everyone knows that the car will spontaneously roll down the hill and not up. The car rolls down a hill because the bottom of the hill represents a region of lower potential energy than the original position of the car. As the car rolls down the hill, its potential energy is converted into the kinetic energy of the moving car and heat from friction. A similar principle applies to chemical reactions. When we have a mixture of reactants and products, the reaction proceeds spontaneously in one direction but not the other. Chemical reactions are favorable (referred to as spontaneous) in one direction because the reactants, like the car on the hill, are at a higher potential energy state (in the form of chemical energy) than the products. (See Figure 14-2.) When the reaction proceeds, the difference in chemical energy between reactants and products can be released as heat energy (exothermic reaction) or, under the proper conditions, converted directly into electrical energy. *The voltaic cell (also called the galvanic cell) uses a favorable or spontaneous redox reaction to generate electrical energy through an external circuit.*

When a strip of metallic zinc is immersed in a $CuSO_4$ solution it is obvious that a favorable and spontaneous chemical reaction occurs. The gray-colored zinc quickly becomes covered with a brown coating of elemental copper. (If a strip of copper is placed in a solution of $ZnSO_4$, however, the opposite reaction does not occur. This would be the "uphill" reaction.) The spontaneous reaction is illustrated by the equation

$$Zn(s) + CuSO_4(aq) \longrightarrow ZnSO_4(aq) + Cu(s)$$

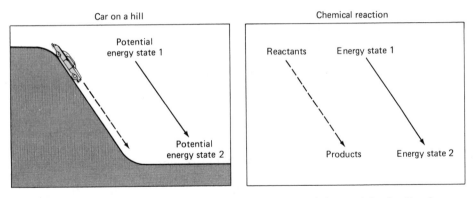

Figure 14-2   ENERGY STATES. A chemical reaction proceeds in a certain direction for the same reason that a car rolls *down* a hill.

This redox reaction occurs in what is known as the Daniell cell. In the early days of the use of electricity, it had very important applications such as generating current for the new telegraph and doorbells in homes. Before such a reaction is useful in generating electricity, however, the oxidation and reduction processes must be separated so that the electrons that are exchanged can be detoured through an external wire where they can be put to use. The Daniell cell is illustrated in Figure 14-3. A zinc strip is immersed in a Zn^{2+} solution, and, in a separate compartment, a copper strip is immersed in a Cu^{2+} solution. A wire connects the two metal strips. The two metal strips are called electrodes. *The* **electrodes** *are the surfaces in a cell at which the reactions take place. The electrode at which oxidation takes place is called the* **anode**. *Reduction takes place at the* **cathode**.

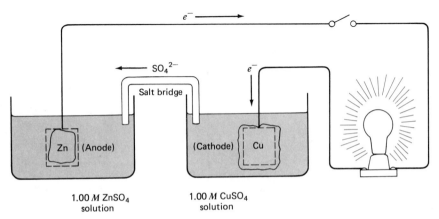

Figure 14-3   THE DANIELL CELL. This chemical reaction produced electricity for the first telegraphs.

In the compartment on the left, the strip of Zn serves as the anode, since the following reaction occurs when the circuit is connected.

$$Zn \longrightarrow Zn^{2+} + 2e^-$$

When the circuit is complete, the two electrons travel in the external wire to the Cu electrode, which serves as the cathode. The reduction reaction occurs at the cathode.

$$2e^- + Cu^{2+} \longrightarrow Cu$$

To maintain neutrality in the solution, some means must be provided for the movement of a SO_4^{2-} ion (or some other negative ion) from the right compartment where a Cu^{2+} has been removed to the left compartment where a Zn^{2+} has been produced. The *salt bridge* is an aqueous gel that allows ions to migrate between compartments but does not allow the mixing of solutions. (If Cu^{2+} ions wandered into the left compartment, they would form a coating of Cu on the Zn electrode, thus short-circuiting the cell.)

As the cell discharges (generates electrical energy), the Zn electrode becomes smaller but the Cu electrode becomes larger. (See the solid lines on the electrodes in Figure 14-3.) An important feature of this cell is that the reaction can be stopped by interrupting the external circuit with a switch. If the circuit is open the electrons can no longer flow, so no further reaction occurs until the switch is again closed.

Two of the most common voltaic cells in use today are the dry cell (flashlight battery) and the lead–acid cell (car battery). (*The word* **battery** *means a collection of one or more separate cells joined together in one unit.*)

One cell of a lead–acid storage battery is illustrated in Figure 14-4. Each cell is composed of two grids separated by an inert spacer. One grid of a fully charged battery contains metallic lead. The other contains PbO_2, which is insoluble in H_2O. Both grids are immersed in a sulfuric acid solution (battery

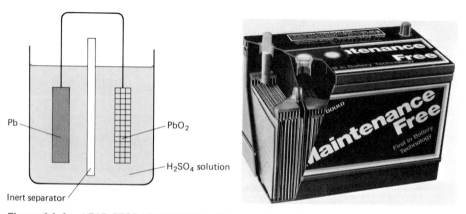

Figure 14-4   LEAD STORAGE BATTERY. The lead storage battery produces electricity to start your car.

acid). When the battery is discharged by connecting the electrodes, the following half-reactions takes place spontaneously.

Anode: $$Pb(s) + H_2SO_4(aq) \longrightarrow PbSO_4(s) + 2H^+(aq) + 2e^-$$

Cathode: $$2e^- + 2H^+(aq) + PbO_2(s) + H_2SO_4(aq) \longrightarrow PbSO_4(s) + 2H_2O$$

Total reaction: $$Pb(s) + PbO_2(s) + 2H_2SO_4(aq) \longrightarrow 2PbSO_4(s) + 2H_2O$$

The electrons released at the Pb anode travel through the external circuit to run lights, starters, radios, or whatever is needed. The electrons return to the PbO_2 cathode to complete the circuit. As the reaction proceeds, both electrodes are converted to $PbSO_4$ and the H_2SO_4 is depleted. Since $PbSO_4$ is also insoluble, it remains attached to the grids as it forms. The degree of discharge of a battery can be determined by the density of the battery acid. Since the density of a fully discharged battery is 1.05 g/mL, the difference in density between this value and the density of a fully charged battery (1.35 g/mL) gives the amount of charge remaining in the battery. The hydrometer discussed in Chapter 2 is used to determine the density of the acid. As the electrodes convert to $PbSO_4$, the battery loses power and eventually becomes "dead."

The convenience of a car battery is that it can be recharged. After the engine starts, an alternator or generator is engaged to push electrons back into the cell in the opposite direction from which they came during discharge. This forces the reverse, nonspontaneous reaction to proceed:

$$2PbSO_4(s) + 2H_2O \longrightarrow Pb(s) + PbO_2(s) + 2H_2SO_4(aq)$$

When the battery is fully recharged, the alternator shuts off, the circuit is open, and the battery is ready for the next start.

The dry cell (invented by Leclanché in 1866) is not rechargeable to any extent but is comparatively inexpensive and easily portable. (In contrast, the lead–acid battery is heavy, is expensive, and must be kept upright.) The dry cell illustrated in Figure 14-5 consists of a zinc anode, which is the outer rim,

Figure 14-5 DRY CELL. The dry cell is comparatively inexpensive, light, and portable.

and an inert graphite electrode. (An inert electrode provides a reaction surface but does not itself react.) In between is an aqueous paste containing NH_4Cl, MnO_2, and carbon. The reactions are as follows:

Anode: $Zn(s) \longrightarrow Zn^{2+}(aq) + 2e^-$

Cathode: $2NH_4^+(aq) + 2MnO_2(s) + 2e^- \longrightarrow$

$$Mn_2O_3(s) + 2NH_3(aq) + H_2O$$

Space travel requires a tremendous source of electrical energy. The requirements are that the source be continuous (no recharging necessary), lightweight, and dependable. Solar energy directly converts rays from the sun into electricity but is not practical for shorter runs such as the space shuttle. Although expensive, a source of power that fills the bill nicely is the fuel cell. A **fuel cell** *uses the direct reaction of hydrogen and oxygen to produce electrical energy.* Figure 14-6 is a picture of a fuel cell. Hydrogen and oxygen gases are fed into the cell where they form water. As long as the gases enter the cell, power is generated. The water that is formed can be removed and used for other purposes in the spacecraft. Since reactants are supplied from external sources, the electrodes are not consummed and the cell does not have to be shut down to be regenerated like a car battery. Fuel cells have had some large-scale application for power generation for commercial purposes but they are quite expensive. The best inert electrode surfaces at which the gases react are made of the extremely expensive metal platinum.

The reactions that take place in the fuel cell are as follows:

Anode: $H_2(g) + 2OH^-(aq) \longrightarrow 2H_2O + 2e^-$

Cathode: $O_2(g) + 2H_2O + 4e^- \longrightarrow 4OH^-(aq)$

Overall: $2H_2(g) + O_2(g) \longrightarrow 2H_2O$

Figure 14-6   FUEL CELL. The fuel cell can generate power without interruption for recharging.

Many other batteries have their special advantages. Nickel–cadmium batteries are popular as replacements for dry cells. Although initially more expensive, they are rechargeable and so last longer. A battery made of silver oxide and zinc can be made very small and supply a small amount of current for a long time. These have applications in digital watches and calculators. More efficient and durable batteries are a subject of much research at the current time. For example, an efficient and relatively inexpensive electrical automobile is the subject of intensive efforts. A small electric car using lead–acid batteries requires at least 18 batteries. These have to be replaced every year or so depending on use. Also, much of the power of these heavy batteries must be used just to move the batteries around, not including the car and passengers. There have been some encouraging possibilities for lighter more durable batteries but production of such an automobile is not planned for the immediate future.

SEE PROBLEMS 14-21 THROUGH 14-25.

14-5 PREDICTING SPONTANE-OUS REDOX REACTIONS

The Daniell cell pitted, in head-to-head competition, the strength of the Cu^{2+} ion versus that of the Zn^{2+} ion as an oxidizing agent. Or, conversely, we could imagine the competition was between the strength of zinc versus that of copper metal as a reducing agent. In either case, the spontaneous reaction written in net ionic form indicates that the winners were Cu^{2+} ion over Zn^{2+} ion and zinc over copper.

$$Zn(s) + Cu^{2+}(aq) \longrightarrow Zn^{2+}(aq) + Cu(s)$$

Let's focus on the relative strengths of Cu^{2+} and Zn^{2+} ions as oxidizing agents. The direction of the spontaneous reaction in the Daniell cell indicates that the half-reaction

$$Cu^{2+} + 2e^- \longrightarrow Cu$$

occurs more readily than

$$Zn^{2+} + 2e^- \longrightarrow Zn$$

Thus, we rate Cu^{2+} as the stronger oxidizing agent.

The competition between the two ions as oxidizing agents is analogous to a tug-of-war between two competitors. In Figure 14-7 two people of unequal strength are pulling on a rope looped around a post. Obviously, the stronger prevails and moves in the desired direction while the other is forced to move in the opposite direction. Likewise, since Cu^{2+} has a stronger tendency to be reduced (is a stronger oxidizing agent) than Zn^{2+}, it is reduced. This means that the Zn^{2+}, Zn half-reaction is forced to go in the opposite direction:

$$Zn \longrightarrow Zn^{2+} + 2e^-$$

Note that when an ion is reduced it forms a species (a metal in this case) that is a potential reducing agent. *The strength of the reduced form as a reducing agent is inversely related to the strength of the original ion as an oxidizing agent.* In other words, the reduced form of a strong oxidizing agent is a weak reducing

Figure 14-7   A TUG-OF-WAR FOR ELECTRONS. The stronger of the two moves in the desired direction. The other moves in the opposite direction.

agent and vice versa. This is the same seesaw relationship that we found with an acid and its conjugate base in the last chapter. Thus, for the reaction discussed above we can now make the following statements: Cu^{2+} is a stronger oxidizing agent than Zn^{2+} which makes Zn a stronger reducing agent than Cu. *A spontaneous reaction occurs when the stronger oxidizing agent reacts with the stronger reducing agent to form weaker oxidizing and reducing agents.* Again, note the similarity to acid–base reactions in which the stronger acid reacts with the stronger conjugate base.

Further experiments can be performed that allow more ions to be ranked as to their strengths as oxidizing agents. For example, a strip of nickel in a Cu^{2+} solution leads to the formation of a coating of copper on the strip of nickel. (See Figure 14-8a.) We can conclude that Cu^{2+} is also a stronger oxidizing agent than Ni^{2+} since the following reaction is apparently spontaneous:

$$Ni(s) + Cu^{2+}(aq) \longrightarrow Ni^{2+}(aq) + Cu(s)$$

Now we can perform experiments to see how the Zn^{2+} ion compares with the Ni^{2+} ion as an oxidizing agent. When a strip of nickel is placed in a Zn^{2+} solution no coating of Zn appears. However, if a strip of Zn is placed in a Ni^{2+} solution a coating of nickel forms on the zinc. (See Figures 14-8b and c.) These experiments suggest that the following reaction is spontaneous.

$$Zn(s) + Ni^{2+}(aq) \longrightarrow Zn^{2+}(aq) + Ni(s)$$

(a)

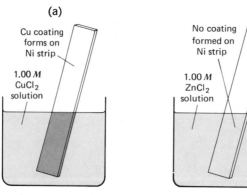

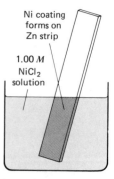

Cu coating forms on Ni strip

1.00 M CuCl₂ solution

No coating formed on Ni strip

1.00 M ZnCl₂ solution

Ni coating forms on Zn strip

1.00 M NiCl₂ solution

Figure 14-8   STRENGTH OF Ni^{2+} AS AN OXIDIZING AGENT. Ni^{2+} is stronger than Zn^{2+} but weaker than Cu^{2+}.

Apparently, the Ni^{2+} ion is a stronger oxidizing agent than Zn^{2+} but weaker than the Cu^{2+} ion. As a result of our experiments we can rank these three oxidizing agents in order of decreasing strength:

$$1 \ Cu^{2+} \quad 2 \ Ni^{2+} \quad 3 \ Zn^{2+}$$

Because of the inverse relationship, the ranking of the three reduced forms of the ions (the metals) as reducing agents is in the reverse order.

$$1 \ Zn \quad 2 \ Ni \quad 3 \ Cu$$

More experiments can provide additional ions and molecular species to our ranking. Eventually, we can construct a table of oxidizing agents ordered according to strength. Such a ranking is given in Table 14-1. In some cases, instruments are required to give quantitative measurements as to oxidizing ability *(known as reduction potentials)* to correctly establish the location of a species in the table. Oxidizing and reducing strengths are compared at the same concentration for all ions involved (1.00 M) and at the same pressure for all gases involved (1.00 atm).

The strongest oxidizing agent is at the top of this table on the left. The species shown on the right act as reducing agents when they undergo reaction. The reducing agents become stronger down the table on the right. A redox reaction takes place between an oxidizing agent on the left and a reducing agent on the right. A favorable or spontaneous reaction occurs between an oxidizing agent in the table and any reducing agent ranked lower in the table. The reaction can be visualized as taking place in a clockwise direction with the

TABLE 14-1   Reducing and Oxidizing Agents

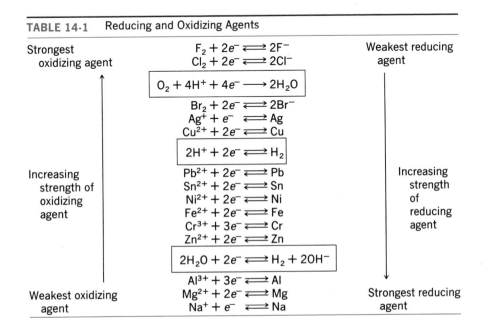

Strongest oxidizing agent	$F_2 + 2e^- \rightleftharpoons 2F^-$ $Cl_2 + 2e^- \rightleftharpoons 2Cl^-$	Weakest reducing agent
	$O_2 + 4H^+ + 4e^- \longrightarrow 2H_2O$	
	$Br_2 + 2e^- \rightleftharpoons 2Br^-$ $Ag^+ + e^- \rightleftharpoons Ag$ $Cu^{2+} + 2e^- \rightleftharpoons Cu$	
	$2H^+ + 2e^- \rightleftharpoons H_2$	
Increasing strength of oxidizing agent	$Pb^{2+} + 2e^- \rightleftharpoons Pb$ $Sn^{2+} + 2e^- \rightleftharpoons Sn$ $Ni^{2+} + 2e^- \rightleftharpoons Ni$ $Fe^{2+} + 2e^- \rightleftharpoons Fe$ $Cr^{3+} + 3e^- \rightleftharpoons Cr$ $Zn^{2+} + 2e^- \rightleftharpoons Zn$	Increasing strength of reducing agent
	$2H_2O + 2e^- \rightleftharpoons H_2 + 2OH^-$	
Weakest oxidizing agent	$Al^{3+} + 3e^- \rightleftharpoons Al$ $Mg^{2+} + 2e^- \rightleftharpoons Mg$ $Na^+ + e^- \rightleftharpoons Na$	Strongest reducing agent

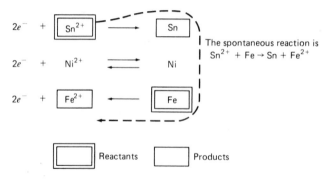

Figure 14-9   SPONTANEOUS REACTION. The stronger oxidizing agent reacts with the stronger reducing agent.

oxidizing agent reacting to the right and the reducing agent reacting in the opposite direction to the left. (See Figure 14-9.)

Table 14-1 can be used to predict spontaneous redox reactions in the same manner that Table 13-1 was used to predict spontaneous acid–base reactions. Of particular interest is the use of Table 14-1 to predict the reactions of certain elements with water. (All of the reactions shown in the table are assumed to occur in aqueous solution.) The highlighted reaction near the top of the table represents the *oxidation* of water. Note that the gases F_2 and Cl_2 spontaneously oxidize water to produce oxygen gas and an acid. (The reaction of Cl_2 with water is quite slow, however.)

$$Cl_2 + 2e^- \rightleftharpoons 2Cl^-$$
$$O_2 + 4H^+ + 4e^- \rightleftharpoons 2H_2O$$

The spontaneous reaction is

$$2Cl_2 + 2H_2O \longrightarrow O_2 + 4H^+ + 4Cl^-$$

The highlighted reaction near the bottom of the table represents the *reduction* of water. Note that the metals Al, Mg, and Na spontaneously reduce water to produce hydrogen gas and a base.

$$2H_2O + 2e^- \rightleftharpoons H_2 + 2OH^-$$
$$Na^+ + e^- \rightleftharpoons Na$$

The spontaneous reaction is

$$2H_2O + 2Na \longrightarrow H_2 + 2Na^+ + 2OH^-$$

These metals are known as *active* metals because of their chemical reactivity with water.

The third highlighted reaction near the middle of the table represents the reduction of aqueous acid solutions (1.00 M H^+) to form hydrogen gas. Note that metals such as Ni and Sn are not oxidized by water but are oxidized by strong acid solutions.

$$2H^+ + 2e^- \rightleftharpoons H_2$$
$$Ni^{2+} + 2e^- \rightleftharpoons Ni$$

The spontaneous reaction is

$$2H^+ + Ni \longrightarrow Ni^{2+} + H_2$$

Acid rain contains a considerably higher H^+ ion concentration than ordinary rain. From this discussion, it becomes understandable why metals such as iron and nickel are sensitive to corrosion by acid rain.

Two of the metals shown in the table — Cu and Ag — are not oxidized by either water or acid solutions. Thus, these metals are relatively unreactive and find use in jewelry and coins.

EXAMPLE 14-5

In which direction will the following reaction be spontaneous?

$$Pb^{2+}(aq) + 2Cl^-(aq) \overset{?}{\longleftrightarrow} Pb(s) + Cl_2(g)$$

PROCEDURE

In Table 14-1 note that Cl_2 is a stronger oxidizing agent than Pb^{2+} and that Pb is a stronger reducing agent than Cl^-. Therefore, the reaction is spontaneous to the left. Metallic lead will dissolve in an aqueous solution of Cl_2. (Cl_2 in swimming pools is hard on many metals, as you can see.)

ANSWER

$$Pb(s) + Cl_2(g) \longrightarrow PbCl_2(aq)$$

EXAMPLE 14-6

A strip of tin metal is placed in a $AgNO_3$ solution. If a reaction takes place, write the equation illustrating the spontaneous reaction.

PROCEDURE

In Table 14-1 note that the oxidizing agent Ag⁺ is above the reducing agent Sn. Therefore, a spontaneous reaction does occur. The balanced equation illustrating this reaction is

ANSWER

$$2Ag^+(aq) + Sn(s) \longrightarrow Sn^{2+}(aq) + 2Ag(s)$$

EXAMPLE 14-7

A length of aluminum wire is placed in water. Does the aluminum react with water?

PROCEDURE

In Table 14-1, note that aluminum is an active metal and should react with water (as an oxidizing agent).

ANSWER

$$6H_2O + 2Al(s) \longrightarrow 2Al^{3+}(aq) + 6OH^-(aq)$$

Since $Al(OH)_3$ is insoluble in water, however, the equation should be

$$6H_2O + 2Al(s) \longrightarrow 2Al(OH)_3(s)$$

Theoretically, aluminum should dissolve in water. Metallic aluminum is actually coated with Al_2O_3 which in fact protects the metal from coming into contact with water. Thus, it is a useful metal even for the hulls of boats despite its high chemical reactivity.

■
*SEE PROBLEMS
14-26 THROUGH
14-39.*

**14-6
ELECTROLYTIC
CELLS**

Chemical reactions, like the car on the hill, move spontaneously in one direction. A car and a chemical reaction can both be made to go in the other direction or uphill, however. In both cases, energy must be supplied from an outside source to make it move in the nonspontaneous or unfavorable direction. In a chemical reaction, if the correct amount of electrical energy is supplied from an outside source, a nonspontaneous reaction can be made to occur. In this case the electrical energy is converted back into chemical energy. *Cells that convert electrical energy into chemical energy are called electrolytic cells.*

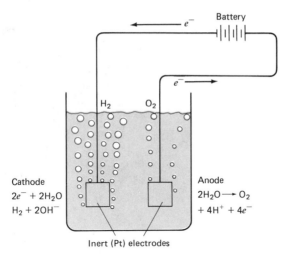

Figure 14-10 ELECTROLYSIS OF WATER. With an input of electrical energy, water can be decomposed to its elements.

An example of an electrolytic cell is shown in Figure 14-10. When sufficient electrical energy is supplied to the electrodes from an outside source, the following nonspontaneous reaction occurs:

$$2H_2O \longrightarrow 2H_2(g) + O_2(g)$$

For this electrolysis to occur, an electrolyte such as Na_2SO_4 must be present in solution. Pure water alone does not have a sufficient concentration of ions to allow conduction of electricity.

Another example of an electrolytic cell is the recharge cycle of the lead–acid battery described in an earlier section. When energy from the engine activates the alternator, electrical energy is supplied to the battery and the nonspontaneous reaction occurs as an electrolysis reaction. This reaction re-forms the original reactants.

Electrolysis has many useful applications. For example, silver or gold can be electroplated onto cheaper metals. In Figure 14-11, the metal spoon is the cathode and the silver bar serves as the anode. When electricity is supplied, the Ag anode produces Ag^+ ions and the spoon cathode reduces Ag^+ ions to give a layer of Ag. The silver-plated spoon can be polished and made to look as good as sterling silver.

Electrolytic cells are used to free elements from their compounds. Such cells are especially useful where metals are held in their compounds by strong chemical bonds. Examples are the metals aluminum, sodium, and magnesium. All aluminum is produced by the electrolysis of molten aluminum salts. Commercial quantities of sodium and chlorine are also produced by electrolysis of

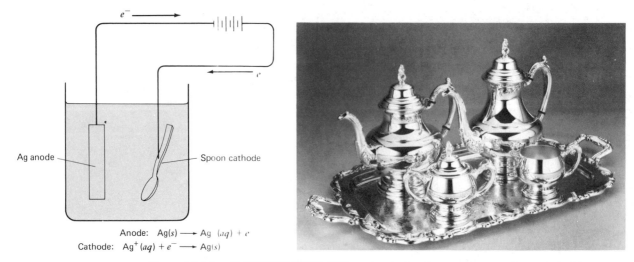

Anode: $Ag(s) \longrightarrow Ag\ (aq) + e$

Cathode: $Ag^+(aq) + e^- \longrightarrow Ag(s)$

Figure 14-11   ELECTROPLATING. With an input of energy, a spoon can be coated with silver. The service has been electroplated with silver.

SEE PROBLEMS
14-40 THROUGH
14-44.

molten sodium chloride. An apparatus used for the electrolysis of molten sodium chloride is illustrated in Figure 14-12. At the high temperature required to keep the NaCl in the liquid state, sodium forms as a liquid and is drained from the top of the cell.

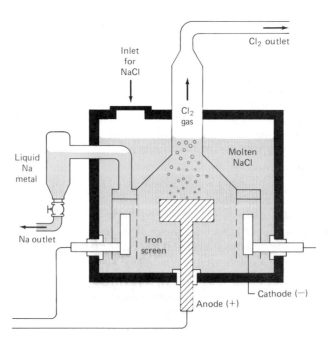

Figure 14-12   ELECTROLYSIS OF NaCl. Elemental sodium and chlorine are prepared by the electrolysis of ordinary table salt in the molten state.

14-7 CHAPTER REVIEW

A common characteristic of a large number of chemical reactions is an exchange of electrons between reactants. These reactions are known as oxidation–reduction or simply redox reactions. In such reactions, the reactant that gives up or loses the electrons is oxidized and the reactant that gains the electrons is reduced. The reactants are also classified as oxidizing and reducing agents as follows:

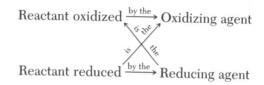

A redox reaction can be divided into two half-reactions: an oxidation and a reduction. These two processes take place so that all electrons lost in the oxidation process are gained in the reduction process. This fact is useful in balancing oxidation–reduction reactions. Most of these reactions would, at best, be very difficult to balance by inspection methods as described in Chapter 9.

Balancing equations by the oxidation state method requires that the atoms undergoing a change in oxidation state be identified. By equalizing electron gain and electron loss with proper coefficients, equations can be balanced. The more useful method is the ion-electron or half-reaction method. In this method, the entire molecule or ion containing the atom undergoing a change is balanced in two half-reactions. The rules for balancing equations by this method are summarized as follows:

1 Identify the ion or molecule containing the atom that has a change in oxidation state and the product species containing the same atom.
2 Balance the atom undergoing the change.
3 Balance oxygens. $\begin{bmatrix} \text{Acidic: add one } H_2O \text{ for each O needed.} \\ \text{Basic: add two } OH^- \text{ ions for each O needed.} \end{bmatrix}$
4 Balance H. $\begin{bmatrix} \text{Acidic add } H^+ \text{ for each H needed.} \\ \text{Basic: add one } H_2O \text{ for each two H's needed.} \end{bmatrix}$
5 Add electrons on the more positive side to balance the charge.
6 Multiply each half-reaction by a whole number so that electrons cancel when the half-reactions are added.
7 Subtract any species common to both sides of the equation.

Spontaneous chemical reactions occur because reactants are higher in chemical energy than products. We are familiar with many exothermic reactions where the difference in energy is released as heat. The energy can also be released as electrical energy in a voltaic cell. In a voltaic cell the two half-reactions are physically separated so that electrons travel in an external circuit or wire. The Daniell cell, the car battery, and the dry cell all involve spontaneous chemical reactions in which the chemical energy is converted directly into electrical energy.

Each substance has its own inherent strength as either an oxidizing agent or a reducing agent. This allows the construction of a table in which oxidizing and reducing agents are ranked by strength. The rankings are arrived at by observations or measurements with electrical instruments. From this table a great many other spontaneous reactions can be predicted. In Table 14-1 stronger oxidizing agents are ranked higher on the left and stronger reducing agents are ranked lower on the right. Reactions can thus be predicted as follows:

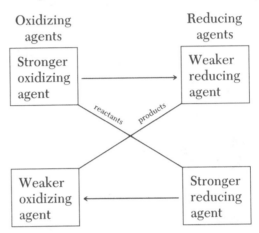

The reactions (or lack of reactions) of certain elements with water can be predicted from Table 14-1. Four groups of elements can be classified:

1   Nonmetals (e.g., F_2 and Cl_2) that are very strong oxidizing agents. These elements *oxidize* water to produce oxygen gas.

2   Metals (e.g., Na and Mg) that are very strong reducing agents. These elements *reduce* water to produce hydrogen gas.

3   Metals (e.g., Cu and Ag) that are very weak reducing agents. These elements are not oxidized by water or a strong acid (1.00 M) solution.

4   Metals (e.g., Ni and Fe) that are moderate reducing agents. These elements are not oxidized by water but are oxidized by aqueous acid.

Many reactions that are predicted to be unfavorable or nonspontaneous can be made to occur if electrical energy is supplied from an outside source. These are referred to as electrolytic cells and are useful in the commercial production of certain elements and in electroplating.

■ EXERCISES

OXIDATION–REDUCTION

14-1   What is the oxidation state of the specified atom?
(a) The S in SO_3
(b) the Co in Co_2O_3
(c) the U in UF_6
(d) the N in HNO_3
(e) the Cr in K_2CrO_4
(f) the Mn in $CaMnO_4$

14-2   What is the oxidation state of the specified atom?
(a) the N in N_2O
(b) the Pb in PbF_4
(c) the H in AlH_3
(d) the O in H_2O_2
(e) the S in $Na_2S_2O_3$
(f) the Hg in Hg_2Cl_2

14-3   Which of the following reactions are oxidation–reduction reactions?
(a) $2H_2 + O_2 \rightarrow 2H_2O$
(b) $CaCO_3 \rightarrow CaO + CO_2$
(c) $2Na + 2H_2O \rightarrow 2NaOH + H_2$
(d) $2HNO_3 + Ca(OH)_2 \rightarrow$
$$Ca(NO_3)_2 + 2H_2O$$
(e) $AgNO_3 + KCl \rightarrow AgCl + KNO_3$
(f) $Zn + CuCl_2 \rightarrow ZnCl_2 + Cu$

14-4   Identify each of the following half-reactions as either oxidation or reduction.
(a) $Na \rightarrow Na^+ + e^-$
(b) $Zn^{2+} + 2e^- \rightarrow Zn$
(c) $Fe^{2+} \rightarrow Fe^{3+} + e^-$
(d) $O_2 + 4H^+ + 4e^- \rightarrow 2H_2O$
(e) $S_2O_8^{2-} + 2e^- \rightarrow 2SO_4^{2-}$

14-5   Identify each of the following changes as either oxidation or reduction.
(a) $P_4 \rightarrow H_3PO_4$
(b) $NO_3^- \rightarrow NH_4^+$
(c) $Fe_2O_3 \rightarrow Fe^{2+}$
(d) $Al \rightarrow Al(OH)_4^-$
(e) $S^{2-} \rightarrow SO_4^{2-}$

14-6   For each of the following unbalanced equations, complete the table below.
(a) $MnO_2 + H^+ + Br^- \rightarrow$
$$Mn^{2+} + Br_2 + H_2O$$
(b) $CH_4 + O_2 \rightarrow CO_2 + H_2O$
(c) $Fe^{2+} + MnO_4^- + H^+ \rightarrow$
$$Fe^{3+} + Mn^{2+} + H_2O$$

14-7   For the following two unbalanced equations, construct a table such as that in Problem 14-6.

(a) $Al + H_2O \rightarrow AlO_2^- + H_2 + H^+$
(b) $Mn^{2+} + Cr_2O_7^{2-} + H^+ \rightarrow$
$$MnO_4^- + Cr^{3+} + H_2O$$

14-8   For the following equations identify the reactant oxidized, the reactant reduced, the oxidizing agent, and the reducing agent.
(a) $Sn + HNO_3 \rightarrow SnO_2 + NO_2 + H_2O$
(b) $IO_3^- + SO_2 + H_2O \rightarrow$
$$I_2 + SO_4^{2-} + H^+$$
(c) $CrI_3 + OH^- + Cl_2 \rightarrow$
$$CrO_4^{2-} + IO_4^- + Cl^- + H_2O$$
(d) $I^- + H_2O_2 \rightarrow I_2 + H_2O + OH^-$

14-9   Identify the product or products containing the elements oxidized and reduced in Problem 14-8.

BALANCING EQUATIONS BY THE OXIDATION STATE METHOD

14-10   Balance each of the following equations by the oxidation state method.
(a) $NH_3 + O_2 \rightarrow NO + H_2O$
(b) $Sn + HNO_3 \rightarrow SnO_2 + NO_2 + H_2O$
(c) $Cr_2O_3 + Na_2CO_3 + KNO_3 \rightarrow$
$$CO_2 + Na_2CrO_4 + KNO_2$$
(d) $Se + BrO_3^- + H_2O \rightarrow$
$$H_2SeO_3 + Br^-$$
(e) $MnO_2 + HBr \rightarrow$
$$MnBr_2 + Br_2 + H_2O$$

14-11   Balance the following equations by the oxidation state method.
(a) $I_2O_5 + CO \rightarrow I_2 + CO_2$
(b) $Al + H_2O \rightarrow AlO_2^- + H_2 + H^+$
(c) $HNO_3 + HCl \rightarrow NO + Cl_2 + H_2O$
(d) $I_2 + Cl_2 + H_2O \rightarrow HIO_3 + HCl$

BALANCING EQUATIONS BY THE ION-ELECTRON METHOD

14-12   Balance each of the following half-reactions in acid solution.
(a) $Sn^{2+} \rightarrow SnO_2$
(b) $CH_4 \rightarrow CO_2$

Reaction	Reactant Oxidized[a]	Product of Oxidation	Reactant Reduced	Product of Reduction	Oxidizing Agent	Reducing Agent
(a)						
(b)						
(c)						

[a] Element, molecule, or ion.

(c) $Fe^{3+} \rightarrow Fe^{2+}$

(d) $I_2 \rightarrow IO_3^-$

(e) $NO_3^- \rightarrow NO_2$

14-13 Balance the following half-reactions in acid solution.

(a) $P_4 \rightarrow H_3PO_4$

(b) $ClO_3^- \rightarrow Cl^-$

(c) $S_2O_3^{2-} \rightarrow SO_4^{2-}$

(d) $NO_3^- \rightarrow NH_4^+$

(e) $H_2O_2 \rightarrow H_2O$

14-14 Balance each of the following by the ion-electron method. All are in acid solution.

(a) $S^{2-} + NO_3^- + H^+ \rightarrow$
$$S + NO + H_2O$$

(b) $I_2 + S_2O_3^{2-} \rightarrow S_4O_6^{2-} + I^-$

(c) $SO_3^{2-} + ClO_3^- \rightarrow Cl^- + SO_4^{2-}$

(d) $Fe^{2+} + H_2O_2 + H^+ \rightarrow Fe^{3+} + H_2O$

(e) $AsO_4^{3-} + I^- + H^+ \rightarrow$
$$I_2 + AsO_3^{3-} + H_2O$$

(f) $Zn + H^+ + NO_3^- \rightarrow$
$$Zn^{2+} + NH_4^+ + H_2O$$

14-15 Balance each of the following by the ion-electron method. All occur in acid solution.

(a) $Mn^{2+} + BiO_3^- + H^+ \rightarrow$
$$MnO_4^- + Bi^{3+} + H_2O$$

(b) $IO_3^- + SO_2 + H_2O \rightarrow$
$$I_2 + SO_4^{2-} + H^+$$

(c) $Se + BrO_3^- + H_2O \rightarrow$
$$H_2SeO_3 + Br^-$$

(d) $P_4 + HOCl + H_2O \rightarrow$
$$H_3PO_4 + Cl^- + H^+$$

(e) $Al + Cr_2O_7^{2-} + H^+ \rightarrow$
$$Al^{3+} + Cr^{3+} + H_2O$$

(f) $ClO_3^- + I^- + H^+ \rightarrow$
$$Cl^- + I_2 + H_2O$$

(g) $As_2O_3 + NO_3^- + H_2O \rightarrow$
$$AsO_4^{3-} + NO + H^+$$

14-16 Balance each of the following half-reactions in basic solution.

(a) $SnO_2^{2-} \rightarrow SnO_3^{2-}$

(b) $ClO_2^- \rightarrow Cl_2$

(c) $Si \rightarrow SiO_3^{2-}$

(d) $NO_3^- \rightarrow NH_3$

14-17 Balance the following half-reactions in basic solution.

(a) $Al \rightarrow Al(OH)_4^-$

(b) $S^{2-} \rightarrow SO_4^{2-}$

(c) $N_2H_4 \rightarrow NO_3^-$

14-18 Balance each of the following by the ion-electron method. All occur in basic solution.

(a) $S^{2-} + OH^- + I_2 \rightarrow$
$$SO_4^{2-} + I^- + H_2O$$

(b) $MnO_4^- + OH^- + I^- \rightarrow$
$$MnO_4^{2-} + IO_4^- + H_2O$$

(c) $BiO_3^- + SnO_2^{2-} + H_2O \rightarrow$
$$SnO_3^{2-} + OH^- + Bi(OH)_3$$

(d) $CrI_3 + OH^- + Cl_2 \rightarrow$
$$CrO_4^{2-} + IO_4^- + Cl^- + H_2O$$

(*Hint:* In (d), two ions are oxidized; include both in one half-reaction.)

14-19 Balance each of the following by the ion-electron method. All are in basic solution.

(a) $ClO_2 + OH^- \rightarrow$
$$ClO_2^- + ClO_3^- + H_2O$$

(b) $OH^- + Cr_2O_3 + NO_3^- \rightarrow$
$$CrO_4^{2-} + NO_2^- + H_2O$$

(c) $Cr(OH)_4^- + BrO^- + OH^- \rightarrow$
$$Br^- + CrO_4^{2-} + H_2O$$

(d) $Mn^{2+} + H_2O_2 + OH^- \rightarrow$
$$H_2O + MnO_2$$

(e) $Ag_2O + Zn + H_2O \rightarrow$
$$Zn(OH)_2 + Ag$$

14-20 Balance the following two equations by the ion-electron method first in acidic solution and then in basic solution.

(a) $H_2 + O_2 \rightarrow H_2O$

(b) $H_2O_2 \rightarrow O_2 + H_2O$

VOLTAIC CELLS

14-21 What is the function of the salt bridge in the voltaic cell?

14-22 In the alkaline battery the following two half-reactions occur.

$$Zn(s) + 2OH^-(aq) \rightarrow$$
$$Zn(OH)_2(s) + 2e^-$$
$$2MnO_2(s) + 2H_2O + 2e^- \rightarrow$$
$$2MnO(OH)(s) + 2OH^-(aq)$$

Which reaction takes place at the anode and which at the cathode? What is the total reaction?

14-23 The following overall reaction takes place in the silver oxide battery.

$$Ag_2O(s) + H_2O + Zn(s) \rightarrow$$
$$Zn(OH)_2(s) + 2Ag(s)$$

The reaction takes place in basic solution. Write the half-reaction that takes place at the anode and the half-reaction that takes place at the cathode.

14-24 The nickel–cadmium battery (nicad) is used as a replacement for a dry cell because it is rechargeable. The overall reaction that takes place is

$$NiO_2(s) + Cd(s) + 2H_2O \rightarrow$$
$$Ni(OH)_2(s) + Cd(OH)_2(s)$$

Write the half-reactions that take place at the anode and the cathode.

14-25 Sketch a galvanic cell in which the following overall reaction occurs.

$$Ni^{2+}(aq) + Fe(s) \rightarrow Fe^{2+}(aq) + Ni(s)$$

(a) What reactions take place at the anode and the cathode?
(b) In what direction do the electrons flow in the wire?
(c) In what direction do the anions flow in the salt bridge?

PREDICTING REDOX REACTIONS

14-26 Given the following information concerning metal strips immersed in certain solutions, write the net ionic equations representing the reactions that occur.

Metal Strip	Solution	Reaction
Cd	$NiCl_2$	Ni coating formed
Cd	$FeCl_2$	N.R. (no reaction)
Zn	$CdCl_2$	Cd coating formed
Fe	$CdCl_2$	N.R.

Where does Cd^{2+} rank as an oxidizing agent in Table 14-1?

14-27 A hypothetical metal (M) forms a coating of Sn when placed in a $SnCl_2$ solution. However, when a strip of Ni is placed in an MCl_2 solution, a coating of the metal M forms on the nickel. Write the net ionic equations representing the reactions that occur. Where does M^{2+} rank as an oxidizing agent in Table 14-1?

14-28 A solution of gold ions (Au^{3+}) spontaneously reacts with water to form metallic gold. Metallic gold does not react with chlorine but does react with fluorine. Write the equations illustrating the two spontaneous reactions and locate Au^{3+} in Table 14-1 as an oxidizing agent.

14-29 Using Table 14-1, predict whether the following reactions occur in aqueous solution. If not, write N.R. (no reaction).
(a) $2Na + 2H_2O \rightarrow H_2 + 2NaOH$
(b) $Pb + Zn^{2+} \rightarrow Pb^{2+} + Zn$
(c) $Fe + 2H^+ \rightarrow Fe^{2+} + H_2$
(d) $Fe + 2H_2O \rightarrow Fe^{2+} + 2OH^- + H_2$
(e) $Cu + 2Ag^+ \rightarrow 2Ag + Cu^{2+}$
(f) $2Cl_2 + 2H_2O \rightarrow 4Cl^- + O_2 + 4H^+$
(g) $3Zn^{2+} + 2Cr \rightarrow 2Cr^{3+} + 3Zn$

14-30 Using Table 14-1, predict whether the following reactions occur in aqueous solution. If not, write N.R.
(a) $Sn^{2+} + Pb \rightarrow Pb^{2+} + Sn$
(b) $Ni^{2+} + H_2 \rightarrow 2H^+ + Ni$
(c) $Cu + F_2 \rightarrow CuF_2$
(d) $Ni^{2+} + 2Br^- \rightarrow Ni + Br_2$
(e) $3Ni^{2+} + 2Cr \rightarrow 2Cr^{3+} + 3Ni$
(f) $2Br_2 + 2H_2O \rightarrow 4Br^- + O_2 + 4H^+$

14-31 Tell whether a reaction occurs in each of the following situations. Write a balanced equation indicating the spontaneous reaction if one occurs.
(a) Some iron nails are placed in a $CuCl_2$ solution.
(b) Silver coins are dropped into an acid solution [$H^+(aq)$].
(c) A copper penny is placed in a $Pb(NO_3)_2$ solution.
(d) Ni metal is placed in a $Pb(NO_3)_2$ solution.
(e) Sodium metal is placed in water.
(f) Zinc metal is placed in water.
(g) Bromine is dissolved in water.
(h) Zinc strips are placed in a $Cr(NO_3)_3$ solution.

14-32 Tell whether a reaction occurs in each of the following situations. Write a balanced equation illustrating the spontaneous reaction if one occurs.
(a) Magnesium metal makes contact with water.

(b) Iron nails are placed in a $ZnBr_2$ solution.

(c) H_2 gas is bubbled into an acid solution of Cu^{2+}.

(d) Copper metal is placed in an acid $[H^+(aq)]$ solution.

(e) Tin metal is placed in an acid solution.

(f) A piece of chromium is placed in a $SnCl_2$ solution.

(g) Chlorine is dissolved in water.

14-33 Which of the following elements react with water: (a) Pb (b) Ag (c) F_2 (d) Br_2 (e) Mg? Write the balanced equation illustrating any reaction of these elements with water.

14-34 Which of the following species will be reduced by hydrogen gas in aqueous solution: (a) Br_2 (b) Cr (c) Ag^+ (d) Ni^{2+}? Write the balanced equation illustrating any reaction of these species with hydrogen.

14-35 In Chapter 13, we mentioned the corrosiveness of acid rain. Why does rain containing a high $H^+(aq)$ concentration cause more damage to iron exposed in bridges and buildings than pure H_2O? Write the reaction between Fe and $H^+(aq)$.

14-36 Br_2 can be prepared from the reaction of Cl_2 with NaBr dissolved in sea water. Explain. Write the reaction. Can Cl_2 be used to prepare F_2 from NaF solutions?

14-37 Describe how a voltaic cell could be constructed from a strip of iron, a strip of lead, a $Fe(NO_3)_2$ solution, and a $Pb(NO_3)_2$ solution. Write the anode reaction, the cathode reaction, and the total reaction.

*14-38 Judging from the relative difference in the strengths of the oxidizing agents (Fe^{2+} vs Pb^{2+}) and (Zn^{2+} vs Cu^{2+}), which do you think would be the more powerful cell, the one in Problem 14-37 or the Daniell cell? Why?

*14-39 The greater the difference in the strength of two oxidizing agents, the more powerful the cell that can be constructed using the two oxidizing agents. Write the equation illustrating the most powerful redox reaction possible between an oxidizing agent and a reducing agent *in aqueous solution*. Consider only the species shown in Table 14-1.

ELECTROLYTIC CELLS

14-40 Chrome plating is an electrolytic process. Write the reaction that occurs when an iron bumper is electroplated from a $CrCl_3$ solution. Are there any metals shown in Table 14-1 that would spontaneously form a chromium layer on the metal?

14-41 Why can't elemental sodium be formed in the electrolysis of an aqueous NaCl solution? Write the reaction that does occur at the cathode. How is elemental sodium produced by electrolysis?

14-42 Why can't elemental fluorine be formed by electrolysis of an aqueous NaF solution? Write the reaction that does occur at the anode. How is elemental fluorine produced?

14-43 A "tin can" is made by forming a layer of tin on a sheet of iron. Is electrolysis necessary for such a process or does it occur spontaneously? Write the equation illustrating this reaction. Is electrolysis necessary to form a layer of tin on a sheet of lead? Write the relevant equation.

14-44 Certain metals can be purified by electrolysis. For example, a mixture of Ag, Zn, and Fe can be dissolved so that their metal ions are present in aqueous solution. If a solution containing these ions is electrolyzed, which metal ion would be reduced to the metal the easiest?

REVIEW TEST ON CHAPTERS 11–14

The following multiple-choice questions have one correct answer.

1. In which of the following compounds would there be dipole–dipole interactions in the liquid state?
   - (a) NaCl
   - (b) H_2
   - (c) CCl_4
   - (d) COS
   - (e) CO_2

2. Which of the following has hydrogen bonding in the liquid state?
   - (a) H_2S
   - (b) NaH
   - (c) CH_4
   - (d) CH_3OH
   - (e) NCl_3

3. Which of the following nonpolar molecules should have the highest boiling point?
   - (a) H_2
   - (b) SF_6
   - (c) CO_2
   - (d) CH_4
   - (e) N_2

4. Based on interactions in the liquid state, which of the following has the highest heat of vaporization?
   - (a) H_2O
   - (b) H_2S
   - (c) CH_4
   - (d) H_2
   - (e) NH_3

5. What happens when heat is applied to a liquid at its boiling point?
   - (a) The heat energy is converted into potential energy.
   - (b) The heat energy is converted into kinetic energy.
   - (c) The heat energy is converted into both potential and kinetic energy.
   - (d) It makes the molecules of the vapor move faster than those in the liquid.

6. The normal boiling point of a liquid is defined as the temperature at which
   - (a) Bubbles form in the liquid.
   - (b) The vapor pressure equals the atmospheric pressure.
   - (c) The vapor pressure equals one atmosphere.
   - (d) The vapor and the liquid exist in equilibrium.

7. A compound has a melting point of 950 °C. Which of the following statements is most likely true?

   - (a) Its molecules are nonpolar.
   - (b) The compound has a low heat of fusion.
   - (c) Its molecules are polar covalent.
   - (d) The compound has a low boiling point.
   - (e) The compound has a high heat of vaporization.

8. When a liquid in an insulated container is allowed to evaporate,
   - (a) The temperature of the liquid rises.
   - (b) The temperature of the liquid does not change.
   - (c) No liquid evaporates unless heat is supplied from the outside.
   - (d) The vapor that escapes is warmer than the liquid.

9. Which of the following equations illustrates the solution of Na_2SO_4 in water?

$$Na_2SO_4(s) \xrightarrow{H_2O}$$

   - (a) $Na_2^{2+}(aq) + SO_4^{2-}(aq)$
   - (b) $2Na^+(aq) + SO_4^{2-}(aq)$
   - (c) $Na^{2+}(aq) + SO_4^-(aq)$
   - (d) $2Na^{2+}(aq) + SO_4^-(aq)$

10. Which of the following equations represents a precipitation reaction?
    - (a) $CaCO_3 \rightarrow CaO + CO_2$
    - (b) $HCl + KOH \rightarrow H_2O + KCl$
    - (c) $Na_2S + Ni(NO_3)_2 \rightarrow NiS + 2NaNO_3$
    - (d) $Zn + CuCl_2 \rightarrow ZnCl_2 + Cu$
    - (e) $CH_2 + 2O_2 \rightarrow CO_2 + 2H_2O$

11. Which of the following is a net ionic equation?
    - (a) $CaCl_2 + Na_2CO_3 \rightarrow CaCO_3 + 2NaCl$
    - (b) $Ca^{2+} + 2Cl^- + 2Na^+ + CO_3^{2-} \rightarrow CaCO_3 + 2Cl^- + 2Na^+$
    - (c) $2Cl^- + 2Na^+ \rightarrow 2Na^+ + 2Cl^-$
    - (d) $Ca^{2+} + CO_3^{2-} \rightarrow CaCO_3$

12. A 2.0-L quantity of 0.10 M solution of HCl contains:
    - (a) 1.0 mol of HCl
    - (b) 0.20 mol of H_2O
    - (c) 20 mol of HCl
    - (d) 0.05 mol of H_2O
    - (e) 0.20 mol of HCl

13. Which of the following is not a colligative property?
    - (a) the boiling point of a pure liquid
    - (b) freezing point depression

(c) vapor pressure lowering

(d) osmotic pressure

14. What is the freezing point of a 0.100 m aqueous solution of a nonelectrolyte? For water $K_f = 1.86$.

(a) $-1.86\ °C$ (d) $18.6\ °C$

(b) $0.186\ °C$ (e) $-18.6\ °C$

(c) $-0.186\ °C$

15. If solutions of each of the following have the same concentration, which has the lowest freezing point?

(a) CH_3OH
    (a nonelectrolyte)

(b) $NaCl$

(c) Na_2SO_4

(d) CaS

(e) all are the same

16. Which of the following is the formula for perchloric acid?

(a) HCl (d) $HClO_3$

(b) $HClO$ (e) $HClO_4$

(c) $HClO_2$

17. Which of the following is a balanced neutralization reaction that leads to the formation of an acid salt?

(a) $HCl + NaOH \rightarrow NaCl + H_2O$

(b) $H_3PO_4 + KOH \rightarrow KHPO_4 + H_2O$

(c) $H_2SO_4 + 2KOH \rightarrow K_2SO_4 + H_2O$

(d) $NaOH + H_2SO_3 \rightarrow NaHSO_3 + H_2O$

18. Which of the following is a strong acid?

(a) HNO_2 (d) $HClO$

(b) HNO_3 (e) KOH

(c) H_2SO_3

19. The concentration of an aqueous HCl solution is 0.01 M. The pH is

(a) 1   (b) 2   (c) 3   (d) 12   (e) 7

20. A solution has a pH of 8.5. The solution is

(a) strongly basic   (d) weakly acidic

(b) weakly basis   (e) strongly acidic

(c) neutral

21. Which of the following is the conjugate acid of the HS^- ion?

(a) H_2S^- (c) S^- (e) H_2S

(b) S^{2-} (d) H_3O^+

22. Which of the following conjugate bases does not exhibit basic nature in water?

(a) S^{2-} (c) ClO_4^- (e) HS^-

(b) NO_2^- (d) PO_4^{3-}

23. A 0.10 M solution of a compound has a pH of 5.4. The compound is

(a) HCl (c) NH_4Cl (e) $NaCl$

(b) KOH (d) K_2S

24. Which of the following pairs of compounds does not form a buffer when aqueous solutions of the two are mixed?

(a) $NaC_2H_3O_2$ and $HC_2H_3O_2$

(b) KBr and HBr

(c) $LiNO_2$ and HNO_2

(d) NH_3 and NH_4Cl

25. Which of the following oxides is the anhydride of $HClO_4$?

(a) CO_2 (c) Cl_2O_7 (e) ClO_4^-

(b) ClO (d) ClO_4^-

26. Which of the following is an oxidation process?

(a) $H_2CO_3 \rightarrow H_2O + CO_2$

(b) $2Br^- \rightarrow Br_2$

(c) $Ca^{2+} + SO_4^{2-} \rightarrow CaSO_4$

(d) $H^+ + OH^- \rightarrow H_2O$

(e) $SO_4^{2-} \rightarrow S^{2-}$

27. What is the oxidation state of the Br in $Ca(BrO_3)_2$?

(a) $+10$ (c) $+5$ (e) -1

(b) $+6$ (d) $+12$

28. In the Daniell cell, what allows for the migration of anions between compartments?

(a) a salt bridge   (c) the external wire

(b) the electrodes   (d) the electrons

29. In the following electrolytic cell, _____ is _____ at the anode.

$$2Cl^-(aq) + Fe^{2+}(aq) \rightarrow Fe(s) + Cl_2(g)$$

(a) Fe^{2+}, reduced   (d) Cl^-, oxidized

(b) Cl^-, reduced   (e) Fe, oxidized

(c) Cl_2, reduced

30. The following equation represents a spontaneous reaction:

$$Br_2(aq) + Ni(s) \rightarrow Ni^{2+}(aq) + 2Br^-(aq)$$

Which of the following statements is correct?

(a) Br_2 is a stronger reducing agent than Ni^{2+}.

(b) Br^- is a stronger reducing agent than Ni.

(c) Ni is an oxidizing agent.

(d) Br_2 is oxidized.

(e) Br_2 is a stronger oxidizing agent than Ni^{2+}.

PROBLEMS

1. The following compounds have nearly the same formula weights: $(CH_3)_2O$ (polar

covalent), C_2H_5OH (polar covalent), C_3H_8 (nonpolar covalent), and BeF_2 (ionic). The melting points of these compounds are 800 °C, −117 °C, −138 °C and −189 °C. Match the melting point with the compound and explain what forces must be overcome in the solid to allow melting.

2. How many calories are required to change 25.0 g of ice at −20.0 °C to steam at 100 °C? (The specific heat of ice is 0.492 cal/g · °C, the heat of fusion is 79.8 cal/g, and the heat of vaporization is 540 cal/g.)

3. An aqueous solution is made by dissolving 40.0 g of H_2SO_4 in enough water to make 850 mL of solution.
   (a) What is the molarity of the H_2SO_4 solution?
   (b) If 100 mL of this solution is diluted to 550 mL, what is the resulting molarity?
   (c) What volume of the original solution is needed to form 2.00 L of a 0.150 M solution?

4. (a) When aqueous solutions of lead(II) nitrate and sodium iodide are mixed, a precipitate of lead(II) iodide forms. Write the balanced equation representing this reaction.
   (b) What volume of 0.40 M lead nitrate is needed to react with 2.50 L of 0.55 M sodium iodide?

5. (a) Write the balanced equation representing the complete neutralization of sulfuric acid with lithium hydroxide.
   (b) What volume (in mL) of 0.25 M sulfuric acid is needed to completely react with 1.80 g of lithium hydroxide?

6. (a) When elemental tin is placed in a nitric acid solution, a spontaneous redox reaction occurs producing SnO_2 and NO_2. Write the balanced equation representing this reaction.
   (b) What mass of NO_2 is produced from the complete reaction of 350 mL of 0.20 M nitric acid?

7. (a) Write the total ionic and net ionic equations for the reaction in Problem 4.

(b) Write the total ionic and net ionic equations for the reaction in Problem 5.

8. From the equation in Problem 6, answer the following.
   (a) What is the reactant oxidized?
   (b) What is the reactant reduced?
   (c) What is the product of oxidation?
   (d) What is the product of reduction?
   (e) What is the oxidizing agent?
   (f) What is the reducing agent?

9. Complete the following equations.
   (a) $HOCl + H_2O \rightleftharpoons$
   (b) $NH_3 + H_2O \rightleftharpoons$
   (c) $KOH + HNO_2 \rightarrow$
   (d) $H_3AsO_3 + 2NaOH \rightarrow$

10. Write equations representing reactions for the following. If no reaction occurs, write N.R.
   (a) $K^+ + H_2O$ (d) $CO_2 + H_2O$
   (b) $CaO + H_2O$ (e) $ClO^- + H_2O$
   (c) $NH_4^+ + H_2O$ (f) $Br^- + H_2O$

11. In a solution of a weak base, $[OH^-] = 6.5 \times 10^{-4}$. What is the value of (a) $[H_3O^+]$ (b) pH (c) pOH?

12. Given the following table of oxidizing and reducing agents, answer the questions below.

Strongest $\longrightarrow$ $I_2 + 2e^- \rightleftharpoons 2I^-$
oxidizing $Cr^{3+} + 3e^- \rightleftharpoons Cr$
agent $Mn^{2+} + 2e^- \rightleftharpoons Mn$ Strongest
    $Ca^{2+} + 2e^- \rightleftharpoons Ca \longleftarrow$ reducing
    agent

   (a) Is the following reaction spontaneous or nonspontaneous?

$$Mn + Ca^{2+} \longrightarrow Ca + Mn^{2+}$$

   (b) Write a balanced equation representing a spontaneous reaction involving the Cr, Cr^{3+} and the Mn, Mn^{2+} half-reactions.
   (c) If an aqueous solution of I_2 (an antiseptic) is spilled on a chromium-coated bumper on an automobile, will a reaction occur? If so, write the balanced equation for the reaction.

FRESH FISH
Fish won't keep for long unless they have been kept on ice. The cold
temperature slows the chemical reactions that cause spoiling.

PURPOSE OF CHAPTER

In Chapter 15 we discuss how molecules of reactants change into products in a reaction at equilibrium, factors that affect the rate of a reaction, and the quantitative implications of the point of equilibrium.

OBJECTIVES FOR CHAPTER

After completion of this chapter, you should be able to:

1 Explain how equilibrium is demonstrated in the examples discussed in previous chapters. (15-1)

2 Describe how reactant molecules are transformed into product molecules according to collision theory. (15-2)

3 Illustrate how a reversible reaction demonstrates dynamic equilibrium in terms of reaction rates. (15-2)

4 Describe how the following factors affect the rate of a reaction. (15-3)

(a) Activation energy

(b) Temperature

CHEMICAL KINETICS
AND EQUILIBRIUM

15

(c) Concentration of reactants

(d) Particle size of solids

(e) Presence of a catalyst

5   Write the law of mass action for any homogeneous (one-phase) reaction. (15-4)

6   Demonstrate how the law of mass action relates to the rate laws for the hypothetical reaction discussed. (15-4)

7   Calculate the value of an equilibrium constant given appropriate experimental data. (15-4)

8   Use the value of the equilibrium constant to calculate the concentration of a species in the mass action expression. (15-4)

9   Describe how the values of K_a and K_b relate to the strength of a weak acid or base. (15-5)

10   Use the value of K_a or K_b and the initial concentration to calculate the pH of a weak acid, a weak base, or a buffer solution. (15-5)

11   Apply the principle of Le Châtelier to show how changes in concentrations, temperature, and volume affect a system at equilibrium. (15-6)

12   Describe the role of a catalyst in a reaction that reaches a point of equilibrium. (15-6)

Who among us has not discovered what was thought to be a new and perhaps alien life form in the refrigerator? Way in the back, perhaps on a long forgotten slice of bologna, there grows a ghastly and green mold of some sort, slowly devouring the meat and perhaps neighboring foods. It could be worse. Without a refrigerator the food would decay within hours and the strange life form could have evolved rapidly compared with the weeks it had in the fridge. At room temperature, meat and dairy products immediately begin to deteriorate as a result of chemical reactions. In the refrigerator, food lasts for weeks and, in the colder environment of the freezer, it lasts for months. Decay, growth of alien life forms, and all other chemical reactions slow as the temperature decreases.

One of the topics of this chapter is how temperature and other factors affect the rate of a reaction. First, however, we review the examples of systems that are reversible and reach a point of equilibrium. These examples were all discussed in previous chapters. We then see how a typical chemical change occurs and how this leads to a system at equilibrium. After discussion of reaction rates we focus on the point of equilibrium and how it is expressed quantitatively. Finally, we see how certain factors change the point of equilibrium.

As background for this chapter, review the kinetic molecular theory discussed in Section 10-1, the distribution of molecular kinetic energies discussed in Section 11-5, and weak acids and bases discussed in Sections 13-3 and 13-6.

15-1 EXAMPLES OF CHEMICAL AND PHYSICAL EQUILIBRIA

The definition of equilibrium is simply *balance.* In our everyday lives we try to maintain an equilibrium between the demands of our job or school and the need to relax. Of course, we are more interested here in the meaning of equilibrium in a chemical sense. As mentioned in previous chapters, equilibrium refers to the balance between two competing chemical reactions. One reaction forms products from reactants while the other is the reverse reaction and re-forms reactants from products. *The **point of equilibrium** in a reversible process is when both reactions proceed at the same rate so that the concentrations of reactants and products remain constant.* Before we discuss what goes on in such a reaction we will review some of the important chemical or physical phenomena that are examples of reversible processes and reach a point of equilibrium. In fact, so far we have discussed four equilibrium situations.

Our first introduction to chemical equilibrium came in Chapter 9 when we discussed an incomplete reaction occurring in the gaseous state. The specific reaction discussed involved a combination reaction between the elements nitrogen and hydrogen, forming the industrially important compound ammonia. When the two reactants are mixed, eventually the amount of ammonia present does not increase even though an appreciable amount of reactants remains. This is because ammonia also undergoes a decomposition reaction to re-form the elements. A reversible reaction that reaches a point of equilibrium is indicated with a double arrow:

$$N_2(g) + 3H_2(g) \rightleftharpoons 2NH_3(g)$$

In Chapter 11 we were again exposed to the concept of equilibrium in a physical context. When water is allowed to evaporate in a closed container at a constant temperature, the partial pressure due to the water (vapor pressure) slowly increases. Eventually, however, there comes a point where the vapor pressure no longer increases. At this point, equilibrium is established between the water molecules that are escaping to the vapor state and the water molecules that are condensing to the liquid state. This equilibrium is illustrated as

$$H_2O(l) \rightleftharpoons H_2O(g)$$

In Chapter 12, we discussed saturated solutions of ionic compounds. When an ample amount of a compound such as table salt, NaCl, is added to water at a certain temperature, the solution eventually becomes saturated and no additional salt seems to dissolve. In fact, some salt is still dissolving but it is balanced by the solid salt that is being re-formed from ions in solution. As in other equilibrium situations, the amounts of solid salt and dissolved salt are constant but the identities of specific ions in the solid or dissolved state are not constant. (See Figure 15-1.) The following equation represents the equilibrium of solid sodium chloride in a saturated solution:

$$NaCl(s) \rightleftharpoons Na^+(aq) + Cl^-(aq)$$

A common but interesting chemical phenomenon that is often displayed at science fairs is the science (and art) of crystal growing. Crystal growing demonstrates the principle of equilibrium. If a comparatively large crystal of a water-soluble compound such as $CuSO_4$ is suspended in a saturated solution of

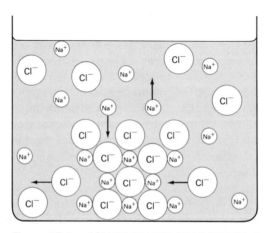

Figure 15-1   SOLUBILITY AND EQUILIBRIUM. At equilibrium the rate of solution of ions equals the rate at which ions are forming the solid.

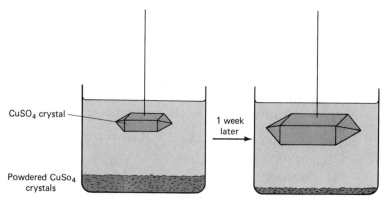

CuSO$_4$ crystal

1 week later

Powdered CuSo$_4$ crystals

Figure 15-2   CRYSTAL GROWTH. The size of the crystals changes, but the total mass does not.

that compound, the small crystals at the bottom get smaller and the large one grows larger. The concentrations of Cu^{2+} and SO_4^{2-} in solution remain constant as does the total mass of solid crystals. The changing size of the crystals, however, indicates that an equilibrium exists in which the identities of ions in the solid and dissolved states are changing. (See Figure 15-2.)

The equilibrium reactions discussed most recently were in Chapter 13 and involved the weak acids and bases. Although the ionization of a weak acid in water is unfavorable, it does occur to a small but measurable extent. At equilibrium, most molecules are present in water in the molecular rather than the ionized state as represented by the following equilibrium equation:

$$HF(aq) + H_2O \rightleftharpoons H_3O^+(aq) + F^-(aq)$$

The quantitative aspects of the reactions of weak acids and bases in water are discussed later in this chapter.

15-2 THE NATURE OF A REACTION AT EQUILIBRIUM

We are now ready to look deeper into how a reversible reaction occurs. So far, we have just had to accept that the atoms in the reactants get shuffled around into some new arrangement in the products. We haven't mentioned how this might occur. It is actually quite simple. We know that molecules are all moving in some way or another since motion relates to kinetic energy which is related to the temperature. Molecules in the gaseous or liquid states have translational motion and collide with each other. When two molecules that undergo a chemical reaction collide, the collision may lead to the breaking of chemical bonds in the reactant molecules as the new bonds of product molecules are forming. *The assumption that chemical reactions are brought about by the collisions of molecules is known as the* **collision theory.** Before we proceed further let us illustrate the collision theory with a simple but hypothetical example of a chemical reaction. We choose a hypothetical reaction rather than a real one because it can be "custom designed" to clearly illustrate the basic principles involved. Real-life reactions (as everything else in real life) are generally more complicated and explanations are required for each complication or exception.

Our reaction involves the reaction of two hypothetical diatomic gaseous molecules, A_2 and B_2, to form the gaseous molecule AB. We will assume that the reaction is reversible and reaches a point of equilibrium where there are appreciable quantities of A_2, B_2, and AB present. The reaction is illustrated by

$$A_2(g) + B_2(g) \rightleftharpoons 2AB(g)$$

We first concentrate on how A_2 and B_2 molecules react to form a product. *The path or method whereby reactant molecules transform into product molecules is known as the* **mechanism** *of the reaction.* The mechanism of our hypothetical reaction in the forward direction is quite simple. It involves a collision between the two reactant molecules that leads to the breaking of the bonds between the atoms in the two elemental molecules and the formation of new bonds to yield two product molecules, AB.

We learned in Chapter 10 that gas molecules are in rapid and random motion constantly undergoing collisions. If all collisions led to products, however, every reaction would be essentially instantaneous. As we know, this does

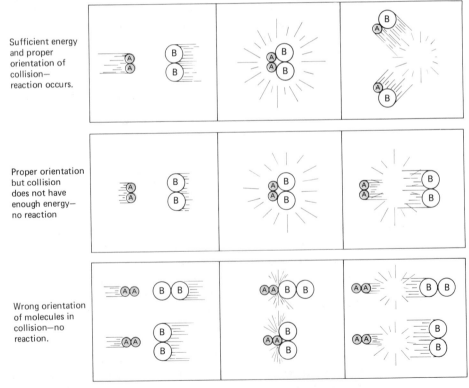

Sufficient energy and proper orientation of collision— reaction occurs.

Proper orientation but collision does not have enough energy— no reaction

Wrong orientation of molecules in collision—no reaction.

Figure 15-3   REACTION OF A_2 AND B_2. Reactions occur only when colliding molecules have the minimum amount of energy and the right orientation.

not happen often. Collisions that lead to the formation of products are actually subject to some important constraints. First, the collisions between two reactant molecules must take place in the right geometric orientation. As illustrated in Figure 15-3 the two reactant molecules must meet in a side-to-side manner rather than end-to-end or side-to-end for new bonds to form. Second, the collision must occur with enough energy to disrupt the bonds in the reactants so that new bonds can form in the products. *This minimum energy needed for the reaction is known as the* **activation energy** and is discussed further in the next section. The activation energy is analogous to the energy required to break a water glass. If the glass is dropped a few inches it doesn't break. If dropped a few feet, it shatters. Just as the collision of the glass must have a minimum energy for it to shatter, so must reactants have a minimum energy to form products. If these two conditions are not met, the colliding molecules simply recoil from each other unchanged. Obviously then, not all collisions lead to products. (See Figure 15-3.)

In the hypothetical reaction, we assume that at least some of the collisions have the right orientation and sufficient energy for the formation of products. If the reaction mixture initially contains only A_2 and B_2 the concentrations of these two molecules begin to decrease as the concentration of the product, AB, increases. As the reaction proceeds, the rate of buildup of AB begins to decrease until, eventually, the concentration of AB no longer changes despite the presence of reactant molecules. To understand this, we turn our attention to the product molecule, AB. If we had started with pure AB we would find that AB slowly decomposes to form A_2 and B_2, just the reverse of the original reaction. This reaction occurs by a mechanism in which two AB molecules collide with the proper orientation and sufficient energy to form A_2 and B_2, (See Figure 15-4.)

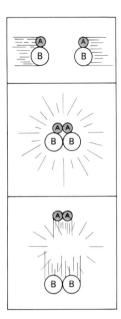

Figure 15-4   REACTION OF 2AB. Collisions between AB molecules having the minimum energy and correct orientation lead to the formation of reactants.

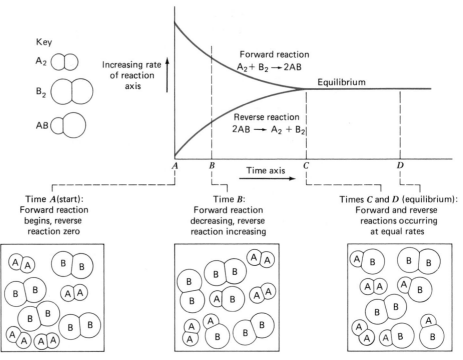

Figure 15-5 A_2, B_2, AB EQUILIBRIUM. Equilibrium is achieved when the rates of the forward and reverse reactions are equal.

Now we put these two reactions together. If we again start with pure A_2 and B_2, at first the only reaction leads to the formation of AB. As the concentration of AB increases, however, the reverse reaction becomes important. Eventually, the rate of formation of products (the forward reaction) is exactly offset by the rate of formation of reactants (the reverse reaction). At this point the concentrations of all species (A_2, B_2, and AB) remain constant. This phenomenon is referred to as a **dynamic equilibrium,** *which emphasizes the changing identities of reactants and products despite the fact that the total amounts of each do not change.* In Figure 15-5 the reaction and the point of equilibrium are illustrated graphically.

If we could shrink ourselves to the molecular level, we would immediately become aware of this dynamic (changing) situation. If one A atom in an A_2 molecule was marked so that it could be traced, we would notice that at one moment it is present in an A_2 molecule, later it is part of an AB molecule, and still later it is part of another A_2 molecule.

$$\begin{array}{ccccc} *A & & *A & & *A \\ | & & | & & | \\ A & \xrightarrow{+B_2} & B & \xrightarrow{+AB} & A & \xrightarrow{+B_2} & \text{etc.} \end{array}$$

15-3 THE RATES OF CHEMICAL REACTIONS

Every chemical reaction takes place at a specific rate. *The rate of a reaction is measured by the rate of disappearance of a reactant or the rate of appearance of a product as a function of time.* Some reactions are essentially instantaneous such as the reaction that occurs with detonation of a stick of dynamite. Others take place very slowly such as the rusting of an iron nail. Why one reaction is rapid and another comparatively slow depends on several factors that affect the rate of a reaction. These include the magnitude of the activation energy, the temperature at which the reaction takes place, the concentration of reactants, the size of the particles of solid reactants, and the presence of a catalyst. We will discuss these factors individually.

Activation Energy

The total energy of a molecule is the sum of its kinetic energy (energy of motion) and its potential energy (energy of position and composition). Let us consider the energy of our two hypothetical molecules that are far away from each other but moving on a course that will eventually lead to a proper collision and formation of products. Before the two molecules come close enough to interact they already contain a certain amount of potential energy in the form of chemical energy. This is represented by the energy of the A_2 and B_2 molecules on the left side of the graph in Figure 15-6. By moving to the right along the horizontal axis in the graph we can follow the change in potential energy of the reactant molecules as they approach each other, collide, and eventually form products. This horizontal axis is referred to as the reaction coordinate and can be viewed as a time axis. The potential energy of the molecules does not change until they begin to collide and thus slow down. The loss of velocity means a loss of kinetic energy, which is converted into potential energy of position. Thus, the total potential energy of molecules begins to increase as they begin to collide. At maximum impact, the motion completely stops for an instant as the two molecules compress together. This corresponds to the maximum potential energy of the collision because all energy is now in this form. This is analogous to hitting a tennis ball with a racquet. At maximum impact, the ball has no kinetic energy but maximum potential energy caused

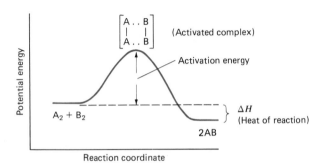

Figure 15-6 ACTIVATION ENERGY. The activation energy is the potential energy barrier between reactants and products.

by the compression of the ball on the racquet. *At maximum impact, both molecules are compressed together, forming what is known as an* **activated complex.** This activated complex has no stability and instantly decomposes to product molecules. The activation energy for the forward reaction is represented by the difference in potential energy of the activated complex (at the peak) and the reactants. As product molecules recoil, potential energy decreases as it is converted into the kinetic energy of the recoiling molecules. Note that the AB molecules end up with lower potential energy in the form of chemical energy than the reactants. *This difference in potential energy represents the heat of the reaction (ΔH).* When the potential energy of the products is lower than that of the reactants, the reaction is exothermic.

The activation energy for a reaction is the potential energy barrier between reactants and products. For a reaction to occur, colliding molecules must contain enough kinetic energy to overcome this barrier. If they don't, then the activated complex simply decomposes into the original reactants. It is much like trying to roll a bowling ball over a small hill. When we hurl the ball, we impart kinetic energy to the ball. As it goes up the hill, it slows as kinetic energy is converted into potential energy. If the ball has kinetic energy that is at least equal to the potential energy of the top of the hill, then the ball goes over and down the other side. If it does not, the ball rolls back.

Perhaps one of the most unique characteristics of any chemical reaction is its activation energy. Consider, for example, the reactions of two nonmetals with oxygen. An allotrope of phosphorus known as white phosphorus reacts almost instantly with the oxygen in the air, even at room temperature, producing an extremely hot flame. As a result, this form of phosphorus must be stored under water to prevent exposure to the air. Hydrogen also reacts with oxygen to form water. This is also a highly exothermic reaction that is used in space rockets. A mixture of hydrogen and oxygen, however, does not react at room temperature. In fact, no appreciable reaction occurs unless the temperature is raised to at least 400 °C or the mixture is ignited with a spark. The difference in the rates of combustion of these two elements at room temperature lies in the activation energies of the two reactions. Figure 15-7 shows the activation

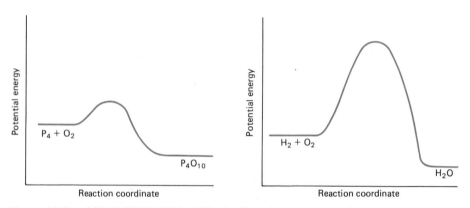

Figure 15-7   COMBUSTION OF P_4 AND H_2. P_4 undergoes combustion at room temperature because of its lower activation energy.

energies for the combustion reactions of hydrogen and phosphorus. Note that the phosphorus reaction has a much lower activation energy than the hydrogen reaction. This explains why collisions between phosphorus molecules and oxygen have enough energy for reaction at room temperature, whereas collisions of the same energy between hydrogen and oxygen do not have the required energy to overcome the activation energy barrier.

Temperature   From the preceding discussion, we established that the rate of a reaction depends on the kinetic energy of the reactant molecules:

$$r \text{ (rate)} \propto \text{energy of colliding molecules}$$

In Chapter 11 we discussed the distribution of kinetic energies of molecules at a certain temperature. This is again illustrated in Figure 15-8. Note that since temperature is related to the average kinetic energy, the higher the temperature, the higher the average kinetic energy of the molecules. In the figure, the dashed line represents the activation energy (E_a). As mentioned, this is the minimum amount of kinetic energy that colliding molecules must have for products to form. Note that at the higher temperature (T_2) the curve intersects the dashed line at a higher point than T_1, meaning that a greater fraction of molecules can overcome the activation energy in a collision. As a result, the rates of *all* chemical reactions increase as the temperature increases.

There is another reason why an increase in temperature increases the rate of a reaction. In addition to the energy of the collisions, the rate of a reaction depends on the frequency of collisions (the number of collisions per second):

$$r \text{ (rate)} \propto \text{frequency of collisions}$$

Recall that the kinetic energy (K.E.) of a moving object is given by the relationship

$$\text{K.E.} = \tfrac{1}{2}mv^2 \quad (m = \text{mass}, v = \text{velocity})$$

This indicates that the more kinetic energy a moving object has, the faster it is moving.

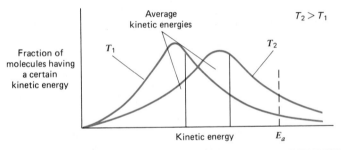

Figure 15-8   DISTRIBUTION OF KINETIC ENERGIES AT TWO TEMPERATURES. The dashed line is the minimum kinetic energy needed for molecules in a collision to form products.

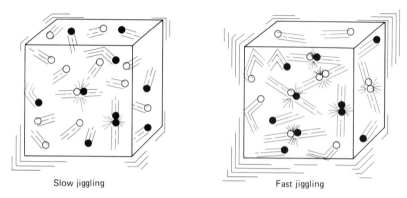

Slow jiggling   Fast jiggling

Figure 15-9   EFFECT OF VELOCITY ON COLLISIONS. Collisions occur more frequently (and with more force) as the velocity increases.

As the temperature increases, the average velocity of the molecules increases. This means that the frequency of collisions also increases. As an analogy, imagine a box containing Ping-Pong balls (four white and four dark balls) as shown in Figure 15-9. If we jiggle the box, the balls move around and collide. If we jiggle the box faster (analogous to a higher temperature), the balls move around faster and there are more frequent collisions. The increased noise we hear tells us that, indeed, collisions are not only more energetic but more frequent.

All chemical reactions occur at a faster rate as the temperature increases. In fact, the rates of most reactions approximately double for every 10 °C rise in temperature. Most of the increase in rate is due to the increased energy of the collisions, with a lesser contribution from the increased rate of collisions.

Concentration of reactants

Let's look again at the box of Ping-Pong balls. The frequency of collisions depends not only on the velocity but on the number of Ping-Pong balls in the box. Shaking a box containing 20 balls at the same rate as one with 10 balls would obviously create a much greater racket. In Figure 15-10 we have doubled the number of dark balls but still have only four white balls. This doubles the chances for a dark–white collision for any one white ball. Similarly, if we tripled the number of dark balls, the rate of dark–white collisions would be tripled. This relationship can be written mathematically:

frequency of dark–white collisions ∝ concentration of dark balls

Conversely,

frequency of dark–white collisions ∝ concentration of white balls

In the case of a chemical reaction, the frequency of collisions (and thus the rate of the reaction) is also dependent on the number of reactant molecules in a fixed volume (which is the concentration). In the reaction of A_2 and B_2, the

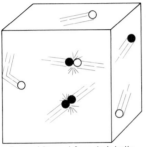
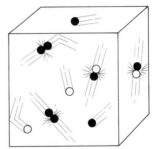

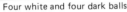
Four white and four dark balls   Four white and eight dark balls

Figure 15-10 EFFECT OF CONCENTRATION ON COLLISIONS. Collisions occur more frequently when there are more balls in the box.

rate of formation of AB doubles if the concentration of A_2 doubles (and $[B_2]$* is held constant). Mathematically, this can be expressed as

$$r_f \propto [A_2] \quad [B_2] \text{ constant} \quad (r_f = \text{rate of forward reaction})$$

The result is similar if $[B_2]$ doubles and $[A_2]$ is held constant:

$$r_f \propto [B_2] \quad [A_2] \text{ constant}$$

These two relations can be combined into one relation known as the rate law for the forward reaction

$$r_f \propto [A_2][B_2]$$

or, as an equality,

$$r_f = k_f[A_2][B_2]$$

where k_f is called the **rate constant** for the forward reaction. *The* **rate law** *for a reaction expresses the effect that changes in concentrations of reactants have on the rate of the reaction.*

Note in this rate law that if both $[A_2]$ and $[B_2]$ are doubled, the rate doubles twice, or fourfold. In the reverse reaction, both colliding molecules are identical. Therefore, the doubling of $[AB]$ in effect doubles the concentrations of both colliding molecules, and the rate increases fourfold. The rate law for the reverse reaction (r_r) is thus

$$r_r = k_r[AB][AB] = k_r[AB]^2$$

where k_r is the rate constant for the reverse reaction.

In our example, we assumed a simple one-step mechanism for the reaction and then rationalized a rate law from this proposed mechanism. In fact, chemists follow the opposite sequence. From a laboratory study of how changes in concentrations of reactants affect the rate of a reaction, they determine a rate law. A mechanism is then proposed that is consistent with the experimentally

---

* $[B_2]$ symbolizes the numerical value of the concentration in units of moles per liter (mol/L).

determined rate law. As one may expect, few reactions are as simple as our hypothetical example. Other reactions have rate laws that are more complicated and indicate more involved mechanisms. For example, few reactions actually occur in one step. Most involve two or more steps, each involving collisions between molecules or certain species that are intermediate between reactants and products. *The study of reaction rates and their relation to the mechanism of a reaction is a fascinating area of chemistry known as* **chemical kinetics.**

Particle Size

If we wish to burn an old dead tree there is a slow way and a quick way. Trying to burn it as one big log could take days. If we cut the tree up into smaller logs, we could burn it in hours. In fact, if we changed the tree into a pile of sawdust and spread it around, we might burn it in a matter of minutes. Related to this is the inherent danger in the storage of grain in large silos. Normally, a pile of grain burns slowly. However, small particles of grain dust can become suspended in air, forming a dangerous mixture that can actually detonate. Such explosions have occurred. In the reaction between a solid and a gas, a solid and a liquid, or a liquid and a gas the surface area of the solid or liquid obviously affects the rate of the reaction. The more area that is exposed, the faster the rate of the reaction.

Catalysts

A **catalyst** *is a substance that is not consumed in a reaction but whose presence increases the rate of the reaction.* Catalysts work by providing an alternate reaction mechanism with a lower activation energy. (See Figure 15-11.) The presence of a catalyst is analogous to having a tunnel through the mountains rather than having to go over the top of the pass. It provides an easier and thus quicker route. A lower activation energy for a reaction means a faster reaction rate. Catalysts work in different ways in different reactions. They may be

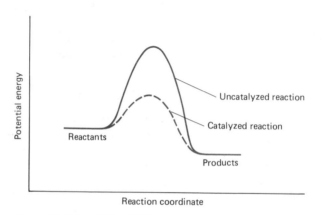

Figure 15-11   ACTIVATION ENERGY AND CATALYSIS. A catalyst speeds a reaction by lowering the activation energy.

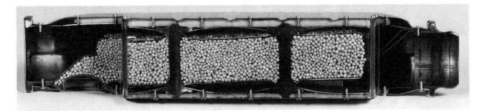

Figure 15-12 THE CATALYTIC CONVERTER. This device on the automobile helps to reduce air pollution.

intimately mixed with the reactants (homogeneous catalysts), as in an aqueous solution, or they may simply provide a surface on which reactions take place (heterogeneous catalysts).

Perhaps the most familiar application of catalysts concerns the catalytic converter in an automobile. The exhaust from the engine contains poisonous carbon monoxide and unburned fuel (mainly C_8H_{18}). Both contribute to air pollution, which is still a serious environmental concern in many localities. The catalytic converter, which is attached to the exhaust pipe, contains finely divided platinum and/or palladium. These metals provide a surface for the following reactions, which occur only at a very high temperature in the absence of a catalyst.

■
SEE PROBLEMS
15-1 THROUGH
15-9.

$$2CO(g) + O_2(g) \longrightarrow 2CO_2(g)$$
$$2C_8H_{18}(g) + 25O_2(g) \longrightarrow 16CO_2(g) + 18H_2O(g)$$

Both reactions are exothermic, which explains why the catalytic converter becomes quite hot when the engine is running. (See Figure 15-12.)

15-4 THE POINT OF EQUILIBRIUM AND THE LAW OF MASS ACTION

We are now ready to consider the point of equilibrium for a specified reaction. *The* **point of equilibrium** *is reached when the concentrations of reactants and products are constant.* In some reactions, a point of equilibrium is reached in which products are favored. In others, such as the ionization of a weak acid, very little product forms. One factor that determines the point of equilibrium is illustrated by the heat of the reaction in Figure 15-6. For an exothermic reaction, note that the activation energy for the forward reaction is less than the activation energy for the reverse reaction (the energy difference from the peak to the energy of the products). This means that the forward reaction occurs more easily than the reverse, suggesting that products are favored over reactants at equilibrium. The heat of the reaction is not the only factor that determines the point of equilibrium, but it is usually the most important.

Actually, long before much was known about rate laws and mechanisms, experimental results indicated a quantitative relationship between reactants and products present in a system at equilibrium. It was found that the relative proportions of reactants and products were distributed in a predictable man-

TABLE 15-1   The Value of K_{eq}

For the reaction $H_2(g) + I_2(g) \rightleftharpoons 2HI(g)$ at 450 °C,

$$K_{eq} = \frac{[HI]^2}{[H_2][I_2]} \qquad [X] = \text{concentration in mol/L}$$

	Initial Concentration			Equilibrium Concentration			
Expt	[H_2]	[I_2]	[HI]	[H_2]	[I_2]	[HI]	K_{eq}
1	2.000	2.000	0	0.428	0.428	3.144	53.9
2	0	0	2.000	0.214	0.214	1.572	54.0
3	1.000	1.000	1.000	0.321	0.321	2.358	54.0

ner at a specified temperature for a particular reaction. This is illustrated with the following general reaction at equilibrium.

$$aA + bB \rightleftharpoons cC + dD$$

For a reaction at equilibrium, the following relationship exists and is known as the **law of mass action**.

$$K_{eq} = \frac{[C]^c[D]^d}{[A]^a[B]^b}$$

Note that the coefficients (a, b, c, and d) of the compounds (A, B, C, and D) become exponents of the molar concentrations (mol/L) of the compounds. The products are written in the numerator, and the reactants are written in the denominator. The law of mass action for a particular reaction consists of two parts. K_{eq} *is called the equilibrium constant and has a definite numerical value at a certain temperature. The ratio to the right of the equality is called the* **mass action expression.**

In Table 15-1, the results of three experiments are listed. In each experiment we start with a different distribution of reactants and products. At equilibrium, the concentrations of all species are distributed so that solution of the mass action expression yields approximately the same constant.

We can now see that the mass action expression is consistent with our previous discussion of rate laws. In the discussion in the previous section we indicated (Figure 15-5) that the point of equilibrium is reached (time C) when the rates of the forward and reverse reactions are equal. This can be expressed mathematically as

$$r_f = r_r$$

For the hypothetical reaction the two rate laws can now be set equal to each other:

$$k_f[A_2][B_2] = k_r[AB]^2$$
$$\frac{k_f}{k_r} = \frac{[AB]^2}{[A_2][B_2]}$$

Since the ratio of two constants is another constant, the expression can be simplified:

$$\frac{k_f}{k_r} = K_{eq} = \frac{[AB]^2}{[A_2][B_2]}$$

We derived this law of mass action by assuming a mechanism, and from that a rate law. Fortunately, an expression can be written as indicated earlier for any reaction at equilibrium regardless of the details of how the reaction takes place.

We are now ready to write the laws of mass action for some gas phase reactions at equilibrium.

EXAMPLE 15-1

Write the law of mass action for each of the following reactions.

(a) $N_2(g) + 3H_2(g) \rightleftharpoons 2NH_3(g)$
(b) $4NH_3(g) + 5O_2(g) \rightleftharpoons 4NO(g) + 6H_2O(g)$
(c) $PCl_3(g) + Cl_2(g) \rightleftharpoons PCl_5(g)$

SOLUTIONS

(a) $$K_{eq} = \frac{[NH_3]^2}{[N_2][H_2]^3}$$

(b) $$K_{eq} = \frac{[NO]^4[H_2O]^6}{[NH_3]^4[O_2]^5}$$

SEE PROBLEMS
15-10 AND 15-11.

(c) $$K_{eq} = \frac{[PCl_5]}{[PCl_3][Cl_2]}$$

We can now illustrate with some examples how the equilibrium constant is calculated from experimental data and then how it is used in calculating equilibrium concentrations.

EXAMPLE 15-2

What is the value for K_{eq} for the following system at equilibrium?

$$2NO(g) + O_2(g) \rightleftharpoons 2NO_2(g)$$

At a certain temperature, the equilibrium concentrations of the gases are $[NO] = 0.890$, $[O_2] = 0.250$, and $[NO_2] = 0.0320$.

SOLUTION

For this reaction,

$$K_{eq} = \frac{[NO_2]^2}{[NO]^2[O_2]}$$

$$= \frac{[0.0320]^2}{[0.890]^2[0.250]} = \frac{1.024 \times 10^{-3}}{0.198}$$

$$= \underline{5.17 \times 10^{-3}}$$

EXAMPLE 15-3

For the equilibrium

$$N_2(g) + 3H_2(g) \rightleftharpoons 2NH_3(g)$$

complete the table and compute the value of the equilibrium constant.

Initial Concentration			Equilibrium Concentration		
[H$_2$]	[N$_2$]	[NH$_3$]	[H$_2$]	[N$_2$]	[NH$_3$]
0.200	0.200	0	___	___	0.0450

PROCEDURE

1 As in other stoichiometry problems, find the [H$_2$] and [N$_2$] that reacted to form the 0.0450 mol/L of NH$_3$ (Section 9-3).

2 Find the [H$_2$] and [N$_2$] remaining at equilibrium by subtracting the concentration that reacted from the initial concentration; that is,

$$[N_2]_{eq} = [N_2]_{initial} - [N_2]_{reacted}$$

3 Substitute the concentrations of all compounds present at equilibrium into the mass action expression and solve to find the value of K_{eq}.

SOLUTION

1 If 0.0450 mol of NH$_3$ is formed, calculate the number of moles of N$_2$ that reacted (per liter).

$$0.0450 \text{ mol NH}_3 \times \frac{1 \text{ mol N}_2}{2 \text{ mol NH}_3} = 0.0225 \text{ mol N}_2 \text{ reacted}$$

The number of moles of H_2 that reacted is

$$0.0450 \text{ mol NH}_3 \times \frac{3 \text{ mol } H_2}{2 \text{ mol NH}_3} = 0.0675 \text{ mol } H_2 \text{ reacted}$$

2 At equilibrium,

$$[N_2]_{eq} = 0.200 - 0.0225 = 0.178$$
$$[H_2]_{eq} = 0.200 - 0.0675 = 0.132$$

3 $$K_{eq} = \frac{[NH_3]^2}{[N_2][H_2]^3} = \frac{(0.0450)^2}{(0.178)(0.132)^3} = \underline{\underline{4.95}}$$

EXAMPLE 15-4

In the preceding equilibrium, what is the concentration of NH_3 at equilibrium if the equilibrium concentrations of N_2 and H_2 are 0.22 and 0.14 mol/L, respectively?

PROCEDURE

In this problem, use the value of K_{eq} found in the previous problem. The concentration of NH_3 can be found by substituting the concentrations of the species given and solving for the one unknown.

SOLUTION

$$K_{eq} = \frac{[NH_3]^2}{[N_2][H_2]^3} \quad \begin{array}{l} [N_2] = 0.22 \\ [H_2] = 0.14 \end{array} \quad K_{eq} = 4.95$$

$$= \frac{[NH_3]^2}{[0.22][0.14]^3} = 4.95$$

$$[NH_3]^2 = 2.99 \times 10^{-3}$$

$$[NH_3] = 5.5 \times 10^{-2} = 0.055$$

Thus, the concentration of $NH_3 = \underline{0.055 \text{ mol/L}}$.

■
SEE PROBLEMS
15-12 THROUGH
15-29.

The numerical value of K_{eq} is a measure of the extent of the reaction to the right, so it tells us about the *position* of the equilibrium. A large value for K_{eq} (e.g., 10^3) tells us that the numerator is much larger than the denominator in the mass action expression. This means that the concentration of at least one of the products in the numerator is considerably larger than that of the reactants in the denominator. Thus the position of equilibrium lies to the right. Conversely, a small value for K_{eq} (e.g., 10^{-3}) indicates that the reactants are favored and the position of equilibrium lies significantly to the left. Intermediate values for K_{eq} indicate appreciable concentrations of both reactants and products present at equilibrium.

**15-5
EQUILIBRIA
OF WEAK
ACIDS AND
BASES IN
WATER**

We are now ready to apply the law of mass action to the equilibria of a typical weak acid in water. As an example of a weak acid we use hypochlorous acid. The ionization of this acid in water is illustrated by the following equation.

$$HOCl(aq) + H_2O \rightleftharpoons H_3O^+(aq) + OCl^-(aq)$$

The law of mass action can be written as

$$K_{eq} = \frac{[H_3O^+][OCl^-]}{[H_2O][HOCl]}$$

In this equilibrium, the concentrations of H_3O^+, OCl^-, and HOCl can all be varied. H_2O is the solvent, however, and is present in a very large excess compared with the other species. Since the amount of H_2O actually reacting is very small compared with the total amount present, the concentration of H_2O is essentially a constant. The $[H_2O]$ can therefore be included with the other constant, K_{eq}, to produce another constant labeled K_a, *which is known as an* **acid ionization constant.**

$$K_{eq}[H_2O] = K_a = \frac{[H_3O^+][OCl^-]}{[HOCl]}$$

Ionization of the weak base NH_3 is illustrated as

$$NH_3(aq) + H_2O \rightleftharpoons NH_4^+(aq) + OH^-(aq)$$

For this reaction, the *equilibrium constant is labeled* K_b, *which is the* **base ionization constant.**

$$K_b = \frac{[NH_4^+][OH^-]}{[NH_3]}$$

The values of the equilibrium constants for weak acids and bases can be determined experimentally as illustrated in the following examples.

EXAMPLE 15-5

In a 0.20 M solution of HNO_2 it is found that 0.009 mol/L of the HNO_2 dissociates. What are the concentrations of H_3O^+, NO_2^-, and HNO_2 at equilibrium, and what is the value of K_a?

PROCEDURE

SIMILAR PROBLEMS ARE ILLUSTRATED IN APPENDIX B (EXAMPLES B-13 AND B-14).

1 Write the equilibrium equation.

2 Calculate the concentration of undissociated HNO_2 present at equilibrium. Remember that the initial concentration given (0.20 M) represents the concentration of undissociated HNO_2 present *plus* the concentration that dissociates. Therefore, $[HNO_2]$ at equilibrium is the initial concentration minus the concentration that dissociates.

$$[HNO_2]_{eq} = [HNO_2]_{initial} - [HNO_2]_{dissociated}$$

3 Note that one H_3O^+ and one NO_2^- are formed for each HNO_2 that dissociates (from the equation stoichiometry).

$$[H_3O^+]_{eq} = [NO_2^-]_{eq} = [HNO_2]_{dissociated}$$

4 Calculate K_a.

SOLUTION

1 $$HNO_2 + H_2O \rightleftharpoons H_3O^+ + NO_2^-$$

2 $$[HNO_2]_{initial} = 0.20 \quad [HNO_2]_{dissociated} = 0.009$$
   $$[HNO_2]_{eq} = 0.20 - 0.009 = \underline{0.19}$$

3 $$\underline{[H_3O^+] = [NO_2^-] = 0.009}$$

4 $$K_a = \frac{[H_3O^+][NO_2^-]}{[HNO_2]} = \frac{[0.009][0.009]}{[0.19]}$$
   $$= \underline{4 \times 10^{-4}}$$

EXAMPLE 15-6

A 0.25 M solution of HCN has a pH of 5.00. What is K_a?

PROCEDURE

1 Write the equilibrium reaction.
2 Convert pH to $[H_3O^+]$.
3 Note that $[CN^-] = [H_3O^+]$ at equilibrium.
4 Calculate $[HCN]$ at equilibrium, which is

$$[HCN]_{eq} = [HCN]_{initial} - [HCN]_{dissociated}$$

$[HCN]_{dissociated} = [H_3O^+]_{eq}$, since one H_3O^+ is produced at equilibrium for every HCN that dissociates.

5 Use these values to calculate K_a.

SOLUTION

$$HCN + H_2O \rightleftharpoons H_3O^+ + CN^-$$

$$[H_3O^+] = \text{antilog } 5.00$$

$$[H_3O^+] = 1.0 \times 10^{-5} = [CN^-]$$

$$[HCN]_{eq} = 0.25 - (1.0 \times 10^{-5}) \approx 0.25$$

$$(\approx \text{ means approximately equal})$$

Note that the amount of HCN that dissociates (10^{-5} M) is negligible compared with the initial concentration of HCN (0.25 M).* Therefore, $[HCN]_{eq} = [HCN]_{initial}$.

$$K_a = \frac{[H_3O^+][CN^-]}{[HCN]}$$

$$= \frac{(1.0 \times 10^{-5})(1.0 \times 10^{-5})}{(0.25)}$$

$$= \underline{4.0 \times 10^{-10}}$$

---

* For the purposes of these calculations (two significant figures), a number is considered negligible compared with another if it is less than 10% of the larger number.

In Table 13-1, several acids were listed in order of decreasing acid strength. The value of K_a is the quantitative measure of this acid strength. The lower the position in the table the smaller the value of K_a.

The K_a values for some weak acids are shown in Table 15-2 and the K_b values for some weak bases, in Table 15-3.

TABLE 15-2　K_a for Some Weak Acids

$$K_a = \frac{[H_3O^+][X^-]}{[HX]}$$　(HX symbolizes a weak acid, X^- its conjugate base)

Acid	Formula	K_a	
Hydrofluoric	HF	6.7×10^{-4}	
Nitrous	HNO_2	4.5×10^{-4}	
Formic	$HCHO_2$	1.8×10^{-4}	Decreasing
Acetic	$HC_2H_3O_2$	1.8×10^{-5}	acid
Hypochlorous	HOCl	3.2×10^{-8}	strength
Hypobromous	HOBr	2.1×10^{-9}	
Hydrocyanic	HCN	4.0×10^{-10}	

TABLE 15-3 K_b for Some Weak Bases

$$K_b = \frac{[HB^+][OH^-]}{[B]}$$ (B symbolizes a weak base, HB^+ its conjugate acid)

Base	Formula	K_b	
Dimethylamine	$NH(CH_3)_2$	7.4×10^{-4}	Decreasing
Ammonia	NH_3	1.8×10^{-5}	base
Hydrazine	N_2H_4	9.8×10^{-7}	↓ strength

EXAMPLE 15-7

A 0.10 M solution of NH_3 is 1.34% ionized. What is the value of K_b (the base ionization constant)?

PROCEDURE

Find the concentrations of all species in the mass action expression at equilibrium. Substitute these values into the expression and solve for K_b.

SOLUTION

$$NH_3 + H_2O \rightleftharpoons NH_4^+ + OH^-$$

$$K_b = \frac{[NH_4^+][OH^-]}{[NH_3]}$$

At equilibrium, 1.34% of the NH_3 is ionized, or

$$0.0134 \times 0.10 = 1.34 \times 10^{-3} \text{ mol/L}$$

According to the equation, for every NH_3 ionized, one NH_4^+ and one OH^- form. Therefore, if 1.34×10^{-3} mol/L ionizes, at equilibrium

$$[NH_4^+] = [OH^-] = 1.34 \times 10^{-3}$$

The concentration of NH_3 at equilibrium is the initial concentration (0.10) minus the concentration that ionizes.

$$[NH_3] = 0.10 - (1.34 \times 10^{-3}) \approx 0.10$$

Substituting these values into the expression,

$$K_b = \frac{(1.34 \times 10^{-3})(1.34 \times 10^{-3})}{(0.10)}$$

$$= \underline{1.8 \times 10^{-5}}$$

SEE PROBLEMS 15-30 THROUGH 15-41.

The values for K_a or K_b can now be used to calculate the pH from a known concentration of weak acid or base, as illustrated in the following examples.

EXAMPLE 15-8

What is the pH of a 0.155 M solution of HOCl?

PROCEDURE

1   Write the equilibrium involved.
2   Write the appropriate mass action expression.
3   Let $x = [H_3O^+]$ at equilibrium; since $[H_3O^+] = [OCl^-]$, $x = [OCl^-]$.
4   At equilibrium, $[HOCl]_{eq} = [HOCl]_{initial} - [HOCl]_{dissociated}$.
5   Using the value for K_a, solve for x.
6   Convert x to pH.

In summary:

	[HOCl]	[H$_3$O$^+$]	[OCl$^-$]
Initial	0.155	0	0
Equilibrium	0.155 − x	x	x

SOLUTION

1
$$HOCl + H_2O \rightleftharpoons H_3O^+ + OCl^-$$

2
$$K_a = \frac{[H_3O^+][OCl^-]}{[HOCl]} = 3.2 \times 10^{-8}$$

3   At equilibrium, $[H_3O^+] = [OCl^-] = x$.
4   At equilibrium, $[HOCl] = 0.155 - x$.

5
$$\frac{[x][x]}{[0.155 - x]} = 3.2 \times 10^{-8}$$

The solution of this equation appears to require the quadratic equation. However, a simplification of this calculation is possible. Note that K_a is a small number, indicating that the degree of dissociation is small (the equilibrium lies far to the left). This means that x is a very small number. Since very small numbers make little or no difference when added to or subtracted from large numbers, they can be ignored with little or no error. (Refer to the example of the large crowd in Section 2-1.)

$$0.155 - x \approx 0.155$$

Therefore, the expression can now be simplified:

$$\frac{[x][x]}{[0.155 - x]} = \frac{x^2}{0.155} = 3.2 \times 10^{-8}$$
$$x^2 = 5.0 \times 10^{-9}$$

To solve for x, take the square root of both sides of the equation:

$$\sqrt{x^2} = \sqrt{5.0 \times 10^{-9}} = \sqrt{50 \times 10^{-10}}$$
$$x = [H_3O^+] = 7.1 \times 10^{-5}$$

(Note that the x is indeed much smaller than 0.155.)

6
$$pH = -\log[H_3O^+] = \underline{4.15}$$

EXAMPLE 15-9

What is the pH of a 0.245 M solution of N_2H_4?

SOLUTION

1
$$N_2H_4 + H_2O \rightleftharpoons HN_2H_4^+ + OH^-$$

2
$$K_b = \frac{[HN_2H_4^+][OH^-]}{[N_2H_4]} = 9.8 \times 10^{-7}$$

3   Let $x = [OH^-] = [HN_2H_4^+]$ (at equilibrium).

4   $[N_2H_4] = 0.245 - x$ (at equilibrium). Since K_b is very small, x is very small. Therefore,

$$[0.245 - x] \approx [0.245]$$

5
$$\frac{[x][x]}{[0.245]} = 9.8 \times 10^{-7}$$
$$x^2 = 2.4 \times 10^{-7}$$
$$x = 4.9 \times 10^{-4} = [OH^-]$$

6
$$pOH = -\log[OH^-] = -\log(4.9 \times 10^{-4}) = 3.31$$
$$pH = 14 - pOH = 14 - 3.31 = \underline{10.69}$$

EXAMPLE 15-10

What is the pH of a buffer solution that is made by dissolving 0.50 mol of $HC_2H_3O_2$ and 0.50 mol of $NaC_2H_3O_2$ in enough water to make 1.00 L of solution?

PROCEDURE

1, 2   As in previous examples.

3   Let $x = [H_3O^+]$ at equilibrium. In this case, $[C_2H_3O_2^-] = 0.50 + x$ (the concentration from the dissolved salt plus that from the dissociation of the acid).

4 $[HC_2H_3O_3] = 0.50 - x$

5, 6   As in previous examples.

SOLUTION

1 $HC_2H_3O_2(aq) + H_2O \rightleftharpoons H_3O^+(aq) + C_2H_3O_2^-(aq)$

2 $K_a = \dfrac{[H_3O^+][C_2H_3O_2^-]}{[HC_2H_3O_2]}$

3, 4   Let $\ x = [H_3O^+], \quad [C_2H_3O_2^-] = 0.50 + x, \quad$ and $\quad [HC_2H_3O_2] = 0.50 - x.$

Since K_a is small, x is small. Therefore,

$$[C_2H_3O_2^-] = 0.50 + x \approx 0.50$$
$$[HC_2H_3O_2] = 0.50 - x \approx 0.50$$

5 $\dfrac{[x][\cancel{0.50}]}{[\cancel{0.50}]} = 1.8 \times 10^{-5}(K_a)$

$$x = [H_3O^+] = 1.8 \times 10^{-5}$$

6 $pH = -\log(1.8 \times 10^{-5}) = \underline{4.74}$

Note that for buffers made up of equal molar concentrations of the acid and the salt, the pH of the solution is simply the negative log of the constant K_a. This is defined as the pK_a.

When $[HX] = [X^-]$, $pH = pK_a$ and $pK_a = -\log K_a$. pK_a values are useful in determining the pH values of various buffer solutions.

SEE PROBLEMS 15-42 THROUGH 15-54.

15-6 THE POINT OF EQUILIBRIUM AND LE CHÂTELIER'S PRINCIPLE

When a chemical system is at equilibrium, any change in conditions may affect the point of equilibrium. Such changes include a change in temperature, pressure, or concentration of a reactant or product. How a system at equilibrium reacts to a change in conditions is summarized by **Le Châtelier's principle**, which states: *When stress is applied to a system at equilibrium, the system reacts in such a way to counteract the stress.*

This principle can be illustrated by a long-distance jogger. When jogging it is desirable to maintain a steady pace, which is analogous to equilibrium. At first the body maintains the pace rather easily, with little change in respiration or heartbeat. After a couple of miles the body begins to tire, which is analogous to stress on the system. If nothing changed, the jogger would slow and finally stop. The body counteracts the stress, however, by faster breathing, faster heartbeat, and increased perspiration. Thus the human body counteracts the stress so that the steady pace (equilibrium) is maintained. (See Figure 15-13.)

We illustrate Le Châtelier's principle with the following important industrial process that reaches an equilibrium:

$$2SO_2(g) + O_2(g) \rightleftharpoons 2SO_3(g)$$

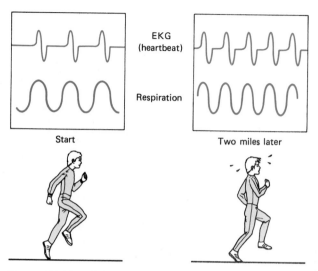

Figure 15-13   STRESS AND THE JOGGER. To compensate for the increased rate of metabolism of a jogger, the heartbeat and respiration increase.

The SO_2 is produced by the combustion of sulfur or sulfur compounds. The SO_3 produced from this reaction is then hydrated to form sulfuric acid (H_2SO_4):

$$SO_3(g) + H_2O \longrightarrow H_2SO_4(aq)$$

In 1988, 40 *million tons* of H_2SO_4 with a value of $3.8 billion was produced by this process in the United States alone. (See Figure 15-14.)

It is important to convert the maximum amount of SO_2 to SO_3 by forcing the point of equilibrium to the right as far as possible.

1   If the system is at equilibrium and additional O_2 is introduced, the system counteracts that stress by removing some of the added O_2. It can do this by reaction with SO_2. An increase in the concentrations of compounds on one side of a reaction at equilibrium ultimately leads to an increase on the other side as the system shifts to remove the stress of the added compound.

The selective removal of a compound from a reaction mixture, such as formation of a precipitate or evolution of a gas, has a similar effect. The system shifts in such a way so as to replace the loss. If the loss is complete, the reaction in the direction of that loss is complete (an irreversible reaction).

The changes in the point of equilibrium can be illustrated by a simple example. In the following hypothetical reaction, equilibrium is established when [A] = 1.00 mol/L and [B] = 2.00 mol/L.

$$A \rightleftharpoons B \quad K_{eq} = \frac{[B]}{[A]} = \frac{2.00}{1.00} = 2.00$$

Figure 15-14   A SULFURIC ACID PLANT. This large plant converts sulfur into some of the millions of tons of sulfuric acid manufactured in this country each year.

If an additional 1.00 mol/L of A is now introduced into the reaction container, note that the system is no longer at equilibrium.

$$\frac{[B]}{[A]} = \frac{2.00}{(1.00 + 1.00)} = 1.00 \neq K_{eq}$$

For the system to return to equilibrium, the numerator becomes larger (more B is formed) and the denominator smaller (some A must react). Thus we see mathematically that introducing additional components present on one side of an equation leads to a shift to the other side. In the example above, equilibrium is reestablished when

$$[A] = 1.33$$
$$[B] = 2.67$$
$$\frac{2.67}{1.33} = 2.00 = K_{eq}$$

This indicates that the added A increases the concentration of B by 0.67 mol/L.

In our industrial example, the conversion of SO_2 to SO_3 is aided by a large excess of O_2 in the reaction container.

2   A change in the volume of the container also affects the point of equilibrium. Note that the reaction produces a volume reduction when it goes to the right. Three moles of gas ($2SO_2 + 1O_2$ or 67.2 L at STP) react to form two moles of gas ($2SO_3$ or 44.8 L at STP). Recall from Boyle's law that when the volume of a container of gas is decreased this corresponds to an increase in pressure. *Thus, when a mixture of gases at equilibrium is compressed, the volume is lowered and the pressure raised.* The chemical system can counteract this stress by going to a lower volume. *Therefore, when a mixture of gases is compressed, the equilibrium shifts in the direction of the fewest number of molecules of gas.* In the reaction discussed, compression of the gaseous mixture results in the formation of a larger proportion of SO_3. When gases are not involved in a reaction at equilibrium, pressure changes have little effect since solids and liquids are virtually incompressible.

3   The reaction of SO_2 with O_2 produces not only SO_3 as a product but also heat energy since this is an exothermic reaction.

$$2SO_2(g) + O_2(g) \rightleftharpoons 2SO_3(g) + \text{heat}$$

The heat is considered a component of the reaction and is produced in an amount directly proportional to the amount of SO_3 produced. Cooling the reaction mixture at equilibrium removes heat from the system. According to Le Châtelier's principle, the system shifts to the right to replace the lost heat. Since SO_3 is formed along with the heat, cooling the mixture produces more SO_3. On the other hand, heating the mixture forces the equilibrium to shift to the left. By shifting to the left, the system attempts to remove the added heat. This phenomenon is reflected in the value for K_{eq}. *For exothermic reactions, K_{eq} becomes larger as the temperature decreases.* This means that, for exothermic reactions, formation of products (on the right) becomes more favorable at lower temperatures.

In all reactions, however, the rate of reaction decreases as the temperature decreases (i.e., the refrigerator effect). Since time is money in industry, the process of maximizing the yield of SO_3 must be compromised to an extent. The reaction must be run at a high enough temperature so that it proceeds to equilibrium in a reasonable time. This reaction is actually carried out at about 400 °C.

Catalysts also play a very significant role in many chemical reactions of industrial significance. A catalyst in an equilibrium reaction increases the rates of the forward and reverse reactions but does not change the point of equilibrium. Reactants and products come to the point of equilibrium faster in the presence of an appropriate catalyst. As a consequence, the use of a catalyst allows the reaction to be run at a considerably lower temperature than would otherwise be possible. In the reaction discussed, heterogeneous catalysts such as V_2O_5 are effective. In other industrial processes, the exact formulation of the catalyst for a specific reaction may be a closely guarded secret.

In summary, the production of SO_3 from SO_2 is aided by

1   A large excess concentration of O_2

2   Compression of the gaseous mixture

■
SEE PROBLEMS
15-55 THROUGH
15-59.

3   A low temperature (but this slows the reaction)

4   The presence of a catalyst (which speeds up the reaction)

15-7 CHAPTER REVIEW

This chapter actually concerns two important topics in chemistry: chemical kinetics and the point of equilibrium. Chemical kinetics is concerned with the rate of a chemical reaction and what that tells us about how a reaction takes place or what is referred to as the mechanism of the reaction. The two topics were tied together by first examining the mechanism of a hypothetical, reversible reaction of gaseous compounds that reaches a point of equilibrium. We found that chemical reactions occur because of collisions between reacting molecules. For reactions at equilibrium, this happens in both forward and reverse reactions. In the collision, the molecules must have the proper geometric orientation and kinetic energy for products to form.

If we accept the idea that collisions of a certain minimum energy (called the activation energy) are important in the formation of products, it is obvious that the rate of a reaction increases with both the energy and the frequency of collisions. The following five factors affect the rate.

1	Activation energy	All reactions require a certain minimum energy for reactants to form products. This is called the activation energy. It is different for each set of potential reactants.
2	Temperature	Higher temperatures increase both the energy of colliding molecules and the frequency of collisions.
3	Concentration of reactants	Increasing the concentrations of reactants increases the frequency of collisions. The mathematical relationship between rate and concentration is known as the rate law. The constant of proportionality in the rate law is called the rate constant.
4	Particle size	Finely divided solids react more rapidly than compact solids because of increased surface area.
5	Catalyst	The presence of a catalyst provides a reaction mechanism with a lower activation energy so the rate of the reaction is increased.

The distribution of concentrations of reactants and products can be predicted from the law of mass action. The mass action expression is constructed from a balanced equation and is equal to a constant, K_{eq}, at a certain temperature. Examples for the two types of equilibria discussed are as follows:

Gaseous

$$A_2(g) + B_2(g) \rightleftharpoons 2AB(g)$$

$$K_{eq} = \frac{[AB]^2}{[A_2][B_2]}$$

Ionic

$$HX(aq) + H_2O \rightleftharpoons H_3O^+(aq) + X^-(aq)$$

$$K_a = \frac{[H_3O^+][X^-]}{[HX]}$$

The value of the equilibrium constant at a certain temperature is obtained by experimental measurements of the distribution of reactants and products present at equilibrium. The constant and initial concentrations can then be used to calculate the concentration of a certain gas or ion present at equilibrium. The ionization constants for weak acids (K_a) and weak bases (K_b) are useful for calculation of the pH of these solutions.

Finally, there was discussion of how the point of equilibrium could be altered. The factors that can affect a system at equilibrium are summarized as follows:

1 Concentration An increase in concentrations on one side of the reaction results in an increase in concentrations on the other side. A decrease on one side leads to an increase on the same side.

2 Volume For gas phase reactions, compression of the mixture leads to a lower volume and a higher pressure. The equilibrium shifts to the side with the smaller total number of molecules of gas.

3 Temperature For exothermic reactions an increase in temperature leads to a decrease in the proportion of products. Just the opposite occurs for an endothermic reaction.

Many important industrial reactions do not take place readily or even at all without a catalyst present. A catalyst brings a reaction mixture to the point of equilibrium faster but does not change the actual point of equilibrium.

■ EXERCISES

THE COLLISION THEORY AND REACTION RATES

15-1 Explain why all collisions between reactant molecules do not lead to products.

15-2 Explain in terms of reaction rates what is occurring when a reaction has reached a point of equilibrium.

15-3 The equation

$$NO_2(g) + CO(g) \rightleftharpoons NO(g) + CO_2(g)$$

represents a reaction in which the mechanism of the forward reaction involves a collision between the two reactant molecules. Describe the probable orientation of this collision.

15-4   In the hypothetical reaction between A_2 and B_2, when was the rate of the forward reaction at a maximum? In the same reaction, when was the rate of the reverse reaction at a maximum? (See Figure 15-5.)

15-5   Besides reaching a point of equilibrium, give another reason why certain reactions do not proceed completely to the right?

15-6   Explain the following facts from your knowledge of collision theory and the factors that affect the rates of reactions.
(a) The rates of chemical reactions approximately double for each 10 °C rise in temperature.
(b) Eggs cook slower in boiling water at higher altitudes, where water boils at temperatures less than 100 °C.
(c) H_2 and O_2 do not start to react to form H_2O except at very high temperatures (unless initiated by a spark).
(d) Wood burns explosively in pure O_2 but slowly in air, which is about 20% O_2.
(e) Coal dust burns faster than a single lump of coal.
(f) Milk sours if left out for a day or two but will last 2 weeks in the refrigerator.
(g) H_2 and O_2 react smoothly at room temperature in the presence of finely divided platinum.

15-7   Explain the following facts from your knowledge of collision theory and the factors that affect the rates of reactions.
(a) A wasp is lethargic at temperatures below 65 °F.
(b) Charcoal burns faster if you blow on the coals.
(c) A 0.10 M boric acid solution ($[H_3O^+] = 7.8 \times 10^{-6}$) can be used for eyewash, but a 0.10 M hydrochloric acid solution ($[H_3O^+] = 0.10$) would cause severe damage.
(d) To keep apple juice from ferment-

ing to apple cider, the apple juice must be kept cold.
(e) Coal does not spontaneously ignite at room temperature but phosphorus does.
(f) Potassium chlorate decomposes at a much lower temperature when mixed with MnO_2.

15-8   Compare the activation energy for the reverse reaction with that for the forward reaction in Figure 15-6. Which are easier to form, reactants from products or products from reactants? Which system would come to equilibrium faster, a reaction starting with pure products or one starting with pure reactants?

15-9   Write the energy diagram for an endothermic reaction similar to that shown in Figure 15-6. In this case, which is greater, the activation energy for the forward or the reverse reaction? Which system would come to equilibrium faster, a reaction starting with pure products or one starting with pure reactants?

THE LAW OF MASS ACTION

15-10   Write the law of mass action for each of the following equilibria.
(a) $CO(g) + Cl_2(g) \rightleftharpoons COCl_2(g)$
(b) $CH_4(g) + 2H_2O(g) \rightleftharpoons$
$$CO_2(g) + 4H_2(g)$$
(c) $4HCl(g) + O_2(g) \rightleftharpoons$
$$2Cl_2(g) + 2H_2O(g)$$
(d) $CH_4(g) + Cl_2(g) \rightleftharpoons$
$$CH_3Cl(g) + HCl(g)$$

15-11   Write the law of mass action for each of the following equilibria.
(a) $3O_2(g) \rightleftharpoons 2O_3(g)$
(b) $N_2(g) + 2O_2(g) \rightleftharpoons 2NO_2(g)$
(c) $C_2H_2(g) + 2H_2(g) \rightleftharpoons C_2H_6(g)$
(d) $4H_2(g) + CS_2(g) \rightleftharpoons$
$$CH_4(g) + 2H_2S(g)$$

THE VALUE OF K_{eq}

15-12   For the hypothetical reaction $A_2 + B_2 \rightleftharpoons 2AB$, $K_{eq} = 1.0 \times 10^8$. Are reactants or products favored at equilibrium?

15-13 For the hypothetical reaction $2C + 3B \rightleftharpoons 2D + F$, $K_{eq} = 5 \times 10^{-7}$. Are reactants or products favored?

15-14 For the reaction $H_2(g) + I_2(g) \rightleftharpoons 2HI(g)$, $K_{eq} = 45$ at a certain temperature. Are reactants or products favored at equilibrium?

15-15 Given the system

$$3O_2(g) \rightleftharpoons 2O_3(g)$$

at equilibrium, $[O_2] = 0.35$ and $[O_3] = 0.12$. What is K_{eq} for the reaction at this temperature?

15-16 Given the system

$$N_2(g) + 2O_2(g) \rightleftharpoons 2NO_2(g)$$

at a certain temperature there are 1.25×10^{-3} mol of N_2, 2.50×10^{-3} mol of O_2, and 6.20×10^{-4} mol of NO_2 in a 1.00-L container. What is K_{eq} for this reaction at this temperature?

15-17 Given the system

$$2SO_3(g) + CO_2(g) \rightleftharpoons$$
$$CS_2(g) + 4O_2(g)$$

what is K_{eq} if, at equilibrium, $[SO_3] = 2.0 \times 10^{-2}$ mol/L, $[CO_2] = 4.5 \times 10^{-3}$ mol/L, $[CS_2] = 6.2 \times 10^{-4}$ mol/L, and $[O_2] = 1.0 \times 10^{-4}$ mol/L?

15-18 Given the system

$$CH_4(g) + 2H_2O(g) \rightleftharpoons$$
$$CO_2(g) + 4H_2(g)$$

at equilibrium we find 2.20 mol of CO_2, 4.00 mol of H_2, 6.20 mol of CH_4, and 3.00 mol of H_2O in a 30.0-L container. What is K_{eq} for the reaction?

15-19 Given the system

$$C_2H_2(g) + 2H_2(g) \rightleftharpoons C_2H_6(g)$$

at equilibrium we find 296 g of C_2H_6 present along with 3.50 g of H_2 and 21.0 g of C_2H_2 in a 400-mL container. What is K_{eq}?

15-20 Consider the following system:

$$2HI(g) \rightleftharpoons H_2(g) + I_2(g)$$

(a) If we start with $[HI] = 0.60$, what would be $[H_2]$ and $[I_2]$ if all of the HI reacts?

(b) If we start with $[HI] = 0.60$ and $[I_2] = 0.20$, what would be $[H_2]$ and $[I_2]$ if all of the HI reacts?

(c) If we start with only $[HI] = 0.60$ and 0.20 mol/L of HI reacts, what are $[HI]$, $[I_2]$, and $[H_2]$ at equilibrium?

(d) From the information in (c) calculate K_{eq}.

(e) What is the K_{eq} for the reverse reaction? How does this value differ from the value given in Table 15-1?

15-21 In the reaction

$$N_2(g) + 3H_2(g) \rightleftharpoons 2NH_3(g)$$

initially 1.00 mol of N_2 and 1.00 mol H_2 are mixed in a 1.00-L container. At equilibrium it is found that $[NH_3] = 0.20$.

(a) What is the concentration of N_2 and H_2 at equilibrium?

(b) What is the K_{eq} for the system at this temperature?

*15-22 At a certain temperature N_2, H_2, and NH_3 are mixed so that the initial concentration of each is 0.50 mol/L. At equilibrium the concentration of N_2 is 0.40 mol/L.

(a) Calculate the concentration of H_2 and NH_3 at equilibrium.

(b) What is the K_{eq} at this temperature?

15-23 At the start of the reaction

$$4NH_3(g) + 5O_2(g) \rightleftharpoons$$
$$4NO(g) + 6H_2O(g)$$

$[NH_3] = [O_2] = 1.00$. At equilibrium it is found that 0.25 mol/L of NH_3 has reacted.

(a) What concentration of O_2 reacts?

(b) What are the concentrations of all species at equilibrium?

(c) Write the total mass action expression and substitute the proper values for the concentrations of reactants and products.

15-24　At the start of the reaction

$$CO(g) + Cl_2(g) \rightleftharpoons COCl_2(g)$$

$[CO] = 0.650$ and $[Cl_2] = 0.435$. At equilibrium it is found that 10.0% of the CO has reacted. What is $[CO]$, $[Cl_2]$, and $[COCl_2]$ at equilibrium? What is the K_{eq}?

15-25　For the reaction

$$PCl_3(g) + Cl_2(g) \rightleftharpoons PCl_5(g)$$

$K_{eq} = 0.95$ at a certain temperature. If $[PCl_3] = 0.75$ and $[Cl_2] = 0.40$ at equilibrium, what is the concentration of PCl_5 at equilibrium?

15-26　At a certain temperature, $K_{eq} = 46.0$ for the reaction

$$4HCl(g) + O_2(g) \rightleftharpoons$$
$$2Cl_2(g) + 2H_2O(g)$$

At equilibrium, $[HCl] = 0.100$, $[O_2] = 0.455$, and $[H_2O] = 0.675$. What is $[Cl_2]$ at equilibrium?

15-27　Using the value for K_{eq} calculated in Problem 15-18, what is the concentration of H_2O at equilibrium if, at equilibrium, $[CH_4] = 0.50$, $[CO_2] = 0.24$, and $[H_2] = 0.20$?

*15-28　For the following equilibrium, $K_{eq} = 56$ at a certain temperature.

$$CH_4(g) + Cl_2(g) \rightleftharpoons$$
$$CH_3Cl(g) + HCl(g)$$

At equilibrium, $[CH_4] = 0.20$ and $[Cl_2] = 0.40$. What are the equilibrium concentrations of CH_3Cl and HCl if they are equal?

*15-29　Using the equilibrium in Problem 15-15, calculate the equilibrium concentration of O_3 if the equilibrium concentration of O_2 is 0.80 mol/L.

EQUILIBRIA OF WEAK ACIDS AND BASES

15-30　Write the expression for K_a or K_b for each of the following equilibria. Where necessary, complete the equilibrium.
(a) $HOBr + H_2O \rightleftharpoons H_3O^+ + OBr^-$
(b) $NH_3 + H_2O \rightleftharpoons NH_4^+ + OH^-$
(c) $H_2SO_3 + H_2O \rightleftharpoons H_3O^+ + HSO_3^-$
(d) $HSO_3^- + H_2O \rightleftharpoons \underline{} + SO_3^{2-}$
(e) $H_3PO_4 + H_2O \rightleftharpoons H_3O^+ + \underline{}$
(f) $HN(CH_3)_2 + H_2O \rightleftharpoons$
$$H_2N(CH_3)_2^+ + \underline{}$$

15-31　Write the expression for K_a or K_b for each of the following equilibria. Where necessary, complete the equilibrium.
(a) $N_2H_4 + H_2O \rightleftharpoons N_2H_5^+ + OH^-$
(b) $HCN + H_2O \rightleftharpoons H_3O^+ + CN^-$
(c) $H_2C_2O_4 + H_2O \rightleftharpoons H_3O^+ + \underline{}$
(d) $H_2PO_4^- + H_2O \rightleftharpoons$
$$\underline{} + HPO_4^{2-}$$

15-32　A 0.10 M solution of a weak acid HX has a pH of 5.0. A 0.10 M solution of another weak acid HB has a pH of 5.8. Which is the weaker acid? Which has the larger value for K_a?

15-33　A hypothetical weak acid HZ has a K_a of 4.5×10^{-4}. Rank the following 0.10 M solutions in order of increasing pH: HZ, $HC_2H_3O_2$, and HOCl.

15-34　Given 0.10 M solutions of the compounds KOH, NaCl, HNO_2, NH_3, HNO_3, $HCHO_2$, and N_2H_4, rank them in order of increasing pH.

15-35　The bicarbonate ion (HCO_3^-) can act as either an acid or a base in water. For the reaction

$$HCO_3^- + H_2O \rightleftharpoons H_3O^+ + CO_3^{2-}$$
$$K_a = 4.7 \times 10^{-11}$$

And for the base reaction

$$HCO_3^- + H_2O \rightleftharpoons OH^- + H_2CO_3$$
$$K_b = 2.2 \times 10^{-8}$$

Is a solution of $NaHCO_3$ acidic or basic?

15-36　In a 0.20 M solution of cyanic acid, HOCN, $[H_3O^+] = [OCN^-] = 6.2 \times 10^{-3}$.
(a) What is [HOCN] at equilibrium?
(b) What is K_a?
(c) What is the pH?

15-37　A 0.58 M solution of a weak acid (HX) is 10.0% dissociated.
(a) Write the equilibrium and the mass action expression.
(b) What are $[H_3O^+]$, $[X^-]$, and [HX] at equilibrium?

(c) What is K_a?

(d) What is the pH?

15-38 Nicotine (Nc) is a nitrogen base in water in the same manner as ammonia. Write the equilibrium illustrating this base behavior. In a 0.44 M solution of nicotine, $[OH^-] = [NcH^+] = 5.5 \times 10^{-4}$. What is K_b for nicotine? What is the pH of the solution?

15-39 In a 0.085 M solution of carbolic acid, HC_6H_5O, the pH = 5.48. What is K_a?

15-40 Novacaine (Nv) is a nitrogen base in water, like ammonia. Write the equilibrium. In a 1.25 M solution of novacaine, pH = 11.46. What is K_b?

15-41 In a 0.300 M solution of chloroacetic acid, HOC_2Cl_3, $[HOC_2Cl_3] = 0.277$ M at equilibrium. What is the pH? What is K_a?

15-42 What is the pH of a 0.65 M solution of HOBr?

15-43 What is the pH of a 1.50 M solution of HNO_2?

15-44 What is $[OH^-]$ in a 0.55 M solution of NH_3?

15-45 What is $[H_3O^+]$ in a 0.25 M solution of $HC_2H_3O_2$?

15-46 What is the pH of a 1.00 M solution of $NH(CH_3)_2$?

15-47 What is the pH of a 0.567 M solution of N_2H_4?

15-48 What is the pH of a buffer made by mixing 0.45 mol of NaCN and 0.45 mol of HCN in 2.50 L of solution?

15-49 What is the pH of a buffer made by dissolving 1.20 mol of NH_3 and 1.20 mol of NH_4ClO_4 in 13.5 L of solution?

15-50 What is the pH of a buffer made by dissolving 0.20 mol of KOBr and 0.60 mol of HOBr in 850 mL of solution?

15-51 What is the pH of a buffer that contains 150 g of HNO_2 and 150 g of $LiNO_2$?

*15-52 A buffer of pH = 7.50 is desired. Which of the following two compounds should be dissolved in water in equimolar amounts to provide this buffer solution: HNO_2, HOCl, HCN, $NaNO_2$, NH_3, KCN, KOCl, NH_4Cl?

*15-53 A buffer of pH = 9.25 is desired. Which of the following two compounds should be dissolved in water in equimolar amounts to provide this buffer solution: $HC_2H_3O_2$, HOCl, NH_3, N_2H_4, $KC_2H_3O_2$, NH_4Cl, NaOCl, N_2H_5Br?

*15-54 What is the pH of a buffer made by adding 0.20 mol of NaOH to 1.00 L of a 1.00 M solution of $HC_2H_3O_2$? Assume no volume change. (*Hint:* First consider the partial neutralization of $HC_2H_3O_2$ by NaOH.)

LE CHÂTELIER'S PRINCIPLE AND EQUILIBRIUM

15-55 The following equilibrium is an important industrial process used to convert N_2 to NH_3. The ammonia is used mainly for fertilizer. This method for the production of NH_3 is called the Haber process.

$$N_2(g) + 3H_2(g) \underset{500\,°C}{\overset{catalyst}{\rightleftharpoons}}$$
$$2NH_3(g) + \text{heat energy}$$

Determine the direction in which the equilibrium will be shifted by the following changes.

(a) increasing the concentration of N_2

(b) increasing the concentration of NH_3

(c) decreasing the concentration of H_2

(d) decreasing the concentration of NH_3

(e) compressing the reaction mixture

(f) removing the catalyst

(g) How will the yield of NH_3 be affected by raising the temperature to 750 °C?

(h) How will the yield of NH_3 be affected by lowering the temperature to 0 °C? How will this affect the rate of formation of NH_3?

15-56 Consider the following equilibrium:

$$4NH_3(g) + 5O_2(g) \rightleftharpoons$$
$$4NO(g) + 6H_2O(g) + \text{heat energy}$$

How will this system at equilibrium be affected by

(a) increasing the concentration of O_2

(b) removing all of the H_2O as it is formed

(c) increasing the concentration of NO

(d) compressing the reaction mixture
(e) increasing the volume of the reaction container
(f) increasing the temperature
(g) decreasing the concentration of O_2
(h) adding a catalyst

15-57 The following equilibrium takes place at high temperatures.

$$N_2(g) + 2H_2O(g) + \text{heat energy} \rightleftharpoons 2NO(g) + 2H_2(g)$$

How will the yield of NO be affected by
(a) increasing $[N_2]$
(b) decreasing $[H_2]$
(c) compressing the reaction mixture
(d) decreasing the volume of the reaction container
(e) decreasing the temperature
(f) adding a catalyst

15-58 Consider the following equilibrium:

$$2SO_3(g) + CO_2(g) + \text{heat energy} \rightleftharpoons CS_2(g) + 4O_2(g)$$

How will the yield of CS_2 at equilibrium be affected by
(a) decreasing the volume of the reaction vessel

(b) adding a catalyst
(c) increasing the temperature
(d) increasing the original concentration of CO_2
(e) removing some O_2 as it forms

15-59 Fortunately for us, the major components of air, N_2 and O_2, do not react under ordinary conditions. At very high temperatures, however, like those found in an automobile engine, the equilibrium constant for the following reaction becomes larger.

$$N_2(g) + O_2(g) \rightleftharpoons 2NO(g)$$

(a) Does the information given indicate that the formation of NO is an endothermic or an exothermic process?
(b) Since NO is a serious pollutant, it is desirable to minimize its formation in automobile engines. Would a cooler- or hotter-running engine increase the yield of NO?
(c) Would a lower pressure in the engine affect the formation of NO?

CARBON-BASED LIFE FORMS
Plants and animals are composed of complex organic molecules.
Carbon is the key element in these molecules.

AN INTRODUCTION
TO ORGANIC
CHEMISTRY

PURPOSE OF CHAPTER

In Chapter 16 we introduce some important classes of organic compounds, their unique chemical structures, and their common uses.

OBJECTIVES FOR CHAPTER

After completion of this chapter, you should be able to:

1   Write the Lewis structures of selected organic compounds. (16-1)

2   Illustrate with Lewis structures the isomers of a specified compound. (16-1)

3   Illustrate why the compound C_3H_7Cl has only two isomers. (16-1)

4   Write condensed and fully condensed formulas for specified hydrocarbons. (16-1)

5   Write the formulas of simple alkanes from the general formula of the homologous series. (16-2)

6   Determine the IUPAC name of simple branched alkanes. (16-2)

7 Describe how petroleum is refined and the uses for the various components that are obtained. (16-2)

8 Write the formulas of simple alkenes from the general formula. (16-3)

9 Describe how alkenes are used to make polymers and the names and uses of representative polymers of this type. (16-3)

10 Write the formulas of simple alkynes from the general formula. (16-4)

11 Illustrate the difference in structure and chemical reactivity between the aromatic compound benzene and a simple alkene. (16-5)

12 Distinguish between compounds containing a functional group and an alkane. (16-6)

13 Give the unique structural feature, an example, and a common use of organic compounds in the following classes.
(a) alcohol (16-7)
(b) ether (16-8)
(c) amine (16-9)
(d) aldehyde (16-10)
(e) ketone (16-10)
(f) carboxylic acid (16-11)
(g) ester (16-11)
(h) amide (16-11)

Perfectly positioned about 90 million miles from a medium-sized star, protected by a thick atmosphere, our planet glides silently in space. Under these perfect conditions the formation and stable existence of chemical bonds that compose living systems are possible. These living systems grow and reproduce on the earth, move gracefully in the oceans, and float effortlessly in the air. These creatures and plants are all composed of compounds that include the element carbon as the central character. We, of course, consider ourselves to be the very ultimate carbon-based life form. Could there be life forms based on other elements? Science fiction stories would have us think so, but it is actually quite unlikely if not impossible. In this chapter we see how this unique element forms the types of molecules that are part of living systems.

Traditionally, all chemical compounds have been relegated to two categories: organic and inorganic. **Organic compounds** *include most of the compounds of carbon, especially those containing carbon–hydrogen bonds.* However, much of the chemistry that we have discussed so far concerns *compounds derived from minerals and other noncarbon compounds and these are known as* **inorganic compounds.** A few compounds of carbon such as $CaCO_3$ and LiCN are considered inorganic compounds since they are derived from minerals and do not contain C—H bonds. Originally, the division of compounds into the two categories was determined by whether they were derived from (or part of) a living system (organic) or from a mineral (inorganic). In fact, before 1828 it was thought that organic compounds and their decay products could be synthesized from inorganic compounds (e.g., H_2O, CO_2, and NH_3) only by living matter (organisms). Only life had that magic ingredient of nature called the "vital force" that allowed the miracle of organic synthesis. In 1828, however, a German scientist named Friedrich Wohler synthesized urea (H_2NCONH_2), which had been classified as organic, from the inorganic compound ammonium cyanate (NH_4CNO). Urea is a waste product of the metabolism of proteins and until 1828 was thought to result only from this source. Although it was just one compound, the concept of the "vital force" was doomed. Since that time millions of organic compounds have been synthesized from basic minerals in the laboratory.

Organic compounds are certainly central to our lives. Hydrocarbons (compounds composed only of carbon and hydrogen) are used as fuel to power our cars and to heat our homes. Our bodies are fueled with organic compounds obtained from the food we eat in the form of sugars (carbohydrates), fats, and proteins. This food is made more palatable by organic flavorings, is wrapped in organic plastic, and is kept from spoiling with organic preservatives. Our clothes are made of organic compounds, whether these compounds come from plant and animal sources (cotton and wool) or are synthetic (nylon and Dacron®). These fabrics are made colorful with organic dyes. When we are ill, we take drugs that may also be organic: aspirin relieves headaches, codeine suppresses coughs, and diazepam (Valium®) calms nerves. These are only a few examples of how we use organic chemicals daily.

The introduction of such a large topic as organic chemistry in one chapter is not an easy assignment. Synthesis, structure, nomenclature, properties, and the reactions of each class of compound are important topics, and they can't all be emphasized. In this text, we will present the following information for each class of compound: the unique feature of the chemical structure, an introduction to the nomenclature, how the compounds are synthesized, and some familiar uses of the compounds.

As background to this chapter, you should review:

1　Lewis structures (Section 6-8)
2　Polarity of covalent bonds and molecules (Sections 6-10 and 6-11)
3　Solubility of polar and nonpolar compounds in solvents (Section 12-2)

16-1 THE NATURE OF ORGANIC COMPOUNDS

Of all of the elements of the periodic table, only one has the properties that can lead to the formation of millions of compounds some of which are parts of living systems. This, of course, is carbon. Two important properties of carbon make it capable of forming large, stable molecules. First, a carbon atom forms strong chemical bonds to other carbon atoms, thus allowing almost infinite chains of carbon atoms chemically bound to each other. No other element even comes close to carbon's ability to form long chains. Second, the carbon–hydrogen bond that is the major part of most organic molecules is not chemically reactive (usually). One reason for the lack of reactivity relates to the polarity of the C—H bond. Since the electronegativities of carbon and hydrogen are nearly the same the bond is essentially nonpolar. Compounds containing highly polar bonds are generally much more reactive than compounds containing nonpolar bonds. Otherwise, organic molecules might be much less stable around molecules such as oxygen or water. For example, science fiction stories suggest that silicon (also in group 14 (IVA), under carbon) may be analogous to carbon as a basis for life. However, the silicon–hydrogen bond is polar and most compounds containing this bond react spontaneously with air and water at room temperature.

Most organic compounds are typical molecular compounds. Thus they tend to be low-melting-point solids or even gases or liquids at room temperature. Also, the molecules of most organic compounds have low polarity. This means that they have low solubilities in polar liquids such as water. Inorganic compounds, on the other hand, are likely to be high-melting-point solids with an appreciable solubility in water. This behavior is typical of ionic compounds or at least of highly polar molecular compounds.

Since carbon is in group 14 (IVA) and has four valence electrons, it must form a total of *four bonds* to achieve an octet of outer electrons. Carbon forms single, double, and triple bonds. Let's review some simple organic compounds to see how these molecules appear when written as Lewis structures.

Single Bonds

Double Bond

C_2H_4

$$\begin{array}{cc} H & H \\ \diagdown & \diagup \\ & C{=}C \\ \diagup & \diagdown \\ H & H \end{array}$$

CH_2O

$$\begin{array}{cc} H \\ \diagdown \\ & C{=}\ddot{\underset{\displaystyle\cdot\cdot}{O}} \\ \diagup \\ H \end{array}$$

Triple Bond

$$C_2H_2 \qquad H{-}C{\equiv}C{-}H$$

EXAMPLE 16-1

Give the Lewis structures for the following compounds:

(a) C_3H_8 (b) C_3H_6 (c) H_3C_2N (all H's on one C)

ANSWERS

(a)
$$\begin{array}{ccc} H & H & H \\ | & | & | \\ H{-}C{-}C{-}C{-}H \\ | & | & | \\ H & H & H \end{array}$$

(b)
$$\begin{array}{cc} H & H \\ | & | \\ H{-}C{-}C{=}C{-}H \\ | & | \\ H & H \end{array} \quad \text{or} \quad \begin{array}{cc} H & H \\ | & | \\ H{-}C\!-\!\!-\!\!-\!C{-}H \\ \diagdown \quad \diagup \\ C \\ \diagup \quad \diagdown \\ H \quad H \end{array}$$

(c)
$$\begin{array}{c} H \\ | \\ H{-}C{-}C{\equiv}N\!: \\ | \\ H \end{array}$$

Many times we find that there is more than one Lewis structure for a given formula. Such is the case with butane, C_4H_{10}, since two correct structures can be drawn. *Compounds with different structures but the same molecular formula are called* **isomers.**

$$\begin{array}{cccc} H & H & H & H \\ | & | & | & | \\ H{-}C{-}C{-}C{-}C{-}H \\ | & | & | & | \\ H & H & H & H \end{array} \xleftarrow{\;} \begin{array}{c} \text{isomers} \\ \text{of} \\ \text{each other} \end{array} \xrightarrow{\;} \begin{array}{ccc} H & H & H \\ | & | & | \\ H{-}C{-}C{-}C{-}H \\ | & | & | \\ H & \underset{\displaystyle\begin{array}{c} | \\ H{-}C{-}H \\ | \\ H \end{array}}{} & H \end{array}$$

n-butane   isobutane

Figure 16-1   GEOMETRY OF CH_4 AND C_2H_6. In these two molecules the carbon is at the center of a tetrahedron.

The Lewis structures represented above imply that the molecules are flat and two dimensional. In fact, whenever carbon is chemically bound to four different atoms, it lies at the center of a tetrahedron. In this environment, the H—C—H bond angle is actually about 109°. A more realistic representation of the structure of *n*-butane may be

However, this representation still implies only two dimensions, so long-chain structures are usually shown in a straight line. A true representation of the structure requires the use of molecular models. In Figure 16-1, the CH_4 and C_2H_6 molecules are represented by "ball-and-stick" models. In a three-dimensional representation we can see that all the hydrogens in the C_2H_6 molecule are identical. Thus if one of the hydrogens is replaced by a chlorine we could confirm that only one isomer is possible. The four Lewis representations of this compound (C_2H_5Cl) that follow are all identical and superimposable.

On the other hand, if a Cl is substituted for an H in C_3H_8, to form C_3H_7Cl, two (and only two) isomers exist:

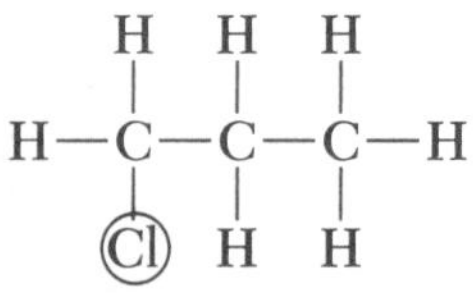

$$\begin{array}{cccc} & H & H & H \\ & | & | & | \\ H- & C- & C- & C-H \\ & | & | & | \\ & \textcircled{Cl} & H & H \end{array} \quad \text{and} \quad \begin{array}{cccc} & H & H & H \\ & | & | & | \\ H- & C- & C- & C-H \\ & | & | & | \\ & H & \textcircled{Cl} & H \end{array}$$

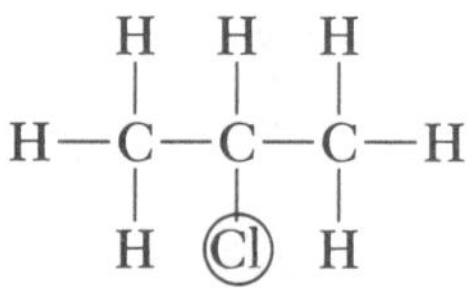

Although actual geometry is important in the study of organic chemistry, we will continue to represent structures in the simplified manner, but it should be understood that the molecules actually have a three-dimensional nature.

Since the molecular formula by itself does not differentiate among the various isomers, it is necessary to refer to isomers by name and by a detailed structure. Drawing the structure can become tedious if all hydrogens and carbons are written out, especially for large molecules. **A condensed formula in which separate bonds are not written out is helpful.** For example,

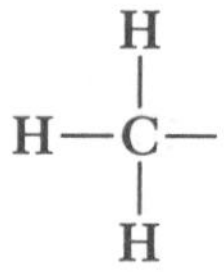

$$\begin{array}{c} H \\ | \\ H-C- \\ | \\ H \end{array}$$

is represented as

$$CH_3-$$

and

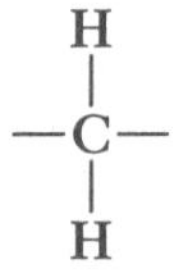

$$\begin{array}{c} H \\ | \\ -C- \\ | \\ H \end{array}$$

is represented as

$$-CH_2-$$

Depending on what we are trying to show, the structure may be partially or fully condensed.

	Partially Condensed	**Fully Condensed**
n-Butane*	$CH_3-CH_2-CH_2-CH_3$	$CH_3(CH_2)_2CH_3$
Isobutane*	$\begin{array}{c} H \\ | \\ CH_3-C-CH_3 \\ | \\ CH_3 \end{array}$	$(CH_3)_3CH$

---

* The n is an abbreviation of normal and refers to the isomer in which all the C's are bound consecutively in a continuous chain. The iso refers to the isomer in which there is a branch involving three carbons at the end of a continuous chain of carbons.

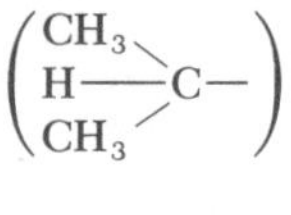

$$\left(\begin{array}{c} CH_3 \diagdown \\ H-\!\!-C- \\ CH_3 \diagup \end{array} \right)$$

TABLE 16-1   Isomers

Formula	Isomer, Name		
CH_5H_{12}	$CH_3CH_2CH_2CH_2CH_3$ n-pentane	$CH_3CH_2CH{-}CH_3$ $\overset{\displaystyle CH_3}{\vert}$ isopentane	$CH_3{-}\overset{\displaystyle CH_3}{\underset{\displaystyle CH_3}{\vert\atop\vert}}{-}CH_3$ neopentane
C_3H_6	$H_2C{=}CHCH_3$ propene	$\overset{H_2C{-}{-}CH_2}{\underset{CH_2}{\diagdown\diagup}}$ cyclopropane	(Note that the carbons can also be arranged in a ring or cyclic structure.)
C_2H_6O	CH_3CH_2OH ethanol	$H_3C{-}O{-}CH_3$ dimethyl ether	
C_3H_6O	$CH_3CH_2\overset{\displaystyle O}{\overset{\displaystyle \|}{C}}H$ propanol	$H_3C{-}\overset{\displaystyle O}{\overset{\displaystyle \|}{C}}{-}CH_3$ propanone	$H_2C{=}CHCH_2OH$ ally alcohol and others

SEE PROBLEMS 16-1 THROUGH 16-6.

In a fully condensed structure, it is understood that the CH_2's in parentheses are in a continuous chain and the CH_3's in parentheses are attached to the same atom.

A few other compounds and their isomers are shown in Table 16-1. As you can see, the number of isomers increases as the number of carbons increases, and addition of a **hetero atom** *(any atom other than carbon or hydrogen)* also increases the number of isomers.

16-2 ALKANES

Natural gas, gasoline, and candle wax are all composed of compounds that are closely related. First, they are all hydrocarbons. **Hydrocarbons** *are compounds that contain only carbon and hydrogen.* Second, they are all alkanes. **Alkanes** *are hydrocarbons containing only single covalent bonds.* Alkanes can be described by the general formula C_nH_{2n+2}. This refers only to *open-chain alkanes* meaning that the carbons do not form a ring. The simplest alkane ($n = 1$) is methane (CH_4), in which the carbon shares *four pairs of electrons* with four different hydrogen atoms. The next alkane ($n = 2$) is ethane (C_2H_6) and the third is propane (C_3H_8). These alkanes are members of a homologous series. *In a* **homologous series,** *the next member differs from the previous one by a constant structure unit, which is one carbon and two hydrogens (CH_2).*

$$H{-}\overset{\displaystyle H}{\underset{\displaystyle H}{\vert\atop\vert}}C{-}H \qquad H{-}\overset{H}{\underset{H}{\vert\atop\vert}}C{-}\overset{H}{\underset{H}{\vert\atop\vert}}C{-}H \qquad H{-}\overset{H}{\underset{H}{\vert\atop\vert}}C{-}\overset{H}{\underset{H}{\vert\atop\vert}}C{-}\overset{H}{\underset{H}{\vert\atop\vert}}C{-}H$$

methane   ethane   propane

TABLE 16-2 Alkanes

Formula	Name	Formula	Name
CH_4	methane	C_6H_{14}	hexane
C_2H_6	ethane	C_7H_{16}	heptane
C_3H_8	propane	C_8H_{18}	octane
C_4H_{10}	butane	C_9H_{20}	nonane
C_5H_{12}	pentane	$C_{10}H_{22}$	decane

The homologous series of alkanes up to 10 carbons, together with the names of the compounds, is given in Table 16-2.

The names in Table 16-2 are the basis for the names of all organic compounds. By altering them slightly, we can name other classes of organic compounds that are discussed later. Two systems of nomenclature are used in organic chemistry. The most systematic is the one devised by the International Union of Pure and Applied Chemistry (the IUPAC system). Although the rules for naming complex molecules can be extensive, we will be concerned with just the basic concepts. Compounds are also known by *common* or *trivial* names. Sometimes these names follow a pattern; sometimes they do not. They have been used for so many years that it is hard to break the habit of using them. When a chemical that is frequently known by its common name is encountered, that name is given in parentheses.

The name of a simple organic compound has two parts; the *prefix* gives the number of carbons in the longest carbon chain and the *suffix* tells what kind of a compound it is. The underlined portions of the names in Table 16-2 are the prefixes used for compounds containing 1 through 10 carbons in the longest chain; *meth-* stands for one carbon, *eth-* for two carbons, and so on. The ending used for alkanes is *-ane*. Therefore, the one-carbon alkane is methane, the two-carbon alkane is ethane, and so on.

Organic compounds can exist as unbranched compounds (all carbons bound to each other in a continuous chain) or as branched compounds. Previously, we indicated that *n*-butane is an unbranched alkane and that isobutane is a branched alkane. The IUPAC system bases its names on the longest carbon chain in the molecule, whereas the common names frequently include all of the carbons in the name (e.g., isobutane). The longest carbon chain in isobutane is three carbons long and is therefore considered a propane in the IUPAC system. Note in isobutane that there is a CH_3— group attached to the propane chain. In this system of nomenclature, the branches are named separately. *Since the branches can be considered as groups of atoms substituted for a hydrogen, the branches are called* **substituents.** *Substituents that contain one less hydrogen than an alkane are called* **alkyl groups.** Alkyl groups are not compounds by themselves; they must always be attached to some other group or atom. They are named by taking the alkane name, dropping the *-ane* ending, and substituting *-yl*. The most common alkyl groups are given in Table 16-3.

TABLE 16-3   Alkyl Groups

Alkyl Group[a]	Name	Alkyl Group[a]	Name
$CH_3—$	methyl	$(CH_3)_2CH—$	isopropyl
$CH_3CH_2—$	ethyl	$CH_3CH_2CH_2CH_2—$	n-butyl
$CH_3CH_2CH_2—$	n-propyl	$(CH_3)_3C—$	tert-butyl (or t-butyl)

[a] The dash (—) shows where the alkyl group is attached to a carbon chain or to another atom such as a halogen, an oxygen, or a nitrogen.

Thus the alkyl substituent in isobutane is a methyl group. The IUPAC name for isobutane is *methylpropane*.

The longest carbon chain is three carbons long and is therefore a *propane* chain.

A substituent on the longest carbon chain called a *methyl* group.

$$CH_3—CH—CH_3$$
$$CH_3$$

The location of the substituent on the longest carbon chain is designated by a number. The carbons are numbered from the end that is closest to the substituent. Thus, the following compound is named 3-ethylheptane rather than 5-ethylheptane.

$$\begin{array}{c} CH_3 \\ | \\ CH_2 \end{array}\Big\} \text{ ethyl group}$$

$$CH_3—CH_2—CH_2—CH_2—CH—CH_2—CH_3 \quad \text{heptane chain}$$

	1	2	3	4	5	6	7
or	7	6	5	4	③	2	1

When the carbons form a ring, the name is prefixed with *cyclo-*. Therefore, the three-carbon ring compound

$$\begin{array}{ccc} H & & H \\ \diagdown & & \diagup \\ & C & \\ \diagup & \diagdown & \\ H—C & — & C—H \\ | & & | \\ H & & H \end{array}$$

is called cyclopropane. Cycloalkanes have the general formula C_nH_{2n}.

EXAMPLE 16-2

Draw a line through the longest carbon chain in the following compounds and circle the substituents. Name the longest chain.

(a) 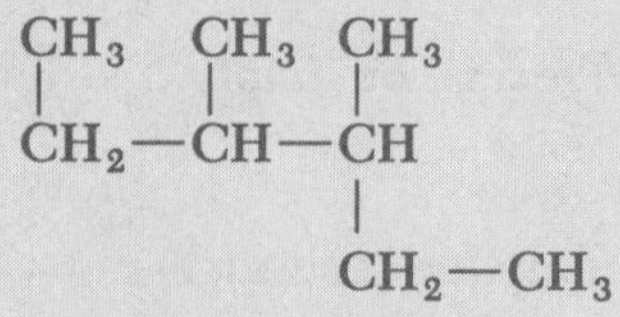
$$CH_3 \quad CH_3 \quad CH_3$$
$$CH_2-CH-CH$$
$$|$$
$$CH_2-CH_3$$

(b) 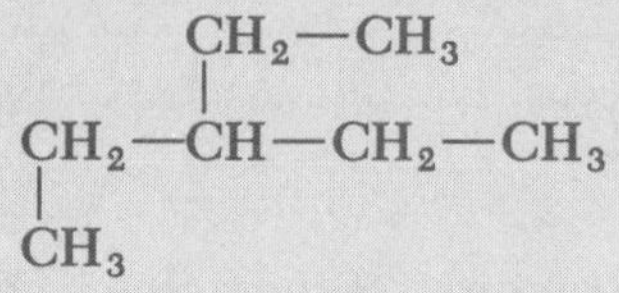
$$CH_2-CH_3$$
$$|$$
$$CH_2-CH-CH_2-CH_3$$
$$|$$
$$CH_3$$

(c) 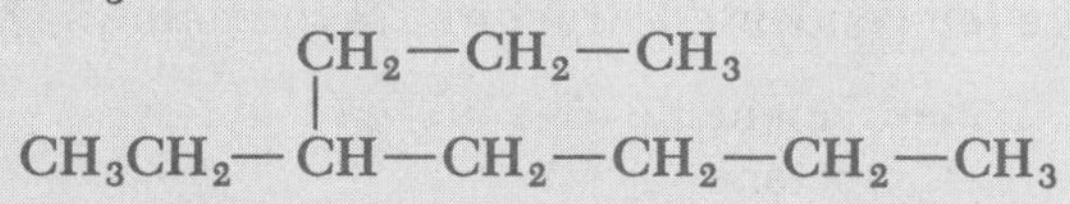
$$CH_2-CH_2-CH_3$$
$$|$$
$$CH_3CH_2-CH-CH_2-CH_2-CH_2-CH_3$$

(d) 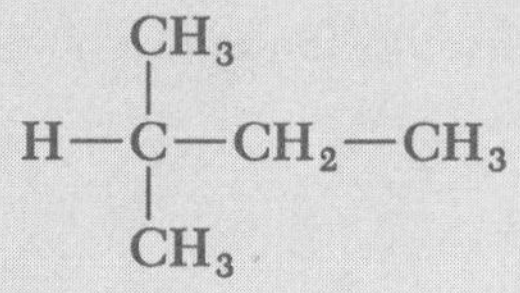
$$CH_3$$
$$|$$
$$H-C-CH_2-CH_3$$
$$|$$
$$CH_3$$

ANSWERS

(a) 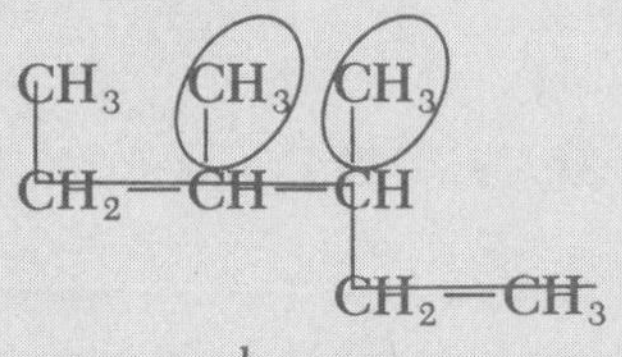
$$CH_3 \quad CH_3 \quad CH_3$$
$$CH_2-CH-CH$$
$$CH_2-CH_3$$
hexane

(b) 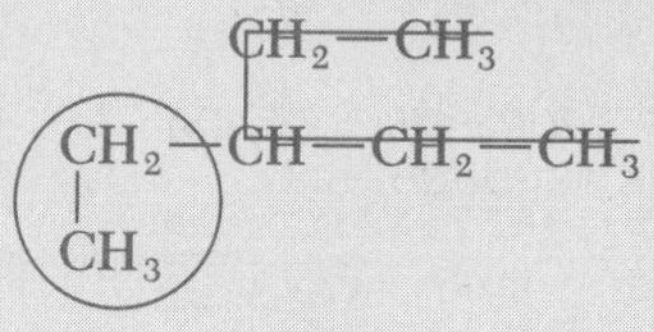
$$CH_2-CH_3$$
$$CH_2-CH-CH_2-CH_3$$
$$CH_3$$
pentane

(c) 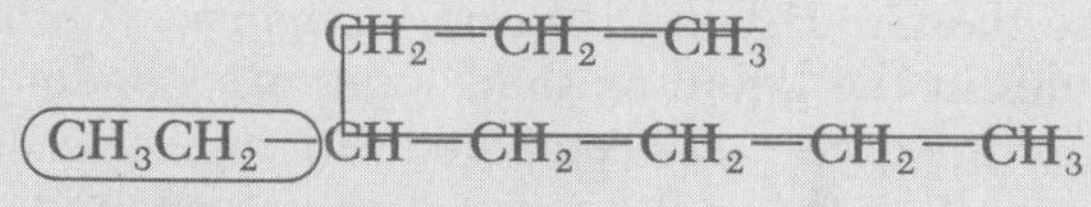
$$CH_2-CH_2-CH_3$$
$$CH_3CH_2-CH-CH_2-CH_2-CH_2-CH_3$$
octane

(d) 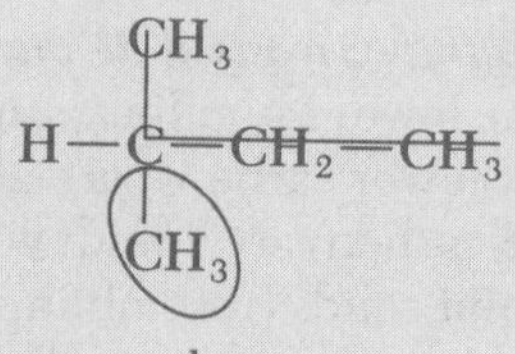
$$CH_3$$
$$H-C-CH_2-CH_3$$
$$CH_3$$
butane

EXAMPLE 16-3

Name the compounds in Example 16-2 by the IUPAC method.

ANSWERS
(a) 3-methyl-4-methylhexane or simply 3,4-dimethylhexane
(b) 3-ethylpentane
(c) 4-ethyloctane
(d) methylbutane (no number necessary)

EXAMPLE 16-4

Write the condensed structures of the following:

(a) 2-methylpentane
(b) 3-ethyloctane
(c) 3-isopropyl-2-methylhexane (Substituent groups are listed in alphabetical order.)

ANSWERS

(a) 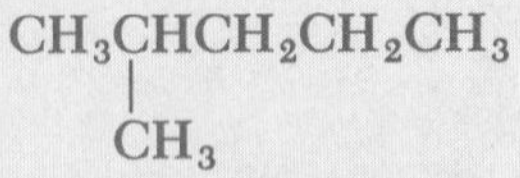 $CH_3CHCH_2CH_2CH_3$
         |
        CH_3

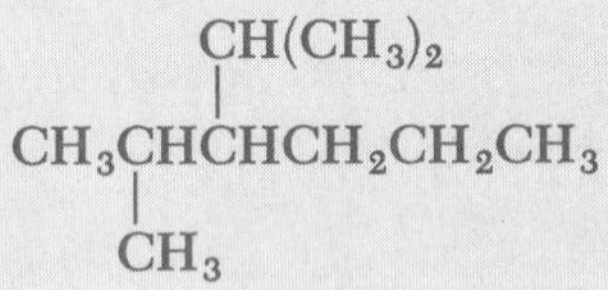

(b) 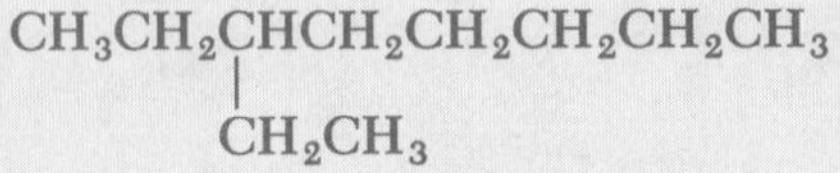 $CH_3CH_2CHCH_2CH_2CH_2CH_2CH_3$
           |
        CH_2CH_3

(c) $CH(CH_3)_2$
           |
    $CH_3CHCHCH_2CH_2CH_3$
          |
        CH_3

SEE PROBLEMS 16-7 THROUGH 16-21.

Alkanes are all essentially nonpolar compounds because of geometry and because the C—H bond is nearly nonpolar. The only attractions between molecules in the liquid or solid states are London forces. These forces of attraction are quite weak compared with forces between ions but are roughly proportional to the molar mass of the compound. Thus, starting with the simplest and lightest member of the series (CH_4), we find that the alkanes become less volatile as the molar mass increases. **Volatility** *refers to the tendency for a liquid to vaporize to the gaseous state and is related to the boiling point.* The lighter the alkane, the lower its boiling point and the more volatile it is. The lightest alkanes (four or fewer carbons) are all gases at room temperature. Those alkanes with 5 to 18 carbons are liquids while those with more than 18 carbons are low-melting-point solids (resembling candle wax). All alkanes are odorless and colorless. They are also extremely flammable.

There are two major sources of alkanes, natural gas and crude oil (petroleum). Natural gas is mainly methane with smaller amounts of ethane, pro-

Figure 16-2 AN OIL REFINERY. The crude oil is separated into various fractions in large refineries.

pane, and butanes. Unlike natural gas, petroleum contains hundreds of compounds, the majority of which are open-chain and cyclic alkanes. Before we can make use of petroleum, it must be separated into groups of compounds with similar properties. Further separation may or may not be carried out depending on the final use of the hydrocarbons.

Crude oil is separated into groups of compounds according to boiling points by distillation in a refinery. (See Figure 16-2.) In such a distillation, the liquid is boiled and the gases move up a large column that becomes cooler and cooler toward the top. Compounds condense (become liquid) at different places in the column, depending on their boiling points. As the liquids condense, they are drawn off, providing a rough separation of the crude oil. Some of the material is too high boiling to vaporize and remains in the bottom of the column. A drawing showing this process and the various fractions obtained is shown in Figure 16-3. Note that the fewer the carbon atoms in the alkane, the lower the boiling point (the more volatile it is).

The composition of crude oil itself varies somewhat depending on where it is found. Certain crude oil, such as that found in Nigeria and Libya, is called "light" oil because it is especially rich in the hydrocarbons that are present in gasoline. Otherwise, one fraction can be converted into another by three processes: cracking, reforming, and alkylation. **Cracking** *changes large molecules into small molecules.* **Reforming** *removes hydrogens from the carbons and/or changes unbranched hydrocarbons into branched hydrocarbons.* (Branched hydrocarbons perform better in gasoline; that is, they have a higher "octane" rating.) **Alkylation** *takes small molecules and puts them together to make larger molecules.* In all of these processes, catalysts are used, but there is a different catalyst for each process.

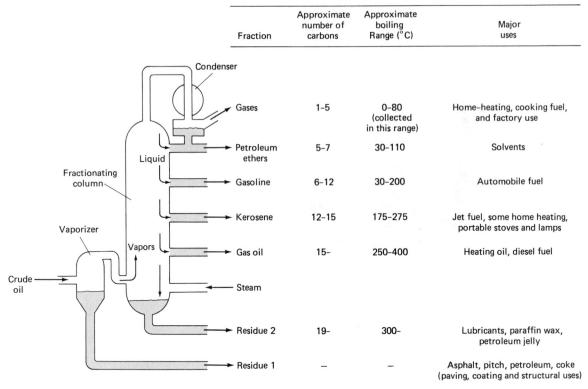

Fraction	Approximate number of carbons	Approximate boiling Range (°C)	Major uses
Gases	1–5	0–80 (collected in this range)	Home-heating, cooking fuel, and factory use
Petroleum ethers	5–7	30–110	Solvents
Gasoline	6–12	30–200	Automobile fuel
Kerosene	12–15	175–275	Jet fuel, some home heating, portable stoves and lamps
Gas oil	15–	250–400	Heating oil, diesel fuel
Steam			
Residue 2	19–	300–	Lubricants, paraffin wax, petroleum jelly
Residue 1	–	–	Asphalt, pitch, petroleum, coke (paving, coating and structural uses)

Figure 16-3 REFINING OF PETROLEUM. Oil is separated into fractions according to boiling points.

About 96% of all oil and gas is burned as fuel, whereas only 4% is used to make other organic chemicals. As a fuel, hydrocarbons burn to give carbon dioxide, water, and a great deal of heat energy.

$$CH_4 + 2O_2 \xrightarrow{\text{spark}} CO_2 + 2H_2O + \text{heat}$$

$$2C_8H_{18} + 25O_2 \xrightarrow{\text{spark}} 16CO_2 + 18H_2O + \text{heat}$$

Industrially, most synthetic organic chemicals have their ultimate origin in *the alkanes obtained from crude oil.* These **petrochemicals** are put to a wide range of uses in the manufacture of fibers, plastics, coatings, adhesives, synthetic rubber, some flavorings, perfumes, and pharmaceuticals.

16-3 ALKENES   *Alkenes are hydrocarbons that contain a double bond. Organic compounds with multiple bonds are said to be* **unsaturated.** The general formula for an open-chain alkene with one double bond is C_nH_{2n}. The simplest alkene is ethene (common name ethylene), which has the structure

$$
\begin{array}{c}
\overset{\displaystyle H}{\diagdown}\overset{\displaystyle H}{\diagup} \\
C\!=\!C \\
\overset{\displaystyle}{\diagup}\overset{\displaystyle}{\diagdown} \\
HH
\end{array}
$$

Alkenes are named by dropping the *-ane* ending of the corresponding alkane and substituting *-ene*. The location of the double bond is indicated by a number in a manner similar to the naming of branched alkanes by the IUPAC method. The double bond is located by numbering the C—C bonds starting from the end closest to the double bond in the main hydrocarbon chain.

$$CH_3—CH_2—CH_2—CH=CH—CH_3$$

$543②1$ 2-hexene

Only small amounts of alkenes are found naturally in crude oil; the majority are made from alkanes by the reforming process during the refining of crude oil. When alkanes are heated over a catalyst, hydrogen is lost from the molecule and alkenes together with hydrogen are formed.

$$C_nH_{2n+2} \xrightarrow[\text{heat}]{\text{catalyst}} C_nH_{2n} + H_2$$

$$C_2H_6 \xrightarrow[\text{heat}]{\text{catalyst}} CH_2=CH_2 + H_2$$

A large amount of alkenes are produced industrially to make polymers. When certain compounds called initiators are added to an alkene or a mixture of alkenes, the double bond is broken and the alkenes become joined to each other by single bonds. This produces a high-molar-mass molecule called a **polymer,** *which has repeating units of the original alkene (called the monomer).*

$$CH_2=CH_2 \xrightarrow{\text{initiator}}$$
monomer (ethylene)

$$\text{etc.} -\!\!\boxed{CH_2—CH_2}\!-\!\!\boxed{CH_2—CH_2}\!-\!\!\boxed{CH_2—CH_2}\!-\!\!\boxed{CH_2—CH_2}\!-\!\text{etc.}$$

written as $-(CH_2—CH_2)_n$
polymer (polyethylene)

Polymers are named by adding *poly-* to the name of the alkene used to form the polymer. In the example shown above, the polymer was made from ethylene (usually common names are used for polymers), and so the polymer is called polyethylene. Since it would be impossible to write out the structure of a polymer, which may contain thousands of carbons, we abbreviate the structure by giving the repeating unit in parentheses along with a subscript n to indicate that the monomer is repeated many times. Groups attached to the double bond affect the properties of the polymer, and by varying the group, we can vary the uses for which a polymer is suited. (See Figure 16-4.) Some commonly used polymers and their uses are given in Table 16-4.

Figure 16-4   POLYMERS. All of these containers are made of plastics composed of polymers.

TABLE 16-4 Polymers

Monomer Name, Structure	Polymer Name, Structure	Some Common Trade Names	Uses
ethylene $CH_2{=}CH_2$	polyethylene $\text{+}CH_2{-}CH_2\text{+}_n$	Polyfilm[a] Marlex[b]	Electrical insulation, packaging (plastic bags), floor covering, plastic bottles, pipes, tubing
propylene $CH_2{=}CH$ $\quad\ \ \ \vert$ $\quad\ \ \ CH_3$	polypropylene $\left(CH_2{-}CH\right)$ $\qquad\ \ \vert$ $\qquad\ \ CH_3{}_n$	Herculon[c]	Pipes, carpeting, artificial turf, molded auto parts, fibers
vinyl chloride $CH_2{=}CH$ $\quad\ \ \ \vert$ $\quad\ \ \ Cl$	polyvinyl chloride (PVC) $\left(CH_2{-}CH\right)$ $\qquad\ \ \vert$ $\qquad\ \ Cl\ \ \ _n$	Tygon[d]	Wire and cable coverings, pipes, rainwear, shower curtains, tennis court playing surfaces.
styrene $CH_2{=}CH$	polystyrene $\left(CH_2{-}CH\right)_n$	Styrofoam[a] Styron[a]	Molded objects (combs, toys, brush and pot handles), refrigerator parts, insulating material, phonograph records, clock and radio cabinets
tetrafluoroethylene $CF_2{=}CF_2$	polytetrafluoroethylene $\text{+}CF_2{-}CF_2\text{+}_n$	Teflon[e] Halon[f]	Gaskets, valves, tubing, coatings for cookware
methyl methacrylate $\qquad CH_3$ $\qquad /$ $CH_2{=}C$ $\qquad \backslash$ $\qquad CO_2CH_3$	polymethyl methacrylate $\qquad CH_3$ $\qquad \vert$ $\text{+}CH_2{-}C\text{+}_n$ $\qquad \vert$ $\qquad CO_2CH_3$	Plexiglas[g] Lucite[e]	Glass substitute, lenses, aircraft glass, dental fillings, artificial eyes, braces
acrylonitrile $CH_2{=}CH$ $\quad\ \ \ \vert$ $\quad\ \ \ CN$	polyacrylonitrile $\left(CH_2{-}CH\right)$ $\qquad\ \ \vert$ $\qquad\ \ CN\ \ _n$	Orlon[e] Acrilan[k]	Fibers for clothing, carpeting

[a] *Dow Chemical Co.*
[b] *Phillips Petroleum Co.*
[c] *Hercules, Inc.*
[d] *U.S. Stoneware Co.*

[e] *E. I. du Pont de Nemours & Co.*
[f] *Allied Chemical Corp.*
[g] *Rohm & Haas Co.*
[h] *Monsanto Industrial Chemicals, Inc.*

Chemicals can add to the double bond to form new single bonds and therefore new compounds. A test based on such a reaction, the addition of bromine to an alkene, is used to show the presence of alkenes. It is a very simple test to perform. The disappearance of the red color of bromine when it is added to a liquid means that multiple bonds are present (the dibromide formed is color-less). We will talk more about this test when we discuss aromatic compounds.

$$CH_2\!=\!CH_2 + Br_2 \longrightarrow \underset{\substack{| \\ Br}}{CH_2}\!-\!\underset{\substack{| \\ Br}}{CH_2}$$

red   colorless

16-4 ALKYNES   Alkynes *are hydrocarbons that contain a triple bond.* The general formula for an open-chain alkyne with one triple bond is C_nH_{2n-2}. The simplest alkyne is ethyne (acetylene), which has the structure

$$H\!-\!C\!\equiv\!C\!-\!H$$

Alkynes are named just like alkenes except that *-yne* is substituted for the alkane ending instead of *-ene*. Thus, $CH_3CH_2CH_2CH_2C\!\equiv\!CCH_2CH_3$ is named 3-octyne.

Alkynes are not found in nature, but can be prepared synthetically. Acety-lene can be made from coal by first reacting the coal with calcium oxide at high temperature and then treating the calcium carbide formed with water:

$$\underset{\substack{coal}}{C} + \underset{\substack{calcium \\ oxide}}{CaO} \overset{\Delta}{\longrightarrow} \underset{\substack{calcium \\ carbide}}{CaC_2} \overset{H_2O}{\longrightarrow} \underset{\substack{acetylene}}{H\!-\!C\!\equiv\!C\!-\!H} + \underset{\substack{calcium \\ hydroxide}}{Ca(OH)_2}$$

A more common method of making acetylene is the reforming process. Meth-ane is heated in the presence of a catalyst, forming acetylene and hydrogen:

$$2CH_4 \xrightarrow[\Delta]{catalyst} H\!-\!C\!\equiv\!C\!-\!H + 3H_2$$

Although acetylene has some use in oxyacetylene torches, its biggest use is in the manufacture of other organic compounds used as monomers for poly-mers. For example, acetylene is used to make vinyl chloride (which is then used to make polyvinyl chloride) and acrylonitrile [which is used to make Acrilan® and Orlon® (see Table 16-4)].

$$H\!-\!C\!\equiv\!C\!-\!H + HCl \longrightarrow \underset{H}{\overset{H}{\diagdown}}C\!=\!C\underset{Cl}{\overset{H}{\diagup}}$$

vinyl chloride

$$H\!-\!C\!\equiv\!C\!-\!H + HCN \longrightarrow \underset{H}{\overset{H}{\diagdown}}C\!=\!C\underset{CN}{\overset{H}{\diagup}}$$

acrylonitrile

16-5 AROMATIC COMPOUNDS

There is another class of hydrocarbons known as the aromatics. *An aromatic compound is a cyclic hydrocarbon containing alternating single and double bonds between adjacent carbon atoms in a six-member ring.** This discussion concentrates on one aromatic compound, benzene (C_6H_6), and its derivatives. *(A derivative of a compound is produced by the substitution of a group or hetero atom on the molecules of the original compound.)* At first glance, benzene looks like a cyclic alkene with three double bonds. It turns out, however, that the three double bonds do not act like alkene double bonds.

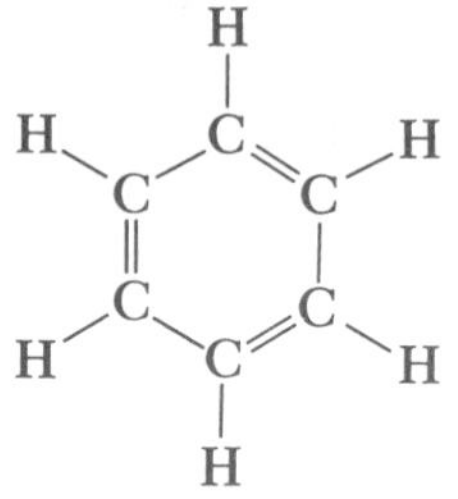

[Simplified structures can be written by omitting carbons (B) or by omitting carbons and hydrogens (C).]

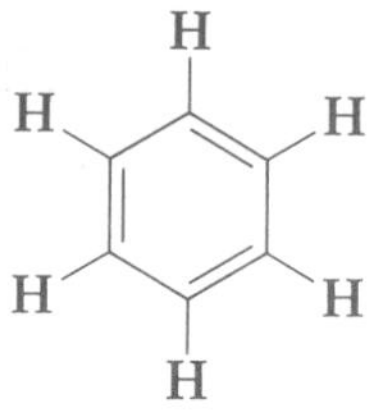

(A)   (B)   (C)

For example, benzene does not "decolorize" a solution of bromine as do simple alkenes.

$$CH_2{=}CH_2 + Br_2 \longrightarrow \underset{\underset{Br \quad Br}{|\quad\;\;|}}{CH_2{-}CH_2}$$

red   colorless

$+ Br_2 \longrightarrow$ no reaction
red   still red

As the preceding reactions illustrate, benzene is less chemically reactive than simple alkenes. It is this property that distinguishes aromatics from alkenes.

Compounds that contain the benzene ring can usually be recognized by name because they are named as derivatives of benzene.

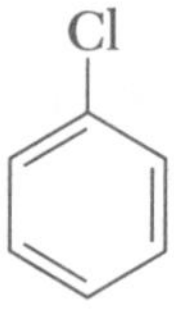

Cl   CH$_2$CH$_3$ CH$_3$

chlorobenzene 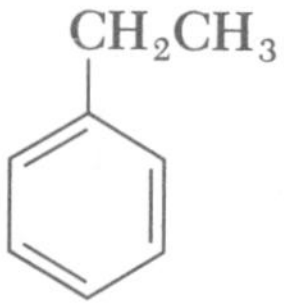   ethylbenzene 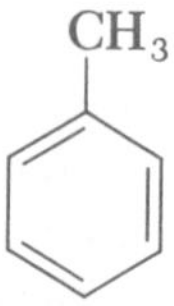   methyl benzene
                                 (toluene)

---

* While there are other types of aromatic compounds, this definition includes the largest group.

OH   NH$_2$ COOH

hydroxybenzene   aminobenzene   benzoic acid
(phenol)   (aniline)

Although some benzene and toluene are present in crude oil, additional quantities can be obtained by the reforming process. When cyclohexane and/ or hexane are heated with a catalyst, benzene is formed. Coal is another source of benzene. When coal is heated to high temperatures in the absence of air, some benzene is formed.

$$\text{cyclohexane} \;\; \text{and/or } CH_3(CH_2)_4CH_3 \;\xrightarrow{\text{catalyst}}\; \text{benzene} + 3H_2 \text{ or } 4H_2$$

SEE PROBLEMS
16-22 THROUGH
16-34.
Benzene and toluene are used mainly as solvents and as starting materials to make other aromatic compounds. Benzene must be used with care, however, because it has been found to be a potent carcinogen (it can cause cancer). Phenol and its derivatives have been used as disinfectants, in the manufacture of dyes, explosives, drugs, and plastics, and as preservatives.

16-6 ORGANIC FUNCTIONAL GROUPS

Organic chemistry extends well beyond the hydrocarbons. When atoms other than carbon and hydrogen (hetero atoms) are part of the molecule the properties are altered drastically from the original hydrocarbon. For example, if a halogen such as Cl is substituted for a hydrogen in methane, the resulting compound, CH_3Cl (chloromethane) has properties very different from those of methane. In most compounds, the nature and bonding of the hetero atom (or atoms) controls the chemistry or the function of the molecule. *The atom or group of atoms taken together in an organic molecule that determines the chemical nature of the molecule is known as the* **functional group.** The functional group of alkenes is the double bond and that of alkynes is the triple bond. Each functional group has a strong influence on the chemistry of the compounds that contain it and thus establishes a specific class of compounds. In the following sections we examine some of these classes of compounds that are determined by the presence of a particular functional group.

Figure 16-5   ALCOHOLS. A major ingredient in each of these products is an alcohol.

**16-7
ALCOHOLS
(R—OH)***

The first and perhaps most familiar class of hydrocarbon derivatives is the alcohols. You are probably well aware that one alcohol is the active ingredient in certain beverages and another is used in rubbing alcohol. Industrially, they are very important solvents. **Alcohols** *are a class of organic compounds that contain the OH functional group (known as a* **hydroxyl group**) *in place of a hydrogen on a carbon chain.*

Alcohols are named by taking the alkane name, dropping the *-e* and substituting *-ol*. Common names are obtained by just naming the alkyl group attached to the —OH followed by *alcohol*. Some very useful alcohols (see Figure 16-5) have more than one hydroxyl group. Two are shown below.

$$CH_3CH_2OH \qquad CH_3CH_2CH_2OH \qquad \begin{matrix} CH_2-CH_2 \\ | \quad\quad | \\ OH \quad OH \end{matrix} \qquad \begin{matrix} CH_2-CH-CH_2 \\ | \quad\quad | \quad\quad | \\ OH \quad OH \quad OH \end{matrix}$$

ethanol	propanol	1,2-ethane*diol*	1,2,3-propane*triol*
(ethyl alcohol)	(*n*-propyl alcohol)	(ethylene glycol)	(glycerol)
		[*diol* means two	[*triol* means three
		hydroxyl groups]	hydroxyl groups]

---

* *The R or R' represents a hydrocarbon group such as an alkyl or an aromatic group.*

Methanol and ethanol can both be obtained from natural sources. Methanol can be prepared by heating wood in the absence of oxygen to about 400 °C; at such high temperatures, methanol, together with other organic compounds, is given off as a gas. Since methanol was once made exclusively by this process, it is often called wood alcohol. Currently, methanol is prepared from synthesis gas, a mixture of CO and H_2. When CO and H_2 are passed over a catalyst at the right temperature and pressure, methanol is formed:

$$CO + 2H_2 \xrightarrow[\Delta]{\text{catalyst}} CH_3OH$$

Ethanol (often known simply as alcohol) is formed in the fermentation of various grains (therefore, it is also known as grain alcohol). In fermentation, the sugars and other carbohydrates in grains are converted to ethanol and carbon dioxide by the enzymes in yeast.

$$\underset{\text{glucose (a sugar)}}{C_6H_{12}O_6} \xrightarrow[\text{enzymes}]{\text{yeast}} 2CH_3CH_2OH + 2CO_2$$

Industrially, ethanol is prepared by reacting ethylene with water in the presence of an acid catalyst.

$$CH_2{=}CH_2 + H_2O \xrightarrow{\text{H}^+} CH_3CH_2OH$$

Isopropanol (isopropyl alcohol or rubbing alcohol) can be prepared from propene (propylene) in the same way.

$$CH_3CH{=}CH_2 + H_2O \xrightarrow{\text{H}^+} CH_3-\underset{\underset{OH}{|}}{\overset{\overset{CH_3}{|}}{C}}-H$$

Methanol has been used as a solvent for shellac, as a denaturant for ethanol (it makes the ethanol undrinkable), and as an antifreeze for automobile radiators. It is very toxic when ingested. In small doses it causes blindness, and in large doses it can cause death.

Ethanol is present in alcoholic beverages such as beer, wine, and liquor. The "proof" of an alcoholic beverage is two times the percent by volume of alcohol. If a certain brand of bourbon is 100 proof, it contains 50% ethanol. Ethanol can also be mixed with gasoline to form a mixture called "gasohol," which is used as a fuel. Ethanol is an excellent solvent and has been used as such in perfumes, medicines, and flavorings. It has also been used as an antiseptic and as a rubbing compound to cleanse the skin and lower a feverish person's temperature. While ethanol is not as toxic as methanol, it can cause coma or death when ingested in large quantities.

Ethylene glycol is the major component of antifreeze and coolant used in automobiles. It is also used to make polymers, the most common of which is Dacron®, a polyester.

Glycerol, which can be obtained from fats, is used in many applications where a lubricant and/or softener is needed. It has been used in pharmaceuticals, cosmetics, foodstuffs, and some liqueurs. When glycerol reacts with nitric acid, it produces nitroglycerin. Nitroglycerin is a powerful explosive,

$$CH_2-CH-CH_2 + 3HNO_3 \longrightarrow CH_2{-}{-}CH{-}{-}CH_2 + 3H_2O$$
$$\underset{\text{glycerol}}{\overset{|}{OH}\ \ \overset{|}{OH}\ \ \overset{|}{OH}} \qquad\qquad \underset{\text{nitroglycerin}}{\overset{|}{ONO_2}\ \ \overset{|}{ONO_2}\ \ \overset{|}{ONO_2}}$$

but it is also a strong smooth-muscle relaxant and vasodilator and has been used to lower the blood pressure and to treat angina pectoris.

16-8 ETHERS (R—O—R′)

An **ether** *contains an oxygen bonded to two hydrocarbon groups* (rather than one hydrocarbon group and one hydrogen as in alcohols). The simplest ether, dimethyl ether, is an isomer of ethanol and has very different properties.

Ethers are named by giving the alkyl groups on either side of the oxygen and adding *ether*.

$$\underset{\text{dimethyl ether}}{H_3C-O-CH_3} \qquad\qquad \underset{\text{ethyl methyl ether}}{CH_3OCH_2CH_3}$$

$$\underset{\text{diethyl ether}}{CH_3CH_2OCH_2CH_3} \qquad \underset{\text{ethyl propyl ether}}{CH_3CH_2OCH_2CH_2CH_3}$$

Diethyl ether is made industrially by reacting ethanol with sulfuric acid. In this reaction, two ethanol molecules are joined together with the loss of an H_2O molecule.

$$2CH_3CH_2OH \xrightarrow{\text{H}_2\text{SO}_4} CH_3CH_2-O-CH_2CH_3 + H_2O$$

The most commonly known ether, diethyl ether (or ethyl ether or just ether), has, in the past, been used extensively as an anesthetic. It has the advantages of being an excellent muscle relaxant that doesn't affect the blood pressure, pulse rate, or rate of respiration greatly. On the other hand, ether has an irritating effect on the respiratory passages and often causes nausea. Its flammability is also a drawback because of the danger of fire and explosions. Diethyl ether is rarely used now as an anesthetic. Other anesthetics that do not have its disadvantages have taken its place.

16-9 Amines (R—NH₂)

An **amine** *contains a nitrogen with single bonds to a hydrocarbon group and two other hydrocarbon groups or hydrogens*. The nitrogen in amines has one pair of unshared electrons similar to ammonia, NH_3. As we learned in Chapter 13, NH_3 utilizes the unshared pair of electrons to form weakly basic solutions in water. In a similar manner, amines are characterized by their ability to act as bases.

$$\overset{..}{N}H_3 + H_2O \rightleftharpoons NH_4^+ + OH^-$$
$$\underset{}{\overset{..}{C}H_3NH_2 + H_2O \rightleftharpoons CH_3NH_3^+ + OH^-} \Bigg\}\ \text{basic solutions}$$

Amines are named by listing the alkyl groups attached to the nitrogen and adding *amine*.

$$CH_3-\overset{..}{\underset{\overset{|}{H}}{N}}-H \qquad CH_3-\overset{..}{\underset{\overset{|}{H}}{N}}-CH_3 \qquad (CH_3CH_2)-\overset{..}{\underset{\overset{|}{CH_3}}{N}}-(CH_2CH_2CH_3)$$

methylamine　　　　dimethylamine　　　　ethyl　　methyl　　propylamine
　　　　　　　　　　(*di* = two methyls)

Simple amines are prepared by the reaction of ammonia with alkyl halides (e.g., CH_3Cl). In the reaction the alkyl group substitutes for the hydrogen on ammonia. The hydrogen and chlorine combine to form HCl which reacts with a second NH_3 to form NH_4Cl.

$$H-\overset{..}{\underset{\overset{|}{H}}{N}}-H + Cl-CH_3 \longrightarrow H-\overset{..}{\underset{\overset{|}{H}}{N}}-CH_3 + HCl$$

　ammonia　　　methyl chloride　　　　　methylamine

$$HCl + NH_3 \longrightarrow NH_4^+\ Cl^-$$

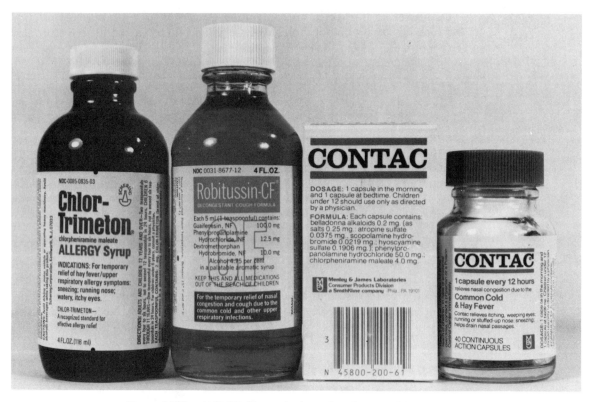

Figure 16-6　AMINES. The active ingredient in an antihistamine is an organic amine.

Further reaction with the same or different alkyl halides leads to substitution of one or both remaining hydrogens on the methylamine.

Amines are used in the manufacture of dyes, drugs, disinfectants, and insecticides. They also occur naturally in biological systems and are important in many biological processes.

Amine groups are present in many synthetic and naturally occurring drugs. (See Figure 16-6.) They may be useful as antidepressants, antihistamines, antibiotics, antiobesity preparations, antinauseants, analgesics, antitussives, diuretics, and tranquilizers, among others. Frequently, a drug may have more than one use [codeine is both an analgesic (pain reliever) and an antitussive agent (cough depressant).] Often a drug may be obtained from plants or animals. One such class of compounds is the alkaloids, which are amines found in plants. Some common drugs are shown below.

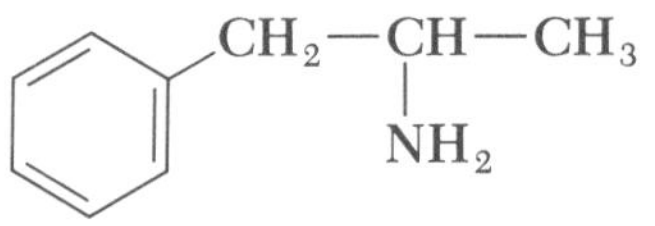

amphetamine
(Benzedrine®;
synthetic
appetite depressant,
stimulant)

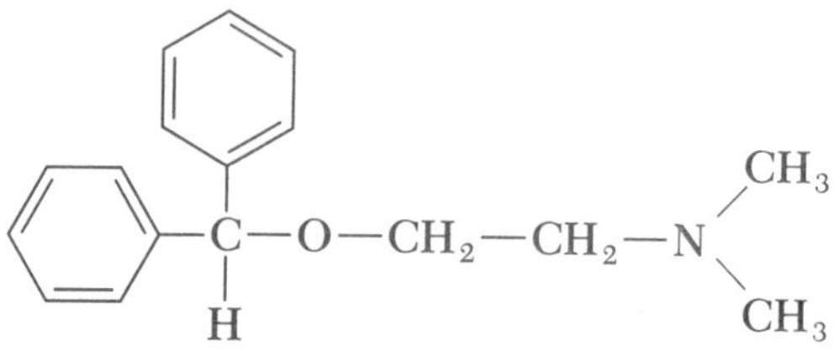

diphenhydramine
(synthetic
antihistamine)

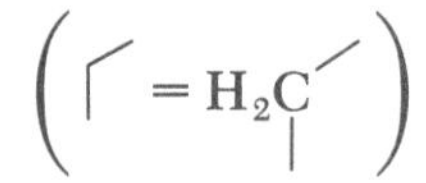

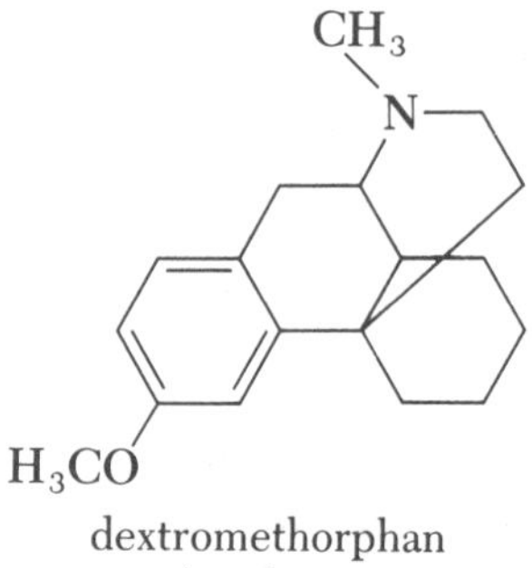

$$\left(\big\lceil = H_2C\big\langle \atop | \right)$$

dextromethorphan
(synthetic
analgesic, antitussive)

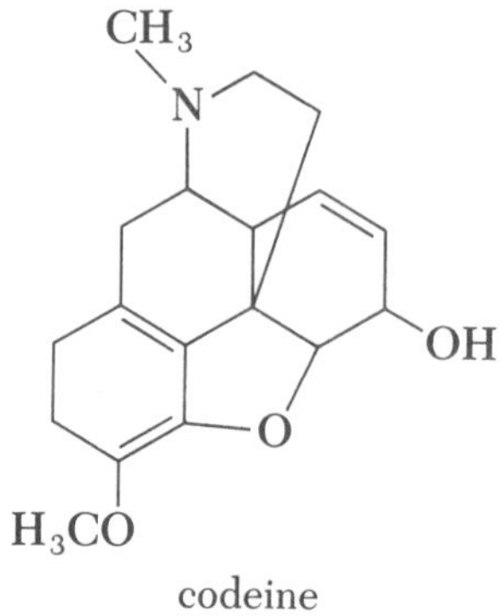

codeine
(from opium;
analgesic, antitussive)

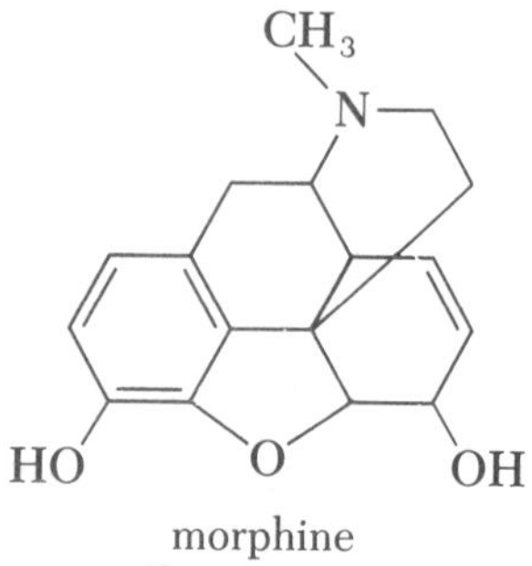

morphine
(from opium;
analgesic)

**16-10
ALDEHYDES**

$$\overset{O}{\underset{\|}{(R-C-H)}}$$

AND KETONES

$$\overset{O}{\underset{\|}{(R-C-R')}}$$

The functional group of aldehydes and ketones is a carbonyl group. A **carbonyl group** *is a carbon with a double bond to an oxygen:*

$$\overset{O}{\underset{\|}{-C-}}$$

In **aldehydes** *the carbonyl group is bound to at least one hydrogen, whereas in* **ketones** *the carbonyl group is bound to two hydrocarbon groups.*

Aldehydes are named by dropping the -e of the corresponding alkane name and substituting -al. Therefore the two-carbon aldehyde is ethanal. Ketones are named by taking the alkane name, dropping the -e, and substituting -one. The common names of ketones are obtained by naming the alkyl groups on either side of the C=O and adding *ketone*. Thus, the four-carbon ketone can be called butanone (**IUPAC**) or methyl ethyl ketone (common).

Aldehydes

$$\overset{O}{\underset{\|}{H-C-H}} \qquad \overset{O}{\underset{\|}{CH_3-C-H}}$$

methanal   ethanal
(formaldehyde)   (acetaldehyde)

Ketones

$$\overset{O}{\underset{\|}{CH_3-C-CH_3}} \qquad \overset{O}{\underset{\|}{CH_3-C-CH_2-CH_2-CH_3}}$$

propanone   2-pentanone
(acetone or   (methyl propyl ketone)
dimethyl ketone)

Many aldehydes and ketones are prepared by the oxidation of alcohols. Industrially, the alcohol is oxidized by heating it in the presence of oxygen and a catalyst. About half of the methanol made industrially is used to make formaldehyde by oxidation. Acetone can be prepared in the same fashion by oxidation of 2-propanol.

$$O_2 + H-\overset{\overset{H}{|}}{\underset{\underset{H}{|}}{C}}-\overset{|}{\underset{|}{O}} \xrightarrow{\text{catalyst}} H-C=O + H_2O$$

methanol   formaldehyde

$$O_2 + CH_3-\overset{\overset{CH_3}{|}}{\underset{\underset{H}{|}}{C}}-\overset{|}{\underset{|}{O}} \xrightarrow{\text{catalyst}} CH_3-C=O + H_2O$$

2-propanol   acetone

Aldehydes and ketones have uses as solvents, in the preparation of polymers, as flavorings, and in perfumes. The simplest aldehyde, formaldehyde, has been used as a disinfectant, antiseptic, germicide, fungicide, and embalming fluid (as a 37% by weight water solution). It has also been used in the preparation of polymers such as Bakelite® (the first commercial plastic) and Melmac® (used to make dishes). Formaldehyde polymers have also been used as coatings on fabrics to give "permanent press" characteristics.

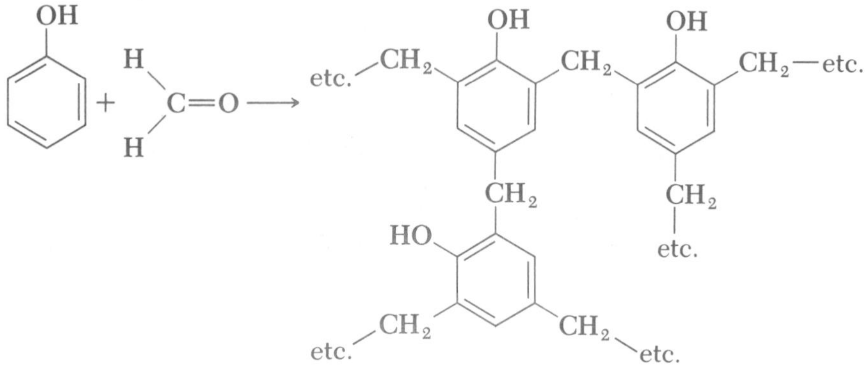

Bakelite

The simplest ketone, acetone, has been used mainly as a solvent. It is soluble in water and dissolves relatively polar and nonpolar molecules. It is an excellent solvent for paints and coatings.

More complex aldehydes and ketones, such as the following two examples, are used in flavorings and perfumes.

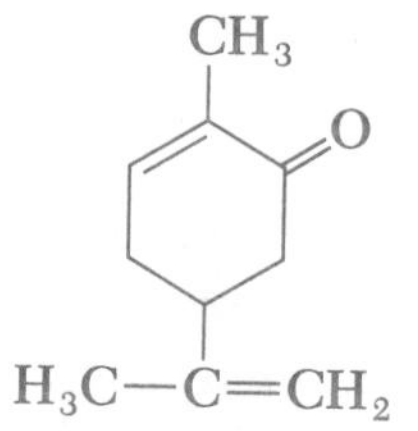

carvone
(in oil of spearmint;
a flavoring)

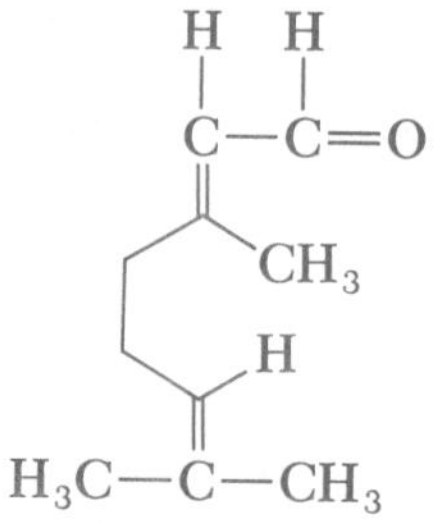

citral
(in oil of lemon grass;
a fragrance)

16-11 CARBOXYLIC ACIDS

$$(R-\overset{\displaystyle O}{\overset{\displaystyle \|}{C}}-O-H),$$

ESTERS

$$(R-\overset{\displaystyle O}{\overset{\displaystyle \|}{C}}-O-R'),$$

AND AMIDES

$$(R-\overset{\displaystyle O}{\overset{\displaystyle \|}{C}}-NH_2)$$

Carboxylic acids *contain the functional group*

$$-\overset{\displaystyle O}{\overset{\displaystyle \|}{C}}-O-H \;(\textit{also written} -COOH)$$

known as a **carboxyl group.** They are acidic organic compounds but not as acidic as inorganic acids such as nitric or sulfuric acid. Carboxylic acids are easily changed into two derivatives: amides and esters. **Amides** *have an amine group substituted for the hydroxyl group of the acid.* **Esters** *have a hydrocarbon group in place of the hydrogen in the carboxyl group.*

Carboxylic acids are named by dropping the *-e* from the alkane name and substituting *-oic acid.* Esters are named by giving the name of the alkyl group attached to the oxygen, followed by the acid name minus the *-ic* acid ending and substituting -*ate*. Amides are named by dropping the -*oic acid* portion of the carboxylic acid name and substituting -*amide*.

The following are a few examples of carboxylic acids, esters, and amides.

Acids

$$
\underset{\substack{\text{methanoic acid}\\ \text{(formic acid)}}}{H-\overset{\displaystyle O}{\overset{\|}{C}}-OH}
\qquad
\underset{\substack{\text{ethanoic acid}\\ \text{(acetic acid)}}}{CH_3-\overset{\displaystyle O}{\overset{\|}{C}}-OH}
\qquad
\underset{\text{benzoic acid}}{\bigcirc\!-\overset{\displaystyle O}{\overset{\|}{C}}-OH}
\qquad
\underset{\text{oxalic acid}}{HO-\overset{\displaystyle O}{\overset{\|}{C}}-\overset{\displaystyle O}{\overset{\|}{C}}-OH}
$$

Esters

$$
\underset{\substack{\text{ethyl methanoate}\\ \text{(ethyl formate)}}}{H-\overset{\displaystyle O}{\overset{\|}{C}}-O-CH_2CH_3}
\qquad
\underset{\substack{\text{ethyl ethanoate}\\ \text{(ethyl acetate)}}}{CH_3-\overset{\displaystyle O}{\overset{\|}{C}}-O-CH_2CH_3}
\qquad
\underset{\text{methyl benzoate}}{\bigcirc\!-\overset{\displaystyle O}{\overset{\|}{C}}-O-CH_3}
$$

Amides

$$
\underset{\substack{\text{methanamide}\\ \text{(formamide)}}}{H-\overset{\displaystyle O}{\overset{\|}{C}}-NH_2}
\qquad
\underset{\substack{\text{ethanamide}\\ \text{(acetamide)}}}{CH_3-\overset{\displaystyle O}{\overset{\|}{C}}-NH_2}
\qquad
\underset{\text{benzamide}}{\bigcirc\!-\overset{\displaystyle O}{\overset{\|}{C}}-NH_2}
$$

Carboxylic acids are made by oxidizing either alcohols or aldehydes. Formic acid, which was first isolated by distilling red ants, can be made by oxidizing either methanol or formaldehyde.

$$
CH_3-OH \xrightarrow[\text{agent}]{\text{oxidizing}} H-\overset{\displaystyle O}{\overset{\|}{C}}-OH \xleftarrow[\text{agent}]{\text{oxidizing}} \overset{\displaystyle O}{\overset{\|}{\underset{H\quad H}{C}}}
$$

Acetic acid can be made from the oxidation of ethanol. In fact, this is what happens when wine becomes "sour." Wine vinegar is produced by air oxidation of the alcohol in ordinary wine. The sharp, tart flavor of vinegar is due to the acetic acid, which is present in a concentration of about 4 to 5%.

Esters are made from the reaction of alcohols and carboxylic acids. In the reaction, H_2O is split off from the two molecules (OH from the acid and H from the alcohol), leading to the union of the remnants of the two molecules.

$$H-\overset{\overset{\displaystyle O}{\|}}{C}-\boxed{OH]+[H}-O-CH_3 \xrightarrow{H^+} H-\overset{\overset{\displaystyle O}{\|}}{C}-O-CH_3 + H_2O$$

Amides are made in a similar fashion, except that the acid is reacted with ammonia or an amine instead of an alcohol.

$$H-\overset{\overset{\displaystyle O}{\|}}{C}-\boxed{OH]+[H}-NH_2 \xrightarrow{\Delta} H-\overset{\overset{\displaystyle O}{\|}}{C}-NH_2 + H_2O$$

$$H-\overset{\overset{\displaystyle O}{\|}}{C}-\boxed{OH]+[H}-NHCH_3 \xrightarrow{\Delta} H-\overset{\overset{\displaystyle O}{\|}}{C}-NHCH_3 + H_2O$$

Carboxylic acids, esters, and amides are frequently present in compounds that have medicinal uses. Salicyclic acid has both antipyretic (fever-reducing) and analgesic (pain-relieving) properties. It has the disadvantage, however, of causing severe irritation of the stomach lining. Acetylsalicyclic acid (aspirin), which is both an acid and an ester, doesn't irritate the stomach as much. Aspirin is broken up in the small intestine, to form salicyclic acid, where it is absorbed. Some people are allergic to aspirin and must take "aspirin substitutes." The common aspirin substitutes are acetaminophen and ibuprofen. Acetaminophen is the active ingredient in Tylenol® and Datril®. Ibuprofen is the active ingredient in Advil® and Nuprin®.

salicylic acid   acetylsalicylic acid

ibuprofen   acetaminophen

We all distinguish the various fruits by their unique odors. These pleasant odors are all caused by esters. Four of these esters are shown below with their familiar odors.

$$H{-}\overset{\displaystyle O}{\overset{\|}{C}}OCH_2CH_3$$

ethyl formate
(rum)

$$CH_3\overset{\displaystyle O}{\overset{\|}{C}}O(CH_2)_4CH_3$$

pentyl acetate
(bananas)

$$CH_3CH_2CH_2\overset{\displaystyle O}{\overset{\|}{C}}OCH_2CH_3$$

ethyl butanoate
(pineapple)

methyl salicylate
(oil of wintergreen)

Figure 16-7   NYLON. Nylon was first synthesized by DuPont chemists. It was one of the first synthetic fibers.

One of the biggest uses for carboxylic acids, esters, and amides is in the formation of condensation polymers. *In* **condensation polymers,** *a small molecule (usually H₂O) is given off during formation of the polymer.* These are different from the polymers discussed earlier that resulted from the joining of alkenes (called *addition* polymers). Two of the most widely known condensation polymers are nylon 6,6 (see Figure 16-7) and Dacron®. The first is a polyamide, made from a diacid and a diamine; the second is a polyester, made from a diacid and a dialcohol (diol). Both of these polymers are used to make fibers.

■
*SEE PROBLEMS
16-35 THROUGH
16-51.*

Nylon 6,6

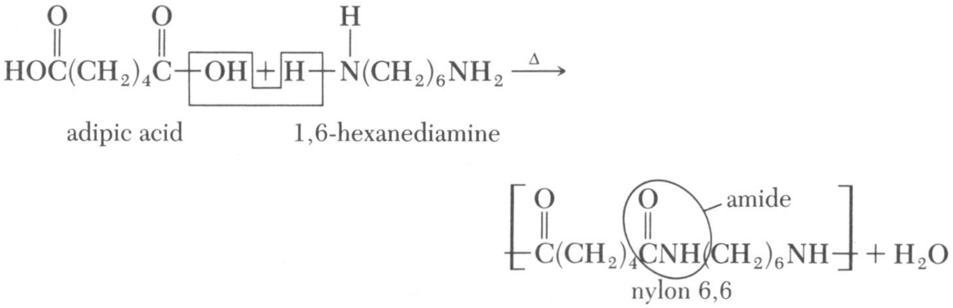

adipic acid   1,6-hexanediamine

nylon 6,6

Dacron®

terephthalic acid   ethylene glycol

Dacron®

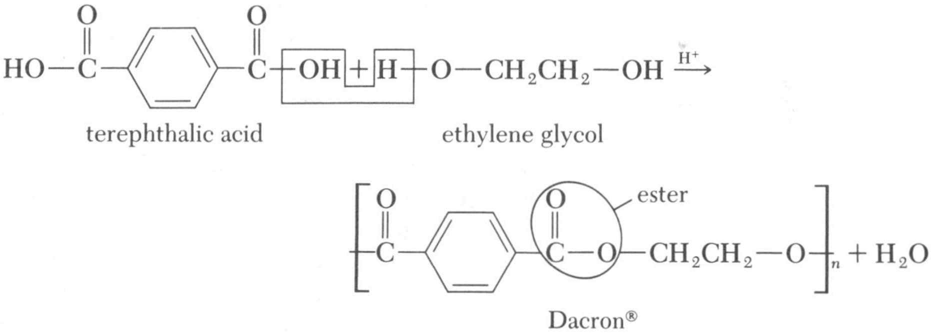

| 16-12 CHAPTER REVIEW | To introduce the chemistry of about six million organic compounds, we have to look for common properties so as to classify the compounds into groups. The first groupings possible are organic compounds containing only carbon and hydrogen (hydrocarbons) and organic compounds containing other (hetero) atoms. First let us consider hydrocarbons. These compounds can also be classified into one of the several groups depending on the nature of the bonds in the molecules: |

Name of Series	General Formula	Example	Name	Comments
Alkane	C_nH_{2n+2}	C_3H_8	propane	Alkanes contain only single bonds and are saturated.
Alkene	C_nH_{2n}	C_3H_6	propene propylene	Alkenes contain double bonds and are unsaturated.
Alkyne	C_nH_{2n-2}	C_3H_4	propyne	Alkynes contain triple bonds and are unsaturated.
Aromatic	—	C_6H_6	benzene	Aromatic compounds are cyclic, with alternating double bonds, and are less reactive than alkenes.

A significant feature of organic compounds is the existence of isomers. Isomers of a compound have the same molecular formula but differ in the arrangement of the bonds. Isomers may be the same type of compound, such as alkanes, and differ only in some structural characteristic (e.g., butane and isobutane) or may actually have different functional groups and have very different properties (e.g., ethanol and dimethyl ether).

Petroleum and natural gas are the sources of hydrocarbons. By use of the processes of cracking, reforming, and alkylation the hydrocarbons found in natural sources can be used to make many other useful hydrocarbons. The nomenclature of hydrocarbons is somewhat complicated because of the use of formal IUPAC names as well as the more popular common names.

Hydrocarbons all have their important uses: alkanes for fuel, alkenes to make polymers, alkynes to make alkenes for various polymers, and aromatics to make solvents, drugs, and flavors.

The focal point of chemical reactivity in an organic compound is the functional group. In alkenes and alkynes the double and triple bonds serve as functional groups (alkanes are thus comparatively unreactive). Hetero atoms attached to hydrocarbons also serve as functional groups. The nature and arrangement of the hetero atoms in a functional group impose certain common properties on a group of compounds; thus, this group of compounds can be treated as a class. Eight classes of compounds and their functional group, formula, example, and name are summarized as follows:

Name of Class	Functional Group	General Formula[a]	Example	IUPAC Name (Common Name)
Alcohols	$-\overset{\vert}{\underset{\vert}{C}}-\overset{\cdot\cdot}{\underset{\cdot\cdot}{O}}-H$	$R-OH$	CH_3CH_2OH	ethanol (ethyl alcohol)
Ethers	$-\overset{\vert}{\underset{\vert}{C}}-\overset{\cdot\cdot}{\underset{\cdot\cdot}{O}}-\overset{\vert}{\underset{\vert}{C}}-$	$R-O-R'$	$CH_3OCH_2CH_3$	(ethyl methyl ether)

Name of Class	Functional Group	General Formula[a]	Example	IUPAC Name (Common Name)
Amines	—C—N—	R—NH₂	$CH_3CH_2NH_2$	(ethyl amine)
Aldehydes	—C—H (with =O)	R—C(=O)—H	CH_3CH_2CHO	propanal (propionaldehyde)
Ketones	—C—C—C— (with =O)	R—C(=O)—R'	$CH_3COCH_2CH_3$	butanone (methyl ethyl ketone)
Carboxylic acid	—C—OH (with =O)	RCOH (with =O)	$CH_3CH_2CH_2COOH$	butanoic acid (butyric acid)
Esters	—C—O—C— (with =O)	R—C(=O)—OR'	CH_3COOCH_3	methyl ethanoate (methyl acetate)
Amides	—C—N— (with =O)	R—C(=O)—NH₂	CH_3CONH_2	ethanamide (acetamide)

[a] R and R' stand for hydrocarbon groups (e.g., alkyl). They may be different groups or the same group.

■ EXERCISES

LEWIS STRUCTURE AND ISOMERS

16-1 Draw Lewis structures for each of the following compounds. (More than one structure may be possible.)
(a) CH_3Br
(b) C_3H_4
(c) C_4H_8
(d) CH_5N
(e) C_2H_7N
(f) C_3H_8O

16-2 Draw Lewis structures for each of the following compounds. (More than one structure may be possible.)
(a) C_5H_{10}
(b) C_2H_4O
(c) C_4H_9Cl
(d) C_3H_9N

16-3 Write all of the isomers for C_6H_{14}.

16-4 Write all of the isomers for C_6H_{12}.

16-5 Which of the following pairs of compounds are isomers of each other? Why or why not?
(a) $CH_3CH_2OCH_2CH_3$ and $CH_3CH_2CH_2CH_3$
(b) $CH_3CH_2CH_2OH$ and $CH_3CH_2OCH_2CH_3$
(c) $CH_3CH_2NH_2$ and CH_3NHCH_3
(d) $CH_3CH_2CH_2NHCH_3$ and $CH_3CH_2CH_2NHCH_2CH_3$
(e) $CH_2CH_2CH_2CH_3$ and $CH_3CH_2CH_2CH_2$ with OH groups

    $CH_2CH_2CH_2CH_3$
    |
    OH

    $CH_3CH_2CH_2CH_2$
    |
    OH

16-6 Which of the following pairs of compounds are isomers of each other? Why or why not?
(a) $CH_3(CH_2)_3CH_3$ and $CH_3CH_2CH(CH_3)_2$
(b) CH_3CHCH_3 and $CH_3CH_2CHCH_3$
    | |
    Cl   Cl

(c) $CH_3CH_2CHCH_3$ and

           |

          NH_2

    $CH_3CHCH_2CH_3$

        |

        NH_2

(d) $(CH_3)_3N$ and $CH_3CH_2CH_2NH_2$

HYDROCARBONS

16-7 Which of the following are formulas of open-chain alkanes?
(a) C_4H_{10} (e) C_7H_{15}
(b) C_3H_7 (f) C_7H_{14}
(c) C_8H_{14} (g) $C_{18}H_{38}$
(d) $C_{10}H_{22}$ (h) C_9H_{15}

16-8 Write the formulas of any compound listed in Problem 16-7 that could be a cyclic alkane.

16-9 Write the formulas of any alkyl groups listed in Problem 16-7.

16-10 Which of the following could be the formula of a cyclic alkane?
(a) C_8H_{18} (c) C_6H_6
(b) C_5H_8 (d) C_5H_{10}

16-11 Which of the following are formulas of alkyl groups?
(a) C_4H_8 (d) C_6H_6
(b) C_4H_9 (e) C_6H_{13}
(c) C_6H_{10} (f) C_6H_{11}

16-12 The following skeletal structures represent alkanes. Fill in the proper number of hydrogens on each carbon atom.

(a) C—C—C

(b) C—C—C—C—C—C
        | |
        C   C—C—C

(c) C—C—C—C
      | |
      C C

(d) C—C—C—C—C
        | |
        C   C
         \ /
          C

16-13 The following skeletal structures represent alkanes. Fill in the proper

number of hydrogens on each carbon atom.

(a) C—C—C—C
         |
         C

(b) C—C—C
      |
      C
      |
      C

(c)
         C—C   C
        /   \ / \
  C—C   C
       \   /   \
        C   C

(d)
             C
             |
  C—C—C—C—C
        | |
        C—C C

16-14 Write condensed structures for the following compounds.
(a) 3-methylpentane
(b) n-hexane
(c) 2,4,5-trimethyloctane
(d) 4-ethyl-2-methylheptane
(e) 3-isopropylhexane
(f) 2,2-dimethyl-4-t-butylnonane

16-15 Write condensed structures for the following compounds.
(a) cyclopentane
(b) n-pentane
(c) 3,3-dimethylpentane
(d) 3-ethyl-3-methylhexane
(e) 3,5-dimethyl-4-isopropyloctane

16-16 Write the structure and the **correct** IUPAC name for (a) 2-ethyl-3-methyl-pentane and (b) 5-n-propyl-5-isopropylhexane.

16-17 How many carbons are in the longest chain of each of the following compounds? Write the name of the longest alkane chain for each.

               $CH_3—CH—CH_3$
                      |
(a) $CH_3CH—CH_2—CH—CH_3$
        |
       CH_3

     CH_3 $CH_2CH_2CHCH_3$
     | | |
(b) $CH_2—CH_2—CH$ CH_3
               |
             CH_2CH_3

(c)
$$CH_2-\underset{\underset{CH_3}{|}}{\overset{\overset{CH_2CH_2CH_2}{|}}{C}}-CH_3 \quad CH_2CH_2CH_3$$

(d)
$$CH_2-\underset{\underset{CH_2-CH_3}{|}}{\overset{\overset{CH_3}{|}}{CH}}-\underset{\underset{CH_2-CH_3}{|}}{CH_2}$$

16-18 Name the substituents attached to the longest chains in the previous problem and give the IUPAC names for the compounds.

16-19 Give the longest chain, the substituents, and the proper IUPAC name for the following compounds.

(a)
$$CH_3CH_2\underset{\underset{CH_2CH_3}{|}}{\overset{\overset{CH_3}{|}}{CH}}-CHCH_2CH_2CH_3$$

(b)
$$CH_3CH_2CH_2\underset{\underset{CH_3-CH-CH_3}{|}}{CH}-(CH_2)_2CH_3$$

(c)
$$CH_3CH_2\underset{\underset{CH_2CH_2CH_3}{|}}{CH}-\underset{\underset{CH_3CCH_3}{|}}{\overset{\overset{CH_3}{|}}{CH}}-CH_2CH_2CH_3$$

(d)
$$CH_3CH_2CH_2CH_2-\underset{\underset{CH_2CH_2CH_2CH_3}{|}}{\overset{\overset{CH_2CH_2CH_3}{|}}{C}}-CH_2CH_2CH_2CH_3$$

16-20 Give the proper IUPAC names for (a), (b), and (c) in Problem 16-12.

16-21 Give the proper IUPAC names for (a), (b), and (d) in Problem 16-13.

16-22 C_nH_{2n} can represent an alkane or an alkene. Explain.

16-23 What is the general formula of a cyclic alkene that contains one double bond?

16-24 Identify the following as an alkane, alkene, or alkyne. Assume that the compounds are not cyclic.

(a) C_8H_{16} (e) CH_4
(b) C_5H_{12} (f) $C_{10}H_{20}$
(c) C_4H_8 (g) $C_{18}H_{38}$
(d) $C_{20}H_{38}$

16-25 The following skeletal structures represent alkenes or alkynes. Fill in the proper number of hydrogens on each carbon.

(a) $C-C-C=C$
(b) $C-C\equiv C-C$
(c) $C-C-C=C-C$
      $\underset{\underset{C}{|}}{}$
(d) $C-C-C\equiv C-C-C$
          $\underset{\underset{C-C}{|}}{}$

16-26 The following skeletal structures represent alkenes or alkynes. Fill in the proper number of hydrogens on each carbon.

(a) $C-C=C-C-C$

(b)

(c) $C-C-C-C\equiv C$
      $\underset{\underset{C}{|}}{}$
(d) $C-C-C-C$
     $\underset{\underset{C\equiv C-C}{|}}{}$

16-27 Give the condensed structures of the following:
(a) 2-butyne
(b) 3-metyl-1-butene
(c) 3-propyl-2-hexene
(d) 5-methyl-3-octyne

16-28 Give the proper IUPAC names for the compounds in Problem 16-25.

16-29 Give the proper IUPAC names for the compounds in Problem 16-26.

16-30 Give the abbreviated structure of the polymer that would be formed from each of the following alkenes. Name the polymer.

(a) $CH_2{=}\underset{\underset{CH_3}{|}}{CH_3}$ propylene

(b) $CH_2{=}\underset{\underset{F}{|}}{CH}$ vinyl fluoride

(c) $CH_2{=}CF_2$ difluoroethylene

(d) $CH_2{=}CH$ methyl acrylate
$\quad\quad\ \ \ |$
$\quad\quad\ CO_2CH_3$

16-31 A certain polymer has repeating units represented by the formula

$$\left(\!\!\begin{array}{cc} CH{-}CH \\ |\quad\ | \\ F\quad CH_3 \end{array}\!\!\right)_n$$

Write the formula of the monomer.

16-32 Naphthalene has the formula $C_{10}H_8$ and is composed of two benzene rings that share two carbons. Write the Lewis structure, including hydrogens, for naphthalene.

16-33 Anthracene and phenanthrene are isomers with the formula $C_{14}H_{10}$. They are composed of two benzene rings that share two carbons with a third benzene ring (not the same two). Write Lewis structures for the two isomers including hydrogens.

16-34 There are three isomers for dichlorobenzene. In this molecule, two chlorines are substituted for two hydrogens on a benzene rings. Write the Lewis structures for the three isomers.

DERIVATIVES OF HYDROCARBONS

16-35 Circle and name the functional groups in each of the following compounds:

(a)
$$\quad\quad\quad\quad\quad O$$
$$\quad\quad\quad\quad\quad \|$$
$$CH_3CH{=}CHCH_2CH$$

(b)

(c) $CH_3OCH_2CH_2\overset{\displaystyle O}{\overset{\|}{C}}NH_2$

(d) $H{-}C{\equiv}C{-}CH_2CH_2CH_2OH$

(e)

(f) $CH_3CH_2CH_2COOH$

16-36 Aspartame is an effective sweetener that is used as a low-calorie substitute for sugar. Identify the functional groups present in an aspartame molecule.

16-37 Tell how the following differ in structure.
(a) alcohols and ethers
(b) aldehydes and ketones
(c) amines and amides
(d) carboxylic acids and esters.

16-38 Glycine has the formula $C_2H_5O_2N$. It is a member of a class of compounds called amino acids that combine to form proteins. Glycine contains both an amine and a carboxylic acid group. Write the structure for glycine. Can glycine function as an acid, base, neither, or both in water?

16-39 Which of the eight classes of compounds containing hetero atoms contains a carboxyl group? Which contains a hydroxyl group?

16-40 What two classes of compounds combine to form esters? What two classes combine to form amides?

16-41 To what class does each of the following compounds belong?
(a) 3-heptanone
(b) 3-nonene
(c) 2-methylpentane
(d) 2-ethylhexanal
(e) ethylbenzene
(f) propanol

16-42 To what class does each of the following compounds belong?

(a) 2-octanol
(b) 2-butyne
(c) ethyl pentanoate
(d) 3-ethylhexene
(e) dimethyl ether
(f) aminobenzene

16-43 Write condensed structures for the following compounds.
(a) n-butyl alcohol
(b) di-n-propyl ether
(c) trimethylamine
(d) propanal
(e) 3-pentanone
(f) propanoic acid
(g) methyl acetate
(h) propanamide

16-44 Write the condensed structure of a compound with five carbon atoms containing each of the eight herero atom functional groups studied.

*16-45 Name the following compounds.

(a) $CH_3CCH_2CH_2CH_3$
       $\underset{O}{\|}$

(b) $CH_3CH_2CH_2-\underset{\underset{O}{\|}}{C}-OH$

(c) $CH_3-\underset{\underset{O}{\|}}{C}-NH_2$

(d) $(CH_3)C-OH$

(e) $CH_3CH_2CH_2CH_2-\underset{\underset{O}{\|}}{C}-H$

*16-46 Name the following compounds.

(a) $CH_3CH_2-\underset{\underset{O}{\|}}{C}-OC_2H_2$

(b) $(C_2H_5)_2NH$
(c) $CH_3CH_2CH_2CH_2OH$
(d) $CH_3CH_2-\underset{\underset{O}{\|}}{C}-CH_2CH_2CH_3$

(e) $CH_3-\underset{\underset{O}{\|}}{C}-H$

16-47 Complete the following equations.

(a) $C_3H_8 + \text{excess } O_2 \longrightarrow$
(b) $CH_2=CHCH_3 + H_2O \xrightarrow{H^+}$
(c) $C_2H_5Cl + 2NH_3 \longrightarrow$
(d) $NH(CH_3)_2(aq) + HCl(aq) \longrightarrow$
(e) $CH_3CH_2COOH + NH(CH_3)_2 \longrightarrow$
(f) $HC\equiv CH + HBr \longrightarrow$
(g) $CH_3CH_2COOH + H_2O \longrightarrow$

16-48 Complete the following equations

(a) $C_3H_7OH + \text{excess } O_2 \longrightarrow$
(b) $CH_2=CHCH_3 + Br_2 \longrightarrow$
(c) $NH(CH_3)_2 + H_2O \longrightarrow$

(f) $CH_3CH_2CH_2OH + O_2 \xrightarrow{\text{catalyst}}$
(g) $CH_3CH_2COOH + CH_3OH \longrightarrow$
(h) $CH_3CH_2COOH(aq)$
$\qquad\qquad + NaOH(aq) \longrightarrow$

16-49 Give a general method for making each of the following classes of compounds.
(a) alcohols   (d) esters
(b) ketones   (e) amides
(c) carboxylic   (f) aldehydes
    acids   (g) amines

16-50 Tell how the following can be obtained from natural sources.
(a) ethanol   (d) methane
(b) acetic acid   (e) gasoline
(c) methanol

16-51 Give one possible use for each of the following compounds or class of compounds.
(a) formaldehyde
(b) acetic acid
(c) ethylene glycol
(d) acetylsalicyclic acid
(e) amines
(f) esters
(g) alkenes
(h) alkanes

The successful athlete must be ''in shape'' physically. The successful
chemistry student must be ''in shape'' mathematically.

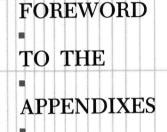

FOREWORD
TO THE
APPENDIXES

It is amazing how students who have *confidence* enjoy and do well in chemistry. Confidence, of course, comes from being prepared with the proper background. Just as a professional football player has the proper background of a successful college career, the serious chemistry student needs a proper mathematical background. Many, if not most, students who take this course need to give some early attention to reviewing math. The reason—whether a deficient secondary school background, a change in career intentions, or a lengthy interruption of studies—is not important. If the deficiencies are serious, the student should consider taking math *before* this course. More often, a review like that provided by the following appendixes can patch the weak spots so that math is not a hindrance to an understanding of the chemistry.

Appendixes A, B, and C are intended to aid your review of the math and algebra required in the study of college or university chemistry. These appendixes each begin with a short test, so that you can quickly gauge whether the review and exercises that follow are needed.

Appendix D complements and expands on the introduction to problem solving by the factor-label method in Chapter 2. Students who can become comfortable with the types of problems in this appendix will find themselves in excellent condition for the quantitative aspects of chemistry. Appendix E is an introduction to the use of common logarithms, which are needed to understand pH in Chapter 13. Appendix F is a discussion of the interpretation and construction of graphs, a topic of prime importance in both the social and the natural sciences. Appendix G contains instructions on the use of calculators specifically for problems in this text. Appendix H is a glossary of terms used throughout the text, and finally, Appendix I contains answers to about two-thirds of the problems at the ends of the chapters.

This section reviews the following basic mathematical skills.

APPENDIX

BASIC

MATHEMATICS

A

A TEST OF BASIC MATHEMATICS

Following is a brief test on some basic mathematical skills necessary in chemistry. Work the test as carefully as you can *without a calculator*. Compare your answers with the answers given at the end of the test. If you have 100% of the answers correct, proceed to Appendix B on algebra. If you don't get 100, recheck your calculations. If one or two mistakes were due to minor computational errors, then you should move on to Appendix B. If you made more than two mistakes or can't identify the source of your error, work through Appendix A before proceeding. After you have finished, take the test again. Perhaps you will then get 100. If not, consult with your instructor for further review.

Math Test

Addition and Subtraction

1 Carry out the following calculations:

   (a) $43.76 + 6.90 + 134.88$ 1(a) _____

   (b) $0.668 - 9.232 + 3.445$ 1(b) _____

   (c) $5.65 - (-2.34) + (-0.89)$ 1(c) _____

Multiplication

2 Carry out the following calculations. Express the answer to the proper number of significant figures and include units in (c) and (d).

   (a) 10.7×8.6 2(a) _____

   (b) $(-0.560) \times 234$ 2(b) _____

   (c) 16 in. $\times$ 34 in.   2(c) _____

   (d) 13.6 g/mL $\times$ 187 mL   2(d) _____

Roots of Numbers

3 Express the following roots in decimal fraction form. Include units and the proper number of significant figures in the answer.

   (a) $\sqrt{82 \text{ cm}^2}$ 3(a) _____

   (b) $\sqrt[3]{512 \text{ in.}^3}$ 3(b) _____

Division, Fractions, and Decimal Fractions

4 Express the following answers in decimal fraction form (e.g., $\frac{3}{5} = 0.6$). Include units and the proper number of significant figures in the answer.

   (a) 465 miles $\div$ 5.7 hr   4(a) _____

   (b) $\dfrac{115 \text{ g}}{7.82 \text{ mL}}$ 4(b) _____

   (c) $\dfrac{186 \text{ ft}^3}{4.5 \text{ ft}}$ 4(c) _____

Multiplication and Division of Fractions

5 Express the following answers in decimal fraction form to three significant figures. When a problem has units, include the units in the answer.

(a) $(-\frac{9}{4}) \times \frac{24}{35} \times (-\frac{7}{3})$ 5(a) _____

(b) $\frac{3}{4} \div (\frac{6}{7} \times \frac{1}{2})$ 5(b) _____

(c) $\dfrac{187 \text{ torr}/7.45}{760 \text{ torr}/\text{atm}}$ 5(c) _____

(d) $(\frac{2}{3} \text{ cm})^3$ 5(d) _____

Decimal Fractions and Percent

6 There are 156 apples in a crate. Seventy apples are Jonathans, 45 are Golden Delicious, and the others are rotten. What is the decimal fraction of rotten apples?

                                                   6 _____

7 Express the following fractions as percentages to three significant figures.

(a) $\frac{3}{8}$ 7(a) _____

(b) 0.0875   7(b) _____

8 Express the following percentages as decimal fractions.

(a) 98.9%   8(a) _____

(b) 0.74%   8(b) _____

9 (a) A young man received a score of 74% on a test with 155 total points. How many points did he receive? (Round off to the nearest integer.)

(b) On a different test, a young woman received 85 points which was a score of 74%. How many points were on the test? (Round off to the nearest integer.)

Answers to Test   1(a) 185.54   1(b) −5.119   1(c) 7.10   2(a) 92   2(b) −131
2(c) 540 in.²   2(d) 2540 g   3(a) 9.0 cm   3(b) 8.00 in.
4(a) 82 miles/hr   4(b) 14.7 g/mL   4(c) 41 ft²   5(a) 3.60   5(b) 1.75
5(c) 0.0330 atm   5(d) 0.296 cm³   6 0.263   7(a) 37.5%   7(b) 8.75%
8(a) 0.989   8(b) 0.0074   9(a) 115   9(b) 115

B REVIEW OF BASIC MATHEMATICS

The following is a quick (very quick) refresher of fundamentals of math. This may be sufficient to aid you if you are just a little rusty on some of the basic concepts. For more thorough explanations and practice, however, you are urged to use a more comprehensive math review workbook or consult with your instructor.

1 Addition and Subtraction   Since most calculations in this text use numbers expressed in decimal form, we will emphasize the manipulation of this type of number. In addition and subtraction, it is important to line up the decimal point carefully before doing the math.

EXAMPLE A-1

Add the following numbers: 16.75, 13.31, and 175.67.

$$\downarrow$$

$$
\begin{array}{r}
16.75 \\
13.31 \\
\underline{175.67} \\
\underline{205.73}
\end{array}
$$

Subtraction is simply the addition of a negative number. Remember that subtraction of a negative number changes the sign to a plus (two negatives make a positive). For example, $4 - 7 = -3$, but $4 - (-7) = 4 + 7 = 11$.

EXAMPLE A-2

Carry out the following calculations.

(a) $11.8 + 13.1 - 6.1$

$$
\begin{array}{r}
11.8 \\
+13.1 \\
\underline{24.9}
\end{array}
\qquad
\begin{array}{r}
24.9 \\
-6.1 \\
\underline{18.8}
\end{array}
$$

(b) $47.82 - 111.18 - (-12.17)$

This is the same as $47.82 - 111.18 + 12.17$.

$$
\begin{array}{r}
47.82 \\
+12.17 \\
\underline{59.99}
\end{array}
\qquad
\begin{array}{r}
-111.18 \\
+59.99 \\
\underline{-51.19}
\end{array}
$$

Exercises

A-1 Carry out the following calculations.
(a) $47 + 1672$
(b) $11.15 + 190.25$
(c) $114 + 26 - 37$
(d) $-97 + 16 - 118$
(e) $0.897 + 1.310 - 0.063$
(f) $-0.377 - (-0.101) + 0.975$
(g) $17.489 - 318.112 - (-0.315) + (-3.330)$

Answers: (a) 1719 (b) 201.40 (c) 103 (d) -199 (e) 2.144 (f) 0.699 (g) -303.638

2 Multiplication Multiplication is expressed in various ways as follows:

$$13.7 \times 115.35 = 13.7 \cdot 115.35 = (13.7)(115.35) = 13.7(115.35)$$

If it is necessary to carry out the multiplication in longhand, you must be careful to place the decimal point correctly in the answer. Count the *total* number of digits to the right of the decimal point in both multipliers (three in this example). The answer has that number of digits to the right of the decimal point in the answer. Finally, round off to the proper number of significant figures.

$$13.7 \times 2.15 =$$

$$1 + 2 = \textcircled{3}$$

(decimal places)

$$
\begin{array}{r}
13.7 \\
\times 2.15 \\
\hline
685 \\
137 \\
274 \\
\hline
29455
\end{array}
$$

$$= 29.455 = 29.5^*$$

$$\textcircled{3}$$

When a number (called a *base*) is multiplied by itself one or more times it is said to be raised to a *power*. The power (called the *exponent*) indicates the number of bases multiplied. For example, the exact values of the following numbers raised to a power are

$$4^2 = 4 \times 4 = 16 \qquad \text{(``four squared'')}$$
$$2^4 = 2 \times 2 \times 2 \times 2 = 16 \qquad \text{(``two to the fourth power'')}$$
$$4^3 = 4 \times 4 \times 4 = 64 \qquad \text{(``four cubed'')}$$
$$(14.1)^2 = 14.1 \times 14.1 = 198.81$$

In the calculations used in this book, most numbers have specific units. In multiplication, the units as well as the numbers are multiplied. For example,

$$3.7 \text{ cm} \times 4.61 \text{ cm} = 17 \text{ (cm} \times \text{cm}) = 17 \text{ cm}^2$$
$$(4.5 \text{ in.})^3 = 91 \text{ in.}^3$$

In the multiplication of a series of numbers, grouping is possible:

$$(a \times b) \times c = a \times (b \times c)$$
$$3.0 \text{ cm} \times 148 \text{ cm} \times 3.0 \text{ cm} = (3.0 \times 3.0) \times 148 \times (\text{cm} \times \text{cm} \times \text{cm})$$
$$= \underline{1300 \text{ cm}^3}$$

When multiplying signs, remember:

$$(+) \times (-) = - \qquad (+) \times (+) = + \qquad (-) \times (-) = +$$

For example, $(-3) \times 2 = -6$; $(-9) \times (-8) = +72$.

---

* Rounded off to three significant figures. See Section 2-2.

Exercises

A-2 Carry out the following calculations. For (a) through (d) carry out the multiplication completely. For (e) through (h) round off the answer to the proper number of significant figures and include units.

(a) $16.2 \times (-118)$ (d) $(-47.8) \times (-9.6)$

(b) $(4 \times 2) \times 879$ (e) $3.0 \text{ ft} \times 18 \text{ lb}$

(c) $(-8) \times (-2) \times (-37)$ (f) 17.7 in. $\times (13.2$ in. $\times 25.0$ in.)

(g) What is the area of a circle where the radius is 2.2 cm? (Area $= \pi r^2$, $\pi = 3.14$.)

(h) What is the volume of a cylinder 5.0 in. high with a cross-section radius of 0.82 in.? (Volume $=$ area of cross section $\times$ height.)

Answers: (a) -1911.6 (b) 7032 (c) -592 (d) 458.88
(e) 54 ft $\cdot$ lb (f) 5840 in.^{3} (g) 15 cm^2 (h) 11 in.^{3}

3 Roots of Numbers A root of a number is a fractional exponent. It is expressed as

$$\sqrt[x]{a} = a^{1/x}$$

If x is not shown (on the left), it is assumed to be 2 and is known as the *square root*. The square root is the number that when multiplied by itself gives the base a. For example,

$$\sqrt{4} = 2 \quad (2 \times 2 = 4)$$
$$\sqrt{9} = 3 \quad (3 \times 3 = 9)$$

The square root of a number may have either a positive or a negative sign. Generally, however, we are interested only in the positive root in chemistry calculations.

If the square root of a number is not a whole number, it may be computed on a calculator or found in a table. Without these available, an educated approximation can come close to the answer. For example, the square root of 54 lies between 7 ($7^2 = 49$) and 8 ($8^2 = 64$) but closer to 7. An educated guess of 7.3 would be excellent.

The cube root of a number is expressed as

$$\sqrt[3]{b} = b^{1/3}$$

It is the number multiplied by itself two times that gives b. For example,

$$\sqrt[3]{27} = 3.0 \quad (3 \times 3 \times 3 = 27)$$
$$\sqrt[3]{64} = 4.0 \quad (4 \times 4 \times 4 = 64)$$

The hand calculator (see Appendix G) is the most convenient source of roots of numbers.

Exercises

A-3 Find the following roots. If necessary, first approximate the answer, then check with a calculator.

(a) $\sqrt{25}$ (b) $\sqrt{36 \text{ cm}^2}$ (c) $\sqrt{144 \text{ ft}^4}$ (d) $\sqrt{40}$

(e) $\sqrt{7.0}$ (f) $110^{1/2}$ (g) $100^{1/3}$ (h) $\sqrt[3]{50}$

(i) What is the radius of a circle that has an area of 150 ft²

(j) What is the radius of the cross section of a cylinder if it has a volume of 320 m³ and a height of 6.0 m?

Answers: (a) 5.0   (b) 6.0 cm   (c) 12.0 ft²   (d) 6.3   (e) 2.6   (f) 10.5
(g) 4.64   (h) 3.7   (i) 6.91 ft   (j) 4.1 m

4 Division, Fractions, and Decimal Fractions   Division problems are expressed in fraction form as follows:

Division form	Common fraction form	Decimal fraction form

$$88.8 \div 2.44 = \frac{88.8}{2.44} = 88.8/2.44 = \quad 36.4$$

Generally, in chemistry, answers are expected in decimal fraction form rather than common fraction form. Therefore, to obtain the answer, the *numerator* (the number on the top) is divided by the *denominator* (the number on the bottom). Before doing the actual calculation, it helps to have a feeling for how the fractional number should look in decimal form. When the numerator is smaller than the denominator, the fraction is known as a *proper fraction,* and the decimal fraction is less than one. When the numerator is larger than the denominator, the fraction is known as an *improper fraction,* and the decimal fraction is greater than one.

To carry out the division longhand, it is easier to divide a whole number in the denominator into the numerator. To do this, move the decimal in both numerator and denominator the same number of places. In effect, you are multiplying both numerator and denominator by the same number, which does not change the value of the fraction.

$$\frac{a}{b} = \frac{a \times c}{b \times c}$$

$$\frac{88.8}{2.44} = \frac{88.8 \times 100}{2.44 \times 100} = \frac{8880}{244} = \underline{\underline{36.4}}$$

$$2.44\overline{)88.800} = 244\overline{)8880.00} = 36.4 \text{ (three significant figures)}$$

$$\begin{array}{r} 36.39 \\ \underline{732} \\ 1560 \\ \underline{1464} \\ 960 \\ \underline{732} \\ 2280 \\ 2176 \end{array}$$

Many divisions can be simplified by "cancellation." Cancellation is the elimination of common factors in the numerator and denominator. This is possible because a number divided by itself is equal to unity (e.g., $25/25 = 1$). As in multiplication, all units must also be divided. If identical units appear in both numerator and denominator, they can also be canceled.

$$\frac{a \times \cancel{c}}{b \times \cancel{c}} = \frac{a}{b}$$

$$\frac{\overset{1}{\cancel{190}} \times 4 \, \cancel{\text{torr}}}{\underset{1}{\cancel{190}} \, \cancel{\text{torr}}} = \frac{4}{=}$$

$$\frac{2500 \, \text{cm}^3}{150 \, \text{cm}} = \frac{\overset{1}{\cancel{50}} \times 50 \, \cancel{\text{cm}} \times \text{cm} \times \text{cm}}{\underset{1}{\cancel{50}} \times 3 \, \cancel{\text{cm}}} = \frac{50 \, \text{cm}^2}{3} = \underline{\underline{17 \, \text{cm}^2}}$$

$$\frac{2800 \, \text{miles}}{45 \, \text{hr}} = \frac{\cancel{5} \times 560 \, \text{miles}}{\cancel{5} \times 9 \, \text{hr}} = \frac{62 \, \text{miles}}{1 \, \text{hr}} = \underline{\underline{62 \, \text{miles/hr}}}$$

This is read as 62 miles "per" one hour or simply 62 miles per hour. The word "per" implies a fraction or a ratio with the unit after "per" in the denominator. If a number is not written or read in the denominator with a unit, it is assumed that the number is unity and is known to as many significant figures as the number in the numerator (i.e., 62 miles per 1.0 hr).

Exercises

A-4   Express the following in decimal fraction form.

(a) 892 miles ÷ 41 hr

(b) 982.6 ÷ 0.250

(c) 195 ÷ 2650

(d) $\dfrac{67.5 \, \text{g}}{15.2 \, \text{mL}}$

(e) $\dfrac{1890 \, \text{cm}^3}{66 \, \text{cm}}$

(f) $\dfrac{146 \, \text{ft} \cdot \text{hr}}{0.68 \, \text{ft}}$

(g) $\dfrac{0.8772 \, \text{ft}^3}{0.0023 \, \text{ft}^2}$

(h) $\dfrac{37.50 \, \text{ft}}{0.455 \, \text{sec}}$

Answers: (a) 22 miles/hr   (b) 3930   (c) 0.0736   (d) 4.44 g/mL
(e) 29 cm²   (f) 210 hr   (g) 380 ft   (h) 82.4 ft/sec

5 Multiplication and Division of Fractions   When two or more fractions are multiplied, all numbers *and units* in both numerator and denominator can be combined into one fraction.

Express the following answers to two significant figures.

EXAMPLE A-3

$$\frac{3}{5} \times \frac{75}{4} \times \frac{16}{7} = \frac{3 \times 75 \times 16}{5 \times 4 \times 7} = \frac{3 \times \overset{15}{\cancel{75}} \times \overset{4}{\cancel{16}}}{\underset{1}{\cancel{5}} \times \underset{1}{\cancel{4}} \times 7} = \frac{180}{7} = \underline{\underline{26}}$$

EXAMPLE A-4

$$\frac{42 \text{ miles}}{\text{hr}} \times \frac{3}{7} \text{ hr} \times \frac{5280 \text{ ft}}{\text{miles}} = \frac{\overset{6}{\cancel{42}} \times 3 \times 5280 \text{ } \cancel{\text{miles}} \times \cancel{\text{hr}} \times \text{ft}}{\underset{1}{\cancel{7}} \text{ } \cancel{\text{hr}} \times \cancel{\text{miles}}}$$

$$= \underline{\underline{95,000 \text{ ft}}}$$

EXAMPLE A-5

$$\frac{3}{4} \text{ mol} \times \frac{0.75 \text{ g}}{\text{mol}} \times \frac{1 \text{ mL}}{19.3 \text{ g}} = \frac{3 \times 0.75 \times 1 \times \cancel{\text{mol}} \times \cancel{\text{g}} \times \text{mL}}{4 \times 1 \times 19.3 \times \cancel{\text{mol}} \times \cancel{\text{g}}}$$

$$= \underline{\underline{0.029 \text{ mL}}}$$

The division of one fraction by another is the same as the multiplication of the numerator by the *reciprocal* of the denominator. The reciprocal of a fraction is simply the fraction in an inverted form (e.g., $\frac{3}{5}$ is the reciprocal of $\frac{5}{3}$).

$$\frac{a}{b/c} = a \times \frac{c}{b} \qquad \frac{a/b}{c/d} = \frac{a}{b} \times \frac{d}{c} = \frac{a \times d}{b \times c}$$

EXAMPLE A-6

$$\frac{1650}{3/5} = 1650 \times \frac{5}{3} = \underline{\underline{2750}}$$

EXAMPLE A-7

$$\frac{145 \text{ g}}{7.50 \text{ g/mL}} = 145 \text{ g} \times \frac{1 \text{ mL}}{7.50 \text{ g}} = \underline{\underline{19.3 \text{ mL}}}$$

Exercises

A-5 Express the following in decimal fraction form. If units are not used, round off the answer to three significant figures. If units are included, round off to the proper number of significant figures and include units in the answer.

(a) $\frac{3}{8} \times \frac{4}{7} \times \frac{21}{20}$

(b) $\frac{250}{273} \times \frac{175}{300} \times (-6)$

(c) $\frac{4}{9} \times (-\frac{5}{8}) \times (-\frac{3}{4})$

(d) $195 \text{ g/mL} \times 47.5 \text{ mL}$

(e) $0.75 \text{ mol} \times 17.3 \text{ g/mol}$

(f) $(3.57 \text{ in.})^2 \times 0.85 \text{ in.} \times \frac{16.4 \text{ cm}^3}{\text{in.}^3}$

(g) $\frac{\frac{150}{350}}{\frac{25}{42}}$

(h) $\frac{(-\frac{3}{7})}{(-\frac{4}{9})}$

(i) $\frac{(-\frac{17}{3})}{\frac{8}{9}}$

(j) $\frac{\frac{16}{9} \times \frac{10}{14}}{\frac{5}{6}}$

(k) $\frac{75.2 \text{ torr}}{760 \text{ torr/atm}}$

(l) $\frac{(55.0 \text{ miles/hr}) \times (5280 \text{ ft/mile}) \times (1 \text{ hr/60 min})}{60 \text{ sec/min}}$

(m) $\frac{305 \text{ K} \times 62.4 \frac{\text{L} \cdot \text{torr}}{\text{K} \cdot \text{mol}} \times 0.25 \text{ mol}}{650 \text{ torr}}$

Answers: (a) 0.225 (b) -3.21 (c) 0.208 (d) 9260 g (e) 13 g
(f) 180 cm³ (g) 0.720 (h) 0.964 (i) -6.38 (j) 1.52
(k) 0.0989 atm (l) 80.7 ft/sec (m) 7.3 L

6 Decimal Fractions and Percent In the examples of the fractions thus far we have seen that the units of the numerator can be profoundly different from those of the denominator (e.g., miles/hr, g/mL, etc.). In other problems in chemistry we use fractions to express a component part in the numerator to the total in the denominator. In most cases such fractions are expressed without units and in decimal fraction form.

EXAMPLE A-8

A box of nails contains 985 nails; 415 of these are 6-in. nails, 375 are 3-in. nails, and the rest are roofing nails. What is the fraction of roofing nails?

SOLUTION

Roofing nails = total − others = $985 - (415 + 375) = 195$

$$\frac{\text{component}}{\text{total}} = \frac{195}{375 + 415 + 195} = \underline{\underline{0.198}}$$

EXAMPLE A-9

A mixture contains 4.25 mol of N_2, 2.76 mol of O_2, and 1.75 mol of CO_2. What is the fraction of moles of O_2 present in the mixture? (This fraction is known, not surprisingly, as "the mole fraction." The mole is a unit of quantity, like dozen.)

SOLUTION

$$\frac{\text{component}}{\text{total}} = \frac{2.76}{4.25 + 2.76 + 1.75} = \underline{\underline{0.315}}$$

Exercises

A-6   A grocer has 195 dozen boxes of fruit; 74 dozen boxes are apples, 62 dozen boxes are peaches, and the rest are oranges. What is the fraction of the boxes that are oranges?

A-7   A mixture contains 9.85 mol of gas. A 3.18-mol quantity of the gas is N_2, 4.69 mol is O_2, and the rest is He. What is the mole fraction of He in the mixture?

A-8   The total pressure of a mixture of two gases, N_2 and O_2, is 0.72 atm. The pressure due to O_2 is 0.41 atm. What is the fraction of the pressure due to O_2?

Answers: **A-6** 0.30   **A-8** 0.201   **A-8** 0.57

The decimal fractions that have just been discussed are frequently expressed as percentages. Percent simply means parts "per" 100. Percent is obtained by multiplying a fraction in decimal form by 100%.

EXAMPLE A-10

If 57 out of 180 people at a party are women, what is the percent women?

SOLUTION

The fraction of women in decimal form is

$$\frac{57}{180} = 0.317$$

The percent women is

$$0.317 \times 100\% = \underline{31.7\% \text{ women}}$$

The general method used to obtain percent is

$$\frac{\text{component}}{\text{total}} \times 100\% = \underline{\hspace{1cm}} \% \text{ of component}$$

To change from percent back to a decimal fraction, divide the percent by 100%, which moves the decimal to the left two places.

$$86.2\% = \frac{86.2\%}{100\%} = 0.862 \quad \text{(fraction in decimal form)}$$

Exercises

A-9 Express the following fractions as percents: $\frac{1}{4}$, $\frac{3}{8}$, $\frac{9}{8}$, $\frac{55}{25}$, 0.67, 0.13, 1.75, 0.098.

A-10 A bushel holds 198 apples, 27 of which are green. What is the percent of green apples?

A-11 A basket contains 75 pears, 8 apples, 15 oranges, and 51 grapefruit. What is the percent of each?

Answers: **A-9** $\frac{1}{4} = 25\%$, $\frac{3}{8} = 37.5\%$, $\frac{9}{8} = 112.5\%$, $\frac{55}{25} = 220\%$, 0.67 = 67%, 0.13 = 13%, 1.75 = 175%, 0.098 = 9.8%   **A-10** 13.6%, **A-11** 50.3% pears, 5.4% apples, 10.1% oranges, 34.2% grapefruit

We have seen how the percent is calculated from the total and the component part. We now consider problems where percent is given and we calculate either the component part as in Example A-11 or the total as in Example A-12. Such problems can be solved in two ways. The method we employ here uses the percent as a conversion factor and the problem is solved by the factor-label method. (See Section 2-5 and Appendix D.) They can also be solved algebraically as is done in Appendix B.

EXAMPLE A-11

A certain crowd at a rock concert was composed of about 87% teenagers. If the crowd totaled 586 people, how many were teenagers?

PROCEDURE

Remember that percent means "per 100." In this case it means 87 teenagers per 100 people or, in fraction form,

$$\frac{87 \text{ teenagers}}{100 \text{ people}}$$

If this fraction is then multiplied by the number of people, the result is the component part or the number of teenagers.

$$586 \text{ people} \times \frac{87 \text{ teenagers}}{100 \text{ people}} = (586 \times 0.87) = \underline{\underline{510 \text{ teenagers}}}$$

EXAMPLE A-12

A professional baseball player got a hit 28.7% of the times he batted. If he got 246 hits, how many times did he bat?

PROCEDURE

The percent can be written in fraction form and then inverted. It thus relates the total at bats to the number of hits:

$$28.7\% = \frac{28.7 \text{ hits}}{100 \text{ at bats}} \quad \text{or} \quad \frac{100 \text{ at bats}}{28.7 \text{ hits}}$$

If this is now multiplied by the number of hits, the result is the total number of at bats.

$$246 \text{ hits} \times \frac{100 \text{ at bats}}{28.7 \text{ hits}} = \left(\frac{246}{0.287}\right) = \underline{\underline{857 \text{ at bats}}}$$

Exercises

A-12 In a certain audience, 45.9% were men. If there were 196 people in the audience, how many women were present?

A-13 In the alcohol molecule, 34.8% of the mass is due to oxygen. What is the mass of oxygen in 497 g of alcohol?

A-14 The cost of a hamburger in 1988 is 216% of the cost in 1970. If hamburgers cost $0.75 each in 1970, what do they cost in 1988?

A-15 In a certain audience, 46.0% are men. If there are 195 men in the audience, how large is the audience?

A-16 If a solution is 23.3% by weight HCl and it contains 14.8 g of HCl, what is the total mass of the solution?

A-17 An unstable isotope has a mass of 131 amu. This is 104% of the mass of a stable isotope. What is the mass of the stable isotope?

Answers: **A-12** 106 women   **A-13** 173 g   **A-14** $1.62
A-15 424 people   **A-16** 63.5 g   **A-17** 126 amu

This section reviews the following basic algebra skills.

1   Operations on algebra equations

2   Word problems and algebra equations

3   Direct and inverse proportionalities

APPENDIX

BASIC

ALGEBRA

B

A TEST OF BASIC ALGEBRA	Following is a test on some of the basic algebra skills that are necessary in chemistry. After you take the test, compare your answers with those listed at the end of the test. If you miss more than one question in any one section, you should work through the review following the test. In any case, you will be referred to this review at specific points in the text when certain calculations are first encountered.

Operations on Algebra Equations

Algebra Test

1 Solve the following equations for x. (Isolate all numbers and variables other than x on the right-hand side of the equation.)

(a) $4x + 17 = 73$ 1(a) _____

(b) $\dfrac{y}{3x} + 4 = R$ 1(b) _____

(c) $3x^2 + y = 225 - 2x^2$ 1(c) _____

(d) $\dfrac{1}{2x + 1} = y$ 1(d) _____

Word Problems and Algebra Equations

2 Write an algebra equation that expresses each of the following.

(a) A number x is equal to 6 more than y.

       2(a) _____

(b) Three times a number R is equal to 8 less than Z.

       2(b) _____

(c) A number b is equal to 36 less than 5 times the square of d.

       2(c) _____

(d) Four times the product of two numbers s and t is equal to 14% of the sum of two other numbers p and q.

       2(d) _____

3 Write algebra equations that represent the following (let x = an unknown) and then solve for x.

(a) The sum of the lengths of two wooden planks is 24 ft. If one plank is three times the length of the other, what is the length of the smaller plank?

       3(a) Equation _____   Answer _____

(b) Two students received 65% and 80% on a chemistry quiz. The sum of their scores was 180 points. How many points were on the test?

       3(b) Equation _____   Answer _____

(c) A large brick weighs 1 lb more than twice as much as a smaller brick. The sum of the weights of the two bricks is 28 lb. What is the weight of each brick?

       3(c) Equation _____   Answer _____

(d) There was \$9.87 in a cash register, with one more than twice as many dimes as quarters and three less than four times as many pennies as quarters. How many quarters were in the cash register?

3(d) Equation _____   Answer _____

4   Answer the following:

(a) A small car weighs 100 lb more than two-thirds of the weight of a midsize car. The difference in weight between the two cars is 800 lb. What is the weight of the small car?

4(a) _____

(b) A major leaguer during his career got 32 home runs, 25 triples, 55 doubles, and the rest singles. The fraction of his hits that were doubles was 0.119. How many singles did he get?

4(b) _____

(c) An oil refinery held 175 barrels of oil. When refined, each barrel yields 24 gallons of gasoline. If 3120 gallons of gasoline were produced, what percentage of the original barrels of oil were refined?

4(c) _____

Direct and Inverse Proportionalities

5   Write equations for the following. Use k for a constant of proportionality.

(a) A quantity x is directly proportional to y.

5(a) _____

(b) A quantity z is inversely proportional to the square root of r.

5(b) _____

(c) A quantity q is directly proportional to A and inversely proportional to the square of the sum of B and C.

5(c) _____

(d) The cube root of a quantity t is directly proportional to the square of v and inversely proportional to 6 less than the quantity w.

5(d) _____

Answers to Test

1(a) $x = 14$ 1(b) $x = y/(3R - 12)$ 1(c) $x = \pm\sqrt{45 - y/5}$
1(d) $x = (1 - y)/2y$ 2(a) $x = y + 6$ 2(b) $3R = Z - 8$
2(c) $b = 5d^2 - 36$ 2(d) $4st = 0.14(p + q)$ 3(a) $x + 3x = 24$, $x = 6$ ft
3(b) $0.65x + 0.80x = 180$, $x = 124$ points   3(c) $x + (2x + 1) = 28$,
$x = 9$ lb, 19 lb   3(d) $0.25x + 0.10(2x + 1) + 0.01(4x - 3) = 9.87$,
$x = 20$ quarters   4(a) 1900 lb   4(b) 350   4(c) 74.3%   5(a) $x = ky$
5(b) $z = k/\sqrt{r}$ 5(c) $q = kA/(B + C)^2$ 5(d) $\sqrt[3]{t} = kv^2/(w - 6)$

B   REVIEW OF BASIC ALGEBRA

1 Operations on Algebra Equations   Many of the quantitative problems of chemistry require the use of basic algebra.

As an example of a simple algebra equation we use

$$x = y + 8$$

In any algebraic equation the equality remains valid when identical operations are performed on both sides of the equation. The following operations illustrate this principle.

1   A quantity may be added to or subtracted from both sides of the equation.

$$\text{(add 8)} \qquad x \underline{+\, 8} = y + 8 \underline{+\, 8} \quad x + 8 = y + 16$$
$$\text{(subtract 8)} \quad x \underline{-\, 8} = y + 8 \underline{-\, 8} \quad x - 8 = y$$

2   Both sides of the equation may be multiplied or divided by the same quantity.

$$\text{(multiply by 4)} \quad \underline{4}x = \underline{4}(y + 8) = 4y + 32$$
$$\text{(divide by 2)} \qquad \frac{x}{2} = \frac{(y + 8)}{2} \qquad\qquad \frac{x}{2} = \frac{y}{2} + 4$$

3   Both sides of the equation may be raised to a power, or a root of both sides of an equation may be taken.

$$\text{(equation squared)} \quad x^2 = (y + 8)^2$$
$$\text{(square root taken)} \quad \sqrt{x} = \sqrt{y + 8}$$

4   Both sides of an equation may be inverted.

$$\frac{1}{x} = \frac{1}{y + 8}$$

In addition to operations on both sides of an equation, two other points must be recalled.

1   As in any fraction, identical factors in the numerator and the denominator in an algebraic equation may be canceled.

$$\frac{\cancel{4}x}{\cancel{4}} = x = y + 8 \qquad \text{or} \qquad x = \frac{\cancel{4}(y + 8)}{\cancel{4}} = y + 8$$

2   Quantities equal to the same quantity are equal to each other. Thus, substitutions for equalities may be made in algebraic equations.

$$x = y + 8$$
$$x = 27$$

Therefore, since $x = x$,

$$y + 8 = 27$$

We can use these basic rules to solve algebraic equations. Usually, we need to isolate one variable on the left-hand side of the equation with all other numbers and variables on the right-hand side of the equation. The following examples illustrate the isolation of a variable on the left-hand side of the equation.

EXAMPLE B-1

Solve for x in $x + 5 = 92$.

SOLUTION

Subtract 5 from both sides of the equation.

$$x + 5 - 5 = 92 - 5$$
$$x = 87$$

In practice, a number or a variable may be moved to the other side of an equation with a change of sign. For example,

$$\text{if} \quad x + 5 = y$$
$$\text{then} \quad x = y - 5$$

EXAMPLE B-2

Solve for x in $x + y + 8 = z + 6$.

SOLUTION

Move $+y$ and $+8$ to the right by changing signs.

$$x = z + 6 - y - 8$$
$$= z - y + 6 - 8 = z - y - 2$$

EXAMPLE B-3

Solve for x in

$$\frac{x + 8}{y} = z$$

SOLUTION

First, move y to the right by multiplying both sides by y.

$$\not{y} \cdot \frac{x + 8}{\not{y}} = z \cdot y$$

This leaves

$$x + 8 = zy$$

Subtract 8 from both sides to obtain the final answer.

$$x = zy - 8$$

In practice, the numerator of a fraction on the left becomes the denominator on the right. The denominator of a fraction on the left becomes the numerator on the right. For example, consider the following equations.

(move z to the right)   (move $k + 5$ to the right)

If $xz = y + 8$ If $\dfrac{x}{k+5} = B$

$x \cdot \widehat{z} = y + 8$ $\dfrac{x}{\widehat{k+5}} = B$

then $x = \dfrac{y + 8}{z}$ then $x = B(k + 5)$

In these problems, the numerical value of x can be found if values for the other variables are known. For example, if $y = -2$ and $z = 8$ in Example B-3,

$$x = zy - 8$$
$$= (-2)(8) - 8 = -16 - 8 = -24$$

EXAMPLE B-4

Solve for x in

$$\frac{4x + 2}{3 + x} = 7$$

SOLUTION

First, multiply both sides by $(3 + x)$ to clear the fraction.

$$(3 + x) \cdot \frac{4x + 2}{3 + x} = 7(3 + x)$$

This leaves

$$4x + 2 = 21 + 7x$$

To move integers to the right and the x variable to the left, subtract $7x$ and 2 from both sides of the equation. This leaves

$$-3x = 19$$

Finally, divide both sides by -3 to move the -3 to the right.

$$x = -\tfrac{19}{3} = \underline{-6.33}$$

EXAMPLE B-5

Solve for T_2 in

$$\frac{P_1 V_1}{T_1} = \frac{P_2 V_2}{T_2}$$

SOLUTION

To move T_2 to the left, multiply both sides by T_2.

$$T_2 \cdot \frac{P_1 V_1}{T_1} = \cancel{T_2} \cdot \frac{P_2 V_2}{\cancel{T_2}} = P_2 V_2$$

Move $P_1 V_1$ to the right by dividing by $P_1 V_1$.

$$\frac{T_2}{\cancel{P_1 V_1}} \cdot \frac{\cancel{P_1 V_1}}{T_1} = \frac{P_2 V_2}{P_1 V_1}$$

Finally, move T_1 to the right by multiplying both sides by T_1.

$$T_2 = \frac{T_1 P_2 V_2}{P_1 V_1}$$

EXAMPLE B-6

Solve the following equation for y.

$$\frac{2y}{3} + x = 9z + 4$$

SOLUTION

First, to clear the fraction, multiply both sides by 3. This leaves

$$2y + 3x = 27z + 12$$

Subtract $3x$ from both sides, which leaves

$$2y = 27z + 12 - 3x$$

Finally, divide both sides by 2, which leaves

$$y = \frac{27z + 12 - 3x}{2}$$

EXAMPLE B-7

Solve the following equation for x.

$$4x^2 - 60 = 16y$$

SOLUTION

First, divide both sides by 4 which leaves

$$x^2 - 15 = 4y$$

Next, move the -15 to the right.

$$x^2 = 4y + 15$$

The final answer is

$$x = \pm\sqrt{4y + 15}$$

Exercises

B-1 Solve for x in $17x = y - 87$.

B-2 Solve for x in

$$\frac{y}{x} + 8 = z + 16$$

B-3 Solve for T in $PV = (\text{wt}/MM)RT$.

B-4 Solve for x in

$$\frac{7x - 3}{6 + 2x} = 3r$$

B-5 Solve for x in $18x - 27 = 2x + 4y - 35$. If $y = 3x$, what is the value of x?

B-6 Solve for x in

$$\frac{x}{4y} + 18 = y + 2$$

B-7 Solve for x in $5x^2 + 12 = x^2 + 37$.

B-8 Solve for r in

$$\frac{80}{2r} + \frac{y}{r} = 11$$

What is the value of r if $y = 14$?

Answers: **B-1** $x = (y - 87)/17$ **B-2** $x = y/(8 + z)$ **B-3** $T = PV \cdot MM/\text{wt} \cdot R$ **B-4** $x = 3(6r + 1)/(7 - 6r)$ **B-5** $x = (y - 2)/4$. When $y = 3x$, $x = -2$.   **B-6** $x = 4y(y - 16)$ **B-7** $x = \pm 2.5$
B-8 $r = (40 + y)/11$. When $y = 14$, $r = \frac{54}{11}$

2 Word Problems and Algebra Equations Eventually, a necessary skill in chemistry is the ability to translate word problems into algebra equations and then solve. The key is to assign a variable (usually x) to be equal to a certain quantity and then to treat the variable consistently throughout the equation. Again, examples are the best way to illustrate the problems.

EXAMPLE B-8

Translate each of the following to an equation.

(a) A number x is equal to a number that is 4 larger than y.

$$x = y + 4$$

(b) A number z is equal to three-fourths of u.

$$z = \tfrac{3}{4}u$$

(c) The square of a number r is equal to 16.9% of the value of w.

$$r^2 = 0.169w \quad \text{(change percent to a decimal fraction)}$$

(d) A number t is equal to 12 plus the square root of q.

$$t = 12 + \sqrt{q}$$

Exercises

Write algebraic equations for the following.

B-9   A number n is equal to a number that is 85 smaller than m.
B-10 A number y is equal to one-fourth of z.
B-11 Fifteen percent of a number k is equal to the square of another number d.
B-12 A number x is equal to 14 more than the square root of v.
B-13 Four times the sum of two numbers, q and w, is equal to 68.

B-14 Five times the product of two variables, s and t, is equal to 16 less than the square of s.

B-15 Five-ninths of a number C is equal to 32 less than a number F.

Answers: **B-9** $n = m - 85$ **B-10** $y = z/4$ **B-11** $0.15k = d^2$
B-12 $x = \sqrt{v} + 14$ **B-13** $4(q + w) = 68$ **B-14** $5st = s^2 - 16$
B-15 $\frac{5}{9}C = F - 32$

We now move from the abstract to the real. In the following examples it is necessary to translate the problem into an algebraic expression as in the previous examples. There are two types of examples that we will use. The first you will certainly recognize, but the second type may be unfamiliar, especially if you have just begun the study of chemistry. However, it is *not* important that you understand the units of the chemistry problems at this time. What *is* important is for you to notice that the problems are worked in the same manner regardless of the units.

EXAMPLE B-9

John is 2 years more than twice as old as Mary. The sum of their ages is 86. How old is each?

SOLUTION

Let $x =$ age of Mary. Then $2x + 2 =$ age of John.

$$x + (2x + 2) = 86$$
$$3x = 84$$
$$x = \underline{28} \quad \text{(age of Mary)}$$
$$[2(28) + 2] = \underline{58} \quad \text{(age of John)}$$

EXAMPLE B-10

One mole of SF_6 has a mass 30.0 g less than four times the mass of 1 mol of CO_2. The mass of 1 mol of SF_6 plus the mass of 1 mol of CO_2 is equal to 190 g. What is the mass of 1 mol of each?

SOLUTION

Let $x =$ mass of 1 mol of CO_2. Then $4x - 30 =$ mass of 1 mol of SF_6.

$$x + (4x - 30) = 190$$
$$x = \underline{44 \text{ g}} \quad \text{(mass of 1 mol of } CO_2\text{)}$$
$$[4(44) - 30] = \underline{146 \text{ g}} \quad \text{(mass of 1 mol of } SF_6\text{)}$$

EXAMPLE B-11

Two students took the same test, and their percent scores differed by 10%. If there were 200 points on the test and the total of their point scores was 260 points, what was each student's percent score?

PROCEDURE

Set up an equation relating each person's percent scores to their total points (260).

Let x = percent score of higher test.
Then $x - 10$ = percent score of lower test.

The points that each person scores is the percent in fraction form multiplied by the points on the test.

$$\frac{\% \text{ grade}}{100 \text{ points}} \times (\text{points on test}) = \text{points scored}$$

SOLUTION

$$\left[\frac{x}{100}(200 \text{ points})\right] + \left[\frac{x-10}{100}(200 \text{ points})\right] = 260 \text{ points}$$

$$200x + 200x - 2000 = 26{,}000$$

$$400x = 28{,}000$$

$$x = 70$$

$$\text{higher score} = \underline{70\%} \quad \text{lower score} = 70 - 10 = \underline{60\%}$$

EXAMPLE B-12

If an 8.75-g quantity of sugar represents 65.7% of a mixture, what is the mass of the mixture?

SOLUTION

Let x = mass of the mixture. Then

$$\frac{65.7}{100} x = 0.657 \, x = 8.75$$

$$x = \frac{8.75}{0.657} = \underline{\underline{13.3 \text{ g}}}$$

EXAMPLE B-13

In a certain year 14.5% of a rancher's herd of 876 cows had two calves each. The rest had none.

How many calves were born?
How many cows did not have a calf?
How large was the herd including calves?

SOLUTION

Let x = number of cows having calves. Then $2x$ = number of calves born.

$$x = 0.145 \times 876 = \underline{127 \text{ cows having calves}}$$
$$2x = 2 \times 127 = \underline{254 \text{ calves born}}$$
$$876 - x = 876 - 127 = \underline{749 \text{ cows not having calves}}$$
$$876 + 2x = 876 + 254 = \underline{1130 \text{ total cattle}}$$

EXAMPLE B-14

This example is directly analogous to the preceding example. A certain compound, AX_2, can break apart (dissociate) into one A and two X's. In a 0.250-mol quantity of AX_2, 6.75% of the moles dissociated. (A mole, abbreviated mol, is a unit of quantity, like dozen or gross.)

How many moles f AX_2 dissociated?
How many moles of AX_2 did not dissociate?
How many total moles of particles (AX_2, A, and X) are present?

SOLUTION

Let x = number of moles of AX_2 that dissociate.

$$x = 0.0675 \times 0.250 \text{ mol} = \underline{0.0169 \text{ mol dissociated}}$$

The number of moles of AX_2 that are not dissociated = $0.250 - x$.

$$0.250 - 0.0169 = \underline{0.233 \text{ mol undissociated}}$$

total moles present

$$= \text{moles } AX_2 \text{ (undissociated)} + \text{moles of A} + \text{moles of X}$$
$$\text{moles } AX_2 = 0.233$$

$$\text{moles of A} = x = 0.0169$$
$$\text{moles of X} = 2x = 0.0338$$

Total moles $= 0.233 + 0.0169 + 0.0338 = \underline{0.284 \text{ mol}}$

EXAMPLE B-15

A used car dealer has Fords, Chevrolets, and Plymouths. There are 120 Fords, 152 Chevrolets, and the rest are Plymouths. If the fraction of Fords is 0.310, how many Plymouths are on the lot?

SOLUTION

Let x = number of Plymouths.

$$\text{fraction of Fords} = \frac{\text{number of Fords}}{\text{total number of cars}} = 0.310$$

$$\frac{120}{120 + 152 + x} = 0.310$$

$$120 = 0.310(272 + x)$$
$$120 = 84.3 + 0.310x$$
$$x = \underline{\underline{115 \text{ Plymouths}}}$$

EXAMPLE B-16

There is a 0.605-mol quantity of N_2 present in a mixture of N_2 and O_2. If the mole fraction of N_2 is 0.251, how many moles of O_2 are present?

SOLUTION

Let x = number of moles of O_2. Then

$$\text{mole fraction N}_2 = \frac{\text{mol of N}_2}{\text{total mol present}} = 0.251$$

$$\frac{0.605}{0.605 + x} = 0.251$$

$$0.605 = 0.251(0.605 + x)$$
$$x = \underline{\underline{1.80 \text{ mol}}}$$

Exercises

In the following exercises, a problem concerning an everyday situation is followed by one or two closely analogous problems concerning a chemistry situation. In both cases the mechanics of the solution are similar. Only the units differ.

B-16 The total length of two boards is 18.4 ft. If one board is 4.0 ft longer than the other, what is the length of each board?

B-17 An isotope of iodine has a mass 10 amu less than two-thirds the mass of an isotope of thallium. The total mass of the two isotopes is 340 amu. What is the mass of each isotope?

B-18 An isotope of gallium has a mass 22 amu more than one-fourth the mass of an isotope of osmium. The difference in the two masses is 122 amu. What is the mass of each?

B-19 An oil refinery held 175 barrels of oil. When refined, each barrel yields 24 gallons of gasoline. If 3120 gallons of gasoline were produced, what percentage of the original barrels of oil was refined?

B-20 A solution contained 0.856 mol of a substance A_2X. In solution some of the A_2X's break up into A's and X's (note that each mole of A_2X yields 2 mol of A). If 0.224 mol of A is present in the solution, what percentage of the moles of A_2X dissociated (broke apart)?

B-21 In Las Vegas, a dealer starts with 264 decks of cards. If 42.8% of the decks were used in an evening, how many jacks (four per deck) were used?

B-22 A solution originally contains a 1.45-mol quantity of a compound A_3X_2. If 31.5% of the A_3X_2 dissociates (three A's and two X's per A_3X_2), how many moles of A are formed? How many moles of X? How many moles of undissociated A_3X_2 remain? How many moles of particles (A's, X's, and A_3X_2's) are present in the solution?

B-23 The fraction of kerosene that can be recovered from a barrel of crude oil is 0.200. After a certain amount of oil was refined, 8.90 gal of kerosene, some gasoline, and 18.6 gal of other products were produced. How many gallons of gasoline were produced?

B-24 The fraction of moles (mole fraction) of gas A in a mixture is 0.261. If the mixture contains 0.375 mol of gas B and 0.175 mol of gas C as well as gas A, how many moles of gas A are present?

Answers: **B-16** 7.2 ft, 11.2 ft **B-17** thallium, 210 amu; iodine, 130 amu **B-18** gallium, 70 amu; osmium, 192 amu **B-19** 74.3% **B-20** 13.1% **B-21** 452 jacks **B-22** 1.37 mol of A, 0.914 mol of X, 0.99 mol of A_3X_2, 3.27 mol total **B-23** 17.0 gal **B-24** 0.195 mol

3 Direct and Inverse Proportionalities There is one other point that should be included in a review on algebra-direct and inverse proportionalities. We use these terms often in chemistry.

When a quantity is directly proportional to another, it means that an increase in one variable will cause a corresponding increase of the same percent in the other variable. A direct proportionality is shown as

$$A \propto B \quad (\propto \text{ is the proportionality symbol})$$

which is read "*A* is directly proportional to *B*." A proportionality can be easily converted to an algebraic equation by the introduction of a constant (in our examples designated *k*), called a constant of proportionality. Thus the proportion becomes

$$A = kB$$

or, rearranging,

$$\frac{A}{B} = k$$

Note that *k* is not a variable but has a certain numerical value that does not change as do *A* and *B* under experimental conditions.

A common, direct proportionality that we will study relates Kelvin temperature *T* and volume *V* of a gas at constant pressure. This is written as

$$V \propto T \quad V = kT \quad \frac{V}{T} = k$$

(This is known as Charles' law.)

In a hypothetical case, $V = 100$ L and $T = 200$ K. From this information we can calculate the value of the constant *k*.

$$\frac{V}{T} = \frac{100 \text{ L}}{200 \text{ K}} = 0.50 \text{ L/K}$$

A change in volume or temperature requires a corresponding change in the other *in the same direction*. For example, if the temperature of the gas is changed to 300 K, we can see that a corresponding change in volume is required from the following calculation:

$$\frac{V}{T} = k$$

$$V = kT = 0.50 \text{ L/K} \times 300 \text{ K} = \underline{150 \text{ L}}$$

When a quantity is inversely proportional to another quantity, an increase in one brings about a corresponding *decrease* in the other. An inverse proportionality between *A* and *B* is written as

$$A \propto \frac{1}{B}$$

As before, the proportionality can be written as an equality by the introduction of a constant (which has a value different from the example above).

$$A = \frac{k}{B} \quad \text{or} \quad AB = k$$

A common inverse proportionality that we use relates the volume V of a gas to the pressure P at a constant temperature. This is written as

$$V \propto \frac{1}{P} \quad V = \frac{k}{P} \quad PV = k$$

(This is known as Boyle's law.)

In a hypothetical case, $V = 100$ L and $P = 1.50$ atm. From this information we can calculate the value of the constant k.

$$PV = k = 1.50 \text{ atm} \times 100 \text{ L} = 150 \text{ atm} \cdot \text{L}$$

A change in volume or pressure requires a corresponding change in the other *in the opposite direction.* For example, if the pressure on the gas is changed to 3.00 atm, we can see that a corresponding change in volume is required.

$$PV = k$$
$$V = \frac{k}{P} = \frac{150 \,\text{atm} \cdot \text{L}}{3.00 \,\text{atm}} = \underline{50.0 \text{ L}}$$

When one variable (i.e., x) is directly proportional to two other variables (i.e., y and z), the proportionality can be written as the product of the two.

$$\text{If} \quad x \propto y \quad \text{and} \quad x \propto z$$
$$\text{then} \quad x \propto yz$$

When one variable (i.e., a) is directly proportional to one variable (i.e., b) and inversely proportional to another (i.e., c), the proportionality can be written as the ratio of the two.

$$\text{If} \quad a \propto b \quad \text{and} \quad a \propto \frac{1}{c}$$
$$\text{then} \quad a \propto \frac{b}{c}$$

Quantities can be directly or inversely proportional to the square, square root, or any other function of another variable or number, as illustrated by the following examples.

EXAMPLE B-17

A quantity C is directly proportional to the square of D. Write an equality for this statement and explain how a change in D affects the value of C.

SOLUTION

The equation is

$$C = kD^2$$

Note that a change in D will have a significant effect on the value of C. For example,

If $D = 1$, then $C = k$.
If $D = 2$, then $C = 4k$.
If $D = 3$, then $C = 9k$.

Note that when the value of D is doubled, the value of C is increased *fourfold*.

EXAMPLE B-18

A variable X is directly proportional to the square of the variable Y and inversely proportional to the square of another variable Z. This can be written as two separate equations if it is assumed that Y is constant when Z varies and vice versa.

$$X = k_1 Y^2 \quad (Z \text{ constant})$$
$$X = k_2 / Z^2 \quad (Y \text{ constant})$$
$$(k_1 \text{ and } k_2 \text{ are different constants})$$

This relationship can be combined into one equation when both Y and Z are variables:

$$X = \frac{k_3 Y^2}{Z^2}$$

k_3 is a third constant that is a combination of k_1 and k_2.

Exercises

Write equalities for the following relations.

B-25 X is inversely proportional to $Y + Z$.

B-26 $[H_3O^+]$ is inversely proportional to $[OH^-]$.

B-27 $[H_2]$ is directly proportional to the square root of r.

B-28 B is directly proportional to the square of y and the cube of z.

B-29 The pressure P of a gas is directly proportional to the number of moles n and the temperature T, and inversely proportional to the volume V.

Answers: **B-25** $X = k/(Y + Z)$ **B-26** $[H_3O^+] = k/[OH^-]$
B-27 $[H_2] = k\sqrt{r}$ **B-28** $B = ky^2z^3$ **B-29** $P = knT/V$

This section reviews the following scientific notation skills.

1   Expressing numbers in standard scientific notation

2   Addition and subtraction

3   Multiplication and division

4   Powers and roots

APPENDIX

SCIENTIFIC

NOTATION

C

A TEST OF SCIENTIFIC NOTATION

Following is a short test on the manipulation of numbers expressed in scientific notation. Since you should be comfortable with the use of such numbers in calculations before beginning Chapter 8, you are urged to study the review if you miss more than one problem in any one section. Compare your answers with those that follow the test. In the following problems, zeros to the left of a decimal point but to the right of a digit (e.g., 92<u>00</u>) where no decimal point is showing are assumed not to be significant figures (see Section 2-1). Although the numbers used in this appendix do not have units (for simplicity), treat the numbers as measurements so that the answer to a problem shows the proper number of significant figures.

Expressing Numbers in Standard Scientific Notation

1 Express the following numbers in scientific notation in standard form [one digit to the left of the decimal point in the coefficient (e.g., 7.39×10^6)].

(a) 0.00657   1(a) _____

(b) 14,300   1(b) _____

(c) 0.00030   1(c) _____

(d) 3,457,000,000   1(d) _____

2 Change the following numbers expressed in scientific notation to standard form.

(a) 0.778×10^5 2(a) _____

(b) 145×10^{-8} 2(b) _____

(c) 0.00720×10^{-8} 2(c) _____

(d) 4330×10^{-10} 2(d) _____

Addition and Subtraction

3 Carry out the following calculations. Express your answer to the proper decimal place.

(a) $0.049 + (3.7 \times 10^{-2}) + (76.6 \times 10^{-3})$ 3(a) _____

(b) $(1.62 \times 10^6) + (0.78 \times 10^4) - (0.025 \times 10^7)$ 3(b) _____

(c) $(0.363 \times 10^{-6}) + (71.2 \times 10^{-9}) + (619 \times 10^{-12})$ 3(c) _____

(d) $(481 \times 10^6) - (0.113 \times 10^9) + (3.5 \times 10^5)$ 3(d) _____

Multiplication and Division

4 Carry out the following calculations. Express your answer in standard form to the proper number of significant figures.

(a) $(4.8 \times 10^6) \times (17.2 \times 10^{-2})$ 4(a) _____

(b) $(186 \times 10^{20}) \div (0.044 \times 10^5)$ 4(b) _____

(c) $[(55.3 \times 10^5) \times (1.5 \times 10^6)] \div (2.2 \times 10^{-5})$ 4(c) _____

(d) $0.000879 \div 8.22 \times 10^5$ 4(d) _____

Powers and Roots

5 Carry out the following calculations. Express the answer in standard form.

(a) $(5.0 \times 10^{-5})^2$ 5(a) _____

(b) $\sqrt{4.9 \times 10^5}$ 5(b) _____

(c) $(1.1 \times 10^4)^3$ 5(c) _____

(d) $\sqrt[3]{1.25 \times 10^{-10}}$ 5(d) _____

Answers to Test

1(a) 6.57×10^{-3} 1(b) 1.43×10^4 1(c) 3.0×10^{-4}
1(d) 3.457×10^9 2(a) 7.78×10^4 2(b) 1.45×10^{-6}
2(c) 7.20×10^{-11} 2(d) 4.33×10^{-7} 3(a) 0.163   3(b) 1.38×10^6
3(c) 4.35×10^{-7} 3(d) 3.68×10^8 4(a) 8.3×10^5 4(b) 4.2×10^{18}
4(c) 3.8×10^{17} 4(d) 1.07×10^{-9} 5(a) 2.5×10^{-9} 5(b) 7.0×10^2
5(c) 1.3×10^{12} 5(d) 5.00×10^{-4}

B REVIEW OF SCIENTIFIC NOTATION

Although this topic was first introduced in Section 2-3 in this text, we focus on a review of the mathematical manipulation of numbers expressed in scientific notation in this appendix. Specifically, addition, multiplication, division, and taking the roots of numbers expressed in scientific notation are covered.

As mentioned in Chapter 2, scientific notation makes use of powers of 10 to simplify awkward numbers that employ more than two or three zeros that are not significant figures. The exponent of 10 simply indicates how many times we should multiply or divide a number (called the coefficient) by 10 to produce the actual number. For example, $8.9 \times 10^3 = 8.9$ (the coefficient), multiplied by 10 *three* times, or

$$8.9 \times 10 \times 10 \times 10 = 8900$$

Also, $4.7 \times 10^{-3} = 4.7$ (the coefficient), divided by 10 *three* times, or

$$\frac{4.7}{10 \times 10 \times 10} = 0.0047$$

1 Expressing Numbers in Standard Scientific Notation   The method for expressing numbers in scientific notation was explained in Section 2-3. However, to simplify a number or to express it in the standard form with one digit to the left of the decimal point in the coefficient, it is often necessary to change a number already expressed in scientific notation. Since this is often done in a hurry, a large number of errors may result. Thus, it is worthwhile to practice moving the decimal point of numbers expressed in scientific notation.

EXAMPLE C-1

Change the following numbers to the standard form.

(a) 489×10^4 (b) 0.00489×10^8

PROCEDURE

All you need to remember is to raise the power of 10 one unit for each place the decimal point is moved to the left and lower the power of 10 one unit for each place that the decimal point is moved to the right in the coefficient.

SOLUTION

$$489 \times 10^4 = (4 \; 8 \; 9) \times 10^4 = 4.89 \times 10^{4+2} = \underline{4.89 \times 10^6}$$

$$0.00489 \times 10^8 = (0.0 \; 0 \; 4 \; 8 \; 9) \times 10^8 = 4.89 \times 10^{8-3} = \underline{4.89 \times 10^5}$$

As an aid to remembering whether you should raise or lower the exponent as you move the decimal point, it is suggested that you write (or at least imagine) the coefficient on a slant. For each place that you move the decimal point *up*, add one to the exponent. For each place that you move the decimal point *down*, subtract one from the exponent. Note that the exponent moves up or down with the decimal point.

EXAMPLE C-2

Change the following numbers to the standard form in scientific notation.

(a) 4223×10^{-7} (b) 0.00076×10^{18}

SOLUTION

$$4223 \times 10^{-7} = \begin{bmatrix} 4 & {}^{+3} \\ 2 & {}^{+2} \\ 2 & {}^{+1} \\ 3 & \end{bmatrix} \times 10^{-7} = 4.223 \times 10^{-7+3} = \underline{4.223 \times 10^{-4}}$$

$$0.00076 \times 10^{18} = \begin{bmatrix} 0 \\ 0 & {}^{-1} \\ 0 & {}^{-2} \\ 0 & {}^{-3} \\ 7 & {}^{-4} \\ 6 \end{bmatrix} \times 10^{18} = 7.6 \times 10^{18-4} = \underline{7.6 \times 10^{14}}$$

Exercises   **C-1**   Change the following numbers to standard scientific notation with one digit to the left of the decimal point in the coefficient.

   (a) 787×10^{-6} (d) 0.0037×10^9
   (b) 43.8×10^{-1} (e) 49.3×10^{15}
   (c) 0.015×10^{-16} (f) 6678×10^{-16}

C-2   Change the following numbers to a number with two digits to the left of the decimal point in the coefficient.

   (a) 9554×10^4 (d) 116.5×10^4
   (b) 1.6×10^{-5} (e) 0.023×10^{-1}
   (c) 1×10^6 (f) 0.005×10^{23}

Answers: C-1(a) 7.87×10^{-4} C-1(b) 4.38 C-1(c) 1.5×10^{-18}
C-1(d) 3.7×10^6 C-1(e) 4.93×10^{16} C-1(f) 6.678×10^{-13}
C-2(a) 95.54×10^6 C-2(b) 16×10^{-6} C-2(c) 10×10^5
C-2(d) 11.65×10^5 C-2(e) 23×10^{-4} C-2(f) 50×10^{19}

2 Addition and Subtraction   Addition or subtraction of numbers in scientific notation can be accomplished only when all coefficients have the same exponent of 10. When all the exponents are the same, the coefficients are added and then multiplied by the power of 10. The correct number of places to the right of the decimal point must be shown as discussed in Section 2-2.

EXAMPLE C-3

Add the following numbers.

$$3.67 \times 10^{-4}, 4.879 \times 10^{-4}, \text{ and } 18.2 \times 10^{-4}$$

SOLUTION

$$
\begin{array}{r}
3.67 \ \times 10^{-4} \\
4.879 \times 10^{-4} \\
\underline{18.2 \ \ \times 10^{-4}} \\
26.749 \times 10^{-4} = \underline{26.7 \times 10^{-4}} = \underline{2.67 \times 10^{-3}}
\end{array}
$$

EXAMPLE C-4

Add the following numbers.

$$320.4 \times 10^3, 1.2 \times 10^5, \text{ and } 0.0615 \times 10^7$$

SOLUTION

Before adding, change all three numbers to the same exponent of 10.

$$
\begin{aligned}
320.4 \quad &\times 10^3 = \quad 3.204 \times 10^5 \\
1.2 \quad &\times 10^5 = \quad 1.2 \quad \times 10^5 \\
0.0615 &\times 10^7 = \underline{\quad 6.15 \quad \times 10^5} \\
&\qquad\qquad 10.554 \times 10^5 = \underline{\underline{10.6 \times 10^5}} = \underline{\underline{1.06 \times 10^6}}
\end{aligned}
$$

Exercises

C-3   Add the following numbers. Express the answer to the proper decimal place.

(a) $152 + (8.635 \times 10^2) + (0.021 \times 10^3)$

(b) $(10.32 \times 10^5) + (1.1 \times 10^5) + (0.4 \times 10^5)$

(c) $(1.007 \times 10^{-8}) + (118 \times 10^{-11}) + (0.1141 \times 10^{-6})$

(d) $(0.0082) + (2.6 \times 10^{-4}) + (159 \times 10^{-4})$

C-4   Carry out the following calculations. Express your answer to the proper decimal place.

(a) $(18.75 \times 10^{-6}) - (13.8 \times 10^{-8}) + (1.0 \times 10^{-5})$

(b) $(1.52 \times 10^{-11}) + (17.7 \times 10^{-12}) - (7.5 \times 10^{-15})$

(c) $(481 \times 10^6) - (0.113 \times 10^9) + (8.5 \times 10^5)$

(d) $(0.363 \times 10^{-6}) + (71.2 \times 10^{-9}) + (519 \times 10^{-12})$

Answers: **C-3(a)** 1.036×10^3 **C-3(b)** 1.18×10^6
C-3(c) 1.254×10^{-7} **C-3(d)** 2.44×10^{-2} **C-4(a)** 2.9×10^{-5}
C-4(b) 3.29×10^{-11} **C-4(c)** 3.69×10^8 **C-4(d)** 4.35×10^{-7}

3 Multiplication and Division   When numbers expressed in scientific notation are multiplied, the exponents of 10 are *added*. When the numbers are divided, the exponent of 10 in the denominator (the divisor) is subtracted from the exponent of 10 in the numerator (the dividend).

EXAMPLE C-5

Carry out the following calculation.

$$(4.75 \times 10^6) \times (3.2 \times 10^5)$$

SOLUTION

In the first step, group the coefficients and the powers of 10. Carry out each step separately.

$$= (4.75 \times 3.2) \times (10^6 \times 10^5)$$
$$= 15.200 \times 10^{6+5} = 15 \times 10^{11} = \underline{1.5 \times 10^{12}}$$

EXAMPLE C-6

Carry out the following calculation.

$$(1.62 \times 10^{-8}) \div (8.55 \times 10^{-3})$$

SOLUTION

$$= \frac{1.62 \times 10^{-8}}{8.55 \times 10^{-3}} = \frac{1.62}{8.55} \times \frac{10^{-8}}{10^{-3}} = 0.189 \times 10^{-8-(-3)}$$
$$= 0.189 \times 10^{-5} = \underline{1.89 \times 10^{-6}}$$

Exercises

C-5 Carry out the following calculations. Express your answer to the proper number of significant figures with one digit to the left of the decimal point.

(a) $(7.8 \times 10^{-6}) \times (1.12 \times 10^{-2})$

(b) $(0.511 \times 10^{-3}) \times (891 \times 10^{-8})$

(c) $(156 \times 10^{-12}) \times (0.010 \times 10^4)$

(d) $(16 \times 10^9) \times (0.112 \times 10^{-3})$

(e) $(2.35 \times 10^3) \times (0.3 \times 10^5) \times (3.75 \times 10^2)$

(f) $(6.02 \times 10^{23}) \times (0.0100)$

C-6 Follow the instructions in Problem C-5.

(a) $(14.6 \times 10^8) \div (2.2 \times 10^8)$

(b) $(6.02 \times 10^{23}) \div (3.01 \times 10^{20})$

(c) $(0.885 \times 10^{-7}) \div (16.5 \times 10^3)$

(d) $(0.0221 \times 10^3) \div (0.57 \times 10^{18})$

(e) $238 \div (6.02 \times 10^{23})$

C-7 Follow the instructions in Problem C-5.

(a) $[(8.70 \times 10^6) \times (3.1 \times 10^8)] \div (5 \times 10^{-3})$

(b) $(47.9 \times 10^{-6}) \div [(0.87 \times 10^6) \times (1.4 \times 10^2)]$

(c) $1 \div [(3 \times 10^6) \times (4 \times 10^{10})]$

(d) $1.00 \times 10^{-14} \div [(6.5 \times 10^5) \times (0.32 \times 10^{-5})]$

(e) $[(147 \times 10^{-6}) \div (154 \times 10^{-6})] \div (3.0 \times 10^{12})$

Answers: C-5(a) 8.7×10^{-8} C-5(b) 4.55×10^{-9} C-5(c) 1.6×10^{-8}
C-5(d) 1.8×10^6 C-5(e) 3×10^{10} C-5(f) 6.02×10^{21} C-6(a) 6.6
C-6(b) 2.00×10^3 C-6(c) 5.36×10^{-12} C-6(d) 3.9×10^{-17}
C-6(e) 3.95×10^{-22} C-7(a) 5×10^{17} C-7(b) 3.9×10^{-13}
C-7(c) 8×10^{-18} C-7(d) 4.8×10^{-15} C-7(e) 3.2×10^{-13}

4 Powers and Roots When a number expressed in scientific notation is raised to a power, the coefficient is raised to the power and the exponent of 10 is *multiplied* by the power.

EXAMPLE C-7

Carry out the following calculation.

$$(3.2 \times 10^3)^2 = (3.2)^2 \times 10^{3 \times 2}$$
$$= 10.24 \times 10^6 = \underline{1.0 \times 10^7}$$
$$[(10^3)^2 = 10^3 \times 10^3 = 10 \times 10 \times 10 \times 10 \times 10 \times 10 = 10^6]$$

EXAMPLE C-8

Carry out the following calculation.

$$(1.5 \times 10^{-3})^3 = (1.5)^3 \times 10^{-3 \times 3}$$
$$= \underline{3.4 \times 10^{-9}}$$

To take the root of a number expressed in scientific notation take the root of the coefficient and divide the exponent of 10 by the root. Before taking a root, however, adjust the number so that the exponent of 10 divided by the root produces a whole number. Adjust the exponent in the direction that will leave the coefficient as a number greater than 1 (to avoid mistakes).

EXAMPLE C-9

Carry out the following calculation.

$$\sqrt{2.9 \times 10^5}$$

SOLUTION

First adjust the number so that the exponent of 10 is divisible by 2.

$$\sqrt{2.9 \times 10^5} = \sqrt{29 \times 10^4} = \sqrt{29} \times \sqrt{10^4} = \sqrt{29} \times 10^{4/2}$$
$$= 5.4 \times 10^2$$

EXAMPLE C-10

Carry out the following calculation.

$$\sqrt[3]{6.9 \times 10^{-8}}$$

SOLUTION

Adjust the number so that the exponent of 10 is divisible by 3.

$$\sqrt[3]{6.9 \times 10^{-8}} = \sqrt[3]{69 \times 10^{-9}} = \sqrt[3]{69} \times \sqrt[3]{10^{-9}}$$
$$= 4.1 \times 10^{-9/3} = 4.1 \times 10^{-3}$$

Exercises

C-8   Carry out the following operations.

(a) $(6.6 \times 10^4)^2$

(b) $(0.7 \times 10^6)^3$

(c) $(1200 \times 10^{-5})^2$ (It will be easier to square if you change the number to $1.2 \times 10^?$ first.)

(d) $(0.035 \times 10^{-3})^3$

(e) $(0.7 \times 10^7)^4$

C-9   Take the following roots. Approximate the answer if necessary.

(a) $\sqrt{36 \times 10^4}$ (d) $\sqrt[3]{1.6 \times 10^5}$

(b) $\sqrt[3]{27 \times 10^{12}}$ (e) $\sqrt{81 \times 10^{-7}}$

(c) $\sqrt{64 \times 10^9}$ (f) $\sqrt{180 \times 10^{10}}$

Answers: C-8(a) 4.4×10^9 C-8(b) 3×10^{17} C-8(c) 1.4×10^{-4}
C-8(d) 4.3×10^{-14} C-8(e) 2×10^{27} C-9(a) 6.0×10^2
C-9(b) 3.0×10^4 C-9(c) 2.5×10^5 C-9(d) 54   C-9(e) 2.8×10^{-3}
C-9(f) 1.3×10^6

This section reviews the following problem-solving skills

1 Manipulation of units

2 The conversion factor

3 One-step conversions

4 Multiple-step conversions

APPENDIX

■

PROBLEM SOLVING

■

BY THE

■

FACTOR-LABEL

■

METHOD

■

(DIMENSIONAL

■

ANALYSIS)

■

D

Two of the biggest stumbling blocks to good performance in chemistry are the lack of ability to memorize and the lack of problem-solving skills. The purpose of this section is to supplement Chapter 2 with a more detailed introduction to problem solving. Here we take the time to develop the necessary tools from the basics to simple problems and finally to more challenging problems.

A large number of problems in basic chemistry fit into the category of *conversion* problems. Whether it is from grams to milligrams, grams to pounds, moles to number of molecules, volume of gas to moles, or mass of solute to volume of solution, these problems can all be solved by the same method. Once the method is understood, the actual solution becomes routine. The problems have the same basic feature; a quantity is given in one unit of measurement and you are asked to convert this to another unit of measurement. To solve such problems we use the factor-label method, which employs one or more conversion factors.

The basic procedure for a one-step conversion is

$$\left[\begin{matrix} \text{given} \\ \text{quantity}* \end{matrix} \right] \text{old unit} \times \left[\begin{matrix} \text{conversion} \\ \text{factor} \\ \text{quantity} \end{matrix} \right] \frac{\text{new unit}}{\text{old unit}} = \left[\begin{matrix} \text{requested} \\ \text{quantity} \end{matrix} \right] \text{new unit}$$

If you are asked to convert 24 in. to feet, the answer comes easily because of your familiarity with the units and the relationship of inches to feet. To obtain the answer of 2 ft, most of us probably aren't even aware of what we did mathematically. In this appendix, however, we carefully show what we do with units as well as numbers even for the simple conversions. By getting into the habit of including units in the calculation, the more complex problems can be simplified.

Three types of units are employed in the examples that follow:

1	English units	mass — ounces, pounds distance — feet, yards, miles volume — quarts, gallons quantity — dozen, gross time — hours, minutes
2	Alien units	mass — gerbs distance — frims volume — moks quantity — nums, rims time — bleems
3	Chemical units metric or SI	mass — grams distance — meters, kilometers volume — milliliters, liters quantity — moles time — hours, minutes

---

* This refers to the numerical value of the quantity only.

There is a reason for this approach. Most of the math you need in general chemistry is quite simple and was reviewed in the previous appendixes. Often, however, the units and the awkward numbers cause confusion. It is the goal of this appendix to end the confusion.

First we will work examples and exercises in familiar English units, where you probably have a feeling for the correct answer before doing the actual calculation. By using the factor-label method in these problems we are in effect perfecting the use of a "decoder" that can be used with less familiar systems. Next we travel to an alien planet with our "decoder" to work problems with units that have no meaning to us. In these problems we certainly have no feeling for the approximate answer to a calculation, so there is no way to work the problems without using the "decoder." The factor-label method leads us confidently to the answers in this strange world.

Finally, we return to earth to work examples with the units used in chemistry. To the beginning student these units may also appear alien. However, until these units become familiar, we may rely on the "decoder" to lead us to the correct answers.

In summary, don't be overly concerned at this point if some of the units used in chemistry are unfamiliar. A calculation in the alien or chemical units is like flying an airplane on instruments. Experienced pilots trust the instruments to lead them safely to the destination when the ground can't be seen. Likewise, you can trust the units to lead you safely to the answer when you are not sure whether to multiply or divide. Understandably, pilots prefer to see where they are going with their own eyes. Likewise, you will eventually wish to see where you are going by becoming familiar with the units and calculations.

One can gain confidence in the factor-label method of problem solving by (1) correctly manipulating units, (2) understanding the origin and construction of conversion factors, and (3) successfully carrying out the calculations. We expand on these three points individually.

1 Manipulation of Units The essence of the factor-label method is that the units of the measurement serve as a guide to the correct mathematical manipulation. To do so, the units, as well as the numerical quantities, must be multiplied, divided, squared, or canceled in the calculation. If we arrive at the proper units for the answer, generally the answer is correct unless a mathematical error was made. Before we address the complete measurement including its numerical quantity, however, let's concentrate strictly on the unit or units of the measurement. In the following examples and problems our only goal is to simplify the units in the calculations to see how they would appear in the answer.

EXAMPLE D-1

What are the unit or units of the answers implied by the following calculations?

$$\frac{ft}{mile} \times \frac{mile}{min} \times \frac{min}{hr}$$

Cancel units that occur in both numerator and denominator.

$$\frac{ft}{\cancel{mile}} \times \frac{\cancel{mile}}{\cancel{min}} \times \frac{\cancel{min}}{hr} = \frac{ft}{hr}$$

EXAMPLE D-2

$$\frac{frim^3/mok \times 1/bleem}{frim^2/mok}$$

Rearrange into one quotient and cancel (see Appendix A for manipulation of fractions).

$$\frac{\overset{1}{\cancel{frim^3}} \times \cancel{mok}}{\cancel{frim^2} \times \cancel{mok} \times bleem} = \frac{frim}{bleem}$$

EXAMPLE D-3

$$\frac{\dfrac{L \cdot atm}{K \cdot mol} \times K}{atm}$$

Rearrange into one quotient and cancel.

$$\frac{L \times \cancel{atm} \times \cancel{K}}{\cancel{K} \times mol \times \cancel{atm}} = \frac{L}{mol}$$

Exercises

In the following exercises, write the units of the answers implied from the calculations.

D-1 $\dfrac{ft}{mile} \times lb \times mile$

D-2 $\dfrac{(mile/hr) \times (ft/mile) \times (hr/min)}{sec/min}$

D-3 $\dfrac{(\text{rim/num}) \times (1/\text{bleem})}{1/\text{num}}$

D-4 $\dfrac{(\text{gerb/frim}^3 \times \text{mok} \times \text{frim}}{\text{mok/frim}}$

D-5 $\dfrac{g \times \dfrac{L \cdot \text{torr}}{K \cdot \text{mol}} \times K}{(L \cdot \text{torr})}$

D-6 $\dfrac{1/(\text{atm} \cdot L)}{\text{mol} \times \dfrac{L \cdot \text{atm}}{K \cdot \text{mol}}}$

D-7 $\dfrac{\text{coulomb} \times \text{mol/F}}{(\text{mol/g}) \times \text{coulomb}}$

Answers: **D-1** ft · lb   **D-2** ft/sec   **D-3** rim/bleem   **D-4** gerb/frim
D-5 g/mol   **D-6** K/L² · atm²   **D-7** g/F

The main lesson to be learned from these examples and exercises is to be sure that the units in the calculation lead to the units desired in the answer. For example, if the answer should be mL, but the units come out to be g²/mL, the units indicate that you have made a mistake. Be careful in your manipulation of units to avoid *forcing* a cancellation, such as

$$\frac{g}{\cancel{mL}} \times \frac{1}{\cancel{mL}} = g \quad \text{(wrong)}$$

$$\frac{g}{\cancel{mL}} \times \cancel{mL} = g \quad \text{(right)}$$

2 The Conversion Factor   If we were to make a trip of about 180 miles in a car traveling about 60 miles per hour, most of us know that we would be in the car about 3 hours. We made this quick calculation thanks to a conversion factor that related the miles traveled to the time elapsed. Conversion factors quantitatively relate one unit of measurement to that of another. They may relate units in the same system of measurement such as length (m and km) or they may relate units in entirely different systems such as mass (g) and volume (mL). The relationships that are used as conversion factors originate in two ways:

1   From an *exact* definition (e.g., 10^3 m = 1 km, 12 in. = 1 ft, 60 min = 1 hr): These are usually (but not always) relationships within one system of measurement. Exact relationships have unlimited numbers of significant figures.

2   From a measurement (e.g., 1 lb = 454 g, 1.00 mL has a mass of 4.64 g, 1.00 lb costs $1.75). These are usually between different systems of measurement. The precision of the relationship is indicated by the number of significant figures shown.

Conversion factors are constructed by expressing the equality or equivalency relationship in fraction or *factor* form. For example, 12 in. = 1 ft can be expressed as

$$\frac{12 \text{ in.}}{1 \text{ ft}} \quad \text{or} \quad \frac{1 \text{ ft}}{12 \text{ in.}}$$

(This reads as 12 in. "per" foot or as 1 ft "per" 12 in.) and 1.00 mL weighs 4.64 g as

$$\frac{1.00 \text{ mL}}{4.64 \text{ g}} \quad \text{or} \quad \frac{4.64 \text{ g}}{1.00 \text{ mL}}$$

These are sometimes called "unit factors" because they relate a quantity in one measurement to "one" of another. When unit factors from a measurement are shown, it is implied that the "one" is known to as many significant figures as the numerical quantity. Also, in cases where the "one" is in the denominator, it is often neither shown nor read. Thus the relationship between mL and g shown above is simplified in this text as

$$\frac{1 \text{ mL}}{4.64 \text{ g}} \quad \text{or} \quad \frac{4.64 \text{ g}}{\text{mL}} \quad \text{or} \quad 4.64 \text{ g/mL} \quad (4.64 \text{ g "per" mL})$$

Many of the conversion factors that are used in calculations are in the form of unity factors or their reciprocals. As we will see, however, this is more a matter of convenience than necessity. Still, it is often easier for beginning students of chemistry to understand the problem when they can relate a certain quantity to one unit of the other. Therefore, it is worth our time to practice the construction of unit factors from measurements that relate the two quantities.

EXAMPLE D-4

If 0.250 miles is found to equal 1320 ft, how many feet are there per mile?

SOLUTION

The problem tells us that 1320 ft = 0.250 mile. We put the relationship in factor form and carry out the math.

$$\frac{1320 \text{ ft}}{0.250 \text{ mile}} = \underline{\underline{5280 \text{ ft/mile}}}$$

EXAMPLE D-5

A substance from the alien planet has a mass of 488 gerbs and has a volume of 14.2 moks. How many gerbs are there per mok?

SOLUTION

$$488 \text{ gerbs} = 14.2 \text{ moks}$$

$$\frac{488 \text{ gerbs}}{14.2 \text{ moks}} = \underline{34.4 \text{ gerbs/mok}}$$

EXAMPLE D-6

A 0.785-mole (abbreviated mol) quantity of a compound has a mass of 175 g. What is the mass in grams per mole?

SOLUTION

$$175 \text{ g} = 0.785 \text{ mol}$$

$$\frac{175 \text{ g}}{0.785 \text{ mol}} = \underline{223 \text{ g/mol}}$$

Exercises

In the following problems, include the units of the answer with the number where appropriate.

D-8 The unit price of groceries is sometimes listed in cost per ounce. Which of the two has the smallest cost per ounce: 16 oz of baked beans costing $1.45, or 26 oz costing $2.10?

D-9 An 82.3-doz quantity of oranges weighs 247 lb. What is the weight per dozen oranges? How many dozen oranges are there per pound?

D-10 If two men can dig an 8-ft trench in 2.5 hr, what is the number of feet dug per man per hour? (The units will be ft/man · hr.)

D-11 There are 1480 years in 3.25 bleems. How many bleems are there per year?

D-12 There are 8.61×10^5 objects in 0.574 nums. How many objects are there per num?

D-13 A gas has a volume of 146 L and a mass of 48.5 g. What is the mass per liter? What is the volume per gram?

D-14 There are 8.85×10^5 coulombs in 9.17 faradays. How many coulombs are there per faraday?

D-15 A quantity of 6.02×10^{23} atoms of tungsten has a mass of 184 g. What is the mass per atom? How many atoms per gram?

Answers: **D-8** The 16-oz can costs \$0.091/oz; the 26-oz can costs \$0.081/oz   **D-9** 3.00 lb/doz, 0.333 doz/lb   **D-10** 1.6 ft/man · hr
D-11 2.20×10^{-3} bleems/year   **D-12** 1.50×10^{6} objects/num
D-13 0.332 g/L, 3.01 L/g   **D-14** 9.65×10^{4} coulombs/faraday
D-15 3.06×10^{-22} g/atom, 3.27×10^{21} atoms/g

3 One-Step Conversions   One drawback to the factor-label method of problem solving is that it can become too mechanical. If this happens, one may not think about the answer; that is, "Is it reasonable?" For example, if one desired to purchase a dozen cheap but tasty hamburgers at \$.60 each, a quick multiplication tells you that you need \$7.20. If you had divided 12 by \$0.60 rather than multiply you would have needed \$20 which sounds like quite a bit for what you got. The point is — we need to think about the answer. It is understood at this point, however, that you may not be able to approximate the answer when working with chemical units. Eventually you will. For example, if the mass of one tiny atom is requested, the answer is very small ($\sim 10^{-23}$ g). An answer of 10 g or even 10^{23} g (about the mass of the moon) is out of the money and should be so recognized. Our purpose here is to set you at ease about the calculation itself. When you study the particular topic, you should answer the very important question: "Does this answer make sense?"

In the following problems we will list what is given and requested, a shorthand procedure, the conversion factor properly set up to follow the procedure, and finally the solution. This effort pays off for the more complex problems later.

Remember that, in most calculations, the conversion factor is set up so that the old unit (which is given) is in the denominator and cancels. The new unit (which is requested) is in the numerator and remains in the answer. Obviously, just the opposite would hold if we wished to make a conversion involving the denominator (e.g., miles/hr to miles/min). This all assumes a simple one-step conversion where a conversion factor is available that relates a given measurement to one that is requested. Such conversions are illustrated by the following examples and exercises.

EXAMPLE D-7

If a car travels 55 miles/hr, how many miles does it travel in 8.5 hr?

Given: 8.5 hr
Requested: ____?____ miles
Procedure: hr → miles
Conversion factor: 55 miles/hr

SOLUTION

$$8.5 \; \text{hr} \times \frac{55 \; \text{miles}}{\text{hr}} = \underline{\underline{470 \; \text{miles}}}$$

EXAMPLE D-8

If the same car as in Example D-7 travels 785 miles, how many hours does it take?

Given: 785 miles

Requested: ___?___ hr

Procedure: miles → hr

Conversion factor: $\dfrac{1 \text{ hr}}{55 \text{ miles}}$

SOLUTION

$$785 \text{ miles} \times \frac{1 \text{ hr}}{55 \text{ miles}} = \underline{\underline{14 \text{ hr}}}$$

Note in Example D-7 that a small number (8.5) is converted to a large number (470). In such a case, the conversion factor must be an improper fraction (larger than one). The opposite is true in Example D-8, where a large number (785) is converted to a small number (14). In such a conversion, the factor must be a proper fraction (less than one). This is where having a feeling for the magnitude of the answer is helpful. We want the conversion factor to change the magnitude of what is given in the appropriate direction.

EXAMPLE D-9

A certain type of chain costs $0.55/ft. What does it cost per mile? (5280 ft = 1 mile)

Given: $0.55/ft

Requested: $\dfrac{\$}{\text{mile}}$

Procedure: $\dfrac{\$}{\text{ft}} \longrightarrow \dfrac{\$}{\text{mile}}$

Conversion factor: 5280 ft/mile

In this case, the unit of the denominator is to be changed. The factor should therefore have the given unit in the *numerator* and the requested unit in the *denominator*.

SOLUTION

$$\frac{\$0.55}{\text{ft}} \times \frac{5280 \text{ ft}}{\text{mile}} = \underline{\underline{\$2904/\text{mile}}}$$

One popular exception to the proper use of significant figures is where money is involved. Traditionally, price is expressed to the nearest cent regardless of how it is obtained. An argument to a salesperson in the above example that the price should only be $2900 because of significant figures would not go very far.

EXAMPLE D-10

A section of cable has a mass of 18.5 gerbs/frim. How many frims are there in 97.2 gerbs?

Given: 97.2 gerbs

Requested: ___?___ frims

Procedure: gerbs → frims

Conversion factor: $\dfrac{1 \text{ frim}}{18.5 \text{ gerb}}$

SOLUTION

$$97.2 \text{ gerb} \times \frac{1 \text{ frim}}{18.5 \text{ gerb}} = \underline{5.25 \text{ frims}}$$

EXAMPLE D-11

A spaceship can travel 1.85×10^6 frims in 0.240 bleems. How far can it travel in 0.525 bleem?

Given: 0.525 bleem

Requested: ___?___ frims

Procedure: bleems → frims

Conversion factor: $\dfrac{1.85 \times 10^6 \text{ frims}}{0.240 \text{ bleem}}$

SOLUTION

$$0.525 \text{ bleem} \times \frac{1.85 \times 10^6 \text{ frims}}{0.240 \text{ bleem}} = \underline{4.05 \times 10^6 \text{ frims}}$$

In the previous example, the conversion factor was not a unity factor. That is, the factor did not relate frims to "one" bleem. It is quite appropriate in conversion problems to use conversion factors that are not unity factors. Still, many students feel more comfortable, at least initially, in using unit factors in conversions. If so, the problem would be solved as follows.

$$\frac{1.85\ 10^6\ \text{frims}}{0.240\ \text{bleem}} = 7.71 \times 10^6\ \text{frim/bleem}$$

$$0.525\ \cancel{\text{bleem}} \times \frac{7.71 \times 10^6\ \text{frim}}{\cancel{\text{bleem}}} = \underline{\underline{4.05 \times 10^6\ \text{frims}}}$$

EXAMPLE D-12

What is the mass of a 3.25-mol quantity of a substance if 1.00 mol has a mass of 80.0 g?

Given: 3.25 mol

Requested: ___?___ g

Procedure: mol → g

Conversion factor: 80.0 g/mol

SOLUTION

$$3.25\ \cancel{\text{mol}} \times \frac{80.0\ \text{g}}{\cancel{\text{mol}}} = \underline{\underline{260\ \text{g}}}$$

EXAMPLE D-13

If there are 6.02×10^{23} molecules/mol, how many moles do 5.65×10^{24} molecules represent?

Given: 5.65×10^{24} molecules

Requested: ___?___ mol

Procedure: molecules → mol

Conversion factor: $\dfrac{1\ \text{mol}}{6.02 \times 10^{23}\ \text{molecules}}$

SOLUTION

$$5.65 \times 10^{24}\ \cancel{\text{molecules}} \times \frac{1\ \text{mol}}{6.02 \times 10^{23}\ \cancel{\text{molecules}}} = \underline{\underline{9.39\ \text{mol}}}$$

Exercises

Practice one-step conversions on the following problems.

D-16 If the unit price of chili is $0.18/oz, what is the cost of a 12-oz can of chili? How many ounces can you buy at this rate for $1.75?

D-17   If a bullet travels 1860 ft/sec. how far does it travel in 4.50 sec?

D-18   If there are 4.55 lb of corn per dozen ears, how many dozen ears of corn are in 265 lb?

D-19   There are 1.50×10^6 commas per num (remember, a num is a unit of quantity, like dozen, gross, or mole). How many nums are 83×10^8 commas? How many commas are in 0.420 nums?

D-20   A resident of an alien planet can lose 0.475 gerb in 2.15 bleems when on a diet. How many gerbs can be lost in 0.880 bleem? How many bleems would it take to lose 187 gerbs?

D-21   If 18.0 g of a gas has a volume of 22.4 L under certain conditions, what is the mass of the gas that would have a volume of 1.35 L?

D-22   A pressure of 1 atm is equal to 760 torr. How many torr are there in 0.66 atm?

D-23   There are 6.02×10^{23} molecules/mol. How many molecules are present in 0.212 mol?

D-24   Using the relationship from the previous problem, determine how many molecules there are per gram of a particular substance if 1.00 mol has a mass of 18.0 g.

D-25   There are 96,500 coulombs/faraday. How many faradays are equivalent to 4820 coulombs?

Answers: **D-16** $2.16, 9.72 oz   **D-17** 8370 ft   **D-18** 58.2 doz
D-19 5.5×10^3 nums, 6.30×10^5 commas   **D-20** 0.194 gerbs, 846 bleems   **D-21** 1.08 g   **D-22** 500 torr   **D-23** 1.28×10^{23} molecules
D-24 3.34×10^{22} molecules/g   **D-25** 0.0499 faraday

4 Multiple-Step Conversions   Hopefully, the one-step conversions did not cause a major problem for you despite the unfamiliar units. In any case, the previous exercises included some fairly sophisticated problems encountered in chemistry. We are now ready to advance to some more complex problems. In what follows a direct relationship between the units of what's given and what's requested is not available. In these problems, conversions must be done in two or more steps, with each step requiring a separate conversion factor. The general procedure for a two-step conversion is as follows:

$$\left[\begin{array}{c} \text{given} \\ \text{quantity} \end{array} \right] \cancel{\text{old unit}} \times \left[\begin{array}{c} \text{conversion factor} \\ \text{quantity (a)} \end{array} \right] \frac{\text{unit X}}{\cancel{\text{old unit}}}$$

$$\times \left[\begin{array}{c} \text{conversion factor} \\ \text{quantity (b)} \end{array} \right] \frac{\text{new unit}}{\cancel{\text{unit X}}} = \left[\begin{array}{c} \text{requested} \\ \text{quantity} \end{array} \right] \text{new unit}$$

Note that the conversion factor for each step leaves a new unit in the numerator while canceling an old unit with the denominator. For example, assume that a procedure is planned as follows for a problem in which mass (in lb) is given and the cost (in $) is requested. If conversion factors are available that relate mass (lb) to quantity (dozen) and quantity (dozen) to cost ($), the procedure is

$$\text{lb} \xrightarrow{\text{(a)}} \text{doz} \xrightarrow{\text{(b)}} \$$$

Step (a) converts lb to doz.

$$\cancel{\text{lb}} \times \frac{\text{doz}}{\cancel{\text{lb}}} = \underline{\qquad} \text{doz}$$

Step (b) converts doz to $.

$$\cancel{\text{lb}} \times \frac{\cancel{\text{doz}}}{\cancel{\text{lb}}} \times \frac{\$}{\cancel{\text{doz}}} = \$ \underline{\qquad}$$

In the examples and exercises that follow, experiment with the units before doing the math. Work with the units of what's given and the units of the conversion factors so that undesired units cancel, leaving only the requested unit or units in the answer. When you are sure of your setup, then do the math. Don't be discouraged if you find the problems with alien and chemical units quite challenging. They are meant to be. The purpose is to present problems for which it is almost necessary to use the factor-label method in which the units lead the way to the answer. If you are able to work your way through these exercises successfully, you should be ready for many if not most quantitative problems facing you in general chemistry. Many of the relationships used in the following problems are found in Tables 2-3 and 2-4.

EXAMPLE D-14

If a train travels 75.0 miles/hr, how many miles does it travel in 416 min?

Given: 416 min
Requested: ___?___ miles

Procedure: min $\xrightarrow{\text{(a)}}$ hr $\xrightarrow{\text{(b)}}$ miles

Conversion factors: (a) $\dfrac{1 \text{ hr}}{60 \text{ min}}$ (b) $\dfrac{75.0 \text{ miles}}{\text{hr}}$

SOLUTION

$$416 \,\cancel{\text{min}} \times \frac{1 \,\cancel{\text{hr}}}{60 \,\cancel{\text{min}}} \times \frac{75.0 \text{ miles}}{\cancel{\text{hr}}} = \underline{\underline{520 \text{ miles}}}$$

EXAMPLE D-15

One gallon of gas costs $1.02. What is the cost of 48.0 L (the volume unit used in Mexico and Canada?

Given: 48.0 L
Requested: $ _____?_____

Procedure: L $\xrightarrow{\text{(a)}}$ qts $\xrightarrow{\text{(b)}}$ gal $\xrightarrow{\text{(c)}}$ $
Conversion factors:

$$\text{(a) } \frac{1.06 \text{ qt}}{\text{L}} \quad \text{(b) } \frac{1 \text{ gal}}{4 \text{ qt}} \quad \text{(c) } \frac{\$1.02}{\text{gal}}$$

SOLUTION

$$48.0 \cancel{\text{L}} \times \frac{1.06 \cancel{\text{qt}}}{\cancel{\text{L}}} \times \frac{1 \cancel{\text{gal}}}{4 \cancel{\text{qt}}} \times \frac{\$1.02}{\cancel{\text{gal}}} = \underline{\underline{\$12.97}}$$

EXAMPLE D-16

The speed limit of a spaceship is 55.0 frims/millibleem (10^3 millibleems per bleem). How many frims can a spaceship travel in 0.205 bleem?

Given: 0.205 bleem
Requested: _____?_____ frims

Procedure: bleems $\xrightarrow{\text{(a)}}$ millibleems $\xrightarrow{\text{(b)}}$ frims
Conversion factors:

$$\text{(a) } \frac{10^3 \text{ millibleems}}{\text{bleem}} \quad \text{(b) } \frac{55.0 \text{ frims}}{\text{millibleem}}$$

SOLUTION

$$0.205 \cancel{\text{bleem}} \times \frac{10^3 \cancel{\text{millibleems}}}{\cancel{\text{bleem}}} \times \frac{55.0 \text{ frims}}{\cancel{\text{millibleem}}} = \underline{\underline{11,300 \text{ frims}}}$$

EXAMPLE D-17

There are 1.50×10^6 numecules in 1.00 num. If the mass of 1.00 num (the "num mass") of a substance is 24.0 gerbs, what is the mass in gerbs of 7.50×10^5 numecules of the substance?

Given: 7.50×10^5 numecules

Requested: ____?____ gerbs

Procedure: numecules $\xrightarrow{(a)}$ nums $\xrightarrow{(b)}$ gerbs

Conversion factors:

$$\text{(a)} \quad \frac{1 \text{ num}}{1.50 \times 10^6 \text{ numecule}} \qquad \text{(b)} \quad \frac{24.0 \text{ gerbs}}{\text{num}}$$

SOLUTION

$$7.50 \times 10^5 \text{ \sout{numecule}} \times \frac{1 \text{ \sout{num}}}{1.50 \times 10^6 \text{ \sout{numecule}}} \times \frac{24.0 \text{ gerbs}}{\text{\sout{num}}}$$

$$= \underline{\underline{12.0 \text{ gerbs}}}$$

EXAMPLE D-18

There are 6.02×10^{23} molecules in 1.00 mol. If the mass of 1.00 mol (the "molar mass") of carbon dioxide is 44.0 g, what is the mass of 8.46×10^{22} molecules of carbon dioxide?

Given: 8.46×10^{22} molecules

Requested: ____?____ g

Procedure: molecules $\xrightarrow{(a)}$ mol $\xrightarrow{(b)}$ g

Conversion factors:

$$\text{(a)} \quad \frac{1 \text{ mol}}{6.02 \times 10^{23} \text{ molecules}} \qquad \text{(b)} \quad \frac{44.0 \text{ g}}{\text{mol}}$$

SOLUTION

$$8.46 \times 10^{22} \text{ \sout{molecules}} \times \frac{1 \text{ \sout{mol}}}{6.02 \times 10^{23} \text{ \sout{molecules}}} \times \frac{44.0 \text{ g}}{\text{\sout{mol}}} = \underline{\underline{6.18 \text{ g}}}$$

EXAMPLE D-19

Using the information given in Example D-18, determine how many molecules are in 892 g of carbon dioxide.

Given: 892 g

Requested: ____?____ molecules

Procedure: g $\xrightarrow{(a)}$ mol $\xrightarrow{(b)}$ molecules

Conversion factors:

$$\text{(a)} \ \frac{1 \text{ mol}}{44.0 \text{ g}} \quad \text{(b)} \ \frac{6.02 \times 10^{23} \text{ molecules}}{\text{mol}}$$

SOLUTION

$$892 \ \cancel{g} \times \frac{1 \ \cancel{\text{mol}}}{44.0 \ \cancel{g}} \times \frac{6.02 \times 10^{23} \text{ molecules}}{\cancel{\text{mol}}} = \underline{1.22 \times 10^{25} \text{ molecules}}$$

Exercises

The following problems require two or more conversions to obtain the answer.

D-26   If a train travels at a speed of 85 miles/hr, how many hours does it take to travel 17,000 ft? How many yards can it travel in 37 min? (1760 yd = 1 mile.)

D-27   A certain size of nail costs $1.25/lb. What is the cost of 3.25 kg of nails?

D-28   Two men can push a car 110 yd in 15 min. How many minutes would it take three men to push a car 250 yd at the same rate?

D-29   If a car travels 55.0 miles/hr, what is its speed in feet per second?

D-30   Low-tar cigarettes contain 11.0 mg of tar per cigarette. If all the tar gets into the lungs, how many packages of cigarettes (20 cigarettes per package) would have to be smoked to produce 0.500 lb of tar? If a person smoked two packages per day, how many years would it take to accumulate 0.500 lb of tar?

D-31   If a car gets 24.5 miles to the gallon of gasoline and gasoline costs $1.22/gal, what would it cost to drive 350 km?

D-32   A Martian fruit has a mass of 0.820 gerb/rim. What is the volume (in moks) of 425 rims if there are 115 gerbs per mok?

D-33   If there are 1.50×10^6 numecules/num, how many numecules are in 1 rim if there are 135 rims/num?

D-34   If numecules have a mass of 118 gerbs/num, how many gerbs are in 4.50×10^8 numecules (1.50×10^6 numecules/num)? How many numecules are in 45.5 gerbs?

D-35   A liquid solution from the alien planet contains 0.250 num of a substance per mok. If the substance has a mass of 164 gerbs/num, how many gerbs of the substance are in 152 moks of solution?

D-36   A liquid solution on earth contains 0.250 mol of sulfuric acid per liter. How many grams of sulfuric acid are in 6.75 L of solution if the sulfuric acid has a mass of 98.0 g/mol? What is the volume (in liters) occupied by 15.5 g of sulfuric acid?

D-37   If there are 6.02×10^{23} molecules/mol, how many molecules of sulfuric acid are in the two liquid solutions in Problem D-36?

D-38   How many grams of silver are deposited by 46,200 coulombs of electricity if 1 faraday is equal to 96,500 coulombs and 1 faraday deposits 108 g of silver?

D-39   If 1 mol of a gas occupies 22.4 L under certain conditions, how many liters will 75.0 g of this gas occupy if the gas has a mass 60.0 g/mol?

D-40   One atmosphere pressure is equal to 760.0 torr. What is 355.0 torr expressed in kilopascals (kPa) if 1 atm is equal to 101.4 kPa?

D-41   There are 3 mol of hydrogen atoms per mole of phosphoric acid. How many moles of hydrogen atoms are in 16.0 g of phosphoric acid if 1 mol of phosphoric acid weighs 98.0 g?

Answers: **D-26** 0.038 hr, 9.2×10^4 yards   **D-27** $8.94   **D-28** 23 min
D-29 80.7 ft/sec   **D-30** 1030 packages, 1.41 years   **D-31** $10.83
D-32 3.03 moks   **D-33** 1.11×10^4 numecules/rim   **D-34** 3.54×10^4
gerbs, 5.78×10^5 numecules   **D-35** 6230 gerbs   **D-36** 165 g, 0.633 L
D-37 1.02×10^{24} molecules, 9.52×10^{22} molecules   **D-38** 51.7 g
D-39 28.0 L   **D-40** 47.36 kPa   **D-41** 0.490 mol of H atoms

This section reviews the following common logarithm skills.

APPENDIX

LOGARITHMS

E

In Appendix C we learned that scientific notation is particularly useful in expressing very large or very small numbers. In certain areas of chemistry, however, such as in the expression of H_3O^+ concentration, even the repeated use of scientific notation becomes tedious. In this situation, it is convenient to express the concentration as simply the *exponent of 10. The exponent to which 10 must be raised to give a certain number is called its* **common logarithm.** With common logarithms (or just logs) it is possible to express the coefficient and the exponent of 10 as one number.

Since logarithms are simply exponents of 10, logs of numbers such as 100 can be easily determined. Note that 100 can be expressed as 10^2, so that the log of 100 is simply 2. Other examples of simple logs are

$$1 = 10^0 \qquad \log 1 = 0 \qquad 0.1 = 10^{-1} \qquad \log 0.1 = -1$$
$$10 = 10^1 \qquad \log 10 = 1 \qquad 0.01 = 10^{-2} \qquad \log 0.01 = -2$$
$$100 = 10^2 \qquad \log 100 = 2 \qquad 0.001 = 10^{-3} \qquad \log 0.001 = -3$$
$$1000 = 10^3 \qquad \log 1000 = 3 \qquad 0.0001 = 10^{-4} \qquad \log 0.0001 = -4$$

The same mathematical operations that apply to exponents also apply to logs. For example, consider the multiplication and division of exponents of 10 and logs.

	Exponents	Logarithms
Multiplication	$10^4 \times 10^3 = 10^{4+3}$	$\log(10^4 \times 10^3) = \log 10^4 + \log 10^3$
	$= 10^7$	$= 4 + 3 = 7$
Division	$\dfrac{10^{10}}{10^4} = 10^{10-4} = 10^6$	$\log \dfrac{10^{10}}{10^4} = \log 10^{10} - \log 10^4$
		$= 10 - 4 = 6$

By definition, the log of a number whose coefficient is exactly one is simply the value of the exponent. When the coefficient is not exactly "one" (which means the exponent is fractional) then some aid is required to find the value. Most modern hand calculators have a log function and this, of course, is the most convenient method of finding logs. This method is described in Appendix G. If a calculator does not have a log function or a calculator is not available, log tables are available and simple to use if one understands the meaning of a logarithm. The following describes the use of a simple log table to find logs. *It is helpful to work through these sections even if you have an appropriate calculator so as to appreciate the function and meaning of a logarithm.*

1 Taking the Log of a Number between 1 and 10   Note that $\log 1 = 0$ and $\log 10 = 1$. The log of a number between 1 and 10 is therefore between 0 and 1. The log of two-significant-figure numbers can be found in a two-place log table such as shown in this section. Three-, four-, and even five-place log tables are available for more precise calculations. The two-place table is suitable for our purposes. The log of a number between 1 and 10 can be read directly from the body of the table. The first integer in the number is found in the vertical column and the second (the tenths) is located in the horizontal column. The log is where the two intersect. Since we are taking logs of two-significant-figure numbers, the log should be expressed to two significant figures.

EXAMPLE E-1

What is the log of (a) 5.8, (b) 3.7, and (c) 1.1?

SOLUTION

The answer is between 0 and 1 and is expressed as

$$0.\text{log}$$

$$\log 5.8 = \underline{0.76} \qquad \text{(5 down and 8 across)}$$

$$\log 3.7 = \underline{0.57} \qquad \text{(3 down and 7 across)}$$

$$\log 1.1 = \underline{0.04} \qquad \text{(1 down and 1 across)}$$

---

2 **Taking the Logs of Numbers Less than 1 and Greater than 10**   To find the log of a number less than 1 or greater than 10, follow these steps:

1   Write the number in standard scientific notation (one digit to the left of the decimal point in the coefficient).
2   Look up the log of the coefficient and write as described in Section 1.
3   The value of the log of the coefficient is added to the value of the exponent of 10. That is a result of the general rule

$$\log(A \times B) = \log A + \log B$$
$$\log(A \times 10^B) = \log A + \log 10^B = (\log A) + B$$

TABLE E-1   Two-Place Logarithms

Tenths →
Row
Ones ↓
Column

	0.0	0.1	0.2	0.3	0.4	0.5	0.6	0.7	0.8	0.9
1	0.00	0.04	0.08	0.11	0.15	0.18	0.20	0.23	0.26	0.28
2	0.30	0.32	0.34	0.36	0.38	0.40	0.41	0.43	0.45	0.46
3	0.48	0.49	0.51	0.52	0.53	0.54	0.56	0.57	0.58	0.59
4	0.60	0.61	0.62	0.63	0.64	0.65	0.66	0.67	0.68	0.69
5	0.70	0.71	0.72	0.72	0.73	0.74	0.75	0.76	0.76	0.77
6	0.78	0.79	0.79	0.80	0.81	0.81	0.82	0.83	0.83	0.84
7	0.85	0.85	0.86	0.86	0.87	0.88	0.88	0.89	0.89	0.90
8	0.90	0.91	0.91	0.92	0.92	0.93	0.93	0.94	0.94	0.95
9	0.95	0.96	0.96	0.97	0.97	0.98	0.98	0.99	0.99	1.00

The sum of the two numbers (the exponent plus the log of the coefficient) has two parts. *The number to the right of the decimal point is called the **mantissa**.* As mentioned before, the mantissa is expressed to the same number of decimal places as there are significant figures in the coefficient. *The whole number to the left of the decimal point is called the **characteristic**.* For example,

$$\log(5.7 \times 10^6) = \log 5.7 + \log 10^6 = 0.76 + 6$$

$$= \boxed{6} . \boxed{76}$$

characteristic mantissa

EXAMPLE E-2

What are the logs of the following numbers?

(a) 4.7×10^3

$$\log(4.7 \times 10^3) = \log 4.7 + \log 10^3$$
$$= 0.67 + 3 = \underline{3.67}$$

(b) 15

$$\log 15 = \log(1.5 \times 10^1)$$
$$= \log 1.5 + \log 10^1$$
$$= 0.18 + 1 = \underline{1.18}$$

(c) 0.066

$$\log 0.066 = \log(6.6 \times 10^{-2})$$
$$= \log 6.6 + \log 10^{-2}$$
$$= 0.82 + (-2) = \underline{-1.18}$$

Note that logs of numbers less than one have negative values.

(d) 8.7×10^{-7}

$$\log(8.7 \times 10^{-7}) = \log 8.7 + \log 10^{-7}$$
$$= 0.94 + (-7)$$
$$= \underline{-6.06}$$

Exercises

E-1 Find the logs of the following numbers.

(a) 7.4   (e) 85   (i) 4.1×10^{-12} (m) 160×10^6
(b) 5.2   (f) 1700   (j) 4.5   (n) 0.60×10^{-7}

(c) 0.87   (g) 7.3×10^4 (k) 0.057×10^8

(d) 0.087   (h) 7.3×10^{-4} (l) 32×10^{-5}

E-2   pH is defined as the negative of the log of a number [e.g., $-\log 10^{-2} = -(-2) = 2$]. Find the negative logs of the following.

(a) 8.5×10^{-6} (c) 71×10^{-10} (e) 1.1

(b) 3.4×10^{-11} (d) 0.217×10^{-3}

Answers: E-1(a) 0.87   E-1(b) 0.72   E-1(c) -0.06 E-1(d) -1.06
E-1(e) 1.93   E-1(f) 3.23   E-1(g) 4.86   E-1(h) -3.14
E-1(i) -11.39 E-1(j) 0.65   E-1(k) 6.76   E-1(l) -3.49
E-1(m) 8.20   E-1(n) -7.22 E-2(a) 5.07   E-2(b) 10.47   E-2(c) 8.15
E-2(d) 3.66   E-2(e) -0.04

3 Finding the Antilog of a Positive Log   Not only should we be able to find the log of a number, but we must be able to find *the number whose log has a certain value*. This is called the **antilog** of a log. The procedure is simply the reverse of finding the log. Remember, we are asking the question, "What is the number whose exponent of 10 is a certain value?" For example, What is the antilog (x) of 2?

$$\log x = 2$$
$$x = \underline{\underline{10^2}} \qquad \text{since } \log 10^2 = 2$$

When the log is between 0 and 1, the antilog is between 1 and 10. (This is just the reverse of the statement in Section 1.) To find the antilog, locate the given log in the *body* of the log table. Find the first number on the left margin and the second on the top that correspond to this log. Write the two numbers you found with one digit to the left of the decimal point.

EXAMPLE E-3

What are the antilogs of the following?

(a) 0.84

$$\text{antilog } 0.84 = \underline{\underline{6.9}}$$

(b) 0.23

$$\text{antilog } 0.23 = \underline{\underline{1.7}}$$

(c) 0.96

$$\text{antilog } 0.96 = \underline{\underline{9.1}}$$

If the log is a number greater than one, divide the number into two parts: the mantissa, which lies to the right of the decimal, and the characteristic, which lies to the left. For example,

$$2.56 = 0.56 + 2$$
$$11.11 = 0.11 + 11$$

Take the antilog of the mantissa and express the number as explained above. This becomes the coefficient of the number. The characteristic then becomes the exponent of 10.

EXAMPLE E-4

Find the antilog of (a) 4.65 and (b) 3.89.

(a) $4.65 = 0.65 + 4$

$$\text{antilog } 0.65 = 4.5$$
$$\text{antilog } 4 = 10^4$$
$$= \underline{4.5 \times 10^4}$$

(b) $3.89 = 0.89 + 3$

$$\text{antilog } 0.89 = 7.8$$
$$\text{antilog } 3 = 10^3$$
$$= \underline{7.8 \times 10^3}$$

---

4 Finding the Antilog of a Negative Log   A negative log must be adjusted so that the mantissa is positive without changing the actual value of the log itself. For example, -0.18 is the same number as $(+0.82 - 1)$, and -3.28 is the same number as $(+0.72 - 4)$. To do this, first separate and write the mantissa and the characteristic as described before. *Add* 1 to the mantissa and *subtract* 1 from the characteristic. Note that if you add and subtract the same number from a quantity, you have not actually changed the number (e.g., $15 + 1 - 1 = 15$).

	Separate	Add 1 to mantissa	Subtract 1 from characteristic	
$-3.28 =$	$-0.28 - 3 =$	$(-0.28 + 1) -$	$3 - 1$	$= +0.72 - 4$
$-0.18 =$	$-0.18 - 0 =$	$(-0.18 + 1) -$	$0 - 1$	$= +0.82 - 1$

Locate the mantissa in the body of the log table as described before and locate the corresponding numbers in the left-hand column and the top. The negative characteristic becomes the negative exponent of 10.

EXAMPLE E-5

Find the following antilogs.

(a) -0.02

$$-0.02 = (1 - 0.02) - 1 = 0.98 - 1$$
$$\text{antilog } 0.98 = 9.5$$
$$\text{antilog } (-1) = 10^{-1}$$
$$= 9.5 \times 10^{-1} = \underline{0.95}$$

(b) -4.54

$$-4.54 = (1 - 0.54) - 5 = 0.46 - 5$$
$$\text{antilog } 0.46 = 2.9$$
$$\text{antilog } (-5) = 10^{-5}$$
$$= \underline{2.9 \times 10^{-5}}$$

Exercises

E-3 Find the antilogs of the following.

(a) 0.81   (c) -0.70 (e) 10.94   (g) -0.08 (i) 0.34
(b) 3.46   (d) -5.48 (f) -2.60 (h) 8.40   (j) -5.96

Answers: E-3(a) 6.5   E-3(b) 2.9×10^3 E-3(c) 0.20
E-3(d) 3.3×10^{-6} E-3(e) 8.7×10^{10} E-3(f) 2.5×10^{-3} E-3(g) 0.83
E-3(h) 2.5×10^8 E-3(i) 2.2   E-3(j) 1.1×10^{-6}

This section reviews the following graphical skills.

1   Trends, cycles, and graphs

2   Direct linear relationships

3   Nonlinear relationships

APPENDIX
■

GRAPHS
■

F

The graphical representation of data is important in all branches of science, social as well as natural. Trends and cycles in data may be obscure when presented in a table but become immediately apparent in graphical form. For example, consider a graph in which the average monthly temperature (represented on the vertical axis) is plotted as a function of the month (represented on the horizontal axis). The graph is constructed by marking the temperature for each month and drawing a smooth curve through the points. A quick look at the graph tells us what we already knew—the temperature cycles up and down throughout the seasons.

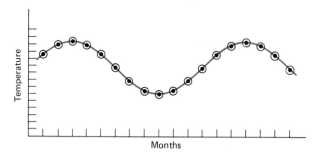

Some useful graphs of this nature are shown in Figures 1-8 and 5-5. In the first, the temperature of a liquid near and at its boiling point is graphed as it varies with time. The graph illustrates the discontinuity in temperature change as the liquid boils. In Figure 5-5, the two graphs illustrate the periodic nature of atomic radius and ionization energy as atomic number increases. These graphs illustrate both short-term and long-term trends.

1 Direct Linear Relationships   Other types of graphs can be useful in predicting data as well as simply illustrating measurements. For example, in Appendix B we discussed proportionalities. In a proportionality a variable (x) is related to another variable (y) by some mathematical operation (i.e., a power, a root, a reciprocal, etc.). A graph translates a proportionality into a line that illustrates the relationship of x to y. There are two types of graphs: linear and nonlinear. A linear graph is represented by a straight line and is a result of a direct, first-power proportionality between the variables x and y. This is expressed as

$$x \propto y$$

Such a graph is illustrated in Figure 10-6, which relates the volume of a gas to the Celsius temperature. Since the data points fall in a straight line, the line can be extended to provide data points beyond the limits of the experiments. (We say the line has been extrapolated.) Using a linear graph, as few as two experiments can be used to predict the results of other measurements.

The relationship between the volume (V) of a gas at constant pressure and the Kelvin temperature (T) can be expressed as

$$V \propto T$$

We will use this direct relationship to construct a linear graph. The following two experiments provide the information necessary to construct such a graph.

Experiment 1   At a temperature of 350 °C, the volume was 12.5 L.

Experiment 2   At a temperature of 250 °C, the volume was 10.5 L.

Actually, it is always risky to assume that only two experiments provide an accurate graph or give all of the needed information, but for simplicity we will assume that it is valid in this case. We will check the graph with an additional experiment later.

To record this information on a graph, we first need some graph paper with regularly spaced divisions. On the vertical axis (called the ordinate or y axis) we put temperature. Pick divisions on the graph paper that spread out the range of temperatures as much as possible. In our example, the temperatures range between 250 and 350 °C, or 100 degrees. Note that the two axes do not need to intersect at $x = y = 0$. We then plot the volume on the horizontal axis (called the abscissa or x axis). The volumes range between 12.5 and 10.5 L, or 2.0 L.

The graph is shown in Figure F-1. An ⊙ is marked at the locations of the two points from the experiments, and a straight line has been drawn between them.

We can now check the accuracy of our graph. A third experiment tells us that at $T = 300$ °C, $V = 11.5$ L. Refer back to the graph. Locate 300 °C on the ordinate and trace that line to where it intersects the vertical line representing the volume of 11.5 L. Since the point of intersection is right on the line, we have confirmation of the accuracy of the original plot. *Normally, four or five measurements are required to provide enough points on the graph so that a line can be drawn through or near as many points as possible.*

A straight-line graph can be expressed by the linear algebraic equation

$$y = mx + b$$

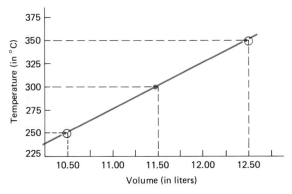

Figure F-1

where y = a value on the ordinate

x = the corresponding value on the abscissa

b = the point on the y axis or ordinate where the straight line intersects the axis

m = the slope of the line

The slope of a line is the ratio of the change on the ordinate to the change on the abscissa. It is determined by choosing two widely spaced points on the line (x_1, y_1 and x_2, y_2). The slope is the difference in y divided by the difference in x.

$$m = \frac{y_2 - y_1}{x_2 - x_1}$$

The larger the slope *for a particular graph,* the steeper the straight line. For the graph shown in Figure F-1, if we pick point 1 to be $x_1 = 10.50$, $y_1 = 250$ and point 2 to be $x_2 = 12.50$, $y_2 = 350$, then

$$m = \frac{350 - 250}{12.50 - 10.50} = \frac{100}{2.00} = 50.0$$

The value for b cannot be determined directly from Figure F-1. The value for b is found by determining the point on the y axis where the line intersects when $x = 0$. If we plotted the data on such a graph, we would find that $b = -273\ ^\circ$C. Therefore, the linear equation for this line is

$$y = 50.0x - 273$$

From this equation we can determine the value for y (the temperature) by substitution of any value for x (the volume).

Exercises

F-1 A sample of water was heated at a constant rate, and its temperature was recorded. The following results were obtained.

Experiment	Temp (°C)	Time (min)
1	10.0	0
2	20.0	4.5
3	30.0	7.5
4	55.0	17.8
5	85.0	30.0

(a) Construct a graph that includes all of the information above with temperature on the ordinate and time on the abscissa.

(b) Calculate the slope of the line and the linear equation $y = mx + b$ for the line.

(c) From the graph find the temperature at 11.0 min and compare it with the value calculated from the equation.

(d) From the graph find the time when the temperature is 65.0 °C and compare it with the value calculated from the equation.

Answers: **F-1(a)** Draw the line touching or coming close to as many points as possible. Note that the straight line does not go through all points. **F-1(b)** $m = 2.5$, $y = 2.5x + 10$ **F-1(c)** From the equation, when $x = 11.0$, $y = 2.5 \cdot 11.0 + 10$, so $y = 37.5$ °C. The graph agrees. **F-1(d)** From the equation, when $y = 65$, $65 = 2.5x + 10$, and $x = 22.0$ min. The graph agrees.

2 Nonlinear Relationships Nonlinear graphs are represented by a curved line and result from an inverse or any type of proportionality other than direct. Examples of such relationships can be expressed as

$$x \propto y^2 \quad \text{and} \quad x \propto \frac{1}{y}$$

An example of an inverse relationship that produces a curve when plotted is the relationship between volume of a gas (V) at constant temperature and pressure (P). The relationship is

$$V \propto \frac{1}{P}$$

In the following experiments the pressure on a volume of gas was varied and the volume was measured at constant temperature.

Experiment	Pressure (torr)	Volume (L)
1	500	15.2
2	600	12.6
3	760	10.0
4	900	8.44
5	1200	6.40
6	1600	4.80

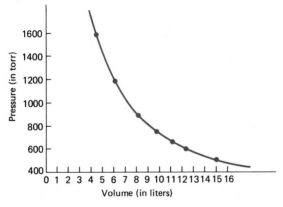

Figure F-2 AN INVERSE RELATIONSHIP.

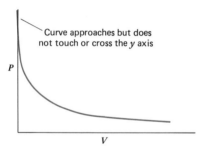

Figure F-3

This information has been graphed in Figure F-2. Note that the line has a gradually changing slope, starting with a very steep slope on the left and going to a small slope on the right of the graph. We can interpret this. In the region of the steep slope the increase in pressure causes little decrease in volume compared with the part of the curve with the small slope to the right. Note that this is a plot of an inverse relationship. That is, the higher the pressure, the lower the volume. If we regraph the information and include much higher pressures and the corresponding volumes, as shown in Figure F-3, we see that *the curve approaches the Y axis but never actually touches it,* no matter how high the pressure. Such a curve is said to approach the axis **asymptotically.** In our example, the curve would eventually appear to change to a vertical line parallel to the y axis.

Exercises

F-2 Construct a graph from the following experimental information. Plot the time on the abscissa and the concentration on the ordinate. Describe what this graph tells you about the change in concentration of HI as a function of (how it varies with) the time.

Experiment	Time (min)	Concentration of HI (mol/L)
1	0.00	2.50×10^{-2}
2	5.0	1.50×10^{-2}
3	10.0	0.90×10^{-2}
4	15.0	0.65×10^{-2}
5	20.0	0.55×10^{-2}
6	35.0	0.50×10^{-2}

(a) From the graph, find the concentration of HI at 12.0 min and at 25 min.

(b) From the graph, find the times at which the concentration of HI is 2.00×10^{-2} and 0.75×10^{-2}.

Answers: **F-2** The concentration decreases rapidly at the start (the curve has a large slope) but then changes little after about 25 min. At that point the curve appears to change to a straight line parallel to the x axis.   (a) At 12.0 min, the concentration is 0.80×10^{-2} mol/L; at 25 min, the concentration is about 0.52×10^{-2} mol/L.   (b) When the concentration is 2.00×10^{-2} mol/L, the time is 2.4 min; when the concentration is 0.75×10^{-2} mol/L, the time is 13.0 min.

This section reviews the following calculator skills.

1 The function, limitations, and types of hand calculators

2 One-step calculations

3 Chain calculations

4 Scientific notation

5 Powers, roots, logs, and antilogs

APPENDIX

CALCULATORS

G

The inexpensive hand calculator has become a way of life for almost all chemistry students. The convenience of this handy instrument is undeniable. Calculations are easy, quick, and accurate. In a way, however, their use creates the risk that we may lose the ability to add, multiply, and divide simple numbers on our own. It is sad for an instructor to see students pull out their calculators when the only exercise is to multiply a number by 10. Obviously, it is easy to become overly dependent on the calculator, but they are here to stay. Besides just adding and subtracting, most calculators divide and multiply, square and take square roots, and take logs and antilogs. Many also handle scientific notation and have several other functions.

There are two basic types of hand calculators. Most students use calculators that have a logic system known as Algebraic Operating System (AOS). These calculators have an "=" key. A few students may have calculators that use a system called Reverse Polish Notation (RPN). Calculators using this system have a key labeled "ENTER". The difference in the two centers on the sequence of instructions. The AOS system basically does the calculations one at a time much as we would write them out. One enters a number, then tells the calculator what he or she plans to do with the next number (e.g., add or multiply). In the RPN system, one basically enters both numbers and then tells the calculator what to do with them.

Different brands of calculators vary as to the number of digits shown in the answer, the labeling of some functions, and, of course, the number of functions available. You should consult the instruction booklet for more details. In this appendix, we concentrate on the types of calculations that one encounters in general chemistry courses. This should save time and confusion in having to read through an often complicated instruction booklet. Calculations encountered in chemistry include powers, logs, and scientific notation. Recall that the calculator usually gives at least six significant figures in an answer. This is almost always far too many for the result of a chemical calculation. The student must know how to express the answer to the proper number of significant figures (see Section 2-2). A chemistry teacher is not impressed with a student who reports a pH of 8.678346 for a calculation made from a two-significant-figure number. Reporting more numbers than necessary makes an answer look worse not better.

1 Addition, Subtraction, Multiplication, and Division   In one-step calculations, one simply types in the numbers and tells the calculator what to do. The sequence is different in the two systems, however. In the following examples, your display will show more digits in the answer. Since we are assuming that the numbers typed into the calculator are results of measurements, we will round off the answer properly. The calculator won't.

For an AOS calculator:
1 Type in the first number.
2 Press the proper function key (+, −, ×, or ÷).
3 Type in the second number.
4 Press =.

For an RPN calculator:
1 Type in the first number.
2 Press ENTER.
3 Type in the second number.
4 Press the proper function key (+, −, ×, or ÷).

EXAMPLE G-1

Carry out the following calculations.

(a) 6.368 + 43.852

AOS system		RPN system	
ENTER	DISPLAY	ENTER	DISPLAY
6.368	6.368	6.368	6.368
+	6.368	ENTER	6.368
43.852	43.852	43.852	43.852
=	50.22	+	50.22

Your answer should be 50.220 to indicate the proper number of decimal places. A calculator does not include zeros that are to the right of the last digit.

(b) 0.0835 ÷ 37.2

ENTER	DISPLAY	ENTER	DISPLAY
0.0835	0.835	0.0835	0.0835
÷	0.0835	ENTER	0.0835
37.2	37.2	37.2	37.2
=	0.00224	÷	0.00224

Some types of calculators (i.e., Casio and Radio Shack) give a display in scientific notation for the answer (i.e., $2.24462 - 03$, which means 2.24462×10^{-3}).

2 Chain Calculations   One of the most convenient features of the modern hand calculator allows us to do a series of calculations without pausing. For example, many gas law problems require successive calculations as illustrated in the following example.

EXAMPLE G-2

Solve the following problem.

$$n = \frac{PV}{RT} = \frac{0.749 \times 7.22}{0.0821 \times 312} \text{ mol} = \underline{\hspace{2cm}} \text{ mol}$$

AOS system		RPN system	
ENTER	DISPLAY	ENTER	DISPLAY
0.749	0.749	0.749	0.749
×	0.749	ENTER	0.749
7.22	7.22	7.22	7.22
÷	5.408	×	5.408
0.0821	0.0821	0.0821	0.0821
÷	65.87	÷	65.87
312	312	312	312
=	0.211	÷	0.211

One can do this problem in any sequence. For example, one could carry out the calculation in the following sequence just as easily: $(0.749 \div 0.0821) \times 7.22 \div 312 = 0.211$.

EXAMPLE G-3

To calculate the formula weight of a compound the atomic weights are multiplied by the number of atoms and the masses totaled. This is illustrated as follows.

Find the formula weight of B_2O_3.
Formula weight $= (2 \times 10.8) + (3 \times 16.0) = $ _____ amu.

AOS system		RPN system	
ENTER	DISPLAY	ENTER	DISPLAY
2	2	2	2
×	2	ENTER	2.00
10.8	10.8	10.8	10.8
+	21.6	×	21.6
3	3	3	3
×	3	ENTER	3.00
16.0	16.0	16.0	16.0
=	69.6	×	48.0
		+	69.6

One other note on the use of the calculator for successive calculations concerns changing signs. For example, in the calculation (2×-6), after the "6" has been entered, "6" appears on the display. To change it to a "-6", type the "$+/-$" or the "CHS" key. The display now reads "-6" and you can proceed with the calculation.

Exercises

G-1 Perform the following calculations.

(a) 0.662×7.99

(b) $73.8 + 0.27 + 45.8$

(c) $43.8 - 178.34$

(d) $7.3 + 8.5 - (-3.4)$

(e) $\dfrac{46.8 \times 23.4}{0.00786}$

(f) $\dfrac{172 + 67 - 89}{55}$

(*Hint:* In problem (f) press the "=" or the "**ENTER**" key after the value of the numerator is calculated. Then carry out the division.)

(g) $(4 \times 14.006) + (3 \times 15.998) \div (6 \times 1.008)$

(h) $(0.810 \times 102.99) + (0.190 \times 104.88)$

(i) $\dfrac{0.233 \times 0.0821 \times 323}{0.645}$

Answers: **G-1(a)** 5.29   **G-1(b)** 119.9   **G-1(c)** -134.5 **G-1(d)** 19.2
G-1(e) 139000   **G-1(f)** 2.7   **G-1(g)** 110.066   **G-1(h)** 103   **G-1(i)** 9.58

3 Scientific Notation   Most hand calculators now have the capacity to easily handle scientific notation. This is done with a key marked "**EE**", "**EXP**", or "**EEX**". The coefficient is entered first; then the exponent key is pressed and the exponent entered. If it is a negative exponent, one must press the $+/-$ or the CHS key to change the sign. The following examples illustrate the use of scientific notation in typical calculations.

EXAMPLE G-4

Carry out the following calculations.

(a) $\dfrac{0.250 \times (6.02 \times 10^{23})}{1.55}$

AOS system			RPN system		
ENTER	DISPLAY		ENTER	DISPLAY	
0.250	0.250		0.250	0.250	
$\times$	0.250		ENTER	0.2500	
6.02	6.02		6.02	6.02	
EXP	6.02	00	EEX	6.02	00
23	6.02	23	23	6.02	23
$\div$	1.505	23	$\times$	1.505	23
1.55	1.55		1.55	1.55	
$=$	9.71	22	$\div$	9.71	22
	(9.71×10^{22})				

(b) $\dfrac{3.76 \times 10^{-18}}{0.862 \times 10^{6}}$

ENTER	DISPLAY		ENTER	DISPLAY	
3.76	3.76		3.76	3.76	
EXP	3.76	00	EEX	3.76	00
18	3.76	18	18	3.76	18
+/−	3.76	−18	CHS	3.76	−18
÷	3.76	−18	ENTER	3.76	−18
0.862	0.862		0.862	0.862	
EXP	0.862	00	EEX	0.862	00
6	0.862	06	6	0.862	06
=	4.36	−24	÷	4.36	−24

4 Powers, Roots, Logs, and Antilogs These are all one-step calculations that require a special function on the calculator. On some calculators a direct key function may not be available. However, an inverse key (INV) is available that provides the desired function. For example, an x^2 key is not available on some calculators but the function is obtained by pressing INV and then the $\sqrt{x}$ key. The square is the inverse of the square root. On another calculator, taking the cube root is the same as raising the number to the $\frac{1}{3}$ power. Again, we note that there are some minor variations in how the functions are presented for different calculators.

EXAMPLE G-5

Carry out the following calculations.

(a) $(6.2)^2$

AOS system		RPN system	
ENTER	DISPLAY	ENTER	DISPLAY
6.2	6.2	6.2	6.2
x^2	38	x^2	38

(b) $(6.2)^3$

ENTER	DISPLAY	ENTER	DISPLAY
6.2	6.2	6.2	6.2
x^y	6.2	ENTER	6.200
3	3	3	3
=	238	y^x	238

(c) $\sqrt{6.2}$

ENTER	DISPLAY		ENTER	DISPLAY
6.2	6.2		6.2	6.2
$\sqrt{x}$	2.49		$\sqrt{x}$	2.49

(d) $\sqrt[3]{6.2}$

ENTER	DISPLAY		ENTER	DISPLAY
6.2	6.2		6.2	6.2
INV*	6.2		ENTER	6.200
x^y	6.2		3	3
3	3		$1/x$	0.3333
=	1.837		y^x	1.837

* $(INV = x^y)$ is the same as $x^{1/y}$

EXAMPLE G-6

Find the following logs (common) and antilogs.*

(a) 35.6

ENTER	DISPLAY
35.6	35.6
log	1.55

(b) 4.88×10^{-9}

ENTER	DISPLAY	
4.88	4.88	
EXP	4.88	00
9	4.88	09
+/−	4.88	−09
log	−8.312	

(c) the antilog of -3.876

ENTER	DISPLAY
3.876	3.876
+/−	−3.876
INV†	−3.876
log	0.000133 or
	1.3304 −04

* Both AOS and RPN systems are the same for these calculations.
† $(INV - log)$ is the same as 10^x.

Exercises

G-2 Perform the following calculations on your calculator.

(a) 0.230^2

(b) $\sqrt{6.80 \times 10^{-3}}$

(c) $\sqrt[3]{0.986}$

(d) $(3.60)^4$

(e) $\sqrt[5]{4.32 \times 10^6}$

(f) the log of 8.33

(g) the log of 0.0257

(h) the log of 567

(i) the antilog of 0.665

(j) the antilog of 2.977

(k) the antilog of -4.005

Answers (given to three significant figures): **G-2(a)** 0.0529
G-2(b) 0.0825 **G-2(c)** 0.995 **G-2(d)** 168 **G-2(e)** 21.2
G-2(f) 0.921 **G-2(g)** -1.590 **G-2(h)** 2.754 **G-2(i)** 4.62
G-2(j) 948 **G-2(k)** 9.89×10^{-5}

APPENDIX

GLOSSARY*

H

A

Accuracy. How close the value of a measurement is to the true value. (2)

Acid. A substance that increases the H_3O^+ ion concentration in water. In the Brønsted definition, an acid is a proton donor. (7, 9, 13)

Acid Anhydride. A nonmetal oxide that reacts with water to form an oxyacid. (13)

Acid Salt. An ionic compound whose anion contains one or more acidic hydrogens. (13)

Acidic Solution. An aqueous solution that has a hydronium ion (H_3O^+) concentration greater than 10^{-7} M at 25 °C (pH less than 7). (13)

Actinide. One of 14 elements in which the $5f$ sublevel is filling (Th through Lr). (5)

Activated Complex. The unstable combination of two reactant molecules at the moment of impact in a collision that leads to products. (15)

Activation Energy. The minimum kinetic energy that colliding reactant molecules must have to overcome the potential energy barrier between reactants and products. (15)

Alcohol. An organic compound containing at least one hydroxyl group (i.e., R—OH). (16)

Aldehyde. An organic compound containing a carbonyl group bound to at least one hydrogen. (16)

$$R-\overset{\overset{\textstyle O}{\|}}{C}-H$$

Alkali Metal. An element in group 1 (IA) of the periodic table. (5)

Alkaline Earth Metal. An element in group 2 (IIA) of the periodic table. (5)

Alkane. A hydrocarbon containing only single bonds (a saturated hydrocarbon). The general formula is C_nH_{2n+2} for open-chain alkanes and C_nH_{2n} for cyclic alkanes. (16)

Alkene. A hydrocarbon containing at least one double bond (an unsaturated hydrocarbon). The general formula for open-chain alkenes is C_nH_{2n}. (16)

Alkylation. The formation of large hydrocarbon molecules from small ones in the refining process. (16)

Alkyl Group. A substituent that is an alkane minus a hydrogen atom [i.e., C_2H_6 (ethane) $-$ H $=$ C_2H_5 (ethyl)]. (16)

Alkyne. A hydrocarbon containing a triple bond. The general formula for open-chain alkynes is C_nH_{2n-2}. (16)

Allotropes. Different forms of the same element in the same physical state (e.g., graphite and diamond are allotropes of carbon). (5)

Alpha (α) Particle. A helium nucleus ($^4_2He^{2+}$) that is emitted from a radioactive nucleus. (3)

Amide. A derivative of a carboxylic acid in which an amine group replaces the hydroxyl group of the acid. (16)

$$R-\overset{\overset{\textstyle O}{\|}}{C}-NH_2$$

Amine. An organic compound containing a nitrogen with a single bond to a hydrocarbon group and a total of two other bonds to hydrocarbon groups or hydrogens. (16)

Amphoteric. A property of a substance indicating that it can act as an acid when a base is present or a base when an acid is present. (13)

Anion. A negatively charged ion. (3)

Anode. The electrode at which oxidation takes place in an electrolytic or voltaic cell. (14)

Aromatic Compounds. The largest class of aromatic compounds are six-membered cyclic hydrocarbons containing alternating single and double bonds. (16)

Arrhenius Acids and Bases. A system that defines acids as substances that produce H^+ ions and bases as substances that produce OH^- ions in aqueous solutions. (13)

Atmosphere. The sea of gases above the solid surface of the earth. Also, the pressure equivalent to 760 torr. (10)

Atom. The smallest particle of an element that can exist and enter into chemical reactions. (3)

Atomic Mass. The mass of a certain isotope compared with that of ^{12}C, which is defined as exactly 12 amu. (3)

Atomic Mass Unit (amu). A mass that is exactly one-twelfth of the mass of an atom of ^{12}C, which is defined as exactly 12 amu. One amu is equal to 1.6605×10^{-24} g. (3)

° The numbers in parentheses at the end of each entry refer to the chapter in which the entry is first discussed plus any chapter in which the topic is discussed in detail.

Atomic Number. The number of protons in the nucleus of an atom of an element. (3)

Atomic Radius. The distance from the nucleus of an atom to the outermost electron. (5)

Atomic Theory. A theory first proposed by John Dalton in 1803 which holds that the basic components of matter are atoms. (3)

Atomic Weight. The weighted average of the atomic masses of all of the naturally occurring isotopes of an element. (3)

Aufbau Principle. The rule that states that electrons fill the lowest available energy levels (sublevels) first. (4)

Autoionization. The partial dissociation of a pure covalent compound to produce a cation and an anion. (13)

Avogadro's Law. Equal volumes of gases at the same temperature and pressure contain equal numbers of molecules. (10)

Avogadro's Number. The number of objects or particles in one mole, equal to 6.02×10^{23}. (8)

B

Balanced Equation. A chemical equation that has the same number of atoms of each element and the same total charge on both sides of the equation. (9)

Barometer. A closed-end glass tube filled with mercury that is used to measure atmospheric pressure. (10)

Base. A substance that increases the OH^- ion concentration in water. In the Brønsted definition, a base is a proton acceptor. (9, 13)

Basic Anhydride. A metal oxide that reacts with water to form an ionic hydroxide. (13)

Basic Solution. An aqueous solution that has a hydroxide ion (OH^-) concentration greater than $10^{-7} M$ at 25 °C (pH greater than 7). (13)

Battery. A combination of one or more voltaic cells. (14)

Beta (β) Particle. A high-energy electron emitted from a radioactive nucleus. (3)

Binary Compounds. Compounds composed of molecules or formula units containing two different elements. (6)

Bohr Model. A model of the atom in which electrons orbit around a central nucleus in discrete energy levels. (4)

Boiling Point. The temperature at which bubbles of vapor form in a liquid. This occurs when the vapor pressure of the liquid equals the restraining or atmospheric pressure. (11)

Boyle's Law. The volume of a quantity of gas is inversely proportional to the pressure at constant temperature. (10)

Brønsted–Lowry Acids and Bases. An acid is a proton (H^+) donor and a base is a proton acceptor. (13)

Buffer Solution. An aqueous solution containing a weak acid and a salt providing its conjugate base (or a weak base and a salt providing its conjugate acid). A buffer solution resists changes in pH. (13)

Buoyancy. The ability of a solid or liquid to float in a certain liquid. (2)

C

Calorie. The amount of heat required to raise the temperature of 1 g of water 1 °C. The precise definition is 1 cal = 4.184 joule (J). (2)

Carbonyl Group. A functional group made up of a carbon with a double bond to an oxygen:

$$\overset{\displaystyle O}{\underset{|}{\overset{\|}{-C-}}}$$

It is the functional group of aldehydes and ketones. (16)

Carboxylic Acid. An organic compound containing a carboxyl group. (16)

Catalyst. A substance that is not consumed in a reaction but whose presence increases the rates of both the forward and the reverse reaction. A catalyst provides an alternate mechanism of lower activation energy but does not shift the point of equilibrium. (9, 15)

Cathode. The electrode at which reduction takes place in an electrolytic or voltaic cell. (14)

Cation. A positively charged ion. (3)

Celsius Scale (°C). A temperature scale with 100 equal divisions between the freezing point of water (defined as 0 °C) and the normal boiling point of water (defined as 100 °C). (2)

Chain Reaction. A self-sustaining reaction. In a nuclear chain reaction one reacting neutron produces an average of three other neutrons that in turn cause reactions. (3)

Chalcogens. Elements in group 16 (VIA) in the periodic table. (5)

Charles' Law. The volume of a quantity of gas is directly proportional to the Kelvin temperature at constant pressure. (10)

Chemical Bond. The forces of attraction that hold atoms or ions together in a compound or element. (3, 6)

Chemical Change. The change of substances into substances of a different composition. (1)

Chemical Equation. The representation of a chemical reaction by means of chemical symbols and coefficients. (8)

Chemical Kinetics. The study of the interrelationships of reaction rates and the mechanism of a reaction. (15)

Chemical Property. A property of a substance that concerns its ability or tendency to change into, or react with, other substances. (1)

Chemical Thermodynamics. The study of heat and its relationship to chemical processes. (9)

Chemistry. The branch of science dealing with the nature, composition, and structure of matter and the changes it undergoes. (1)

Colligative Property. A property that depends on the relative number of particles of solute and solvent in a solution rather than their identities. Vapor pressure lowering, boiling point elevation, freezing point depression, and osmotic pressure are colligative properties of solutions. (12)

Collision Theory. A theory suggesting that chemical reactions result from collisions of reactant molecules having proper orientation and sufficient energy. (15)

Combination Reaction. A chemical reaction in which one compound results from the union of two or more elements and/or compounds. (9)

Combined Gas Law. A combination of Boyle's, Gay-Lussac's, and Charles' laws (i.e., $PV/T = a$ constant). (10)

Combustion Reaction. A chemical reaction in which a compound or element reacts with oxygen. The reaction usually results in the liberation of large amounts of heat and light. (1, 9)

Compound. A pure substance composed of two or more different elements combined in fixed proportions. (1)

Concentrated. A solution containing a relatively large amount of a specified solute per unit volume. (12)

Concentration. The amount of solute present in a specified amount of solvent or solution. (12)

Condensation. The formation of a liquid from the vapor or gaseous state. (11)

Condensed States. The liquid and solid states are referred to as the condensed states because the molecules are relatively close together and these states are essentially incompressible. (11)

Conductor. A substance such as a metal that allows the flow of electricity. (12)

Conjugate Acid (Base). The conjugate acid of a base is formed by the addition of one H^+ to the base. The conjugate base of an acid is formed by the removal of one H^+ from the acid. (13)

Conversion Factor. An equivalency relationship expressed as a ratio. It is used to convert quantities in one unit of measurement into quantities in another. (2, Appendix D)

Covalent Bond. Two atoms held together in an ion or molecule by sharing one to three pairs of electrons. (3, 6)

Cracking. The formation of small hydrocarbon molecules from large ones in the refining process. (16)

D

Dalton's Law of Partial Pressures. The total pressure exerted by a mixture of gases is the sum of the partial pressures of the component gases. (10)

Daniell Cell. A voltaic cell made up of zinc and copper electrodes each immersed in solutions of their ions connected by a salt bridge. (14)

Decomposition Reaction. A chemical reaction in which one compound is decomposed into two or more elements and/or compounds. (9)

Density. The ratio of the mass of a substance (usually in grams) to its volume (usually in milliliters or cubic centimeters), in other words, mass per unit volume. (2)

Derivative. A compound produced by the substitution of a group on the molecules of the original or parent compound. (16)

Diffusion. The mixing of one gas into another. (10)

Dilute. A solution containing a relatively small amount of a specified solute per unit volume. (12)

Dilution. The preparation of a dilute solution from

a concentrated solution by the addition of more solvent. (12)

Dipole. The existence of a partial charge separation in a polar covalent bond or molecule. (6)

Diprotic Acid. An acid containing two acidic hydrogens per molecule. (13)

Discrete Spectra. The definite wavelengths of light (spectral lines) that are emitted by the hot gaseous atoms of an element. (4)

Dissociation. The breakup of one substance into simpler substances. This includes the formation of ions when certain compounds dissolve in water. (13)

Distillation. The process of separation of a solution into its components by vaporization (boiling) followed by condensation of the more volatile component. (1)

Double Bond. A covalent bond in which two pairs of electrons are shared between two atoms. (6)

Double-Replacement Reaction. A reaction involving an exchange of ions in a solution that leads to the formation of a precipitate or a covalent compound. (9, 12, 13)

Dry Cell. A voltaic cell composed of zinc and graphite electrodes immersed in an aqueous paste of NH_4Cl and MnO_2. (14)

Dynamic Equilibrium. The state in which opposing changes occur at equal rates. In a chemical reaction it is the point at which the rate of the forward reaction equals the rate of the reverse reaction. (9, 11, 13, 15)

E

Effusion. The process whereby a gas moves through an opening or hole from a region of high pressure to a region of lower pressure. (10)

Electricity. A flow of electrons through a conductor. (12)

Electrode. A conducting surface at which reactions take place in a voltaic or electrolytic cell. (14)

Electrolysis. The process of forcing electrical energy through an electrolytic cell, thereby causing a chemical reaction. (14)

Electrolyte. A compound whose aqueous solution or pure melt conducts electricity. (12)

Electrolytic Cell. An electrochemical cell that converts electrical energy into chemical energy. (14)

Electron. A subatomic particle with a charge of negative one (-1). In an atom, it exists outside of the nucleus and its mass is very small compared with that of the entire atom. (3)

Electron Affinity. The energy released when a gaseous atom adds an electron to form a gaseous ion. (5)

Electron Configuration. The arrangement of electrons in an atom according to shell and sublevel. (4)

Electronegativity. A measure of the tendency of the atoms of a certain element to attract electrons in a chemical bond. (6)

Electrostatic Forces. Forces of attraction between centers of opposite charge and forces of repulsion between centers of the same charge. (3, 6, 11)

Element. A pure substance that is one of the 108 basic forms of nature. Elements cannot be separated into simpler substances by ordinary chemical means. (1, 3)

Empirical Formula. A formula showing the simplest whole-number ratio of the atoms of the elements that make up a compound. (8)

Endothermic Reaction. A chemical reaction that absorbs heat from the surroundings. (1, 9)

Energy. The ability to do work. It can take the form of light, heat, mechanical, chemical, electrical, or nuclear. It can be of two types: kinetic or potential. (1, 9, 14, 15)

Enthalpy. The heat content or chemical energy of a certain compound. The change in enthalpy is the *heat of reaction*. (9)

Equation Factor. The mole relationships from a balanced equation that serve as conversion factors in stoichiometry problems. (9, 10, 12)

Equilibrium. See *dynamic equilibrium.*

Equilibrium Constant. A number that defines the position of equilibrium for a particular reaction at a specified temperature. (15)

Ester. A derivative of a carboxylic acid in which a hydrocarbon group (R′) replaces the hydrogen on the oxygen in the parent acid. (16)

$$\begin{array}{c} \quad\; O \\ \quad\; \| \\ R-C-O-R' \end{array}$$

Ether. An organic compound containing an oxygen bonded to two hydrocarbon groups. (16)

$$R-O-R'$$

Evaporation. The vaporization of a liquid below its boiling point. (11)

Exact Number. A number that results from a definition or an actual count. (2)

Excited State. An atom is in an excited state when at least one electron is in an energy state higher than the lowest available state. (4)

Exothermic Reaction. A chemical reaction that releases heat energy to the surroundings. (1, 9, 15)

Extrapolate. To extend a line on a graph beyond the experimental points. (10, Appendix F)

F

Factor-label Method. A problem-solving technique in which units and conversion factors are used for calculation purposes. (2, Appendix D)

Fahrenheit Scale (°F). A temperature scale with 180 equal divisions between the freezing point of water (32 °F) and the normal boiling point of water (212 °F). (2)

Family. See *group.*

Filtration. A laboratory procedure whereby a solid phase is separated from a liquid phase in a heterogeneous mixture. (1)

Fission. The splitting of a large, unstable nucleus into two smaller nuclei of similar size with the production of energy. (3)

Formula. The symbols of the elements and the number of atoms of each element that make up a compound. (3)

Formula Unit. The group of atoms represented by the formula of an ionic compound. (3)

Formula Weight. The total mass of all of the atoms in a molecule (or formula unit for an ionic compound) expressed in atomic mass units (amu). (8)

Freezing Point. See *melting point.*

Fuel Cell. A voltaic cell that can generate a continuous flow of electricity from the reaction of hydrogen and oxygen to produce water. (14)

Functional Group. The part of an organic molecule where most of the chemical reactions occur. (16)

Fusion (Nuclear). The combination of two small nuclei to form a larger nucleus with the production of energy. (3)

G

Gamma Ray. A high-energy form of light emitted from a radioactive nucleus. (3)

Gas. One of the three physical states of matter. A gas does not have a fixed shape or volume and thus fills any container uniformly. (1, 10)

Gas Laws. The various laws governing the behavior of gases. Gas laws are consistent with the kinetic molecular theory as applied to gases. (10)

Gay-Lussac's Law. The pressure of a gas is directly proportional to the Kelvin temperature at constant volume. (10)

Graham's Law of Effusion. The rate of effusion of a gas is inversely proportional to the square root of its formula weight. (10)

Ground State. An atom is in the ground state when all of its electrons are in the lowest available energy levels. (4)

Group. All of the elements in a vertical column in the periodic table. (4, 5)

H

Half-Life. The time required for one-half of a given sample of an isotope to undergo radioactive decay. (3)

Half-Reaction. A balanced equation for an oxidation or a reduction process. (14)

Halogen. An element from group 17 (VIIA) of the periodic table. (5)

Heat of Fusion. The heat required to melt a given mass of a substance at its melting point. (11)

Heat of Reaction. The amount of heat energy absorbed or evolved in a specified chemical reaction. (9)

Heat of Vaporization. The heat required to vaporize a given mass of a substance at its boiling point. (11)

Heterogeneous Matter. A nonuniform mixture of substances consisting of two or more phases with definite boundaries between the phases. (1)

Homogeneous Matter. Matter existing in one phase or physical state with the same properties and composition throughout. Homogeneous matter may be either a solution or a pure substance. (1)

Homologous Series. A series of related compounds in which one member differs from the next by a constant number of atoms. (16)

Hund's Rule. Electrons occupy separate orbitals of the same energy level with their spins parallel. (4)

Hydrate. A compound that contains one or more bonded water molecules that can generally be removed by heating. (11)

Hydration. The formation of attractive interactions between polar water molecules and solute ions or molecules. (12)

Hydration Energy. The energy released by the hydration of one mole of a specific molecule or ion. (12)

Hydrocarbon. An organic compound containing only carbon and hydrogen. (16)

Hydrogen Bond. An electrostatic interaction between a hydrogen bonded to a N, O, or F atom on one molecule and an unshared electron pair on an N, O, or F on another molecule. (11)

Hydrolysis. The reaction of an anion with water to produce a basic solution or the reaction of a cation with water to produce an acidic solution. (13)

Hydrometer. A device used to directly measure the density or specific gravity of a liquid. (2)

Hydronium Ion. The H_3O^+ ion, which represents the H^+ ion in aqueous solution. (13)

Hydroxyl Group. The —OH group as found in an organic compound. It is the functional group of alcohols. (16)

Hypothesis. A tentative explanation of related data. It can be used to predict results of experiments. (1)

I

Ideal Gas. A hypothetical gas whose molecules are considered to have no volume or interactions with each other. An ideal gas obeys the ideal gas law at all temperatures and pressures. (10)

Ideal Gas Law. A gas law that results from combination of the combined gas law and Avogadro's law ($PV = nRT$). (10)

Improper Fraction. A fraction whose numerator is larger than the denominator and thus has a value greater than one. (Appendix A)

Infrared Light. Light with wavelengths somewhat longer than those of red light. (4)

Inner Transition Element. Elements in which the f sublevel is filling (the lanthanides and actinides). (5)

Ion. An atom or a group of atoms that has acquired an electrical charge. (3)

Ion Product of Water. The equilibrium product of the H_3O^+ and OH^- molarities in water. At 25 °C the product, K_w, is a constant that is equal to 1.0×10^{-14}. (13)

Ionic Bond. The ion–ion electrostatic attractions between oppositely charged ions. (6)

Ionic Compound. A compound composed of oppositely charged ions. (3)

Ionization. The process of forming an ion or ions from a molecule or atom. (6, 13)

Ionization Constant. The equilibrium constant for the ionization (dissociation) of a weak acid (K_a) or a weak base (K_b). (15)

Ionization Energy. The energy required to remove an electron from a gaseous atom to form a gaseous ion. (5)

Isomers. Compounds with the same molecular formula but different molecular structures. (16)

Isotopes. Atoms with different mass numbers but the same atomic number. (3)

J

Joule. The SI unit for measuring heat energy. (2, 9)

K

Kelvin Scale (K). The temperature scale in which zero Kelvin is known as absolute zero, which is the lowest possible temperature. The Kelvin scale (T) relates to the Celsius scale [t(C)] by $T = [t(C) + 273]$ K. (2, 10)

Ketone. An organic compound containing a carbonyl group bonded to two hydrocarbon groups. (16)

$$\overset{\displaystyle O}{\underset{}{\overset{\displaystyle \|}{R-C-R}}}$$

Kinetic Energy. Energy resulting from motion. K.E. $= \frac{1}{2}$[mass $\times$ (velocity)^2]. (1)

Kinetic Molecular Theory. Matter is composed of molecules that are in constant motion. (10, 11)

L

Lanthanide. One of the 14 elements in which the $4f$ sublevel is filling (Ce through Lu). (5)

Lattice. The regular three-dimensional array of ions or molecules in a solid crystal. (6)

Lattice Energy. The quantitative value of the ion–ion forces of one mole of a specific ionic compound. (12)

Law. A concise statement or mathematical relationship that describes some behavior of matter. (1)

Law of Mass Action. The relationship that describes the relative distribution of reactants and products at equilibrium. It is composed of an equilibrium constant and a ratio called the mass action expression. (15)

Lead–Acid Battery. A rechargeable voltaic battery composed of lead and lead dioxide electrodes in a sulfuric acid solution. (14)

Le Châtelier's Principle. When stress is applied to a system at equilibrium, the system reacts in such a way so as to counteract the stress. (15)

Lewis Dot Symbol. The symbol of an element with dots representing valence electrons. (6)

Lewis Structure. The representation of a molecule or ion showing all outer *s* and *p* electrons as dots and/or dashes. (6)

Limiting Reactant. The reactant that would produce the least amount of products if completely consumed. (9)

Liquid. One of the three physical states of matter. Liquids have a fixed volume but not a fixed shape. Liquids flow and thus take the shape of the bottom of the container. (1, 11)

London Forces. The inherent forces of attraction between the molecules of a compound. The magnitude of these forces depends on the molar mass of the compound. The greater the mass, the greater the forces. (11)

M

Mass. The quantity of matter (usually in grams or kilograms) that a substance contains. (2)

Mass Action Expression. The fraction containing the concentrations of products in the numerator and reactants in the denominator, each concentration raised to the power corresponding to its coefficient in the balanced equation. (15)

Mass Number. The number of nucleons in an atom. (3)

Matter. Anything that has mass and occupies space. (1)

Measurement. The quantity, dimensions, or extent of something, usually in comparison with a standard. (2)

Mechanism. The means by which reactant molecules collide or otherwise behave so as to eventually form products. (15)

Melting Point. The temperature at which a solid changes to a liquid. (11)

Metal. An element that has a relatively low ionization energy and thus forms positive ions in compounds. Metals are solids (except Hg) and are good conductors of electricity. (5)

Metalloid. An element with properties intermediate between those of metals and nonmetals. (5)

Metallurgy. The conversion of metal ores into metals. (1)

Metathesis Reaction. See *double-replacement reaction.*

Metric System. A system of measurement based on multiples of 10. (2)

Molality. A measure of solution concentration calculated by dividing the moles of solute by the kilograms of solvent. (12)

Molar Mass. The mass in grams of one mole of an element or compound. (8)

Molar Number. See *Avogadro's number.*

Molar Volume. The volume of one mole of gas at STP, which is 22.4 L. (10)

Molarity. A measure of solution concentration calculated by dividing the moles of solute by the liters of solution. (12)

Mole. The SI unit of quantity. It is equal to 6.02×10^{23} particles. (8)

Molecular Compound. A compound composed solely of molecules. (3)

Molecular Formula. The formula of a compound showing the exact number of atoms of each element in one molecule of a compound. (8)

Molecular Weight. See *formula weight.*

Molecule. A group of two or more atoms held together by covalent bonds. (3)

Monomer. Single molecules that can be joined together in a chain to form a repeating structure called a polymer. (16)

Monoprotic Acid. An acid with one acidic hydrogen per molecule. (13)

N

Net Ionic Equation. The balanced ionic equation written without spectator ions. (12)

Neutral Solution. An aqueous solution in which $[H_3O^+] = [OH^-] = 1.0 \times 10^{-7} M$. In a neutral solution pH = 7.0 at 25 °C. (13)

Neutralization. The reaction of an acid and a base to form a salt and, generally, water. (9, 13)

Neutron. A neutral subatomic particle that exists in the nucleus of an atom. It has a mass of approximately 1 amu. (3)

Noble Gas. An element in group 18 (0) in the periodic table. (5)

Nonconductor. A substance that does not allow the passage of an electric current under normal conditions. (12)

Nonelectrolyte. A covalent compound that dissolves in water without the formation of ions. A solution of a nonelectrolyte does not conduct electricity. (12)

Nonmetals. Elements in the upper right-hand corner of the periodic table. They are characterized by high ionization energies. (5)

Nonpolar Bond. A covalent bond in which the electrons are shared equally between the two atoms. (6)

Normal Boiling Point. The temperature at which the vapor pressure of a liquid is equal to 1 atm pressure. (11)

Nuclear Equation. A symbolic representation of the changes of a nucleus or nuclei into other nuclei and particles. (3)

Nuclear Reactor. A device that generates electrical energy from a controlled nuclear fission reaction. (3)

Nucleons. The protons and neutrons that make up the nucleus of the atom. (3)

Nucleus. The small, dense center of the atom containing the nucleons. (3)

O

Octet Rule. The generalization that states that the atoms of the representative elements tend to form bonds in such a manner as to have access to eight valence electrons. (6)

Orbital. A region of space in which there is the highest probability of finding an electron. Each orbital of a certain sublevel can hold up to two electrons. (4)

Organic Chemistry. The branch of chemistry that deals with most of the compounds of carbon. (5, 16)

Osmosis. The process whereby there is a net movement of solvent molecules through a semipermeable membrane from a dilute to a concentrated solution. (12)

Osmotic Pressure. The pressure on one side of a membrane that is needed to prevent the net movement of solvent molecules from a dilute to a concentrated solution. (12)

Oxidation. The loss of electrons from a substance that leads to an increase in oxidation state of an element in the substance. (14)

Oxidation–Reduction (Redox) Reactions. A chemical reaction involving an exchange of electrons between two substances. (14)

Oxidation State. The charge that an atom in a molecule would have if the electrons in each bond were assigned to the more electronegative atom. (7)

Oxidizing Agent. The reactant that causes another reactant to be oxidized. The oxidizing agent accepts electrons from the reactant oxidized. (14)

Oxyacid. An acid derived from hydrogen and an oxyanion. (7)

Oxyanion. An anion containing oxygen and at least one other element. (7)

P

Partial Pressure. The pressure of one component in a mixture of gases. (10)

Parts per Billion (ppb). A unit of concentration for extremely trace amounts obtained by multiplying the ratio of the mass of solute to the mass of solution by 10^9. (12)

Parts per Million (ppm). A unit of concentration for trace amounts obtained by multiplying the ratio of the mass of solute to the mass of solution by 10^6. (12)

Pauli Exclusion Principle. No two electrons in an atom can reside in the same orbital with the same spin. (4)

Percent by Weight. A measure of solution concentration by which the mass of solute is expressed as a percent of the mass of the solution. (12)

Percent Composition. The mass of a component substance (i.e., an element) expressed as a percent of the total mass (i.e., the formula weight). (8)

Percent Yield. The ratio of the actual yield to the theoretical yield times 100%. (9)

Period. A row on the periodic table. It corresponds to the elements between noble gases. (5)

Periodic Law. When elements are arranged by atomic number, their properties vary periodically. (4)

Periodic Table. An arrangement of elements in order of increasing atomic number. Elements with the same number of outer electrons are arranged in vertical columns. (4, 5)

pH. An expression of the H_3O^+ concentration in

an aqueous solution as the negative of the logarithm of the H_3O^+ molarity. (13)

Phase. A physically distinct state, either solid, liquid, or gas. (1)

Physical Change. A change that does not alter the chemical composition of a substance. Usually a change in physical state or shape. (1)

Physical Property. A property of a substance that can be observed or measured without changing the substance into some other substance. (1)

Physical Science. A science concerned with the natural laws of matter other than those concerned with life. (2)

pOH. An expression of the OH^- concentration in an aqueous solution as the negative of the logarithm of the OH^- molarity. (13)

Polar Covalent Bond. A covalent bond in which the pair of electrons resides closer to one atom than the other, thereby establishing a separation of charge (a dipole). (6)

Polyatomic Ion. An ion containing more than one atom. (3, 6)

Polymer. A high-molar-mass substance made from repeating units of an alkene or from combinations of other molecules. (16)

Polyprotic Acid. An acid with more than one acidic hydrogen per molecule. (13)

Potential Energy. Energy that is available because of position or composition. (2, 15)

Precipitate. A solid compound that is formed in solution and settles to the bottom of the container. (9, 12)

Precipitation Reaction. A double-replacement reaction in which two soluble compounds produce a precipitate when mixed. (9, 12)

Precision. How close repeated measurements come to each other. The more precise a measurement, the more significant figures that are expressed. (2)

Pressure. Force per unit area. (10)

Principal Quantum Number (n). A number that corresponds to a particular shell or energy level occupied by the electrons in an atom. (4)

Products. The compounds and/or elements that result from a chemical reaction. (9)

Proper Fraction. A fraction whose numerator is smaller than the denominator and thus has a value less than one. (Appendix A)

Property. A description of a particular characteristic or trait of a substance that can be observed or measured. (1)

Proton. A positively charged subatomic particle that exists in the nucleus of an atom. The mass of a proton is approximately 1 amu. (3)

Pure Substance. Matter that has definite and unchanging composition and properties. Pure substances are either elements or compounds. (1)

R

Radiation. Particles or high-energy light rays that are emitted by an atom or a nucleus of an atom. (3)

Radioactive Decay Series. A series of elements formed from the successive emission of alpha and beta particles starting from a long-lived isotope and ending with a stable isotope. (3)

Radioactivity. The emission of energy or particles from an unstable nucleus. (3)

Radionuclide. A radioactive isotope. (3)

Rate Law. A relationship between the rate of a reaction at a certain temperature and the concentrations of any reactants that affect the rate of the reaction. (15)

Reactants. The compounds or elements that undergo a chemical reaction. (9)

Recrystallization. A purification process in which a liquid (e.g., H_2O) is saturated with an impure compound at a high temperature and then allowed to cool. The precipitate that forms is generally more pure than the original solid. (11)

Redox Reaction. See *oxidation–reduction reaction.*

Reducing Agent. The reactant that causes another reactant to be reduced. The reducing agent donates electrons to the reactant reduced. (14)

Reduction. The gain of electrons by a substance that leads to a decrease in oxidation state of an element in the substance. (14)

Reforming. The rearrangement of long-chain hydrocarbons to branched hydrocarbons and/or the removal of hydrogen from hydrocarbons. (16)

Representative Elements. Groups 1, 2, and 13–17 (IA through VIIA) in the periodic table. (5)

Resonance Structures. The representation of all possible Lewis structures of a molecule or ion having the same number of electrons and arrangement of atoms. (6)

S

Salt. An ionic compound formally composed of a positive ion from a base and a negative ion from an acid. (7, 13)

Saturated Hydrocarbon. A hydrocarbon containing only single covalent bonds (an alkane). (16)

Saturated Solution. A solution containing the maximum amount of a specified solute that can dissolve at that temperature. (12)

Scientific Method. The method whereby modern scientists explain the behavior of nature with hypotheses and theories or describe the behavior with laws. (1)

Scientific Notation. A number written in exponential form. It is used to express nonsignificant zeros in very large or small numbers by a power of 10. (2, Appendix C)

Semipermeable Membrane. A thin porous membrane that allows passage of small solvent molecules but blocks passage of solute molecules or ions. (12)

Shell (Energy Level). A principal energy state for an electron in an atom. Its number relates to the most probable distance of the electrons from the nucleus. (4)

SI units. The international system of units of measurement. (2)

Significant Figure. A measured digit in a number. The number of significant figures in a measurement relates to the degree of precision. (2)

Single-Replacement Reaction. A chemical reaction in which an element is substituted for another element in a compound (9, 14)

Solid. One of the three physical states of matter. A solid is characterized by a fixed volume and a fixed shape. (1, 11)

Solubility. The maximum amount of a solute that dissolves in a given amount of solvent at a certain temperature. (12)

Solute. A substance that dissolves in a solvent to form a homogeneous solution. (12)

Solution. A homogeneous mixture of two or more substances with uniform properties and composition. (1, 12)

Solvent. A medium that disperses another substance called a solute to form a homogeneous mixture. The solvent and the solution exist in the same physical state. (12)

Specific Gravity. The ratio of the density of a substance to the density of water at the same temperature. (2)

Specific Heat. The heat required to raise the temperature of one gram of a substance one degree Celsius (1 °C). (2)

Spectator Ion. An ion that is not directly involved in an aqueous reaction and does not appear in the net ionic equation. (12)

Spectrum. The separate color components of a beam of light. (4)

Stock Method. The naming of metal ions in a compound by writing the name of the metal followed by its oxidation state in Roman numerals. (7)

Stoichiometry. The quantitative relationships among reactants and products in a chemical reaction. (9, 10, 12)

STP. Standard temperature and pressure: 1 atm pressure (760 torr) and 0 °C, (273 K). (10)

Strong Electrolyte. A substance that is essentially dissociated completely into ions in aqueous solution. These solutions are good conductors of electricity. (12)

Structural Formula. The formula of a compound written in a way that illustrates the sequence of bonds. (3)

Sublevel. One of the subdivisions (s, p, d, or f) of a shell of electrons in an atom. The orbitals of each sublevel have characteristic shapes. (4)

Sublimation. The vaporization of a solid. (11)

Substance. A form of matter. It is usually thought of as a pure form of matter (i.e., elements or compounds). (1)

Substituent. An atom or group of atoms that can substitute for a hydrogen atom in an organic compound. (16)

Supersaturated Solution. An unstable solution containing more solute in solution than expected from its solubility at that temperature. (11)

Symbol. The letter or letters used to represent an element. (3)

T

Temperature. A measure of the intensity of heat (the average kinetic energy) within any form of matter. (2)

Theoretical Yield. The maximum amount of a product that can form when at least one of the reactants is completely consumed. (9)

Theory. A hypothesis that withstands the test of time and trial is upgraded to a theory. (1)

Thermochemical Equation. A balanced equation that includes the value of the heat of reaction. (9)

Thermometer. A device used to measure temperature. (2)

Torr. A unit of gas pressure equivalent to one millimeter of mercury (1 mm Hg). (10)

Total Ionic Equation. A reaction in aqueous solution involving ionic compounds written so that all ions are shown independently in the equation. (12)

Transition Element. Elements in which a *d* sublevel is filling. These elements are groups 3–12 (the B elements) in the periodic table. (5)

Transmutation. The changing of one element into another by radioactive decay or a nuclear reaction. (3)

Triple Bond. A covalent bond in which three pairs of electrons are shared between two atoms. (6)

Triprotic Acid. An acid containing three acidic hydrogens per molecule. (13)

U

Ultraviolet Light. Light with wavelengths somewhat shorter than those of violet light. (4)

Unit Factor. A conversion factor that relates a quantity of a certain unit to a quantity of "one" of another unit. (2, Appendix D)

Unsaturated Hydrocarbon. A hydrocarbon that contains at least one double or triple bond. (16)

Unsaturated Solution. A solution containing less than the maximum amount of a specified solute that can dissolve at that temperature. (12)

V

Valence Electrons. The outer *s* and *p* electrons of a representative element. (6)

Vapor Pressure. The pressure exerted by a vapor in equilibrium with its liquid or solid form at a specified temperature. (11)

Volatile. Having a significant vapor pressure. (11)

Voltaic Cell. A device that converts the chemical energy from a spontaneous redox reaction directly to electrical energy. (14)

W

Wavelength (λ). The distance between two peaks on a wave. (4)

Weak Electrolyte. A molecular compound that dissolves in water but is only partially dissociated to produce a limited concentration of ions. These solutions are weak conductors of electricity. (12)

Weight. The measure of the attraction of gravity for a portion of matter. (2)

APPENDIX

ANSWERS TO

PROBLEMS

I

Chapter 1

1-1 The following elements occur in nature in the free state in appreciable amounts: carbon, nitrogen, oxygen, sulfur, copper, silver, gold, platinum, palladium, helium, neon, argon, krypton, and xenon.

1-3 It was found that water could be decomposed into simpler substances (hydrogen and oxygen).

1-4 (a) compound (b) element (c) element (d) compound
(e) compound (f) compound

1-6 Not necessarily. For example, the compound, calcium carbonate can be decomposed into two other compounds, calcium oxide and carbon dioxide.

1-7 A liquid (c) has a fixed volume but not a fixed shape.

1-8 The gaseous state is compressible because the basic particles are very far apart and thus the volume of a gas is mostly empty space.

1-10 (a) physical (b) chemical (c) physical (d) chemical (e) chemical
(f) physical (g) physical (h) chemical (i) physical

1-12 (a) chemical (b) physical (c) physical (d) chemical (e) physical

1-14 No. Both compounds and elements have definite and unchanging properties. A compound can be chemically decomposed, however, and an element cannot.

1-16 Original substance: green, solid (physical); can be decomposed (chemical).
  Substance is a compound.
  Gas: gas, colorless (physical); can be decomposed (chemical). Substance is a
  compound since it can be decomposed.
  Solid: shiny, solid (physical); cannot be decomposed (chemical). Substance is
  an element since it cannot be decomposed.

1-18 A compound such as carbon dioxide may be formed from a mixture of elements, but once formed, it is no longer a mixture. It is a unique substance with its own properties.

1-19 (a) liquid phase only. The solid has been dispersed into the water as its basic particles.

1-20 (a) homogeneous (b) heterogeneous (c) heterogeneous
(d) homogeneous (e) homogeneous (f) homogeneous
(g) heterogeneous (h) heterogeneous (i) homogeneous

1-22 (a) liquid (b) various solid phases (c) gas and liquid (d) liquid
(e) solid (f) liquid (g) solid and liquid (h) liquid and gas (i) gas

1-24 (a) solution (b) heterogeneous mixture (c) element (d) compound
(e) solution

1-26 Ocean water is the least pure, followed by drinking water, and then rain water.

1-28 When a log burns, most of the compounds formed in the combustion are gases and dissipate into the atmosphere. Only some solid residue (ashes) is left. When zinc and sulfur (both solids) combine, the only product is a solid so there is no weight change. When iron burns, however, its only product is a solid. It weighs more than the original iron because the iron has combined with the oxygen gas from the air.

1-29 (a) exothermic   (b) endothermic   (c) endothermic   (d) exothermic   (e) exothermic

1-30 Gasoline is converted into *heat* energy when it burns. The heat causes the pistons to move. This movement is *mechanical* energy. The mechanical energy turns the alternator, which generates *electrical* energy. The electrical energy is converted into *chemical* energy in the battery.

1-32 (a) potential   (b) kinetic   (c) potential (It is stored energy because of composition.)   (d) kinetic   (e) kinetic

1-35 The kinetic energy is greatest when the swing is moving the fastest, which occurs when it is closest to the ground. The potential energy is greatest when the swing is at its highest point and has momentarily stopped. If there is no loss of energy the sum of the kinetic energy and the potential energy is the same at all points.

Chapter 2

2-1 (b) 74.212 gal (the most significant figures)

2-2 A device used to produce a measurement may provide a reproducible answer to several significant figures, but if the device itself is inaccurate (such as a ruler with the tip broken off) the measurement is inaccurate.

2-4 (a) three   (b) two   (c) three   (d) one   (e) four   (f) two   (g) two   (h) three

2-6 (a) 16.0   (b) 1.01   (c) 0.665   (d) 4890   (e) 87,600   (f) 0.0272   (g) 301

2-8 (a) 188   (b) 12.90   (c) 2300   (d) 48   (e) 0.84

2-10 37.9 qt

2-12 (a) 120 cm^2 (b) 394 ft · lb   (c) 2 cm   (d) 2.3 in.

2-14 (a) 1.57×10^2 (b) 1.57×10^{-1} (c) 3.00×10^{-2} (d) 4.0×10^7 (e) 3.49×10^{-2} (f) 3.2×10^4 (g) 3.2×10^{10} (h) 7.71×10^{-4} (i) 2.34×10^3

2-15 (a) 4.23×10^5 (b) 4.338×10^2 (c) 2.0×10^{-3} (d) 8.8×10^2 (e) 8×10^{-5} (f) 8.20×10^7 (g) 7.5×10^{13} (h) 1.06×10^{-6}

2-16 (a) 0.000476   (b) 6550   (c) 0.00788   (d) 48,900   (e) 4.75   (f) 0.0000034

2-17 (a) 4.89×10^{-4} (b) 4.56×10^{-5} (c) 7.8×10^3 (d) 5.71×10^{-2} (e) 4.975×10^8 (f) 3.0×10^{-4}

2-19 (a) 1.597×10^{-3} (b) 2.30×10^7 (c) 3.5×10^{-5} (d) 2.0×10^{14}

2-21 (a) 3.1×10^{10} (b) 2×10^9 (c) 4×10^{13} (d) 14   (e) 2.56×10^{-14}

2-23 (a) 9×10^7 (b) 8.7×10^7 (c) 8.70×10^7

2-25 (a) 720 cm, 7.2 m, 7.2×10^{-3} km   (b) 5.64×10^4 mm, 5640 cm, 0.0564 km   (c) 2.50×10^5 mm, 2.50×10^4 cm, 250 m

2-26 (a) 8.9 g, 8.9×10^{-3} kg   (b) 2.57×10^4 mg, 0.0257 kg
(c) 1.25×10^6 mg, 1250 g

2-28 (a) $12 = 1$ doz   (c) 3 ft = 1 yd   (e) 10^3 m = 1 km

2-29 (a) $\dfrac{1 \text{ ft}}{12 \text{ in.}}$ (b) $\dfrac{2.54 \text{ cm}}{\text{in.}}$ (c) $\dfrac{5280 \text{ ft}}{\text{mile}}$

(d) $\dfrac{1.06 \text{ qt}}{\text{L}}$ (e) $\dfrac{1 \text{ qt}}{2 \text{ pt}}$, $\dfrac{1 \text{ L}}{1.06 \text{ qt}}$

Note: Your answer may vary from those below in the last significant figure depending on the conversion factor used.

2-31 (a) 4.8 miles, 2.6×10^4 ft, 7.8 km   (b) 2380 ft, 725 m, 0.725 km
(c) 1.70 miles, 2740 m, 2.74 km   (d) 4.21 miles, 2.22×10^4 ft, 6780 m

2-32 (a) 27.1 qt, 25.6 L   (b) 170 gal, 630 L   (c) 2.08×10^3 gal, 8.33×10^3 qt

2-34 55.4 kg

2-35 $28.0 \text{ m} \times \dfrac{100 \text{ cm}}{\text{m}} \times \dfrac{1 \text{ in.}}{2.54 \text{ cm}} \times \dfrac{1 \text{ ft}}{12 \text{ in.}} \times \dfrac{1 \text{ yd}}{3 \text{ ft}} = 30.6$ yd

The team should look for a new punter.

2-36 0.354 L

2-37 6 ft $10\frac{1}{2}$ in. = 82.5 in.

$$82.5 \text{ in.} \times \dfrac{2.54 \text{ cm}}{\text{in.}} \times \dfrac{1 \text{ m}}{10^2 \text{ cm}} = 2.10 \text{ m}$$
$$212 \text{ lb} = 96.4 \text{ kg}$$

2-38 14.6 gal

2-40 $\frac{1}{5} \text{ gal} \times \dfrac{4 \text{ qt}}{\text{gal}} = 0.800$ qt $0.800 \text{ qt} \times \dfrac{1 \text{ L}}{1.06 \text{ qt}} \times \dfrac{1 \text{ mL}}{10^{-3} \text{ L}} = \underline{\underline{755 \text{ mL}}}$

There is slightly more in a "fifth" than in 750 mL.

2-41 105 km/hr

2-43 $14,608

2-45 $\dfrac{\$0.899}{\text{gal}} \times \dfrac{1 \text{ gal}}{4 \text{ qt}} \times \dfrac{1.06 \text{ qt}}{\text{L}} = \$0.238/\text{L}$

80.0 L costs $19.04

2-46 551 miles costs $23.59

$$482 \text{ km} \times \dfrac{1 \text{ mi}}{1.61 \text{ km}} \times \dfrac{1 \text{ gal}}{21.0 \text{ mi}} \times \dfrac{\$0.899}{\text{gal}} = \$12.82$$

2-47 1690 km

2-50 382 nails

2-51 $28.20

2-53 2100 sec or 0.58 hr

2-55 572 °F

2-56 24 °C

2-58 −38 °F

2-60 95 °F

2-61 (a) 320 K   (b) 296 K   (c) 200 K   (d) 261 K   (e) 291 K   (f) 244 K

2-63 (a) −98 °C   (b) 22 °C   (c) 27 °C   (d) −48 °C   (e) 600 °C

2-64 Since $t(C) = t(F)$ sustitute $t(C)$ for $t(F)$ and set the two equations equal.

$$[t(C) \times 1.8] + 32 = \frac{[t(C) - 32]}{1.8} \quad t(C) = \underline{\underline{-40 \ °C}}$$

2-65 2.60 g/mL

2-67 Density = 1.74 g/mL   The solid could be magnesium.

2-69 $285 \, \cancel{g} \times \dfrac{1 \text{ mL}}{13.6 \, \cancel{g}} = 21.0 \text{ mL}$

2-70 1450 g

2-71 670 g

2-73 625 mL

2-74 0.951 g/mL; 4790 mL. Pumice floats on water but sinks in ethyl alcohol.

2-76 2080 g

2-78 One liter of water weighs 1000 g; 1 L of gasoline weighs 670 g.

2-80 $\left(\dfrac{2.54 \text{ cm}}{\text{in.}}\right)^3 = \dfrac{16.4 \text{ cm}^3}{\text{in.}^3} = \dfrac{16.4 \text{ mL}}{\text{in.}^3} \quad \left(\dfrac{12 \text{ in.}}{\text{ft}}\right)^3 = \dfrac{1728 \text{ in.}^3}{\text{ft}^3}$

$\dfrac{1.00 \, \cancel{g}}{\cancel{mL}} \times \dfrac{1.00 \text{ lb}}{454 \, \cancel{g}} \times \dfrac{16.4 \, \cancel{mL}}{\cancel{\text{in.}^3}} \times \dfrac{1728 \, \cancel{\text{in.}^3}}{\text{ft}^3} = 62.4 \text{ lb/ft}^3$

2-81 2.0×10^5 lb (100 tons)

2-83 0.853 J/(g · °C)

2-84 Specific heat = 0.0310 cal/(g · °C)   The substance could be gold.

2-86 $\dfrac{150 \, \cancel{cal}}{50.0 \, \cancel{g}} \times \dfrac{1 \, \cancel{g} \cdot °C}{0.092 \, \cancel{cal}} = 33 \ °C \text{ rise}$

$25 + 33 = \underline{\underline{58 \ °C}}$ (final temperature)

This compares with a 3 °C rise in temperature for 50.0 g of water.

2-87 506 J

2-88 $58 - 25 = 33 \ °C$ rise in temperature $\dfrac{1 \text{ g} \cdot \cancel{°C}}{0.106 \, \cancel{cal}} \times \dfrac{16.0 \, \cancel{cal}}{33 \, \cancel{°C}} = \underline{\underline{4.6 \text{ g}}}$

2-90 Iron will rise 9.38 °C, gold 32 °C, and water 0.997 °C.

2-92 45 °C

2-93 2910 joules

2-94 140 g of water

2-96 Heat lost by metal = heat gained by water.

$$100.0 \text{ g} \times 68.7 \text{ °C} \times \text{specific heat} = 100.0 \cancel{\text{g}} \times 6.3 \text{ }\cancel{\text{°C}} \times \frac{1.00 \text{ cal}}{\cancel{\text{g}} \cdot \cancel{\text{°C}}}$$

specific heat = 0.092 cal/g · °C (The metal could be copper.)

Chapter 3

3-2 C—carbon, Ca—calcium, Cd—cadmium, Ce—cerium, Cf—californium, Cl—chlorine, Cm–curium, Co—cobalt, Cr—chromium, Cs—cesium, Cu—copper

3-3 (a) barium—Ba (b) neon—Ne (c) cesium—Cs (d) platinum—Pt (e) manganese—Mn (f) tungsten—W

3-5 (a) B—boron (b) Bi—bismuth (c) Ge—germanium (d) U—uranium (e) Co—cobalt (f) Hg—mercury (g) Be—beryllium (h) As—arsenic

3-6 (b) and (e)

3-7 (c)

3-9 $^{16}_{8}\text{O}$, $^{14}_{7}\text{N}$, $^{28}_{14}\text{Si}$, $^{40}_{20}\text{Ca}$

3-11 (a) 21 p, 21 e, 24 n (b) 90 p, 90 e, 142 n (c) 87 p, 87 e, 136 n (d) 38 p, 38 e, 52n

3-13

Isotope Name	Isotope Symbol	Atomic Number	Mass Number	Subatomic Particles		
				Protons	Neutrons	Electrons
molybdenum-96	$^{96}_{42}\text{Mo}$	42	96	42	54	42
(a) silver-108	$^{108}_{47}\text{Ag}$	47	108	47	61	47
(b) silicon-28	$^{28}_{14}\text{Si}$	14	28	14	14	14
(c) potassium-39	$^{39}_{19}\text{K}$	19	39	19	20	19
(d) cerium-140	$^{140}_{58}\text{Ce}$	58	140	58	82	58
(e) iron-56	$^{56}_{26}\text{Fe}$	26	56	26	30	26
(f) tin-110	$^{110}_{50}\text{Sn}$	50	110	50	60	50
(g) iodine-118	$^{118}_{53}\text{I}$	53	118	53	65	53
(h) mercury-196	$^{196}_{80}\text{Hg}$	80	196	80	116	80

3-15 $5.81 \times 12.00 = 69.7$ amu. The element is Ga.

3-16 Atomic weight = 40.0 amu. The element is Ca.

3-17 (a) H: $\frac{1}{12} \times 8.00 = 0.667$ (b) N: $\frac{14}{12} \times 8.00 = 9.33$ (c) Na: 15.3
  (d) Ca: 26.7

3-20 79.9 amu

3-21 28.09 amu

3-23 Let X = decimal fraction of ^{35}Cl and Y = decimal fraction of ^{37}Cl. Since there are only two isotopes present, $X + Y = 1$, $Y = 1 - X$.

$$(X \times 35) + (Y \times 37) = 35.5$$
$$(X \times 35) + [(1 - X) \times 37] = 35.5$$
$$X = 0.75 \quad (75\% \ ^{35}Cl)$$
$$Y = 0.25 \quad (25\% \ ^{37}Cl)$$

3-25 (a) Re: at. no. 75, at. wt. 186.2   (b) Co: at. no. 27, at. wt. 58.9332
  (c) Br: at. no. 35, at. wt. 79.904   (d) Si: at. no. 14, at. wt. 28.086

3-26 (e) N_2, diatomic element   (b) CO, diatomic compound

3-27 (a) 6 carbons, 4 hydrogens, 2 chlorines   (b) 2 carbons, 6 hydrogens, 1 oxygen   (c) 1 copper, 1 sulfur, 18 hydrogens, 13 oxygens   (d) 9 carbons, 8 hydrogens, 4 oxygens   (e) 2 aluminums, 3 sulfurs, 12 oxygens   (f) 2 nitrogens, 8 hydrogens, 1 carbon, 3 oxygens

3-28 (a) 12   (b) 9   (c) 33   (d) 21   (e) 17   (f) 14

3-29 (a) 8   (b) 7   (c) 4   (d) 3

3-30 (a) SO_2 (b) CO_2 (c) H_2SO_4 (d) $Ca(ClO_4)_2$ (e) $(NH_4)_3PO_4$ (f) C_2H_2

3-32 Neon would appear as individual spheres that are widely spaced. Oxygen would appear as molecules with two spheres joined together.

3-34 (a) 19 p, 18 e (b) 35 p, 36 e (c) 16 p, 18 e (d) $7 + 16 = 23$ p, 24 e
  (e) 13 p, 10 e (f) 11 p, 10 e

3-36 (a) $^{90}_{38}Sr^{2+}$ (b) $^{52}_{24}Cr^{3+}$ (c) $^{79}_{34}Se^{2-}$ (d) $^{14}_{7}N^{3-}$ (e) $^{139}_{57}La^{3+}$

3-38 (a) $^{0}_{-1}e$ (b) $^{90}_{38}Sr$ (c) $^{231}_{90}Th$ (d) $^{41}_{20}Ca$ (e) $^{210}_{81}Tl$

3-40 (a) $^{230}_{90}Th \rightarrow ^{226}_{88}Ra + ^{4}_{2}He$ (b) $^{214}_{84}Po \rightarrow ^{210}_{82}Pb + ^{4}_{2}He$
  (c) $^{210}_{84}Po \rightarrow ^{210}_{85}At + ^{0}_{-1}e$ (d) $^{239}_{94}Pu \rightarrow ^{235}_{92}U + ^{4}_{2}He$ (e) $^{14}_{6}C \rightarrow ^{14}_{7}N + ^{0}_{-1}e$

3-41 $^{234}_{90}Th$, $^{234}_{91}Pa$, $^{234}_{92}U$, $^{230}_{90}Th$, $^{226}_{88}Ra$, $^{222}_{86}Rn$, $^{218}_{84}Po$, $^{214}_{82}Pb$, $^{214}_{83}Bi$, $^{210}_{81}Tl$, $^{210}_{82}Pb$, $^{210}_{83}Bi$, $^{210}_{84}Po$, $^{206}_{82}Pb$

3-42 2.5 g

3-44 (a) $^{90}_{39}Y$ (b) 50 yr

3-46 about 11,500 yr old

3-48 (a) $^{87}_{34}Se$ (b) $4^{1}_{0}n$ (c) $^{97}_{38}Sr$

3-49 (a) $^{35}_{16}S$ (b) $^{2}_{1}H$ (c) $^{30}_{15}P$ (d) $8^{0}_{-1}e$ (e) $^{254}_{102}No$ (f) $^{237}_{92}U$ (g) $4^{1}_{0}n$

3-51 Fission occurs when a large nucleus is split into two smaller nuclei and neutrons. Fusion occurs when two small nuclei join to form a larger nucleus. In both processes the mass of the products is less than the mass of the original nuclei. This mass has been converted to energy according to Einstein's relationship, $E = mc^2$.

3-52 $^{239}_{93}Np$, $_{-1}^{0}e$

Review Test for Chapters 1–3

Multiple Choice

1 (a)   2. (d)   3. (b)   4. (d)   5. (a)   6. (a)   7. (b)   8. (c)   9. (c)
10. (a)   11. (d)   12. (a)   13. (b)   14. (c)   15. (c)   16. (c)   17. (d)
18. (c)   19. (a)   20. (a)   21. (c)   22. (d)   23. (a)   24. (a)   25. (b)

Problems

1. (a) 174   (b) 0.00232   (c) 18,400   2. 17.17 cm   3. 69 in.²
4. 13 g/mL   5. 2.0×10^7 cm²   6. 0.014 mL   7. 1.2×10^{12}
8. 2×10^{-9} 9. 1.6×10^{13} 10. 74.4 in.   11. 41 L   12. 317 km/hr
13. 1.00×10^2 g/cm   14. 4.5 g/cm³ = 4.5 g/mL   15. 60.0 mL   16. 44 °C
17. an increase of 49 Fahrenheit degrees   18. 0.288 J/(g · °C)   19. 121.8 amu,
Sb, antimony   20. 114.6 amu

Chapter 4

4-1   (b) Sr   and   (d) Mg

4-3   Ultraviolet light has shorter wavelengths but higher energy than visible light. Ultraviolet light can damage tissue, thus causing a burn.

4-4   Since these two shells are close in energy, transitions of electrons from these two levels to the $n = 1$ shell have similar energy. Thus the wavelengths of light from the two transitions are very close together.

4-5   Since these two shells are comparatively far apart in energy, transitions from these two levels to the $n = 1$ shell have comparatively different energies. Thus the wavelengths of light from the two transitions are quite different. [The $n = 3$ to $n = 1$ transition has a shorter wavelength (more energy) than the $n = 2$ to $n = 1$ transition.]

4-6   32 electrons

4-8   The ground state for a lithium atom is the $n = 2$ shell. Thus the $n = 3$ shell is an excited state, and light can be emitted when the electron falls to the ground state.

4-9 $2d$, $1p$, $3f$

4-10 $s - 2$, $p - 6$, $d - 10$, $f - 14$

4-13   The $n = 5$ shell holds $2n^2$ or 50 electrons. The s, p, d, and f sublevels hold a total of 32 electrons. Therefore, the g sublevel holds 18 electrons.

4-14   (a) $6s$ (b) $5p$ (c) $4p$ (d) $4d$

4-16 $4s, 4p, 5s, 4d, 5p, 6s, 4f$

4-17   (a) Mg: $1s^2 2s^2 2p^6 3s^2$ (b) Ge: $1s^2 2s^2 2p^6 3s^2 3p^6 4s^2 3d^{10} 4p^2$
      (c) Pd: $1s^2 2s^2 2p^6 3s^2 3p^6 4s^2 3d^{10} 4p^6 5s^2 4d^8$ (d) Si: $1s^2 2s^2 2p^6 3s^2 3p^2$

4-19   (a) S: $[\text{Ne}]3s^2 3p^4$ (b) Zn: $[\text{Ar}]4s^2 3d^{10}$ (c) Pu: $[\text{Rn}]7s^2 6d^1 5f^5$
      (d) I: $[\text{Kr}]5s^2 4d^{10} 5p^5$

4-21   (a) F   (b) Ga   (c) Ba   (d) Gd   (e) Cu

4-23   (a) 17 (VIIA)   (b) 13 (IIIA)   (c) 2 (IIA)   (e) 11 (IB)

4-24   15 (VA) [noble gas]$ns^2 np^3$ 5 (VB) [noble gas]$ns^2 (n-1)d^3$

4-26   (b) and (e) group 14 (IVA)

4-28   (a) 14 (IVA)   (b) 18 (0) (noble gases)   (c) 11 (IB)   (d) Pr and Pa

4-29   (a) B   (b) Ar   (c) K, Cr, and Cu   (d) Ga   (e) Hf

4-31   (a) ns^2 (b) $ns^2 (n-1)d^{10}$ (c) $ns^2 np^4$ (d) $ns^2 (n-1)d^2$

4-32   He does not have a filled p sublevel. (There is no $1p$ sublevel.)

4-34   The theoretical order of filling is $6d$, $7p$, $8s$, $5g$. The $6d$ is completed at element 112. The $7p$ and $8s$ fill at element 120. Thus, element 121 would theoretically begin the filling of the $5g$ sublevel. This assumes filling in the order indicated by Figure 4-11.

4-36   (a) three   (b) five   (c) none (There is no $2d$ sublevel.)   (d) one   (e) seven

4-37 $3s$, one; $3p$, three; $3d$, five; total $=$ nine

4-38   If there are 18 electrons in the 5g sublevel, there would be 9 orbitals.

4-40   A $4p$ orbital is shaped roughly like a two-sided baseball bat with two "lobes" lying along one of the three axes. This shape represents the region of highest probability of finding the electrons.

4-42   (a) This is excluded by Hund's rule since electrons are not shown in separate orbitals of the same sublevel with parallel spins.
      (b) This is correct.
      (c) This is excluded by the Aufbau principle because the $2s$ orbital should fill before the $2p$.
      (d) This is excluded by the Pauli exclusion principle since the two electrons in the $2s$ orbital are shown with the same spin.

4-43

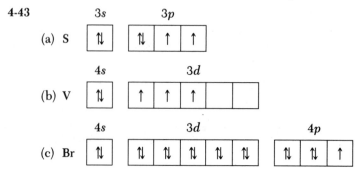

	6s	5d					4f						
(d) Pm	↑↓	↑					↑	↑	↑	↑			

4-45 12 (IIB), none; 5 (VB), three; 16 (VIA), two; 17 (VIIA), one; $[ns^1(n-1)d^5]$, six; Pm, five

Chapter 5

5-1 $Ar = [Ne]3s^23p^6$

5-3 32

5-5 (a) Fe, transition element   (b) Te, representative element   (c) Pm, inner transition element   (d) La, transition element   (e) Xe, noble gas   (f) H, representative element   (g) In, representative element

5-7 (b) Ti,   (e) Pd, and   (g) Ag

5-8 (a) transition   (b) representative   (c) noble gas   (d) inner transition

5-10 (a) Ne   (d) Cl   (f) N

5-12 (a) Ru   (b) Sn   (c) Hf   (h) W

5-13 (d) Te   (f) B

5-14 (a) Ar   (b) Hg   (c) N_2 (d) As   (e) Be   (f) Po

5-16 element 118 in group 18 (0)

5-18 (a) As   (b) Ru   (c) Ba   (d) I

5-20 Cr, 117 pm and Nb, 134 pm

5-22 The outer electron in Hf is in a shell higher in energy than Zr. This alone would make Hf a larger atom. However, in between Zr and Hf lie several sublevels, including the long 4f sublevel (Ce through Lu). The filling of these sublevels, especially the 4f, causes a gradual contraction that offsets the higher shell for Hf.

5-23 (a) V   (b) Cl   (c) Mg   (d) Fe   (e) B

5-24 In, 558 kJ/mol and Ge, 762 kJ/mol

5-26 Ga^+, 578.8 kJ; Ga^{2+}, 2558 kJ; Ga^{3+}, 5521 kJ; Ga^{4+}, 11,700 kJ. The fourth electron must be removed from an inner shell.

5-28 (a) Cs^+ and (e) Sc^{3+} are likely to form. Ne^+ and Te^{2+} would require a large amount of energy to form because they are nonmetals. To form Rb^{2+} and Tl^{4+} an electron would have to be removed from an inner shell.

5-30 (b) Se^{2-} and (d) N^{3-} can form. Be^- is unlikely because Be is a metal. To form F^{2-} and Ar^- an electron would have to be added to an outer shell.

5-32 (b) Al

5-34   (d) I

5-35   (b) Po (It is a metal.),   (c) Cl, and   (d) H

5-37   (d) Mg^{2+} (e) K^+ (c) S   (a) Mg   (b) S^{2-} (f) Se^{2-}

5-39

s^1	s^2	d^1	d^2	d^3	d^4	d^5	d^6	d^7	d^8	p^1	p^2	p^3	p^4
1													2
3	4									5	6	7	8
9	10									11	12	13	14
15	16	17	18	19	20	21	22	23	24	25	26	27	28
29	30	31	32	33	34	35	36	37	38	39	40	41	42
43	44	45*											

*46–50 would be in a $4f$ sublevel.

(a) 6 in second period, 14 in the fourth
(b) Third period, No. 14; fourth period, No. 28
(c) First inner transition element is No. 46 (assuming that one electron goes into the $5d$ and the next in the $4f$ as on earth).
(d) Elements 11 and 17 are most likely to be metals.
(e) Element 12 would have the larger radius in both cases.
(f) Element 7 would have the higher ionization energy in all three cases.
(g) The ions that would be reasonable are 16^{2+} (metal cation), 13^- (nonmetal anion), 15^+ (metal cation), and 1^- (nonmetal anion). The ion 9^{2+} is unreasonable since the second electron would come from a filled inner sublevel; 7^+ because it would be a nonmetal cation; and 17^{4+} because it would be necessary to remove an electron from a filled inner sublevel.

5-40   H is classified as a group 1 (IA) element because it has a ns^1 configuration like the alkali metals. Unlike the alkalis, however, hydrogen is not a metal. It is sometimes classified as a group 17 (VIIA) element because it is one electron short of a noble gas configuration.

5-41   (c) I_2 and (g) Br_2

5-43   (d) F_2

5-45   (d) O_3

5-46   (a) diamond

5-47   (b) Rb

5-49   (d) $[Kr]5s^24d^{10}5p^5$ (iodine)

5-51   Yes. The requirement for a group 3 (IIIB) element is to have a d^1 configuration. The configuration of Lu is $[Xe]6s^25d^14f^{14}$ and that of Lr is $[Rn]7s^26d^15f^{14}$. The only difference between these two elements and La and Ac is a filled f sublevel.

Chapter 6

6-1 (a) Ca· (b) ·Sb· (c) ·Sn· (d) ·Ï: (e) :Ne: (f) ·Bi· (g) ·Vï:

6-2 (a) group 3 (IIIA) (b) group 15 (VA) (c) group 2 (IIA)

6-3 The electrons from filled inner sublevels are not involved in chemical bonding.

6-5 (b) Sr and S (c) H and K (d) Al and F

6-7 (b) S^- (c) Cr^{2+} (e) In^+ (f) Pb^{2+} (h) Tl^{3+}

6-8 They have the noble gas configuration of He, which requires only two electrons.

6-9 (a) K^+ (b) ·Ö:$^-$ (c) :Ï:$^-$ (d) :P̈:$^{3-}$ (e) Ba^+ (f) ·Xë:$^+$ (g) Sc^{3+}

6-11 (b) 0^- (e) Ba^+ (f) Xe^+

6-12 (a) Mg^{2+} (b) Ga^{3+} (pseudo-noble gas) (c) Br^- (d) S^{2-} (e) P^{3-}

6-15 Se^{2-}, Br^-, Rb^+, Sr^{2+}, Y^{3+}

6-16 Tl^{3+} has a full $5d$ sublevel. This is a pseudo-noble gas configuration.

6-17 Cs_2S, Cs_3N, $BaBr_2$, BaS, Ba_3N_2, $InBr_3$, In_2S_3, InN

6-18 (a) CaI_2 (b) CaO (c) Ca_3N_2 (d) $CaTe$ (e) CaF_2

6-20 (a) Cr^{3+} (b) Fe^{3+} (c) Mn^{2+} (d) Co^{2+} (e) Ni^{2+} (f) V^{3+}

6-22 (a) H_2Se (b) GeH_4 (c) ClF (d) Cl_2O (e) NCl_3 (f) CBr_4

6-24 (b) Cl_3, (d) NBr_4, and (e) H_3O

6-25 (a) $SO_4{}^{2-}$ (b) $ClO_3{}^-$ (c) $SO_3{}^{2-}$ (d) $PO_4{}^{3-}$ (e) $C_2{}^{2-}$

6-27 (a)
```
   H  H
   |  |
H—C—C—H
   |  |
   H  H
```
(b) H—Ö—Ö—H (c) :F̈—N̈—F̈:
                        |
                       :F̈:

(d) :Cl̈—S̈—Cl̈: (e)
```
    H       H
    |       |
H—C—Ö—C—H
    |       |
    H       H
```
```
    H  H
    |  |
H—C—C—Ö—H
    |  |
    H  H
```

6-29 (a) :C≡O: (b)
```
.O.    .O.
   \\  //
     S
     |
    :O:
```
(c) $K^+[:C≡N:]^-$

(d)
```
H—Ö:
  |
:S—Ö:
  |
:O—H
```

6-30 (a) $N = 2$ $6(2) + 2 = 14$ valence electrons $= 4 + 6 = 10$ $14 - 10 = 4$ (one triple bond)

(b) $N = 4$ $6(4) + 2 = 26$ valence electrons $= 6 + 3(6) = 24$ $26 - 24 = 2$
(one double bond)

(c) for CN^-: $N = 2$ $6N + 2 = 14$ valence electrons $= 4 + 5 + 1 = 10$
$14 - 10 = 4$ (one triple bond)

(d) $N = 3$ $6N + 2 = 20$ valence electrons $= 6(H) + 8(C) + 6(O) = 20$
$20 - 20 = 0$ (all single bonds)

6-31 (a) N_2O $6N + 2 = 20$ valence electrons $= 16$ (two double or one triple bond)

(b) for the NO_2^- ion in $Ca(NO_2)_2$: $6N + 2 = 20$ valence electrons $= 18$
(one double bond)

(c) for $AsCl_3$: $6N + 2 = 26$ valence electrons $= 26$ (all single bonds)

(d) for H_2S: $6N + 2 = 8$ valence electrons $= 8$ (all single bonds)

(e) for CH_2Cl_2: $6N + 2 = 20$ valence electrons $= 20$ (all single bonds)

(f) for NH_4^+: $6N + 2 = 8$ valence electrons $= 8$ (all single bonds)

6-32 (a) $\overset{..}{N}{=}N{=}\overset{..}{O}$ (b) $Ca^{2+}\left[\ \overset{..}{\underset{O}{N}}\ \right]_2^-$

(c) $:\overset{..}{Cl}{-}\overset{..}{As}{-}\overset{..}{Cl}:$ (d) S (e) $:\overset{..}{Cl}{-}\overset{H}{\underset{H}{C}}{-}\overset{..}{Cl}:$
$\qquad\quad :\overset{..}{Cl}:\qquad\quad H\quad H$

(f) $\left[\ H{-}\overset{H}{\underset{H}{N}}{-}H\ \right]^+$

6-33 (a) $:\overset{..}{Cl}{-}\overset{..}{O}{-}\overset{..}{Cl}:$ (b) $\left[\ :\overset{..}{O}{-}\overset{..}{\underset{\overset{|}{:O:}}{S}}{-}\overset{..}{O}:\ \right]^{2-}$

(c) $\overset{H}{\underset{H}{{\diagdown}}}C{=}C\overset{{\diagup}H}{\underset{{\diagdown}H}{}}$

(d) $\overset{H}{\underset{H}{{\diagdown}}}C{=}\overset{..}{O}$ (e) $\overset{:\overset{..}{F}:}{\underset{:F\quad F:}{B}}$ (f) $:N{\equiv}O:^+$

6-37 (1)(a) 17 (b) 31 (c) 51_3 (d) 97 (e) $7(13)$ (f) $10(13)_2$ (g) 67_2
(h) 3_26
(2) On Zerk, six electrons fill the outer s and p orbitals to make a noble gas structure. Therefore we have a "sextet" rule on Zerk.
(a) $1{-}\overset{..}{7}:$ (b) $3^+1:^-$(ionic) (c) $1{-}\underset{\overset{|}{1}}{5}{-}1$ (d) $9^+:\overset{..}{7}:^-$(ionic)

(e) $:\overset{..}{7}-\overset{..}{13}:$ (f) $10^{2+}(:\overset{..}{13}:^-)_2$(ionic)   (g) $:\overset{..}{7}-\overset{..}{6}-\overset{..}{7}:$

(h) $(3^+)_2:\overset{..}{6}:^{2-}$(ionic)

(3) $4N+2$

6-38 (b) K_2SO_4

6-39 (a)

(b)

(c) $\left[:\overset{..}{O}-\overset{..}{\underset{\underset{\overset{..}{\underset{..}{O}}}{|}}{S}}-\overset{..}{O}:\right]^{2-}$ (only one structure)

6-41

6-42 A resonance hybrid is the actual structure of a molecule or ion that is implied by the various resonance structures. For example, the two structures shown in Problem 6-41 imply that both C—O bonds have properties that are halfway between those of a single bond and a double bond.

6-43 $:\overset{..}{O}-C\equiv O: \leftrightarrow :O\equiv C-\overset{..}{O}:$ The two resonance structures imply that both C—O bonds are halfway between a single and a triple bond, which is a double bond. This is the same as the one structure shown below.

$$\overset{.}{\underset{..}{O}}=C=\overset{..}{\underset{..}{O}}$$

6-44 Cs, Ba, Be, B, C. Cl, O, F

6-45 (a) $\overset{\delta-\quad\delta+}{\underset{\leftarrow}{N-H}}$ (b) $\overset{\delta+\quad\delta-}{\underset{\rightarrow}{B-H}}$ (c) $\overset{\delta+\quad\delta-}{\underset{\rightarrow}{Li-H}}$ (d) $\overset{\delta-\quad\delta+}{\underset{\leftarrow}{F-O}}$ (e) $\overset{\delta-\quad\delta+}{\underset{\leftarrow}{O-Cl}}$

(f) $\overset{\delta-\quad\delta+}{\underset{\leftarrow}{S-Se}}$ (g) $\overset{\delta-\quad\delta+}{\underset{\leftarrow}{C-B}}$ (h) $\overset{\delta+\quad\delta-}{\underset{\rightarrow}{Cs-N}}$ (i) C—S (very low polarity)

6-46 (i) very little polarity, (b) = (f), (d) = (e) = (g), (a), (c), (h) (Those marked as equal are actually not exactly equal as there is always some difference. The precision of the electronegativity scale that we used, however, does not permit us to differentiate.)

6-47 Only (d) is predicted to be ionic on this basis. The electronegativity difference is 2.5.

6-49 (b) I—I

6-51 (a) nonpolar (equal dipoles cancel)   (b) polar (unequal dipoles)   (c) polar (equal dipoles do not cancel)   (d) nonpolar (equal dipoles cancel) (e) polar (unequal dipoles)

6-53 Since the H—S bond is much less polar than the H—O bond, the resultant molecular dipole is much less. The H_2S molecule is less polar than the H_2O molecule.

6-55 The CHF_3 molecule is more polar than the $CHCl_3$ molecule. The C—F bond is more polar, which means that the resultant molecular dipole is larger for CHF_3.

Chapter 7

7-1 (a) Pb $+4$, O -2 (b) P $+5$, O -2,   (c) C -1, H $+1$
(d) N -2, H $+1$ (e) Li $+1$, H -1 (f) B $+3$, Cl -1
(g) Rb $+1$, Se -2 (h) Bi $+3$, S -2

7-3 (a) Li   (e) K   (g) Rb

7-5 $+3$

7-6 (a) P $+5$ (b) C $+3$ (c) Cl $+7$ (d) Cr $+6$ (e) S $+6$
(f) N $+5$ (g) Mn $+7$

7-8 K_3N (-3), N_2H_4 (-2), NH_2OH (-1), $N_2(O)$, N_2O $(+1)$, NO $(+2)$, N_2O_3 $(+3)$, N_2O_4 $(+4)$, $Ca(NO_3)_2$ $(+5)$

7-9 (a) lithium fluoride   (b) barium telluride   (c) strontium nitride
(d) barium hydride   (e) aluminum chloride

7-11 (a) Rb_2Se (b) SrH_2 (c) RaO   (d) Al_4C_3 (e) BeF_2

7-13 (a) bismuth(V) oxide   (b) tin(II) sulfide   (c) tin(IV) sulfide
(d) copper(I) telluride   (e) titanium(IV) carbide

7-15 (a) Cu_2S (b) V_2O_3 (c) AuBr   (d) Ni_2C (e) CrO_3

7-17 (a) Bi_2O_5 (c) SnS_2 (e) TiC   (e) CrO_3

7-19 (c) ClO_3^-

7-21 ammonium, NH_4^+

7-22 (b) permanganate (MnO_4^-) (c) perchlorate (ClO_4^-) (e) phosphate (PO_4^{3-}) (f) oxalate $(C_2O_4^{2-})$

7-23 (a) chromium(II) sulfate   (b) aluminum sulfite   (c) iron(II) cyanide
(d) rubidium hydrogen carbonate   (e) ammonium carbonate
(f) ammonium nitrate   (g) bismuth(III) hydroxide

7-25 (a) $Mg(MnO_4)_2$ (b) $Co(CN)_2$ (c) $Sr(OH)_2$ (d) Tl_2SO_3
(e) $In(HSO_4)_3$ (f) $Fe_2(C_2O_4)_3$ (g) $(NH_4)_2Cr_2O_7$ (h) $Hg_2(C_2H_3O_2)_2$

7-27

	HSO_3^-	Te^{2-}	PO_4^{3-}
NH_4^+	NH_4HSO_3 ammonium bisulfite	$(NH_4)_2Te$ ammonium telluride	$(NH_4)_3PO_4$ ammonium phosphate
Co^{2+}	$Co(HSO_3)_2$ cobalt(II) bisulfite	$CoTe$ cobalt(II) telluride	$Co_3(PO_4)_2$ cobalt(II) phosphate
Al^{3+}	$Al(HSO_3)_3$ aluminum bisulfite	Al_2Te_3 aluminum telluride	$AlPO_4$ aluminum phosphate

7-29 (a) sodium chloride   (b) sodium hydrogen carbonate   (c) calcium carbonate   (d) sodium hydroxide   (e) sodium nitrate   (f) ammonium chloride   (g) aluminum oxide   (h) calcium hydroxide   (i) potassium hydroxide

7-30 (a) Ca_2XeO_6 (b) K_4XeO_6 (c) $Al_4(XeO_6)_3$

7-31 (a) phosphonium fluoride   (b) potassium hypobromite   (c) cobalt(III) iodate   (d) calcium silicate (actual name is calcium metasilicate)   (e) aluminum phosphite   (f) chromium(II) molybdate

7-32 (a) Si   (b) I   (c) B   (d) Kr   (e) H

7-34 (a) carbon disulfide   (b) boron trifluoride   (c) tetraphosphorus decoxide   (d) dibromine trioxide   (e) sulfur trioxide   (f) dichlorine monoxide   (g) phosphorus pentachloride   (h) sulfur hexafluoride

7-36 (a) P_4O_6 (b) CCl_4 (c) IF_3 (d) Cl_2O_7 (e) SF_6 (f) XeO_2

7-38 (a) hydrochloric acid   (b) nitric acid   (c) hypochlorous acid   (d) permanganic acid   (e) periodic acid   (f) hydrobromic acid

7-39 (a) HCN   (b) H_2Se (c) $HClO_2$ (d) H_2CO_3 (e) HI   (f) $HC_2H_3O_2$

7-41 (a) hypobromous acid   (b) iodic acid   (c) phosphorous acid   (d) molybdic acid   (e) perxenic acid

7-42 H_3AsO_3, arsenious acid; H_3AsO_4, arsenic acid

Review Test for Chapters 4–7

Multiple Choice

1. (c)   2. (c)   3. (b)   4. (c)   5. (a)   6. (a)   7. (a)   8. (b)   9. (d)
10. (d)   11. (b)   12. (a)   13. (b)   14. (b)   15. (b)   16. (a)   17. (c)
18. (d)   19. (c)   20. (c)   21. (a)   22. (c)   23. (c)   24. (d)   25. (c)
26. (d)   27. (b)   28. (e)   29. (d)   30. (d)

Problems

1. aluminum (Al)   (a) IIIA, representative element   (b) $[Ne]3s^23p^1$ (c) solid   (d) metal   (e) +3   (f) neon   (g) Al_2S_3, $AlBr_3$, AlN   (h) aluminum sulfide, aluminum bromide, aluminum nitride

2. nitrogen (N)   (a) VA, representative element   (b) $1s^2 2s^2 2p^3$ (c) three
(d) nonmetal   (e) gas   (f) N_2 (g) $:N\equiv N:$ (h) -3 (i) neon   (j) oxygen, positive; boron, negative   (k) Mg_3N_2, Li_3N, NF_3 (l) magnesium
nitride, lithium nitride, nitrogen trifluoride   (m) $:\ddot{F}-\underset{\underset{:\ddot{F}:}{|}}{\ddot{N}}-\ddot{F}:$

3. (a) O   (b) He   (c) Cu   (d) Te   (e) Y   (f) Sb   (g) F   (h) Pt   (i) P
(j) Ba

4. (a) $Mg^{2+}SO_4{}^{2-}$ magnesium sulfate Mg^{2+} $\left[\ :\underset{\underset{:\ddot{O}:}{|}}{\overset{\overset{:\ddot{O}\cdot}{|}}{S}}-\ddot{O}: \right]^{2-}$

(b) no ions   nitrous acid $H-\ddot{O}-\ddot{N}=\ddot{O}\cdot$

(c) $Li^+NO_3{}^-$ lithium nitrate Li^+ $\left[\underset{\ddot{O}\diagdown\ \diagup\ddot{O}}{\overset{\overset{\cdot\ddot{O}\cdot}{\|}}{N}} \right]^-$

(d) $(Co^{3+})_2(CO_3{}^{2-})_3$ cobalt(III) $2(Co^{3+})$ $3\left[\underset{\ddot{O}\diagdown\ \diagup\ddot{O}}{\overset{\overset{\cdot\ddot{O}\cdot}{\|}}{C}} \right]^{2-}$
                     carbonate

(e) no ions   dichlorine monoxide $:\ddot{Cl}-\ddot{O}-\ddot{Cl}:$

5. (a) N_2O_3 no ions $\underset{\ddot{O}\diagup}{\overset{\ddot{O}\diagdown}{N}}-\ddot{N}=\ddot{O}\cdot$

(b) $Cr_2(SO_3)_3$ $(Cr^{3+})_2(SO_3{}^{2-})_3$ $2Cr^{3+}$ $3\left[:\ddot{O}-\underset{\underset{:\ddot{O}:}{|}}{\ddot{S}}-\ddot{O}: \right]^{2-}$

(c) $Fe(OH)_2$ $Fe^{2+}(OH^-)_2$ Fe^{2+} $2\left[:\ddot{O}-H \right]^-$

(d) SrC_2O_4 $Sr^{2+}(C_2O_4{}^{2-})$ Sr^{2+} $\left[\underset{\ddot{O}\diagup\ \ \ \diagdown\ddot{O}}{\overset{\ddot{O}\diagdown\ \ \ \diagup\ddot{O}}{C-C}} \right]^{2-}$

(e) HI   no ions $H-\ddot{I}:$

Chapter 8

8-1 6.02×10^{23} ~~units~~ $\times \dfrac{1 \text{ ~~sec~~}}{2 \text{ ~~units~~}} \times \dfrac{1 \text{ ~~min~~}}{60 \text{ ~~sec~~}} \times \dfrac{1 \text{ ~~hr~~}}{60 \text{ ~~min~~}} \times \dfrac{1 \text{ ~~day~~}}{24 \text{ ~~hr~~}} \times \dfrac{1 \text{ year}}{365 \text{ ~~day~~}}$

$= 9.5 \times 10^{15}$ years (9.5 quadrillion years) for one person

It would take five billion people 1.9×10^6 (1.9 million) years.

8-3 79.4 g

8-5 16.0 ~~g Cu~~ $\times \dfrac{? \text{g X}}{63.5 \text{ ~~g Cu~~}} = 49.1$ g X; ? $= 195$ g

The element is <u>platinum</u>.

8-6 71.5 g of Cu

8-8 Atomic weight is 16.0 g. The element is oxygen. The formula is <u>CO</u>.

8-9 303 g of Br

8-11 (a) 0.468 mol of P, 2.82×10^{23} atoms   (b) 150 g of Rb, 1.05×10^{24} atoms   (c) 27.0 g, 1.00 mol   (d) 5.00 mol, Ge   (e) 7.96×10^{-23} g, 1.66×10^{-24} mol

8-13 (a) 63.5 g   (b) 16 g   (c) 40.1 g

8-15 (a) 1.93×10^{25} (b) 6.02×10^{23} (c) 1.20×10^{24}

8-17 50.0 ~~g Al~~ $\times \dfrac{1.00 \text{ mol}}{27.0 \text{ ~~g Al~~}} = 1.85$ mol Al

50.0 ~~g Fe~~ $\times \dfrac{1.00 \text{ mol}}{55.8 \text{ ~~g Fe~~}} = 0.896$ mol Fe

There are more moles of atoms (or more atoms) in 50.0 g of Al.

8-18 20.0 ~~g Ni~~ $\times \dfrac{1.00 \text{ mol}}{58.7 \text{ ~~g Ni~~}} = 0.341$ mol Ni

2.85×10^{23} ~~atoms~~ $\times \dfrac{1.00 \text{ mol Ni}}{6.02 \times 10^{23} \text{ ~~atoms~~}} = 0.473$ mol Ni

The 2.85×10^{23} atoms contain more Ni.

8-20 (a) 107 amu   (b) 80.1 amu   (c) 108 amu   (d) 98.1 amu   (e) 106 amu   (f) 60.0 amu   (g) 460 amu

8-22 $Cr_2(SO_4)_3$ $2(52.0) + 3(32.0) + 12(16.0) = 392$ amu

8-24 (a) 189 g, 6.32×10^{24} molecules   (b) 0.339 g, 5.00×10^{-3} mol   (c) 0.219 mol, 1.32×10^{23} molecules   (d) 0.0209 g, 7.22×10^{19} molecules   (e) 598 g, 7.48 mol   (f) 7.62×10^{-3} mol, 4.59×10^{21} molecules

8-26 C 5.10 mol, H 15.3 mol, O 2.55 mol, 23.0 mol total
C 61.2 g, H 15.4 g, O 40.8 g, 117.4 g total

8-27 0.135 mol of $Ca(ClO_3)_2$, 0.135 mol of Ca, 0.270 mol of Cl, 0.810 mol of O, 1.215 mol of atoms

8-29 $1.50 \text{ mol } H_2SO_4 \times \dfrac{2 \text{ mol } H}{\text{mol } H_2SO_4} \times \dfrac{1.01 \text{ g } H}{\text{mol } H} = \underline{\underline{3.03 \text{ g } H}}$

48.2 g of S, 72.0 g of O

8-31 $1.20 \times 10^{22} \text{ molecules} \times \dfrac{1.00 \text{ mol } O_2}{6.02 \times 10^{23} \text{ molecules}}$

$$= 0.0199 \text{ mol } O_2 \text{ molecules}$$

$0.0199 \text{ mol } O_2 \times \dfrac{2 \text{ mol } O \text{ atoms}}{\text{mol } O_2} = 0.0398 \text{ mol } O \text{ atoms}$

$0.0199 \text{ mol } O_2 \times \dfrac{32.0 \text{ g}}{\text{mol } O_2} = 0.637 \text{ g } O_2$

same mass as O atoms

8-33   N, 25.89%; O, 74.11%

8-34   Si, 46.7%; O, 53.3%

8-36   (a) C_2H_6O: C, 52.1%; H, 13.1%; O, 34.8%   (b) C_3H_6: C, 85.5%;
H, 14.5%   (c) C_9H_{18}: C, 85.7%; H, 14.3% (essentially the same as C_3H_6)
(d) Na_2SO_4: Na, 32.4%; S, 22.6%; O, 45.0%   (e) $(NH_4)_2CO_3$: N, 29.1%;
C, 12.5%; H, 8.41%; O, 49.9%

8-38   Na, 12.1%; B, 11.3%, O, 71.4%; H, 5.27%

8-40 $(7 \times 12.0) + (5 \times 1.01) + 32.1 + 14.0 + (3 \times 16.0) = 183 \text{ amu}$

$\text{%C:} \dfrac{84.0 \text{ amu}}{183 \text{ amu}} \times 100\% = \underline{\underline{45.9\%}} \quad \text{%H:} \dfrac{5.05 \text{ amu}}{183 \text{ amu}} \times 100\% = \underline{\underline{2.76\%}}$

$\text{%S:} \dfrac{32.1 \text{ amu}}{183 \text{ amu}} \times 100\% = \underline{\underline{17.5\%}} \quad \text{%N:} \dfrac{14.0 \text{ amu}}{183 \text{ amu}} \times 100\% = \underline{\underline{7.65\%}}$

$\text{%O:} = 100 - (45.9 + 2.8 + 17.5 + 7.6) = \underline{\underline{26.2\%}}$

8-42 $125 \text{ g } Na_2C_2O_4 \times \dfrac{1 \text{ mol } Na_2C_2O_4}{134 \text{ g } Na_2C_2O_4} \times \dfrac{2 \text{ mol } C}{\text{mol } Na_2C_2O_4} \times \dfrac{12.0 \text{ g } C}{\text{mol } C} = \underline{\underline{22.4 \text{ g } C}}$

8-43   4.73 lb

8-45   1398 lb

8-46   (a) and (d)

8-47   (a) FeS   (b) SrI_2 (c) $KClO_3$ (d) I_2O_5 (e) Fe_3O_4 (f) $C_3H_5Cl_3$

8-49 N_2O_3

8-51 KO_2

8-53 MgC_2O_4

8-54 CH_2Cl

8-56 $N_2SH_8O_3$

8-57 $C_8H_8O_3$ (*Hint:* $2.67 = \frac{8}{3}$)

8-59 $1.20 \text{ g CO}_2 \times \dfrac{1.00 \text{ mol CO}_2}{44.0 \text{ g CO}_2} \times \dfrac{1 \text{ mol C}}{1 \text{ mol CO}_2} = 0.0273 \text{ mol C}$

$0.489 \text{ g H}_2\text{O} \times \dfrac{1.00 \text{ mol H}_2\text{O}}{18.0 \text{ g H}_2\text{O}} \times \dfrac{2 \text{ mol H}}{1 \text{ mol H}_2\text{O}} = 0.0543 \text{ mol H}$

$\dfrac{0.0273}{0.0273} = 1 \quad \dfrac{0.0543}{0.0273} = 2 \quad \underline{\underline{\text{CH}_2}}$

8-60 $C_9H_{12}Cl_{12}$

8-62 $B_2C_2H_6O_4$

8-63 empirical formula = $KC_2NH_3O_2$
empirical weight = 112 g/emp unit
$\dfrac{224 \text{ g/mol}}{112 \text{ g/emp unit}} = 2$ emp units/mol
$\underline{\underline{K_2C_4N_2H_6O_4}}$ (molecular formula)

8-65 I_6C_6

Chapter 9

9-1 (a) $Cl_2(g)$ (b) $C(s)$ (c) $K_2SO_4(s)$ (d) $H_2O(l)$ (e) $P_4(s)$ (f) $H_2(g)$
(g) $Br_2(l)$ (h) $NaBr(s)$ (i) $S_8(s)$ (j) $Na(s)$ (k) $Hg(l)$ (l) $CO_2(g)$

9-2 (a) $CaCO_3 \rightarrow CaO + CO_2$
(b) $4Na + O_2 \rightarrow 2Na_2O$
(c) $H_2SO_4 + 2NaOH \rightarrow Na_2SO_4 + 2H_2O$
(d) $2H_2O_2 \rightarrow 2H_2O + O_2$
(e) $Si_2H_6 + 8H_2O \rightarrow 2Si(OH)_4 + 7H_2$
(f) $2Al + 2H_3PO_4 \rightarrow 2AlPO_4 + 3H_2$
(g) $Ca(OH)_2 + 2HCl \rightarrow CaCl_2 + 2H_2O$
(h) $Na_2NH + 2H_2O \rightarrow NH_3 + 2NaOH$
(i) $3Mg + N_2 \rightarrow Mg_3N_2$
(j) $CaC_2 + 2H_2O \rightarrow C_2H_2 + Ca(OH)_2$
(k) $2C_2H_6 + 7O_2 \rightarrow 4CO_2 + 6H_2O$

9-4 (a) $2Na(s) + 2H_2O(l) \rightarrow H_2(g) + 2NaOH(aq)$
(b) $2KClO_3(s) \rightarrow 2KCl(s) + 3O_2(g)$
(c) $NaCl(aq) + AgNO_3(aq) \rightarrow AgCl(s) + NaNO_3(aq)$
(d) $2H_3PO_4(aq) + 3Ca(OH)_2(aq) \rightarrow Ca_3(PO_4)_2(s) + 6H_2O(l)$

9-6 For Problem 9-2:   (a) decomposition   (b) combustion, combination
(c) double replacement   (d) decomposition   (e) double replacement (OH
from H_2O replaces H on Si_2H_6)   (f) single replacement   (g) double
replacement   (h) double replacement   (i) combination   (j) double replacement   (k) combustion
For Problem 9-4:   (a) single replacement   (b) decomposition   (c) double
replacement   (d) double replacement

9-8 (a) $2K(s) + Cl_2(g) \rightarrow 2KCl(s)$
(b) $Ca(s) + 2H_2O(l) \rightarrow Ca(OH)_2(aq) + H_2(g)$

(c) $2C_6H_6(l) + 15O_2(g) \rightarrow 12CO_2(g) + 6H_2O(l)$

(d) $Na_2S(aq) + Cu(NO_3)_2(aq) \rightarrow CuS(s) + 2NaNO_3(aq)$

(e) $2Au_2O_3(s) \xrightarrow{\Delta} 4Au(s) + 3O_2(g)$

9-10 (a) $\dfrac{1 \text{ mol } H_2}{1 \text{ mol } Mg}$ (b) $\dfrac{2 \text{ mol } HCl}{1 \text{ mol } Mg}$ (c) $\dfrac{1 \text{ mol } H_2}{2 \text{ mol } HCl}$ (d) $\dfrac{2 \text{ mol } HCl}{1 \text{ mol } MgCl_2}$

9-11 (a) $\dfrac{2 \text{ mol } C_4H_{10}}{8 \text{ mol } CO_2}$ (b) $\dfrac{2 \text{ mol } C_4H_{10}}{13 \text{ mol } O_2}$ (c) $\dfrac{13 \text{ mol } O_2}{8 \text{ mol } CO_2}$ (d) $\dfrac{10 \text{ mol } H_2O}{13 \text{ mol } O_2}$

9-13 (a) $\text{mol } H_2O \rightarrow \text{mol } H_2$

$$0.400 \text{ mol } H_2O \times \frac{2 \text{ mol } H_2}{2 \text{ mol } H_2O} = 0.400 \text{ mol } H_2$$

(b) $g\ O_2 \rightarrow \text{mol } O_2 \rightarrow \text{mol } H_2O$

$$0.640 \text{ g } O_2 \times \frac{1 \text{ mol } O_2}{32.0 \text{ g } O_2} \times \frac{2 \text{ mol } H_2O}{1 \text{ mol } O_2} = 0.0400 \text{ mol } H_2O$$

(c) $g\ O_2 \rightarrow \text{mol } O_2 \rightarrow \text{mol } H_2$

$$0.032 \text{ g } O_2 \times \frac{1 \text{ mol } O_2}{32.0 \text{ g } O_2} \times \frac{2 \text{ mol } H_2}{1 \text{ mol } O_2} = 0.0020 \text{ mol } H_2$$

(d) $g\ H_2 \rightarrow \text{mol } H_2 \rightarrow \text{mol } H_2O \rightarrow g\ H_2O$

$$0.400 \text{ g } H_2 \times \frac{1 \text{ mol } H_2}{2.02 \text{ g } H_2} \times \frac{2 \text{ mol } H_2O}{2 \text{ mol } H_2} \times \frac{18.0 \text{ g } H_2O}{\text{mol } H_2O} = 3.56 \text{ g } H_2O$$

9-14 (a) 1.35 mol of CO_2, 1.80 mol of H_2O, 2.25 mol of O_2

(b) 4.80 g of H_2O

(c) 1.10 g of C_3H_8

(d) 44.0 g of C_3H_8

(e) molecules $O_2 \rightarrow \text{mol } O_2 \rightarrow \text{mol } CO_2 \rightarrow g\ CO_2$

$$1.20 \times 10^{23} \text{ molecules} \times \frac{1 \text{ mol } O_2}{6.02 \times 10^{23} \text{ molecules}} \times \frac{3 \text{ mol } CO_2}{5 \text{ mol } O_2}$$
$$\times \frac{44.0 \text{ g } CO_2}{\text{mol } CO_2} = 5.26 \text{ g } CO_2$$

(f) 0.0997 mol of H_2O

9-16 47.2 g of N_2

9-18 $125 \text{ g } Fe_2O_3 \times \dfrac{1.00 \text{ mol } Fe_2O_3}{160 \text{ g } Fe_2O_3} \times \dfrac{2 \text{ mol } Fe_3O_4}{3 \text{ mol } Fe_2O_3} \times \dfrac{3 \text{ mol } FeO}{1 \text{ mol } Fe_3O_4}$

$$\times \frac{1 \text{ mol } Fe}{1 \text{ mol } FeO} \times \frac{55.85 \text{ g } Fe}{\text{mol } Fe} = 87.3 \text{ g } Fe$$

9-19 0.730 g of HCl

9-21 $\text{mol } FeS_2 \rightarrow \text{mol } H_2S \rightarrow \text{molecules } H_2S$

$$0.520 \text{ mol } FeS_2 \times \frac{1 \text{ mol } H_2S}{1 \text{ mol } FeS_2} \times \frac{6.02 \times 10^{23} \text{ molecules}}{\text{mol } H_2S}$$
$$= 3.13 \times 10^{23} \text{ molecules}$$

9-22 16.9 kg of HNO_3

9-24 2.30×10^3 g (2.30 kg) of C_2H_5OH

9-25 8.14×10^3 g (8.14 kg) of CO

9-27 30.0 g of SO_3 (theoretical yield); 70.7% yield

9-30 theoretical yield $\times \dfrac{\% \text{ yield}}{100\%} =$ actual yield;

theoretical yield $= \dfrac{\text{actual yield}}{\dfrac{\% \text{ yield}}{100\%}}$

theoretical yield $= \dfrac{250 \text{ g}}{0.700} = 357$ g N_2

g $N_2 \rightarrow$ mol $N_2 \rightarrow$ mol $H_2 \rightarrow$ g H_2

$$357 \text{ g } N_2 \times \frac{1 \text{ mol } N_2}{28.0 \text{ g } N_2} \times \frac{4 \text{ mol } H_2}{1 \text{ mol } N_2} \times \frac{2.02 \text{ g } H_2}{\text{mol } H_2} = 103 \text{ g } H_2$$

9-31 86.4%

9-32 If 86.4% is converted to CO_2, 13.6% is converted to CO. Thus 0.136×57.0 g $= 7.75$ g of C_8H_{18} is converted to CO. Note that 1 mol of C_8H_{18} would form 8 mol of CO (because of the eight carbons in C_8H_{18}). Thus,

$$7.75 \text{ g } C_8H_{18} \times \frac{1.00 \text{ mol } C_8H_{18}}{114 \text{ g } C_8H_{18}} \times \frac{8 \text{ mol CO}}{1 \text{ mol } C_8H_{18}} \times \frac{28.0 \text{ g CO}}{\text{mol CO}} = 15.2 \text{ g CO}$$

9-35 $2KClO_3 \rightarrow 2KCl + 3O_2$

Find the mass of $KClO_3$ needed to produce 12.0 g O_2.

g $O_2 \rightarrow$ mol $O_2 \rightarrow$ mol $KClO_3 \rightarrow$ g $KClO_3$

30.6 g $KClO_3$ needed to produce 12.0 g O_2.

Percent purity $= \dfrac{\text{mass of compound}}{\text{mass of sample}} \times 100\%$

$= \dfrac{30.6 \text{ g}}{50.0 \text{ g}} \times 100\% = \underline{61.2\%}$

9-36 4.49% FeS_2

9-38 1.00 mol of H_2 (H_2SO_4 is the limiting reactant.)

$$1.00 \text{ mol } H_2SO_4 \times \frac{2 \text{ mol Al}}{3 \text{ mol } H_2SO_4} = 0.667 \text{ mol of Al used;}$$

$$0.800 - 0.667 = \underline{0.133 \text{ mol Al remains}}$$

9-39

$$3.44 \text{ mol } C_5H_6 \times \frac{10 \text{ mol } CO_2}{2 \text{ mol } C_5H_6} = 17.7 \text{ mol } CO_2$$

$$20.6 \text{ mol } O_2 \times \frac{10 \text{ mol } CO_2}{13 \text{ mol } O_2} = 15.8 \text{ mol } CO_2$$

since O_2 is the limiting reactant:

$$15.8 \text{ mol } CO_2 \times \frac{44.0 \text{ g } CO_2}{\text{mol } CO_2} = \underline{695 \text{ g } CO_2}$$

9-41 NH_3 is the limiting reactant producing 0.750 mol of N_2.

9-42 $20.0 \text{ g AgNO}_3 \times \dfrac{1.00 \text{ mol AgNO}_3}{170 \text{ g AgNO}_3} \times \dfrac{2 \text{ mol AgCl}}{2 \text{ mol AgNO}_3} = 0.118 \text{ mol AgCl}$

$10.0 \text{ g CaCl}_2 \times \dfrac{1.00 \text{ mol CaCl}_2}{111 \text{ g CaCl}_2} \times \dfrac{2 \text{ mol AgCl}}{1 \text{ mol CaCl}_2} = 0.180 \text{ mol AgCl}$

Since $AgNO_3$ produces the smallest yield, it is the limiting reactant.

$$0.118 \text{ mol AgCl} \times \dfrac{143 \text{ g AgCl}}{\text{mol AgCl}} = 16.9 \text{ g AgCl}$$

Convert moles of AgCl (from limiting reactant) to grams of $CaCl_2$ used.

$$0.118 \text{ mol AgCl} \times \dfrac{1 \text{ mol CaCl}_2}{2 \text{ mol AgCl}} \times \dfrac{111 \text{ g CaCl}_2}{\text{mol CaCl}_2} = 6.5 \text{ g CaCl}_2 \text{ used}$$

$$10.0 - 6.5 = \underline{3.5 \text{ g CaCl}_2 \text{ remaining}}$$

9-44   products: 2.94 g H_2O, 2.93 g NO, 4.72 g S
reactants remaining: 3.8 g HNO_3

9-45 NH_3 is the limiting reactant. The theoretical yield based on NH_3 is 141 g of NO. The percent yield is 28.4%.

9-47
$$2Mg(s) + O_2(g) \longrightarrow 2MgO(s) + 1204 \text{ kJ}$$
$$2Mg(s) + O_2(g) \longrightarrow 2MgO(s) \quad \Delta H = -1204 \text{ kJ}$$

9-49
$$CaCO_3(s) + 176 \text{ kJ} \longrightarrow CaO(s) + CO_2(g)$$
$$CaCO_3(s) \longrightarrow CaO(s) + CO_2(g) \quad \Delta H = +176 \text{ kJ}$$

9-50 $1.00 \text{ g C}_8\text{H}_{18} \times \dfrac{1 \text{ mol C}_8\text{H}_{18}}{114 \text{ g C}_8\text{H}_{18}} \times \dfrac{5480 \text{ kJ}}{\text{mol C}_8\text{H}_{18}} = \underline{48.1 \text{ kJ}}$

1.00 g CH_4 yields $\underline{55.6 \text{ kJ}}$ of heat.

CH_4 is a better fuel on a mass basis.

9-52   kJ $\rightarrow$ mol Al $\rightarrow$ g Al

$$35.8 \text{ kJ} \times \dfrac{2 \text{ mol Al}}{850 \text{ kJ}} \times \dfrac{27.0 \text{ g Al}}{\text{mol Al}} = \underline{\underline{2.27 \text{ g Al}}}$$

9-53   69.7 g of glucose

Chapter 10

10-1   The molecules of water are closely packed and thus offer much more resistance. The molecules in a gas are dispersed into what is mostly empty space.

10-3   Since a gas is mostly empty space, more and more molecules can be added. In a liquid, the space is mostly occupied by the molecules so no more can be added.

10-5 Gas molecules are in rapid but random motion. When gas molecules collide with a light dust particle suspended in the air they impart a random motion to the particle.

10-6 When the pressure is high, the gas molecules are forced closer together. In a highly compressed gas, the molecules can occupy an appreciable part of the total volume. When the temperature is low, molecules have a lower average velocity. If there is some attraction, they can momentarily stick together when moving slowly.

10-7 (a) 2.17 atm   (b) 0.0266 torr   (c) 9560 torr   (d) 0.0558 atm
(e) 3.68 lb/in.²   (f) 11 kPa

10-8 (a) 768 torr   (b) 2.54×10^4 atm   (c) 8.40 lb/in.²   (d) 19 torr

10-10 force = 2450, $P = 2450$ g/12.0 cm² = 204 g/cm²

$$1 \text{ atm} = 76.0 \text{ cm Hg} = 1030 \text{ g/cm}^2; \ 204 \ \cancel{\text{g/cm}^2} \times \frac{1 \text{ atm}}{1030 \ \cancel{\text{g/cm}^2}}$$
$$= \underline{\underline{0.198 \text{ atm}}}$$

10-11 0.0102 atm

10-13 For a column of Hg that is 1 cm² in area and 76.0 cm high, weight = 76.0 cm × 1 cm² × 13.6 g/cm³ = 1030 g. For water, volume × density = weight, volume = height × area, height × 1 cm² × 1.00 g/cm³ = 1030 g, $h = 1030$ cm.

$$1030 \ \cancel{\text{cm}} \times \frac{1 \ \cancel{\text{in.}}}{2.54 \ \cancel{\text{cm}}} \times \frac{1 \text{ ft}}{12 \ \cancel{\text{in.}}} = 33.8 \text{ ft}$$

Since the well is more than 33.8 ft deep, suction cannot be used.

10-15 10.2 L

10-17 978 mL

10-18 67.9 torr

10-19 $\dfrac{V_{\text{final}}}{V_{\text{initial}}} = \dfrac{1}{15} = \dfrac{P_{\text{initial}}}{P_{\text{final}}}; \ P_{\text{final}} = 14.2$ atm

10-21

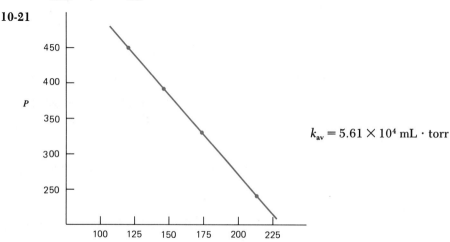

$k_{av} = 5.61 \times 10^4$ mL · torr

10-23 1.94 L

10-25 77 °C

10-26 $T_2 = 341$ K (68 °C)

10-27 2.60×10^4 L

10-30 2.94 atm

10-31 191 K (-82 °C)

10-33 596 K (323 °C)

10-34 30.2 lb/in.^{2}

10-37 (a) and (d)

10-39 Expt 1: T increases, Expt 2: V decreases, Expt 3: P increases, Expt 4: T increases

10-41 1.24 atm

10-42 76.8 L

10-44 88.5 K (-185 °C)

10-47 258 K (-15 °C)

10-49 1260 miles/hr

10-50 SF_6, SO_2, $N_2O = CO_2$, N_2, H_2

10-51 rate $N_2 = 1.19$ rate Ar

10-54 127 g/mol

10-56 $\dfrac{r_{235_U}}{r_{238_U}} = \sqrt{\dfrac{352}{349}} = 1.004$; $r_{235_U} = 1.004 r_{238_U}$

10-57 762 torr

10-59 6.8 torr

10-61 730 torr

10-63 N_2, 756 torr; O_2, 84 torr; SO_2, 210 torr

10-64 N_2, 1.02 atm; CO_2, 1.73 atm

10-65 $P_{N_2} = 300$ torr, $P_{O_2} = 170$ torr (when compressed from 4.00 to 2.00 L) $P_{CO_2} = 225$ torr, $P_{total} = 695$ torr

10-67 1.70 L

10-68 Let X = total moles needed in expanded balloon.

$$188 \text{ L} \times \frac{X \text{ mol}}{8.40 \text{ mol}} = 275 \text{ L} \qquad X = \underline{12.3 \text{ mol}}$$

$$(12.3 - 8.4) = \underline{3.9 \text{ mol must be added}}$$

10-70 $\dfrac{V_1}{V_2} = \dfrac{n_1}{n_2}$; $n_2 = 5.47 \times 10^{-3}$ mol (original gas + added N_2); $n_{N_2} = n_{total} -$
$n_{original} = 5.47 \times 10^{-3} - 2.50 \times 10^{-3} = 2.97 \times 10^{-3}$ mol N_2;

$$2.97 \times 10^{-3} \text{ mol N}_2 \times \frac{28.0 \text{ g N}_2}{\text{mol N}_2} = \underline{\underline{0.0832 \text{ g N}_2}}$$

10-72 98 °C

10-74 13.0 g NH_3

10-75 589 torr

10-76 $n_T = 0.223$ mol (N_2) + 0.286 mol (CO_2) = 0.509 mol

$$P = \frac{nRT}{V} = \frac{0.509 \text{ mol} \times 0.0821 \dfrac{L \cdot \text{atm}}{\text{mol} \cdot K} \times 306 K}{0.825 L} = \underline{\underline{15.5 \text{ atm}}}$$

10-78 3.35 g of Ne

10-79 0.0148 g H_2

10-81 mass of He (4.2×10^6 g) 9.2×10^3 lb, mass of air (3.0×10^7 g) 6.6×10^4 lb, lifting power = 57,000 lb; lifting power with H_2 = 61,000 lb. H_2 is combustible whereas He is not. In 1937 the German airship *Hindenberg* (which used H_2) was destroyed by fire with great loss of life.

10-82 Molar mass is 41.9 g; empirical formula is CH_2; molecular formula is C_3H_6.

10-84 7.64 L

10-86 112 L

10-88 1.39×10^{-3} g

10-89 1.24 g/L

10-91 34.0 g/mol

10-93 $1.00 \text{ L} \times \dfrac{273 K}{298 K} \times \dfrac{1.20 \text{ atm}}{1.00 \text{ atm}} = 1.10 \text{ L (STP)} \quad \dfrac{3.60 \text{ g}}{1.10 \text{ L}} = \underline{\underline{3.27 \text{ g/L (STP)}}}$

10-94 Find moles of N_2 in 1 L at 500 torr and 22 °C. $n = 0.0272$ mol $N_2 = 0.762$ g; density = 0.762 g/L (500 torr, 22 °C).

10-96 25.8 L CO_2

10-97 vol $O_2 \rightarrow$ mol $O_2 \rightarrow$ mol Mg $\rightarrow$ g Mg

$$5.80 \text{ L O}_2 \times \frac{1 \text{ mol O}_2}{22.4 \text{ L O}_2} \times \frac{2 \text{ mol Mg}}{1 \text{ mol O}_2} \times \frac{24.3 \text{ g Mg}}{\text{mol Mg}} = \underline{\underline{12.6 \text{ g Mg}}}$$

10-99 3.47 L

10-101 (a) g $C_4H_{10} \rightarrow$ mol $C_4H_{10} \rightarrow$ mol $CO_2 \rightarrow$ vol CO_2 (STP)

$$85.0 \text{ g C}_4\text{H}_{10} \times \frac{1.00 \text{ mol C}_4\text{H}_{10}}{58.0 \text{ g C}_4\text{H}_{10}} \times \frac{8 \text{ mol CO}_2}{2 \text{ mol C}_4\text{H}_{10}} \times \frac{22.4 \text{ L}}{\text{mol CO}_2} = 131 \text{ L}$$

(b) 96.3 L O_2
(c) 124 L CO_2

10-102   2080 kg Zr (2.29 tons)

10-104 $n_{O_2} = 1.45$ mol O_2; $1.45 \, \text{mol O}_2 \times \dfrac{1 \text{ mol } CO_2}{2 \text{ mol O}_2} = 0.725$ mol CO_2

$V_{CO_2} = \underline{\underline{11.9 \text{ L}}}$

Review Test for Chapters 8 – 10

Multiple Choice

1. (c)   2. (a)   3. (d)   4. (a)   5. (b)   6. (c)   7. (a)   8. (c)   9. (d)
10. (b)   11. (e)   12. (c)   13. (c)   14. (d)   15. (e)   16. (d)   17. (c)
18. (c)   19. (c)   20. (b)   21. (a)   22. (c)   23. (e)   24. (e)   25. (c)
26. (c)   27. (c)   28. (d)   29. (b)   30. (b)

Problems

1. 1.50 mol of K_2SO_4 has a mass of $\underline{\underline{261 \text{ g}}}$, contains $\underline{\underline{9.03 \times 10^{23}}}$ formula units, contains $\underline{\underline{6.00 \text{ mol}}}$ of O atoms, contains $\underline{\underline{117 \text{ g K}}}$, and contains $\underline{\underline{1.81 \times 10^{24}}}$ K atoms.

2. $C_4H_{10}N_2$ has a molar mass of $\underline{\underline{86.1 \text{ g/mol}}}$, is $\underline{\underline{32.5\%}}$ by weight N, and has an empirical formula of $\underline{\underline{C_2H_5N}}$:

3. Fe_3O_4

4. $C_2H_{14}B_{10}$

5. (a) $6K(s) + Al_2Cl_6 \rightarrow 6KCl(s) + 2$ Al
    (b) $2C_6H_6 + 15 \, O_2(g) \rightarrow 12 \, CO_2(g) + 6 \, H_2O$
    (c) $Cl_2O_3 + H_2O \rightarrow 2HClO_2$
    (d) $3K_2S + 2H_3PO_4 \rightarrow 3H_2S(g) + 2K_3PO_4$
    (e) $B_4H_{10} + 12 \, H_2O \rightarrow 4B(OH)_3 + 11 \, H_2(g)$

6. (a) 0.0233 mol of SbF_3 (b) 0.826 mol of $SbCl_3$ (c) 1.10×10^3 g of CCl_4
    (d) 4.75×10^{25} molecules of CCl_4 (e) 3.12 L of freon

7. 10.0 L of O_2 at STP contains $\underline{\underline{0.446 \text{ mol}}}$ of O_2, contains $\underline{\underline{2.69 \times 10^{23}}}$ molecules of O_2, has a volume of $\underline{\underline{8.06 \text{ L}}}$ at -53 °C, has a volume of $\underline{\underline{13.1 \text{ L}}}$ at 580 torr, has a pressure of $\underline{\underline{1.10 \text{ atm}}}$ at 27 °C, and has a volume of $\underline{\underline{21.2 \text{ L}}}$ if 0.500 mol of a gas is added.

8. Molar mass = 147 g/mol (from the ideal gas law); empirical formula = C_3H_2Cl; molecular formula = $C_6H_4Cl_2$.

Chapter 11

11-1 Since gas molecules are far apart they move a comparatively great distance between collisions. Liquid molecules, on the other hand, are close together and so do not move far between collisions. The farther molecules move, the faster they mix.

11-3 The liquid molecules as well as the food coloring molecules are in motion. Through constant motion and collisions the food coloring molecules will eventually become dispersed.

11-5 Generally, solids have greater densities than liquids. (Ice and water are notable exceptions.) Since the molecules of a solid are held in fixed positions, more of them usually fit into the same volume compared with the liquid state. This is similar to being able to get more people into a room if they are standing still than if they are moving around.

11-6 The two equal O—H bond dipoles would cancel if they were at an angle of 180° (linear).

11-8

NH_3 is a polar molecule because the three equal N—H bond dipoles do not cancel.

11-10 (a), (b), and (f)

11-12 (a), (c), (d), and (f)

11-14 All are nonpolar molecules and have only London forces of attraction. The higher the molar mass, however, the greater the forces. The lightest is Cl_2 and is a gas at room temperature. Br_2 is intermediate and is a liquid. The heaviest is I_2 and the forces of attraction are apparently great enough to hold the molecules in the solid state at room temperature.

11-16 SF_6 is a molecular compound and SnO is an ionic compound (i.e., $Sn^{2+}O^{2-}$). All ionic compounds are found as solids at room temperature because of the strong interionic forces.

11-18 The average kinetic energy of the molecules is the same at the same temperature regardless of the physical state.

11-20 Because H_2O molecules have an attraction for each other, moving them apart increases the potential energy. Since the molecules in a gas at 100 °C are farther apart than in a liquid at 100 °C, the potential energy of the gas molecules is greater.

11-21 (b) and (c)

11-23 There is hydrogen bonding in CH_3OH. Hydrogen bonding is a considerably stronger interaction than an ordinary dipole–dipole force. The stronger the interactions, the higher the boiling point (generally).

11-25 Carbon monoxide is a polar molecule whereas nitrogen is not. The added dipole–dipole forces in CO probably account for the slightly higher boiling point.

11-26

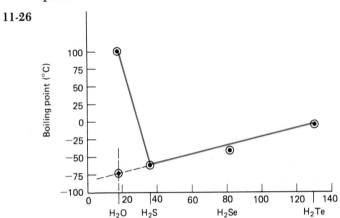

The boiling point of H_2O would be about $-75\ °C$ without hydrogen bonding.

11-28 $PbCl_2$ is an ionic compound (i.e., Pb^{2+}, $2Cl^-$) whereas $PbCl_4$ is a molecular compound. Ionic compounds usually have higher melting points than molecular compounds.

11-30 2.54 g H_2O, 1.64 g NaCl, 6.69 g benzene

11-31 Ethyl ether, 555 cal; H_2O, 2000 cal. Water is more effective than ether in holding heat. 1.30×10^4 J melts 125 g of ethyl alcohol.

11-33 7070 J

11-35 NH_3, 6.12×10^5 J; freon, 7.25×10^4 J. On the basis of mass, NH_3 is more effective than freon.

11-37 Condensation releases 6.22×10^5 J and cooling releases 8.6×10^4 J, for a total of 7.08×10^5 J.

11-39 Heat ice

$$132\ \cancel{g} \times 20\ \cancel{°C} \times \frac{0.492\ cal}{\cancel{g}\cdot \cancel{°C}} = 1{,}300\ cal$$

Melt ice at 0 °C

$$132\ \cancel{g} \times 79.8\ cal/\cancel{g} = 10{,}500\ cal$$

Heat H_2O

$$132\ \cancel{g} \times 100\ \cancel{°C} \times \frac{1.00\ cal}{\cancel{g}\cdot \cancel{°C}} = 13{,}200\ cal$$

Vaporize H_2O at 100 °C

$$132\ \cancel{g} \times 540\ cal/\cancel{g} = \underline{71{,}300\ cal}$$

$$= 96{,}300\ cal\ total$$

11-41 Let $Y =$ the mass of the sample in grams. Then

$$\left(\frac{2260\ J}{g} \times Y\right) + \left(25\ °C \times Y \times \frac{4.18\ J}{g\cdot °C}\right) = 28{,}400\ J$$

$$Y = \underline{12.0\ g}$$

11-43 2.83×10^4 cal

11-44 18 °C

11-46 This is a comparatively high heat of fusion, so the melting point should also be comparatively high.

11-48 This is a comparatively high boiling point, so the compound probably also has a comparatively high melting point.

11-50 Equilibrium refers to a state where opposing forces are balanced. In the case of a liquid in "equilibrium" with its vapor, it means that a molecule escaping to the vapor is replaced by one condensing to the liquid.

11-52 The substance is a gas under these conditions.

11-53 Ethyl alcohol boils at about 52 °C at that altitude. At 10 °C, ethyl ether is a gas at that altitude.

11-55 Figure 11-9 is hard to read at this temperature but it is obviously close. The actual vapor pressure of water at 10 °C is 9.2 torr, so liquid water could theoretically exist. A glass of water would rapidly evaporate, however, and soon freeze due to the rapid evaporation.

11-57 $V = 100$ L, $T = 34 + 273 = 307$ K, $P = 0.70 \times 39.0$ torr $= 27.3$ torr. Using the ideal gas law, $n = 0.143$ mol $H_2O = \underline{2.57 \text{ g } H_2O}$.

11-59 Alcohol is more volatile (i.e., has a higher vapor pressure) than water. It thus evaporates more rapidly and produces a pronounced cooling effect.

11-61 The liquid other than water. The higher the vapor pressure, the faster the liquid evaporates and the faster the liquid cools.

11-63 (a) $2K(s) + 2H_2O(l) \rightarrow 2KOH(aq) + H_2(g)$
(b) $Br_2(l) + H_2O(l) \rightleftharpoons HBr(aq) + HOBr(aq)$
(c) $K_2O(s) + H_2O(l) \rightarrow 2KOH(aq)$
(d) $CO_2(g) + H_2O(l) \rightarrow H_2CO_3(aq)$
(e) $C_2H_4(g) + 3 O_2(g) \rightarrow 2CO_2(g) + 2H_2O(l)$

11-65 (a) magnesium carbonate pentahydrate   (b) sodium sulfate decahydrate
(c) iron(III) bromide hexahydrate

11-67 54.1% H_2O

11-69 Let $x =$ number of H_2O molecules. Then the decimal fraction of H_2O equals

$$\frac{g \ H_2O}{g \ of \ compound}$$

Thus

$$\frac{18x}{120.3 + 18x} = 0.511 \quad x = 7$$

Formula is $MgSO_4 \cdot 7H_2O$.

Chapter 12

12-1   (a) $Na_2S \rightarrow 2Na^+(aq) + S^{2-}(aq)$
         (b) $Li_2SO_4 \rightarrow 2Li^+(aq) + SO_4^{2-}(aq)$
         (c) $K_2Cr_2O_7 \rightarrow 2K^+(aq) + Cr_2O_7^{2-}(aq)$
         (d) $CaS \rightarrow Ca^{2+}(aq) + S^{2-}(aq)$
         (e) $2(NH_4)_2S \rightarrow 4NH_4^+(aq) + 2S^{2-}(aq)$
         (f) $4Ba(OH)_2 \rightarrow 4Ba^{2+}(aq) + 8OH^-(aq)$

12-3   (a)   2 mol of Al^{3+} and 3 mol of SO_4^{2-}
         (b)   6 mol of Mg^{2+} and 4 mol of PO_4^{3-}

12-4

There is hydrogen bonding between solute and solvent.

12-6 $PbSO_4$, $MgSO_3$, Ag_2O

12-8   At 10 °C, Li_2SO_4 is the most soluble; at 70 °C, KCl is the most soluble.

12-10   (a) unsaturated   (b) supersaturated   (c) saturated   (d) unsaturated

12-12   A specific amount of Li_2SO_4 will precipitate unless the solution becomes supersaturated.

$$(\text{solubility at } 0 \text{ °C}) - (\text{solubility at } 100 \text{ °C}) = \text{mass of precipitate}$$
$$(35 \text{ g}/100 \text{ g } H_2O \times 500 \text{ g } H_2O) -$$
$$(28 \text{ g}/100 \text{ g } H_2O \times 500 \text{ g } H_2O) = 35 \text{ g } Li_2SO_4$$

12-13   (a) $2KI(aq) + Pb(C_2H_3O_2)_2(aq) \rightarrow PbI_2(s) + 2KC_2H_3O_2(aq)$
          (b)   no reaction
          (c)   no reaction
          (d) $BaS(aq) + Hg_2(NO_3)_2(aq) \rightarrow Hg_2S(s) + Ba(NO_3)_2(aq)$
          (e) $FeCl_3(aq) + 3KOH(aq) \rightarrow Fe(OH)_3(s) + 3KCl(aq)$

12-15   (a) $2K^+(aq) + S^{2-}(aq) + Pb^{2+}(aq) + 2NO_3^-(aq) \rightarrow$
$$PbS(s) + 2K^+(aq) + 2NO_3^-(aq)$$
            $S^{2-}(aq) + Pb^{2+}(aq) \rightarrow PbS(s)$
          (b) $2NH_4^+(aq) + CO_3^{2-}(aq) + Ca^{2+}(aq) + 2Cl^-(aq) \rightarrow$
$$CaCO_3(s) + 2NH_4^+(aq) + 2Cl^-(aq)$$
            $CO_3^{2-}(aq) + Ca^{2+}(aq) \rightarrow CaCO_3(s)$
          (c) $2Ag^+(aq) + 2ClO_4^-(aq) + 2Na^+(aq) + CrO_4^{2-}(aq) \rightarrow$
$$Ag_2CrO_4(s) + 2Na^+(aq) + 2ClO_4^-(aq)$$
            $2Ag^+(aq) + CrO_4^{2-}(aq) \rightarrow Ag_2CrO_4(s)$

12-16   (a) $2K^+(aq) + 2I^-(aq) + Pb^{2+}(aq) + 2C_2H_3O_2^-(aq) \rightarrow$
$$PbI_2(s) + 2K^+(aq) + 2C_2H_3O_2^-(aq)$$
          (d) $Ba^{2+}(aq) + S^{2-}(aq) + Hg_2^{2+}(aq) + 2NO_3^-(aq) \rightarrow$
$$Hg_2S(s) + Ba^{2+}(aq) + 2NO_3^-(aq)$$
          (e) $Fe^{3+}(aq) + 3Cl^-(aq) + 3K^+(aq) + 3OH^-(aq) \rightarrow$
$$Fe(OH)_3(s) + 3K^+(aq) + 3Cl^-(aq)$$

12-18 (a) $Pb^{2+}(aq) + 2I^-(aq) \rightarrow PbI_2(s)$
(d) $Hg_2^{2+}(aq) + S^{2-}(aq) \rightarrow Hg_2S(s)$
(e) $Fe^{3+}(aq) + 3OH^-(aq) \rightarrow Fe(OH)_3(s)$

12-20 (a) $CuCl_2(aq) + Na_2CO_3(aq) \rightarrow CuCO_3(s) + 2NaCl(aq)$
Filter the solid $CuCO_3$.
(b) $(NH_4)_2SO_3(aq) + Pb(NO_3)_2(aq) \rightarrow PbSO_3(s) + 2NH_4NO_3(aq)$
Filter the solid $PbSO_3$.
(c) $2KI(aq) + Hg_2(NO_3)_2(aq) \rightarrow Hg_2I_2(s) + 2KNO_3(aq)$
Filter the solid Hg_2I_2.
(d) $NH_4Cl(aq) + AgNO_3(aq) \rightarrow AgCl(s) + NH_4NO_3(aq)$
Filter the solid AgCl; the desired product remains after the water is removed by boiling.
(e) $Ca(C_2H_3O_2)_2(aq) + K_2CO_3(aq) \rightarrow CaCO_3(s) + 2KC_2H_3O_2(aq)$
Filter the solid $CaCO_3$; the desired product remains after the water is removed by boiling.

12-21 1.49%

12-22 15.77%

12-23 0.375 mol of NaOH

12-26 8.81%

12-27 100 ppm

12-29 0.542 M

12-30 (a) 0.873 M   (b) 1.40 M   (c) 12.4 L   (d) 41.3 g   (e) 0.294 M
(f) 0.024 mol   (g) 310 mL   (h) 49.0 g   (i) 2.00 mL

12-34 $1.17 \times 10^{-3} M$

12-35 $[Ba^{2+}] = 0.166$, $[OH^-] = 0.332$

12-36 1.84 M

12-38 1.32 g/mL

12-39 0.833 L

12-41 Volume of 0.800 M NaOH needed is 313 mL. Slowly add the concentrated NaOH to about 500 mL of water in a 1-L volumetric flask. Dilute to the 1-L mark with water.

12-43 0.140 M

12-44 720 mL

12-45 $0.250\,\cancel{L} \times \dfrac{0.200\,\cancel{mol}}{\cancel{L}} \times \dfrac{60.0\,\cancel{g}}{\cancel{mol}} \times \dfrac{1.00\ mL}{1.05\,\cancel{g}} = 2.86\ mL$

12-47 $n = (0.150\,\cancel{L} \times 0.250\ mol/\cancel{L}) + (0.450\,\cancel{L} \times 0.375\ mol/\cancel{L}) = 0.206\ mol$
$V = 0.150\ L + 0.450\ L = 0.600\ L$
$n/V = 0.343\ M$

12-48
$$\left(0.500 \, L \times 0.250 \, \frac{mol}{L}\right) \times \frac{1 \, mol \, Cr(OH)_3}{3 \, mol \, KOH} \times \frac{103 \, g \, Cr(OH)_3}{mol \, Cr(OH)_3}$$
$$= 4.29 \, g \, Cr(OH)_3$$

12-50 145 g $BaSO_4$

12-51 4.08 L

12-53 0.653 L

12-55 NaOH is the limiting reactant and produces 1.75 g of $Mg(OH)_2$.

12-57 An aqueous solution of AB is a good conductor of electricity. AB dissociates into A^+ and B^- ions. A solution of AC is a weak conductor of electricity. AC only partially dissociates into ions (i.e., $AC \rightleftharpoons A^+ + C^-$). A solution of AD is a nonconductor of electricity. AD is present as undissociated molecules in solution.

12-59 The solute is 2.50 m in both solvents.

12-61 1.00 m

12-62 15.8 g

12-64 Let x = grams of KBr; then the grams of $H_2O = 1000 - x$.

$$\frac{mol \, solute}{kg \, solvent} = m = 1.00$$

$$\frac{x \, g/119 \, g/mol}{\left(\dfrac{1000 - x}{1000}\right) kg} = 1.00$$

$$x = 106 \, g; \text{ mass of } H_2O = \underline{\underline{894 \, g}}$$

12-65 $-0.37 \, °C$

12-67 834 g

12-68 101.4 °C

12-70 2.93 m

12-71 80.5 °C

12-73 $-14.5 \, °C$

12-75 (a) $-5.8 \, °C$ (b) $-6.4 \, °C$ (c) $-5.0 \, °C$

12-76 The salty water removes water from the cells of the skin by osmosis. After a prolonged period one would dehydrate and become thirsty.

12-79 One can concentrate a dilute solution by boiling away some solvent if the solute is not volatile. If the solution is separated from pure solvent by a semipermeable membrane, reverse osmosis can be applied. Pressure must be exerted on the solution that is greater than its osmotic pressure to cause solvent molecules to migrate to the pure solvent. As the solution becomes more concentrated, the osmotic pressure becomes greater and the corresponding pressure that is applied must be increased.

Chapter 13

13-1   (a) HNO_3, nitric acid   (b) HNO_2, nitrous acid   (c) $HClO_3$, chloric acid   (d) H_2SO_3, sulfurous acid

13-2   (a) CsOH, cesium hydroxide   (b) $Sr(OH)_2$, strontium hydroxide
      (c) $Al(OH)_3$, aluminum hydroxide   (d) $Mn(OH)_3$, manganese(III) hydroxide

13-4   (a) $HNO_3 + H_2O \rightarrow H_3O^+ + NO_3^-$
      (b) $HNO_2 + H_2O \rightarrow H_3O^+ + NO_2^-$
      (c) $HClO_3 + H_2O \rightarrow H_3O^+ + ClO_3^-$
      (d) $H_2SO_3 + H_2O \rightarrow H_3O^+ + HSO_3^-$

13-6   (a) acid   (b) normal salt   (c) acid   (d) acid salt   (e) normal salt
      (f) base

13-7   (a) $HC_2H_3O_2 + KOH \rightarrow KC_2H_3O_2 + H_2O$
      (b) $2HI + Ca(OH)_2 \rightarrow CaI_2 + 2H_2O$
      (c) $H_2SO_4 + Ca(OH)_2 \rightarrow CaSO_4 + 2H_2O$

13-9   (b) $2H^+(aq) + 2I^-(aq) + Ca^{2+}(aq) + 2OH^-(aq) \rightarrow$
$$Ca^{2+}(aq) + 2I^-(aq) + 2H_2O$$
      $H^+ + OH^- \rightarrow H_2O$
      (c) $2H^+(aq) + SO_4^{2-}(aq) + Ca^{2+}(aq) + 2OH^-(aq) \rightarrow CaSO_4(s) + 2H_2O$
      (net ionic equation is the same as the total ionic equation)

13-10   (a) $2HBr + Ca(OH)_2 \rightarrow CaBr_2 + 2H_2O$
       (b) $2HClO_2 + Sr(OH)_2 \rightarrow Sr(ClO_2)_2 + 2H_2O$
       (c) $2H_2S + Ba(OH)_2 \rightarrow Ba(HS)_2 + 2H_2O$
       (d) $H_2S + 2LiOH \rightarrow Li_2S + 2H_2O$

13-12 $LiOH + H_2S \rightarrow LiHS + H_2O$
       $LiOH + LiHS \rightarrow Li_2S + H_2O$

13-14 $H_2S + NaOH \rightarrow NaHS + H_2O$

13-15 $Ca(OH)_2(aq) + 2H_3PO_4(aq) \rightarrow Ca(H_2PO_4)_2(aq) + 2H_2O$

13-17 $NH(CH_3)_2 + H_2O \rightleftharpoons HNH(CH_3)_2^+ + OH^-$

13-18   HX is a weak acid. The percent ionization is $\dfrac{0.010}{0.10} \times 100\% = 10\%$.

13-20 $[H_3O^+] = 0.55$

13-21 $[H_3O^+] = 0.030 \times 0.55 = 0.016$

13-23   From the first ionization, $[H_3O^+] = 0.354$. From the second ionization $[H_3O^+] = 0.25 \times 0.354 = 0.088$. Therefore, the total $[H_3O^+] = 0.354 + 0.088 = 0.442$.

13-25   The system would not be at equilibrium if $[H_3O^+] = [OH^-] = 10^{-2}\ M$. Therefore, H_3O^+ reacts with OH^- until the concentration of each is reduced to $10^{-7}\ M$ (i.e., a neutralization reaction occurs).
$$H_3O^+ + OH^- \longrightarrow 2H_2O$$

13-26 (a) $[H_3O^+] = K_w/[OH^-] = 10^{-14}/10^{-12} = 10^{-2}$
(b) $[H_3O^+] = 10^{-15}$
(c) $[OH^-] = 5.0 \times 10^{-10}$

13-28 $[H_3O^+] = 0.0250$; $[OH^-] = 4.00 \times 10^{-13}$

13-29 In lye $[H_3O^+] = 3.92 \times 10^{-15}$; in ammonia, $[H_3O^+] = 2.5 \times 10^{-12}$.

13-30 (a) acidic   (b) basic   (c) acidic

13-32 (a) acidic   (b) basic   (c) acidic   (d) basic

13-34 (a) 6.00   (b) 9.00   (c) 12.00   (d) 9.40   (e) 10.19

13-36 (a) 1.0×10^{-3} (b) 2.9×10^{-4} (c) 1.0×10^{-6} (d) 2.4×10^{-8}
(e) 2.0×10^{-13}

13-38 For Problem 13-34:   (a) acidic   (b) basic   (c) basic   (d) basic
(e) basic
For Problem 13-16:   (a) acidic   (b) acidic   (c) acidic   (d) basic
(e) basic

13-40 pH = 1.12

13-42 pH = 12.56

13-43 $[H_3O^+] = 0.10 \times 0.10 = 0.010$ pH = 2.00

13-46 (a) strongly acidic   (b) strongly acidic   (c) weakly acidic
(d) strongly basic   (e) neutral   (f) weakly basic   (g) weakly acidic
(h) strongly acidic

13-47 ammonia (pH = 11.4), eggs (pH = 7.8), rain water (pH = 5.7), grape juice
(pH = 4.0), vinegar (pH = 2.6), sulfuric acid (pH = 0.40)

13-49 (a) NO_3^- (b) HSO_4^- (c) PO_4^{3-} (d) CH_3^- (e) OH^- (f) NH_2^-

13-51 (a) $HClO_4$, ClO_4^- and H_2O, OH^- (b) HSO_4^-, SO_4^{2-} and $HOCl$, OCl^-
(c) H_2O, OH^- and NH_3, NH_2^- (d) NH_4^+, NH_3 and H_3O^+, H_2O

13-54 (a) $NH_3 + H_2O \longrightarrow NH_4^+ + OH^-$

      B_1 A_1

          A_2 B_2

(b) $N_2H_4 + H_2O \longrightarrow N_2H_5^+ + OH^-$

      B_1 A_1

          A_2 B_2

(c) $HS^- + H_2O \longrightarrow H_2S + OH^-$

      B_1 A_1

          A_2 B_2

(d) $H^- + H_2O \longrightarrow H_2 + OH^-$

B_1 | A_1

A_2 B_2

(e) $F^- + H_2O \longrightarrow HF + OH^-$

B_1 | A_1

A_2 B_2

13-55 Brønsted-Lowry base: $HS^- + H_3O^+ \rightleftharpoons H_2S + H_2O$
Brønsted-Lowry acid: $HS^- + OH^- \rightleftharpoons S^{2-} + H_2O$

13-57 (b) and (d)

13-59 (a) $H_3O^+ + HS^- \rightarrow H_2S + H_2O$
(b) $HOCl + CN^- \rightarrow HCN + OCl^-$
(c) $H_2SO_3 + OH^- \rightarrow HSO_3^- + H_2O$
(d) $HBr + CH_3^- \rightarrow Br^- + CH_4$

13-61 (a) $H_2S + H_2O \rightleftharpoons HS^- + H_3O^+$
(b) no reaction
(c) $HNO_3 + H_2O \rightarrow NO_3^- + H_3O^+$
(d) $NH_4^+ + H_2O \rightleftharpoons NH_3 + H_3O^+$

13-63 (c) H^- cannot exist in water.

13-65 (a) HS^- (b) H_3O^+ (c) OH^- (d) $NH(CH_3)_2$

13-67 (a) $F^- + H_2O \rightleftharpoons HF + OH^-$
(b) $SO_3^{2-} + H_2O \rightleftharpoons HSO_3^- + OH^-$
(c) $HNH(CH_3)_2^+ + H_2O \rightleftharpoons NH(CH_3)_2 + H_3O^+$
(d) $HPO_4^{2-} + H_2O \rightleftharpoons H_2PO_4^- + OH^-$
(e) $CN^- + H_2O \rightleftharpoons HCN + OH^-$
(f) Li^+ does not hydrolyze.

13-69 $CaC_2(s) + 2H_2O \rightarrow C_2H_2(g) + 2OH^-(aq) + Ca^{2+}(aq)$

13-72 (a) neutral (neither ion hydrolyzes) (b) acidic (cation hydrolysis)
(c) basic (anion hydrolysis) (d) neutral (neither ion hydrolyzes)
(e) acidic (cation hydrolysis) (f) basic (anion hydrolysis)

13-74 Cation: $NH_4^+ + H_2O \rightleftharpoons NH_3 + H_3O^+$
Anion: $CN^- + H_2O \rightleftharpoons HCN + OH^-$
Since the solution is basic, the anion hydrolysis reaction must take place to a greater extent than the cation hydrolysis.

13-76 (a), (b), (e), (h), and (i)

13-77 There is no equilibrium involved when HCl dissolves in water. A reservoir of un-ionized acid must be present to replace any H_3O^+ lost when a strong base is added. Likewise, the Cl^- ion does not exhibit base behavior in water, so it cannot react with any H_3O^+ added to the solution.

13-80
$$NH_3(aq) + H_2O \rightleftharpoons NH_4^+(aq) + OH^-(aq)$$
Added H_3O^+ $H_3O^+ + NH_3 \longrightarrow NH_4^+ + H_2O$
Added OH^- $OH^- + NH_4^+ \longrightarrow NH_3 + H_2O$

13-82
$$H_2PO_4^-(aq) + H_2O \rightleftharpoons HPO_4^{2-}(aq) + H_3O^+(aq)$$
Added H_3O^+ $H_3O^+ + HPO_4^{2-} \longrightarrow H_2PO_4^- + H_2O$
Added OH^- $OH^- + H_2PO_4^- \longrightarrow HPO_4^{2-} + H_2O$

13-83 (a) $Sr(OH)_2$ (b) H_2SeO_4 (c) H_3PO_4 (d) $CsOH$ (e) HNO_2
(f) $HClO_3$

13-85 $LiOH(aq) + CO_2(g) \rightarrow LiHCO_3(aq)$

13-86 $2LiNO_3$

13-88 (a) acidic: $H_2S + H_2O \rightleftharpoons HS^- + H_3O^+$
(b) basic: $ClO^- + H_2O \rightleftharpoons HClO + OH^-$
(c) neutral
(d) basic: $NH_3 + H_2O \rightleftharpoons NH_4^+ + OH^-$
(e) acidic: $HN_2H_4^+ + H_2O \rightleftharpoons N_2H_4 + H_3O^+$
(f) basic: $Ba(OH)_2 \rightarrow Ba^{2+} + 2OH^-$
(g) neutral
(h) basic: $NO_2^- + H_2O \rightleftharpoons HNO_2 + OH^-$
(i) acidic: $H_2SO_3 + H_2O \rightleftharpoons H_3O^+ + HSO_3^-$
(j) acidic: $Cl_2O_3 + H_2O \rightarrow 2HClO_2$; $HClO_2 + H_2O \rightleftharpoons H_3O^+ + ClO_2^-$

13-90 LiOH, strongly basic, $pH = 13.0$; $SrBr_2$, neutral, $pH = 7.0$; KOCl, weakly basic, $pH = 10.2$; NH_4Cl, weakly acidic, $pH = 5.2$; HI, strongly acidic, $pH = 1.0$. The concentration is $0.10\ M$.

Chapter 14

14-1 (a) S $(+6)$ (b) Co $(+3)$ (c) U $(+6)$ (d) N $(+5)$ (e) Cr $(+6)$
(f) Mn $(+6)$

14-3 (a), (c), and (f)

14-4 (a) oxidation (b) reduction (c) oxidation (d) reduction (e) reduction

14-6 (a) Br^-; Br_2; MnO_2; Mn^{2+}; MnO_2; Br^-
(b) CH_4; CO_2; O_2; CO_2 and H_2O; O_2; CH_4
(c) Fe^{2+}; Fe^{3+}; MnO_4^-; Mn^{2+}; MnO_4^-; Fe^{2+}

14-7 (a) Al, AlO_2^-, H_2O, H_2, H_2O, Al
(b) Mn^{2+}, MnO_4^-, $Cr_2O_7^{2-}$, Cr^{3+}, $Cr_2O_7^{2-}$, Mn^{2+}

14-10

$$-5e \times \textcircled{4} = -20e^-$$

$-3 \qquad\qquad +2$

(a) $4NH_3 + 5O_2 \longrightarrow 4NO + 6H_2O$

$\quad\quad 0 \qquad\qquad -2 \qquad -2$

$$+4e^- \times \textcircled{5} = +20e^-$$

$$-4e^- \times \textcircled{1} = -4e^-$$

$\quad 0 \qquad\qquad +4$

(b) $Sn + 4HNO_3 \longrightarrow SnO_2 + 4NO_2 + 2H_2O$

$\quad\quad +5 \qquad\qquad\qquad +4$

$$+1e^- \times \textcircled{4} = +4e^-$$

(c) Before the number of electrons is calculated, note that the Na_2CrO_4 in the products must first be multiplied by 2 since there are two Cr's in Cr_2O_3 in the reactants.

$$+2e^- \times \textcircled{3} = +6e^-$$

$\qquad\qquad +5 \qquad\qquad\qquad\qquad +3$

$Cr_2O_3 + 2Na_2CO_3 + 3KNO_3 \longrightarrow 2CO_2 + 2Na_2CrO_4 + 3KNO_2$

$+6 \qquad\qquad\qquad\qquad\qquad\qquad +12$

$$-6e^- \times \textcircled{1} = -6e^-$$

$$-4e^- \times \textcircled{3} = -12e^-$$

(d) $0 \qquad\qquad\qquad\qquad +4$

$\quad 3Se + 2BrO_3^- + 3H_2O \longrightarrow 3H_2SeO_3 + 2Br^-$

$\qquad\quad +5 \qquad\qquad\qquad\qquad -1$

$$+6e^- \times \textcircled{2} = +12e^-$$

$$+2e \times \textcircled{1} = +2e$$

(e)

$+4 \qquad\qquad +2$

$\quad MnO_2 + 4HBr \longrightarrow MnBr_2 + Br_2 + 2H_2O$

$\qquad\qquad -2 \qquad\qquad 0$

$$-2e \times \textcircled{1} = -2e$$

(*Note:* 2HBr are needed for the Br$^-$ in MnBr$_2$.)

14-12 (a) $2H_2O + Sn^{2+} \rightarrow SnO_2 + 4H^+ + 2e^-$
(b) $2H_2O + CH_4 \rightarrow CO_2 + 8H^+ + 8e^-$
(c) $e^- + Fe^{3+} \rightarrow Fe^{2+}$
(d) $6H_2O + I_2 \rightarrow 2IO_3^- + 12H^+ + 10e^-$
(e) $e^- + 2H^+ + NO_3^- \rightarrow NO_2 + H_2O$

14-14   (a) $$S^{2-} \rightarrow S + 2e^- \qquad \big| \times 3$$
$$\underline{3e^- + 4H^+ + NO_3^- \rightarrow NO + 2H_2O \big| \times 2}$$
$$3S^{2-} + 8H^+ + 2NO_3^- \rightarrow 3S + 2NO + 4H_2O$$

(b) $I_2 + 2S_2O_3^{2-} \rightarrow 2I^- + S_4O_6^{2-}$

(c) $$H_2O + SO_3^{2-} \rightarrow SO_4^{2-} + 2H^+ + 2e^- \big| \times 3$$
$$\underline{6e^- + 6H^+ + ClO_3^- \rightarrow Cl^- + 3H_2O \qquad \big| \times 1}$$
$$3SO_3^{2-} + ClO_3^- \rightarrow Cl^- + 3SO_4^{2-}$$

(d) $2H^+ + 2Fe^{2+} + H_2O_2 \rightarrow 2Fe^{3+} + 2H_2O$

(e) $AsO_4^{3-} + 2I^- + 2H^+ \rightarrow I_2 + AsO_3^{3-} + H_2O$

(f) $4Zn + 10H^+ + NO_3^- \rightarrow 4Zn^{2+} + NH_4^+ + 3H_2O$

14-16   (a) $2OH^- + SnO_2^{2-} \rightarrow SnO_3^{2-} + H_2O + 2e^-$

(b) $6e^- + 4H_2O + 2ClO_2^- \rightarrow Cl_2 + 8OH^-$

(c) $6OH^- + Si \rightarrow SiO_3^{2-} + 3H_2O + 4e^-$

(d) $8e^- + 6H_2O + NO_3^- \rightarrow NH_3 + 9OH^-$

14-18   (a) $$8OH^- + S^{2-} \rightarrow SO_4^{2-} + 4H_2O + 8e^- \big| \times 1$$
$$\underline{I_2 + 2e^- \rightarrow 2I^- \qquad\qquad\qquad \big| \times 4}$$
$$S^{2-} + 8OH^- + 4I_2 \rightarrow SO_4^{2-} + 8I^- + 4H_2O$$

(b) $8MnO_4^- + 8OH^- + I^- \rightarrow 8MnO_4^{2-} + IO_4^- + 4H_2O$

(c) $2H_2O + BiO_3^- + SnO_2^{2-} \rightarrow SnO_3^{2-} + Bi(OH)_3 + OH^-$

(d) $$32OH^- + CrI_3 \rightarrow CrO_4^{2-} + 3IO_4^- + 16H_2O + 27e^- \qquad\qquad \big| \times 2$$
$$\underline{2e^- + Cl_2 \rightarrow 2Cl^- \qquad\qquad\qquad\qquad\qquad\qquad\qquad \big| \times 27}$$
$$2CrI_3 + 64OH^- + 27Cl_2 \rightarrow 2CrO_4^{2-} + 6IO_4^- + 32H_2O + 54Cl^-$$

14-20   (a) $2H_2 + O_2 \longrightarrow 2H_2O$

(b) $$H_2O_2 \longrightarrow O_2 + 2H^+ + 2e^- \big| \times 1$$
$$\underline{2e^- + 2H^+ + H_2O_2 \longrightarrow 2H_2O \qquad \big| \times 1}$$
$$2H_2O_2 \longrightarrow O_2 + 2H_2O$$

$$2OH^- + H_2O_2 \longrightarrow O_2 + 2H_2O + 2e^- \big| \times 1$$
$$\underline{2e^- + H_2O_2 \longrightarrow 2OH^- \qquad\qquad \big| \times 1}$$
$$2H_2O_2 \longrightarrow O_2 + 2H_2O$$

14-22   Zn reacts at the anode and MnO_2 reacts at the cathode.

$$Zn + 2MnO_2 + 2H_2O \longrightarrow Zn(OH)_2 + 2MnO(OH)$$

14-24
$$\text{Anode:} \quad Cd + 2OH^- \longrightarrow Cd(OH)_2 + 2e^-$$
$$\text{Cathode:} \ NiO_2 + 2H_2O + 2e^- \longrightarrow Ni(OH)_2 + 2OH^-$$

14-26
$$Ni^{2+} + Cd \longrightarrow Cd^{2+} + Ni$$
$$Cd^{2+} + Zn \longrightarrow Zn^{2+} + Cd$$

Cd^{2+} is a stronger oxidizing agent than Zn^{2+} but is weaker than Ni^{2+}. It appears to have strength comparable to Fe^{2+}.

14-28   Two spontaneous reactions are

$$4Au^{3+} + 6H_2O \longrightarrow 4Au + 3O_2 + 12H^+$$
$$2Au + 3F_2 \longrightarrow 2Au^{3+} + 6F^-$$

The half-reaction

$$Au^{3+} + 3e^- \longrightarrow Au \text{ belongs between the } F_2 \text{ and } Cl_2 \text{ half-reactions.}$$

14-29 (a) yes (b) N.R. (c) yes (d) N.R. (e) yes (f) yes (g) N.R.

14-31 (a) $CuCl_2(aq) + Fe(s) \rightarrow FeCl_2(aq) + Cu(s)$
(A coating of Cu forms on the nails.)
(b) N.R.
(c) N.R.
(d) $Ni(s) + Pb(NO_3)_2(aq) \rightarrow Ni(NO_3)_2(aq) + Pb(s)$
(e) $2Na(s) + 2H_2O \rightarrow 2NaOH(aq) + H_2(g)$
(f) N.R.
(g) N.R.
(h) $2Cr^{3+} + 3Zn \rightarrow 3Zn^{2+} + 2Cr$

14-33 (c) $2F_2 + 2H_2O \rightarrow 4F^- + O_2 + 4H^+$
(e) $Mg + 2H_2O \rightarrow Mg^{2+} + H_2 + 2OH^-$

14-35 $Fe + 2H_2O \rightarrow$ N.R.
$Fe + 2H^+ \rightarrow Fe^{2+} + H_2$
Acid rain has a higher $H^+(aq)$ concentration, thus making the second reaction possible.

14-37 The spontaneous reaction is

$$Pb(NO_3)_2(aq) + Fe(s) \longrightarrow Fe(NO_3)_2(aq) + Pb(s)$$
$$\text{Anode:} \quad Fe(s) \longrightarrow Fe^{2+}(aq) + 2e^-$$
$$\text{Cathode:} \quad Pb^{2+}(aq) + 2e^- \longrightarrow Pb(s)$$

14-38 The Daniell cell is more powerful. The greater the separation between oxidizing and reducing agent in the table, the more powerful the cell.

14-40
$$3Fe(s) + 2Cr^{3+}(aq) \longrightarrow 3Fe^{2+}(aq) + 2Cr(s)$$

Yes, the following reaction is spontaneous.

$$3Zn(s) + 2Cr^{3+}(aq) \longrightarrow 3Zn^{2+}(aq) + 2Cr(s)$$

14-41 As indicated in Table 14-1, Na reacts spontaneously with water. The cathode reaction is

$$2H_2O + 2e^- \longrightarrow H_2 + 2OH^-$$

Water is more easily reduced than Na^+. Elemental sodium is produced by electrolysis of the molten salt, NaCl.

14-44 The strongest oxidizing agent is reduced the easiest. Thus the reduction of Ag^+ (i.e., $Ag^+ + e^- \rightarrow Ag$) occurs first. This procedure can be used to purify silver.

Review Test for Chapters 11 – 14

Multiple Choice

1. (d) **2.** (d) **3.** (b) **4.** (a) **5.** (a) **6.** (c) **7.** (e) **8.** (d) **9.** (b)
10. (c) **11.** (d) **12.** (e) **13.** (a) **14.** (c) **15.** (c) **16.** (e) **17.** (d)
18. (b) **19.** (b) **20.** (b) **21.** (e) **22.** (c) **23.** (c) **24.** (b) **25.** (c)
26. (b) **27.** (c) **28.** (a) **29.** (d) **30.** (e)

Problems

1. BeF_2—800 °C: Ion–ion forces are the strongest.
    C_2H_5OH——117 °C: Hydrogen bonding is stronger than dipole–dipole forces.
    $(CH_3)_2O$——138 °C: Dipole–dipole forces, but no hydrogen bonding.
    C_3H_8——189 °C: Only London forces since it is nonpolar.

2. 1.82×10^4 cal

3. (a) $0.480\ M$ (b) $0.0873\ M$ (c) 625 mL

4. (a) $Pb(NO_3)_2(aq) + 2NaI(aq) \rightarrow PbI_2(s) + 2NaNO_3(aq)$
    (b) 1.7 L

5. (a) $H_2SO_4(aq) + 2LiOH(aq) \rightarrow Li_2SO_4(aq) + 2H_2O$
    (b) 150 mL

6. (a) $Sn + 4HNO_3 \rightarrow SnO_2 + 4NO_2 + 2H_2O$
    (b) 3.2 g NO_2

7. (a) $Pb^{2+} + 2NO_3^- + 2Na^+ + 2I^- \rightarrow PbI_2 + 2Na^+ + 2NO_3^-$
    $\quad Pb^{2+} + 2I^- \rightarrow PbI_2$
    (b) $2Li^+ + 2OH^- + 2H^+ + SO_4^{2-} \rightarrow 2H_2O + 2Li^+ + SO_4^{2-}$
    $\quad OH^- + H^+ \rightarrow H_2O$

8. (a) Sn   (b) HNO_3 (c) SnO_2 (d) NO_2 (e) HNO_3 (f) Sn

9. (a) $HOCl + H_2O \rightleftharpoons H_3O^+ + OCl^-$
    (b) $NH_3 + H_2O \rightleftharpoons NH_4^+ + OH^-$
    (c) $KOH + HNO_2 \rightarrow KNO_2 + H_2O$
    (d) $H_3AsO_3 + 2NaOH \rightarrow Na_2HAsO_3 + 2H_2O$

10. (a) N.R.   (b) $CaO + H_2O \rightarrow Ca(OH)_2$ (c) $NH_4^+ + H_2O \rightleftharpoons NH_3 + H_3O^+$
    (d) $CO_2 + H_2O \rightarrow H_2CO_3$ (e) $ClO^- + H_2O \rightleftharpoons HClO + OH^-$ (f) N.R.

11. (a) 1.5×10^{-11} (b) pH $= 10.82$ (c) pOH $= 3.18$

12. (a) nonspontaneous
    (b) $2Cr^{3+} + 3Mn \rightarrow 2Cr + 3Mn^{2+}$
    (c) yes: $3I_2 + 2Cr \rightarrow 2Cr^{3+} + 6I^-$

Chapter 15

15-1   Colliding molecules must have the proper orientation relative to each other
at the time of the collision and the colliding molecules must have the
minimum kinetic energy for the particular reaction.

15-3   Most likely, the collision would involve the carbon end of the CO molecule
colliding with one oxygen on the NO_2 as follows:

15-4 The rate of the forward reaction was at a maximum at the beginning of the reaction; the rate of the reverse reaction was at a maximum at the point of equilibrium.

15-6 (a) As the temperature increases, the frequency of collisions between molecules increases as does the average energy of the collisions. Both contribute to the increased rate of reaction.

(b) The cooking of eggs initiates a chemical reaction that occurs more slowly at lower temperatures.

(c) The average energy of colliding molecules at room temperature is not sufficient to initiate a reaction between H_2 and O_2.

(d) A higher concentration of O_2 increases the rate of combustion.

(e) When a solid is finely divided, a greater surface area is available for collisions with oxygen molecules. Thus it burns faster.

(f) The souring of milk is a chemical reaction that slows as the temperature drops. It takes several days in a refrigerator.

(g) The platinum is a catalyst. Since the activation energy in the presence of a catalyst is lower, the reaction can occur at a lower temperature.

15-8 Products are easier to form because the activation energy for the forward reaction is less than that for the reverse reaction. This is true of all exothermic reactions. The system should come to equilibrium faster when it starts with pure reactants.

15-10 (a) $K_{eq} = \dfrac{[COCl_2]}{[CO][Cl_2]}$ (b) $K_{eq} = \dfrac{[CO_2][H_2]^4}{[CH_4][H_2O]^2}$

(c) $K_{eq} = \dfrac{[Cl_2]^2[H_2O]^2}{[HCl]^4[O_2]}$ (d) $K_{eq} = \dfrac{[CH_3Cl][HCl]}{[CH_4][Cl_2]}$

15-12 products

15-14 There will be an appreciable concentration of both reactants and products present at equilibrium.

15-15 $K_{eq} = 0.34$

15-16 $K_{eq} = 49.2$

15-18 $K_{eq} = \dfrac{[CO_2][H_2]^4}{[CH_4][H_2O]^2} = \dfrac{[2.20/30][4.00/30]^4}{[6.20/30][3.00/30]^2} = 0.0112$

15-20 (a) $[H_2] = [I_2] = 0.30$

(b) $[H_2] = 0.30$; $[I_2] = 0.50$

(c) $[HI] = 0.60 - 0.20 = 0.40$; $[H_2] = 0.10$; $[I_2] = 0.10$

(d) $K_{eq} = 6.3 \times 10^{-2}$

(e) For the reverse reaction, $K_{eq} = 16$. This value is smaller than that used in Table 15-1. This indicates that the equilibrium in this problem was established at a temperature different from that in Table 15-1.

15-22 (a) If $[N_2]_{eq} = 0.40$, then 0.10 mol/L of N_2 reacted. This means that 0.30 mol/L H_2 reacted, forming an additional 0.20 mol/L of NH_3.

$[H_2] = 0.50 - 0.30 = 0.20$ mol/L

$[NH_3] = 0.50 + 0.20 = 0.70$ mol/L

(b) $K_{eq} = 150$

15-23 (a) **The concentration of O_2 that reacts is**

$$0.25 \;\cancel{\text{mol NH}_3} \times \frac{5 \text{ mol } O_2}{4 \;\cancel{\text{mol NH}_3}} = 0.31 \text{ mol } O_2$$

(b) $[NH_3] = 0.75$; $[O_2] = 0.69$; $[NO] = 0.25$; $[H_2O] = 0.38$

(c) $K_{eq} = \dfrac{[NO]^4[H_2O]^6}{[NH_3]^4[O_2]^5} = \dfrac{[0.25]^4[0.38]^6}{[0.75]^4[0.69]^5}$

15-25 $[PCl_5] = 0.28$

15-27 $[H_2O] = 0.26$

15-28 Let $x = [HCl] = [CH_3Cl]$, then $\dfrac{x^2}{[CH_4][Cl_2]} = K_{eq} = 56.$

$$\dfrac{x^2}{(0.20)(0.40)} = 56 \qquad x^2 = 4.48 \qquad x = \underline{2.1 \text{ mol/L}}$$

15-30 (a) $K_a = \dfrac{[H_3O^+][OBr^-]}{[HOBr]}$ (b) $K_b = \dfrac{[NH_4^+][OH^-]}{[NH_3]}$

(c) $K_a = \dfrac{[H_3O^+][HSO_3^-]}{[H_2SO_3]}$ (d) $K_a = \dfrac{[H_3O^+][SO_3^{2-}]}{[HSO_3^-]}$

(e) $K_a = \dfrac{[H_3O^+][H_2PO_4^-]}{[H_3PO_4]}$ (f) $K_b = \dfrac{[H_2N(CH_3)_2^+][OH^-]}{[HN(CH_3)_2]}$

15-32 The acid HB is weaker because it produces a smaller H_3O^+ concentration (higher pH). The stronger acid HX has the larger value of K_a.

15-33 The strongest acid produces the lowest pH (at the same initial concentration of acid).
HZ, $HC_2H_3O_2$, HOCl

15-35 A solution of $NaHCO_3$ is slightly basic because the hydrolysis (base) reaction occurs to a greater extent than the acid ionization reaction.

15-36 (a) $[HOCN] = 0.20 - 0.0062 = 0.19$

(b) $K_a = \dfrac{[H_3O^+][OCN^-]}{[HOCN]} = \dfrac{(6.2 \times 10^{-3})(6.2 \times 10^{-3})}{(0.19)} = 2.0 \times 10^{-4}$

(c) pH = 2.21

15-37 (a) $HX + H_2O \rightleftharpoons H_3O^+ + X^-$; $K_a = \dfrac{[H_3O^+][X^-]}{[HX]}$

(b) $[H_3O^+] = 0.100 \times 0.58 = 0.058 = [X^-]$
$[HX] = 0.58 - 0.058 = 0.52$

(c) $K_a = 6.5 \times 10^{-3}$

(d) pH = 1.24

15-40 $Nv + H_2O \rightleftharpoons NvH^+ + OH^-$
$K_b = 6.6 \times 10^{-6}$

15-41 $[H_3O^+] = 0.300 - 0.277 = 0.023$; pH = 1.64; $K_a = 1.9 \times 10^{-3}$

15-42 pH = 4.43

15-44 $[OH^-] = 3.2 \times 10^{-3}$

15-46 $pH = 12.43$

15-48 $pH = 9.40$

15-50 $pH = 8.20$

15-52 In a buffer solution, when the acid and its conjugate base are present in equimolar amounts

$$K_a = \frac{[H_3O^+][\text{anion}]}{[\text{acid}]} \quad \text{or } [H_3O^+] = K_a$$
$$\text{likewise } [OH^-] = K_b$$

When $pH = 7.5$, $[H_3O^+] = 3.2 \times 10^{-8}$ and $[OH^-] = 3.1 \times 10^{-7}$. Since $K_a = 3.2 \times 10^{-8}$ for HOCl, an equimolar mixture of <u>HOCl and KOCl</u> produces the required buffer.

15-54 $pH = 4.14$

15-55 (a) right   (b) left   (c) left   (d) right   (e) right   (f) no effect
(g) decrease   (h) increase the yield but greatly decrease the rate of formation of NH_3

15-57 (a) increase   (b) increase   (c) decrease   (d) decrease   (e) decrease
(f) no effect

15-59 (a) endothermic   (b) hotter   (c) no effect, since there are the same number of moles of gas on both sides of the equation

Chapter 16

16-1

(a) 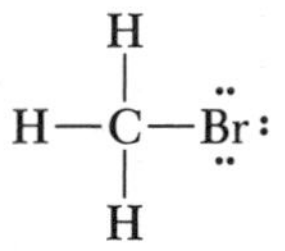
```
    H
    |    ..
H — C — Br :
    |    ..
    H
```

(b)
```
 H              H
  \            /
   C = C = C
  /            \
 H              H
```

(c)
```
 H    H  H
  \   |  |
   C = C — C — C — H
  /   |  |  |
 H    H  H  H
```
```
 H  H  H  H
 |  |  |  |
 H — C — C = C — C — H
 |       |       |
 H       H       H
```
```
    H  H
    |  |
 H — C — C — H
    |  |
 H — C — C — H
    |  |
    H  H
```
```
      H   H
       \ /
        C
       / \
 H — C     C — H
    |       | \
    H       H  H
  H—C        C—H
```

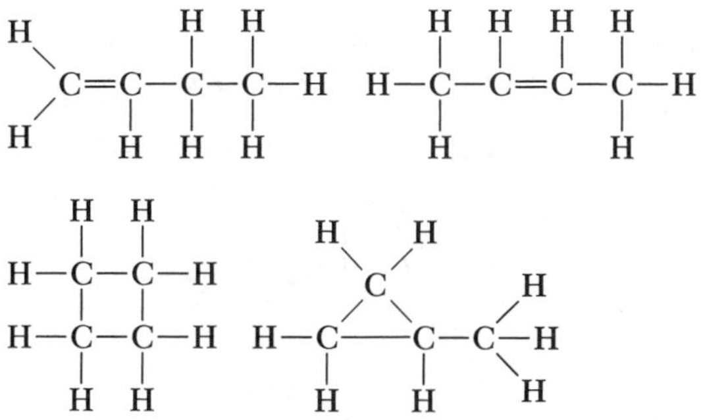

(d) 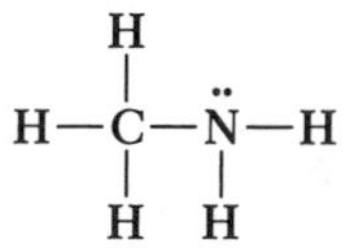
```
    H
    |    ..
H — C — N — H
    |    |
    H    H
```

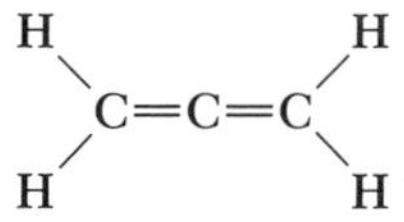

$$(e) \quad H-\overset{\overset{\displaystyle H}{|}}{\underset{\underset{\displaystyle H}{|}}{C}}-\overset{\overset{\displaystyle H}{|}}{\underset{\underset{\displaystyle H}{|}}{C}}-\overset{..}{N}-H, \quad H-\overset{\overset{\displaystyle H}{|}}{\underset{\underset{\displaystyle H}{|}}{C}}-\overset{..}{\underset{\underset{\displaystyle H}{|}}{N}}-\overset{\overset{\displaystyle H}{|}}{\underset{\underset{\displaystyle H}{|}}{C}}-H$$

$$(f) \quad H-\overset{\overset{\displaystyle H}{|}}{\underset{\underset{\displaystyle H}{|}}{C}}-\overset{\overset{\displaystyle H}{|}}{\underset{\underset{\displaystyle H}{|}}{C}}-\overset{\overset{\displaystyle H}{|}}{\underset{\underset{\displaystyle H}{|}}{C}}-\overset{..}{\underset{..}{O}}-H \qquad H-\overset{\overset{\displaystyle H}{|}}{\underset{\underset{\displaystyle H}{|}}{C}}-\overset{\overset{\displaystyle :O:}{|}}{\underset{\underset{\displaystyle H}{|}}{C}}-\overset{\overset{\displaystyle H}{|}}{\underset{\underset{\displaystyle H}{|}}{C}}-H$$

$$H-\overset{\overset{\displaystyle H}{|}}{\underset{\underset{\displaystyle H}{|}}{C}}-\overset{\overset{\displaystyle H}{|}}{\underset{\underset{\displaystyle H}{|}}{C}}-\overset{..}{\underset{..}{O}}-\overset{\overset{\displaystyle H}{|}}{\underset{\underset{\displaystyle H}{|}}{C}}-H$$

16-3 $CH_3-CH_2-CH_2-CH_2-CH_2-CH_3$ $CH_3-CH_2-CH_2-\underset{\underset{\displaystyle CH_3}{|}}{CH}-CH_3$

$CH_3-CH_2-\underset{\underset{\displaystyle CH_3}{|}}{CH}-CH_2-CH_3$ $CH_3-CH_2-\overset{\overset{\displaystyle CH_3}{|}}{\underset{\underset{\displaystyle CH_3}{|}}{C}}-CH_3$

$CH_3-\underset{\underset{\displaystyle CH_3}{|}}{CH}-\underset{\underset{\displaystyle CH_3}{|}}{CH}-CH_3$

16-5 (a) no (no oxygen in second compound)   (b) no (unequal number of carbons)   (c) yes   (d) no (unequal number of carbons)   (e) These two structures represent the same compound.

16-7 (a) C_4H_{10} (d) $C_{10}H_{22}$ and   (g) $C_{18}H_{38}$

16-8 (f) C_7H_{14}

16-9 (b) C_3H_7 and   (e) C_7H_{15}

16-12 (a) $CH_3-CH_2-CH_3$

(b) $CH_3-\underset{\underset{\displaystyle CH_3}{|}}{CH}-CH_2-\underset{\underset{\underset{\displaystyle CH_2-CH_2-CH_3}{}}{|}}{CH}-CH_2-CH_3$

(c) $CH_3-\underset{\underset{\displaystyle CH_3}{|}}{CH}-\overset{\overset{\displaystyle CH_3}{|}}{\underset{\underset{\displaystyle CH_3}{|}}{C}}-CH_3$

(d) $CH_3-\underset{\underset{\displaystyle CH_2}{|}}{CH}-CH_2-\overset{\overset{\displaystyle CH_3}{|}}{\underset{\underset{\displaystyle CH_2}{|}}{C}}-CH_3$

$$\underset{\displaystyle CH_2}{\diagdown \quad \diagup}$$

16-14 (a) $CH_3-CH_2-CH-CH_2-CH_3$
$\quad\quad\quad\quad\quad\quad\quad\quad\ |$
$\quad\quad\quad\quad\quad\quad\quad\ CH_3$

(b) $CH_3CH_2CH_2CH_2CH_2CH_3$

(c) $CH_3-CH-CH_2-CH-CH-CH_2CH_2CH_3$
$\quad\quad\quad\ \ |\quad\quad\quad\quad\ |\quad\ |$
$\quad\quad\quad\ CH_3\quad\quad\ CH_3\ \ CH_3$

(d) $CH_3CH-CH_2-CHCH_2CH_2CH_3$
$\quad\quad\quad\ |\quad\quad\quad\quad\ |$
$\quad\quad\quad CH_3\quad\quad\ CH_2CH_3$

(e) $CH_3CH_2CHCH_2CH_2CH_3$
$\quad\quad\quad\quad\quad\ |$
$\quad\quad\quad\quad\ CH(CH_3)_2$

(f) $\quad\quad\ CH_3$
$\quad\quad\quad\ |$
$\quad CH_3CCH_2-CH-CH_2CH_2CH_2CH_2CH_3$
$\quad\quad\quad\ |\quad\quad\ |$
$\quad\quad\ CH_3\quad\ C(CH_3)_3$

16-16 (a) $CH_3-CH-CH-CH_2-CH_3$ 3,4-dimethylhexane
$\quad\quad\quad\quad\quad\ |\quad\ |$
$\quad\quad\ CH_3-CH_2\ \ CH_3$

(b) $\quad\quad\quad\quad\quad\quad\quad CH_3-CH_2-CH_2$
$\quad\quad\quad\quad\quad\quad\quad\quad\quad\quad\quad\ |$
$\quad CH_3-CH_2-CH_2-CH_2-C-CH_3$ 4-isopropyl-4-
$\quad\quad\quad\quad\quad\quad\quad\quad\quad\quad\ |$ methyloctane
$\quad\quad\quad\quad\quad\quad\quad\quad CH(CH_3)_2$

16-17 (a) six (hexane)   (b) eight (octane)   (c) nine (nonane)   (d) seven (heptane)

16-18 (a) methyl   2,3,5-trimethylhexane
(b) methyl and ethyl   5-ethyl-2-methyloctane
(c) methyl   3,3-dimethylnonane
(d) methyl   4-methylheptane

16-20 (a) propane   (b) 4-ethyl-2-methylheptane
(c) 2,3,3-trimethylbutane

16-22 This general formula is that of a straight-chain alkene with one double bond or it can also apply to a cyclic alkane.

16-24 (a) alkene   (b) alkane   (c) alkene   (d) alkyne   (e) alkane
(f) alkene   (g) alkane

16-25 (a) $CH_3CH_2CH=CH_2$ (b) $CH_3C\equiv CCH_3$
(c) $CH_3CH_2-C=CHCH_3$ (d) $CH_3CH_2C\equiv CCHCH_3$
$\quad\quad\quad\quad\quad\ |$ $\quad\quad\quad\quad\quad\quad\quad\ |$
$\quad\quad\quad\quad\ CH_3$ $\quad\quad\quad\quad\quad\quad CH_2CH_3$

16-27 (a) $CH_3C\equiv CCH_3$ (b) $CH_2=CHCHCH_3$
$\quad\quad\quad\quad\quad\quad\quad\quad\quad\quad\quad\quad\quad\ |$
$\quad\quad\quad\quad\quad\quad\quad\quad\quad\quad\quad\quad\ CH_3$

(c) $CH_3CH=CCH_2CH_2CH_3$ (d) $CH_3CH_2C\equiv CCHCH_2CH_2CH_3$
$\quad\quad\quad\quad\quad\ |$ $\quad\quad\quad\quad\quad\quad\quad\quad\ |$
$\quad\quad\quad CH_2CH_2CH_3$ $\quad\quad\quad\quad\quad\quad\quad CH_3$

16-28 (a) 1-butene   (b) 2-butyne   (c) 3-methyl-2-pentene
(d) 3-methyl-4-heptyne

16-30 (a) $(-CH_2-CH-)_n$ polypropylene
$\qquad\qquad\quad | $
$\qquad\qquad\ CH_3$

(b) $(-CH_2-CH-)_n$ polyvinylfluoride
$\qquad\qquad\quad | $
$\qquad\qquad\ F$

(c) $(-CH_2-CF_2-)_n$ polydifluoroethylene

(d) $(-CH_2-CH-)_n$ polymethyacrylate
$\qquad\qquad\quad | $
$\qquad\qquad\ CO_2CH_3$

16-31 $CHF{=}CHCH_3$

16-32

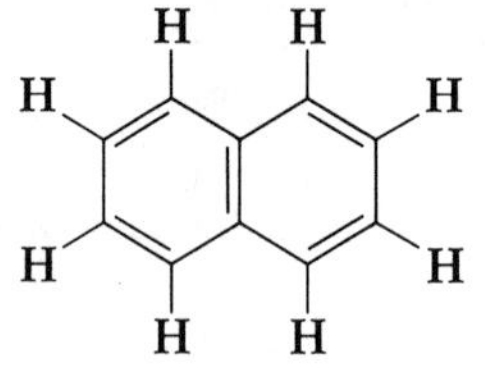

16-34

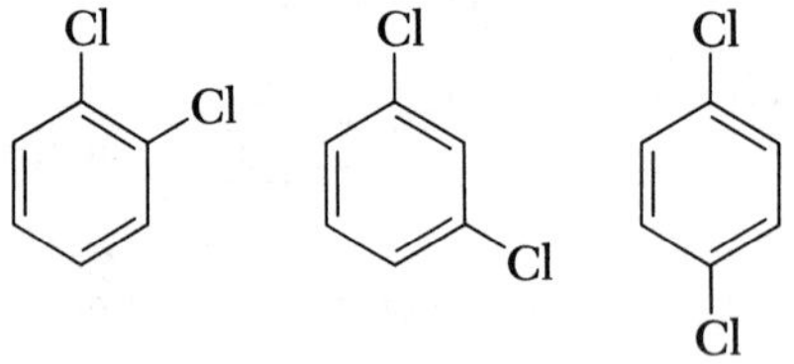

16-35 (a) $-C{=}C-$, alkene; and $\overset{\overset{\textstyle O}{\|}}{C}-H$, aldehyde

(b) $-NH_2$, amine; and $-\overset{\overset{\textstyle \|}{\ }}{\underset{\underset{\textstyle O}{\|}}{C}}-$, ketone

(c) $-O-$, ether; and $-\overset{\overset{\textstyle O}{\|}}{C}-NH_2$, amide

(d) $-C{\equiv}C-$, alkyne; and $-OH$ alcohol

(e) $-COOCH_3$, ester

(f) $-COOH$, carboxylic acid

16-37 (a) In an ether the O is between two C's. In an alcohol the O is between a C and a H.

(b) In an aldehyde the carbonyl (i.e., C=O) is between a C and a H. In a ketone the carbonyl is between two C's.

(c) In an amine the NH_2 group is attached to a hydrocarbon group. In an amide the NH_2 group is attached to a carbonyl group.

(d) Carboxylic acids have a hydrogen attached to an oxygen; in esters the hydrogen is replaced by a hydrocarbon group.

16-39 Carboxylic acids contain a carboxyl group, and alcohols contain a hydroxyl group.

16-41 (a) ketone (b) alkene (c) alkane (d) aldehyde (e) aromatic
(f) alcohol

16-43 (a) $CH_3CH_2CH_2CH_2OH$ (b) $CH_3CH_2CH_2OCH_2CH_2CH_3$
(c) $(CH_3)_3N$ (d) CH_3CH_2CHO (e) $CH_3CH_2CCH_2CH_3$
$$\underset{O}{\|}$$

(f) CH_3CH_2COOH (g) $CH_3-\underset{\underset{O}{\|}}{C}-O-CH_3$ (h) $CH_3CH_2-\underset{\underset{O}{\|}}{C}-NH_2$

16-45 (a) 2-pentanone (b) butanoic acid (c) acetamide
(d) t-butyl alcohol (e) pentanal

16-47 (a) $C_3H_8 + 5O_2 \rightarrow 3CO_2 + 4H_2O$
(b) $CH_2{=}CHCH_3 + H_2O \rightarrow CH_3-\underset{\underset{OH}{|}}{C}HCH_3$

(c) $C_2H_5Cl + 2NH_3 \rightarrow C_2H_5NH_2 + NH_4{}^+Cl^-$
(d) $NH(CH_3)_2(aq) + HCl(aq) \rightarrow HNH(CH_3)_2{}^+(aq) + Cl^-(aq)$
(e) $CH_3CH_2COOH + NH(CH_3)_2 \rightarrow CH_3CH_2-\underset{\underset{O}{\|}}{C}-O-N(CH_3)_2 + H_2O$

(f) $HC{\equiv}CH + HBr \rightarrow CH_2{=}CHBr$
(g) $CH_3CH_2COOH + H_2O \rightleftharpoons CH_3CH_2COO^- + H_3O^+$

PHOTO CREDITS

Figure 5-12: (top left) Kathy Bendo; (top right) Peter M. Lerman. Figure 5-13: (bottom left) Courtesy Linde Division, Union Carbide; (bottom right) Joel E. Arem, PhD. Page 148: Figure 5-14: (left) Life Science Library; (right) Peter M. Lerman.

Chapter 6:
Opener: Dick Raphael for Sports Illustrated © Time Inc. Page 163: Figure 6-1: OPC, Inc.

Chapter 7:
Opener: Edith Reichmann/Monkmeyer Press Photo Service. Page 202: Figure 7-2: (left) Peter M. Lerman; (right) Peter M. Lerman.

Chapter 8:
Opener: Ray Atkeson. Page 216: Figure 8-1: (left) Peter M. Lerman; (center) Peter M. Lerman; (right) Peter M. Lerman. Page 223: Figure 8-2: (left) Peter M. Lerman; (center) Peter M. Lerman; (right) Peter M. Lerman. Page 232: Figure: 8-3: Peter M. Lerman.

Chapter 9:
Opener: Courtesy of United States Steel. Page 245: Figure 9-2: Peter M. Lerman. Page 246: Figure 9-3: Peter M. Lerman. Page 247: Figure 9-4: Peter M. Lerman. Page 248: Figure 9-5: (left) Peter M. Lerman; (right) Peter M. Lerman.

Chapter 10:
Opener: Phaneuf/Gurdziel/The Picture Cube.

Chapter 11:
Opener: Pro Pix/Monkmeyer Press Photo Service. Page 334: Figure 11-10: (left) Peter M. Lerman; (right) Peter M. Lerman.

Chapter 12:
Opener: Russell A. Thompson/Taurus Photos. Page 370: Figure 12-12: (left) Peter M. Lerman; (right) EPA/Art Resources.

Chapter 13:
Opener: Del Mulkey/Photo Researchers, Inc. Page 382: Figure 13-1: Kathy Bendo. Page 396: Figure 13-8: Peter M. Lerman. Page 410: Figure 13-11: (left) Westfälisches Amt Fur Denkmalpflege; (right) Westfälisches Amt Fur Denkmalpflege.

Chapter 14:
Opener: John Spragens, Jr./Photo Researchers, Inc. Page 423: Figure 14-1: (left) Peter M. Lerman; (center) Life Science Library; (right) Kathy Bendo. Page 436: Figure 14-4: Courtesy of Gould, Inc. Page 437: Figure 14-5: Courtesy Duracell, Inc. Page 438: Figure 14-6: NASA. Page 446: Figure 14-11: Peter M. Lerman.

Chapter 15:
Opener: Jan Lukas/Photo Researchers, Inc. Page 471: Figure 15-12: Courtesy of General Motors. Page 484: Figure 15-14: The Sulfer Institute.

Chapter 16:
Opener: Robert E. Largent/Photo Researchers, Inc. Page 500: Figure 16-1: Peter M. Lerman. Page 507: Figure 16-2: Courtesy Standard Oil Company of California. Page 510: Figure 16-4: (top) Exxon Chemical Company, U.S.A.; (bottom) Courtesy Owen-Illinois. Page 515: Figure 16-5: Kathy Bendo. Page 518: Figure 16-6: Kathy Bendo. Page 524: Figure 16-7: Courtesy Dupont Company.

Appendix
Opener: Harvey Stein.

INDEX